Lecture Notes
in Control and Information Sciences   245

Editor: M. Thoma

Springer-Verlag London Ltd.

A. Garulli, A. Tesi and A. Vicino (Eds)

Robustness in Identification and Control

Springer

Editors

A. Garulli, Assistant Professor
Dipartimento di Ingegneria dell'Informazione, Università di Siena, Via Roma, 56-53100 Siena, Italy

A. Tesi, Assistant Professor
Dipartimento di Sistemi e Informatica, Università di Firenze, Via S. Marta, 3-50139 Firenze, Italy

A. Vicino, Professor
Dipartimento di Ingegneria dell'Informazione, Università di Siena, Via Roma, 56-53100 Siena, Italy

British Library Cataloguing in Publication Data
Robustness in identification and control. – (Lecture notes
  in control and information sciences ; 245)
  1.Control theory
  I.Garulli, A. II. Tesi, A. III.Vicino, A.
  629.8'312
  ISBN 978-1-85233-179-5

Library of Congress Cataloging-in-Publication Data
Robustness in identification and control / Andrea Garulli, Alberto
  Tesi, and Antonio Vicino (eds.).
    p.   cm. -- (Lecture notes in control and information sciences
  ; 245)
    Includes bibliographical references.
    ISBN 978-1-85233-179-5   ISBN 978-1-84628-538-7 (eBook)
    DOI 10.1007/978-1-84628-538-7
    1. Automatic control. 2. System identification. I. Garulli,
  Andrea, 1968- . II. Tesi, Alberto, 1957- . III. Vicino,
  Antonio. IV. Series.
  TJ213.R585 1999   99-30419
  629.8--dc21   CIP

Typesetting: Camera ready by contributors

69/3830-543210 Printed on acid-free paper SPIN 10715916

Preface

Automatic control systems operate in highly uncertain environments. The task of a reliable control system is to ensure the achievement of the prescribed goals and guarantee good performance despite the uncertainty. Modeling, analysis and design issues involved in this problem constitute the subject of *Robust Identification and Control.*

Robustness is a keyword which stimulated an unforeseen synergy between people traditionally working in the Identification area and those working in the Robust Control area. The prominence of the research area of Robust Identification and Control is testified by the very intense level of activity developed recently. Quite a number of specific workshops have taken place since the end of the 1980s, some of which have given rise to selected proceedings volumes. A remarkable number of special and invited sessions were organized for the most important conferences and several Special Issues in prestigious journals or volumes have been published in recent years.

The Workshop *Robustness in Identification and Control* held in Siena (Italy) on July 30 - August 2, 1998 was conceived in this context. Its aim was to bring together researchers with an interest in the variety of problems that contribute or utilize the techniques of *Identification for Control* and *Robust Control.* Many leading researchers in the field participated in the event. As an outcome, the forum not only delineated the state of the art in the fundamental theoretical aspects of robustness in identification and control, but also covered the important emerging applications of these areas, such as: aero-elasticity modeling, automotive systems, dynamic vision, electronic microscope control, medical systems, power systems, public health, robotics and turbulence in fluidodynamics.

The aim of the present volume is twofold. First, to publish a selected list of papers presented at the Workshop, collecting the latest contributions to the field from leading researchers worldwide. Second, to provide a comprehensive study of the latest methodological research and an outline of some significant application fields. Authors have been encouraged to write their manuscripts in a tutorial style, departing whenever necessary from the theorem-proof format of journal papers, to make the book pleasant to read for a community much broader than the automatic control community.

The book is organized into two main sections. The first section collects papers in the area of *Identification for Robust Control.* This research area is attracting a growing interest in the control community. While a number of important problems have been solved, several issues are still open. The contributions included in this section cover recent developments in the field and focus on new approaches to solve challenging problems in: mixed stochastic-deterministic information processing, structured/unstructured models of uncertainty, model validation, set membership identification and filtering. The second section gathers papers in the area of *Robust Control*. Although this research area is well assessed at the present stage, a renewed and vigorous interest in the subject has been stimulated by recent advances in worst case and set membership system identification. Contributions in this section deal with: robust synthesis techniques, invariant sets, robust predictive control, limits of performances for linear and nonlinear systems, robust performance with structured and unstructured uncertainty and control of distributed arrays.

Siena,
May 1999

Andrea Garulli
Alberto Tesi
Antonio Vicino

Contents

viii

Robustness in Identification*

R. E. Kalman

Istituto "Asinius Pollio"
Swiss Federal Institute of Technology, Zurich

Our objective is to illuminate the importance of "robustness" in the general scientific context. We view "identification" in the broad sense as covering the basic scientific activity of constructing credible and verifiable scientific models from observational data. So the basic message of this lecture will be

$$\text{robustness} \iff \text{good scientific model}$$

For the purposes of developing this argument, I shall consider three examples, each rather different from the others:

(i) Stability of the solar system;
(ii) Axiomatic "science";
(iii) Statistical origins of robustness.

Stability of the solar system

This problem is as old as human astronomy. It can be attacked at present only on the basis of the (traditional) mathematical model which is based on Newton's law of gravitation and requires treating the planets (and other large bodies, such as moons and large asteroids) as point masses. This model, whose mathematical properties are not yet fully known and which requires advanced mathematical techniques for its study, seems to be, at best, marginally stable and certainly not robust. New astronomical data indicates that the orbit of Pluto (and perhaps that of Neptune) is chaotic but it is not clear if this fact can be deduced from the model by purely mathematical means. In any case, the traditional model is certainly not very "good" and it is surely not robust in regard to newly discovered behavior such as a chaotic orbit. (We are not even sure that Pluto can be called a *planet*; perhaps true planets never have chaotic orbits).

---

* Extended abstract of the lecture given at the Certosa di Pontignano, Siena (Italy), July 30, 1998.

Axiomatic "science"

There have been many examples in the last two hundred years where science has advanced by giant leaps by finding the right axioms, calling the axioms a model, and then verifying the consequences. (Examples are electromagnetic radiation and relativity theory). In such situations, all experiments performed on the model are reproducible but the same may not be true when the experiments are performed on the real system in the real world. So one can never be sure if the axiomatic approach will work, in the sense of leading to new scientific truth. There is reason to be optimistic, however: if we cannot think of other axioms, maybe they do not exist and then Nature is forced to follow the axioms we already have.

But this is dangerous optimism. For a long time, it was thought that the axioms of Euclidean geometry are mandatory for physics, and these axioms are readily verifiable, since all triangles measured (accurately) on earth add up to 180°. Many big minds were dissatisfied with this easy result and looked for alternative axioms.... and they found them. Nowadays no one can deny the physical relevance of nonEuclidean geometry. But that means that Euclidean geometry is not robust and so implies that physics based on it is not a "good" model. Bravo Mandelbrot! Bravo fractals!

Robustness in statistics

Mainstream statistical dogma, since about 1900, begins by postulating a basically probabilistic environmment and then assumes (for simplicity?) a gaussian distribution. This procedure has the basic advantage of having to deal only with means and variances (readily available or easily computed) and the further very real benefit of not needing to worry about the actual distribution present - often an unpleasant task. It is not exaggeration to say that this approach has all the attractions of theft over honest toil.

The difficulty is, of course, lack of robustness. Gauss "derived" his famous distribution by appealing to (what is currently called) the maximum likelihood principle. This is akin to a snake swallowing its own tail since the maximum likelihood principle can be readily applied only in the case of gaussian distributions.

Though there was much grumbling by big minds, it was only around 1960 that Tukey pointed out - by means of the very simple example of adding numbers from two gaussian distributions with unequal means and variances - that the gaussian distribution and related procedures were hopelessly nonrobust. He thereby introduced the word "robust" into statistics, which now finds resonance in many fields, as in the meeting today. (Perhaps the problem with the gaussian distribution is due to its incestuous origin in the mind of Gauss).

In my opinion, Tukey stopped far short of the goal. It is possible to develop a substantial part of classical statistics (that dealing with means and variances) without any appeal to gaussian "randomness" and in fact without any use of probability. These new methods provide very good models for certain kinds of randomness. That they are also robust should be emphasized although that is no news to their maker. (The interested reader may look at the speaker's Nine Lectures in Identification [in econometrics] of which many preliminary copies are in circulation with definitive publication expected before the end of the century).

The shortest way of putting the long story outlined above is to note that the classical probabilistic models of randomness are intrinsically nonrobust. This is because what is commonly called "probability" is an abstract world created by human genius; all attempts to find even faint traces of this abstract world in Nature have (so far) been doomed to failure. So even more succinctly, we may identify "robustness" with "naturalness".

In closing, let me express a rather more personal opinion. The Achilles heel of computer science is its lack of robustness (especially as regards developments in the last 10 years). If computer systems (like Windows) do not include more feedback, on various levels, they will move further away from natural systems and the consequences may be unpleasant indeed.

So let us, by all means, analyze all artificial systems and models from the point of view of robustness and not shy away from publicizing the findings, especially if negative. Robustness is not so far from Darwinian evolution.

Part I

Identification for Robust Control

Comments on Model Validation as Set Membership Identification

Lennart Ljung

Division of Automatic Control
Department of Electrical Engineering
Linköping University, S-581 83 Linköping, Sweden
E-mail: ljung@isy.liu.se
URL: http://www.control.isy.liu.se/

Abstract. We review four basic model validation techniques, one that relates to the "unknown-but-bounded" disturbance assumption, one that has been recently suggested in the "Identification-for-robust-control" context and two more classical statistical tests. By defining the set of models that would pass the chosen model validation test, we may interpret each of these as a set membership identification method. The consequences of such a viewpoint are discussed, and we focus on the important, but perhaps controversial concept of "independence" to make further selections of models within the thus defined sets.

1 Introduction

Model validation has always played a major role in System Identification, as a basic instrument for model structure selection and as the last "quality control" station before a model is delivered to the user.

The issues around model validation have shown renewed interest in connection with the recent discussions on "identification-for-control" or "control-oriented-validation".

In this contribution we shall focus on some classical model validation criteria as well as on some more recently suggested ones. We shall in particular consider the validation process as a screening process that defines a set of models that are not falsified by the test. We shall show the connections between these sets and common identification methods and we shall also discuss if there is any need for further choices within the set of un-falsified models.

We place ourselves in the following situation. A model is given. Let it be denoted by $\hat{G}$ (more specific notation will follow later). We are also given a data set Z^N consisting of measured input-output data from a system. We do not know, or do not care, how the model was estimated, or constructed or given. We might not even know if the data set was used to construct the model.

Our problem is to figure out if the model $\hat{G}$ is any good at describing the measured data, and whether it can be used for robust control design.

A natural start is to consider the model's simulated response to the measured input signal. Let that simulated output be denoted by $\hat{y}$. We would then compare this model output with the actual measured output and contemplate how good the fit is. This is indeed common practice, and is perhaps the most useful, pragmatic way to gain confidence in (or reject) a model. This will be the starting point or our discussion.

Some notation is defined is Section 2, while Section 3 discusses some relevant statistics around the measured and simulated outputs. Note that "statistics" here means some bulk, numerical descriptions of the fit; this has nothing to do with probability theory.

Section 4 deals with split of the residuals and a model error, while Section 5 lists the tests we are considering. The set membership aspects of model validation are outlined in Section 6 while a concluding discussion is given in Section 7.

2   Some Notation

We shall use the following notation. The input will be denoted by $u(t)$ and the output by $y(t)$. The data record thus is

$$Z^N = \{y(1), u(1), \ldots, y(N), u(N)\} \tag{1}$$

The given model $\hat{G}$ will be assumed to be linear, and a function of the shift operator q in the usual way: $\hat{G}(q)$. The simulated output will thus be

$$\hat{y}(t) = \hat{G}(q)u(t) \tag{2}$$

It may be that the model contains a noise assumption, typically in the form of an additive noise or disturbance $v(t)$ with certain properties. It would then be assumed that the actual output is generated as

$$y_m(t) = \hat{G}(q)u(t) + v(t) \tag{3}$$

(We append a subscript m to stress the difference with the measured output.) The model could contain some "prejudice" about the properties of $v(t)$, but this is not at all essential to our discussion. A typical, conventional assumption would be that $v(t)$ is generated from a white noise source through a linear filter:

$$v(t) = \hat{H}(q)e(t) \tag{4}$$

Most of the model validation test are based on simply the difference between the simulated and measured output:

$$\varepsilon(t) = y(t) - \hat{y}(t) = y(t) - \hat{G}(q)u(t) \tag{5}$$

Sometimes prefiltered model errors are studied:

$$\varepsilon(t) = L(q)(y(t) - \hat{y}(t)) = L(q)(y(t) - \hat{G}(q)u(t)) \tag{6}$$

For example, if the model comes with a noise model (4), then a common choice of prefilter is $L(q) = \hat{H}^{-1}(q)$, since this would make $\varepsilon(t)$ equal to the model's *prediction errors*. The prefilter is however not at all essential to our discussion, and we shall cover the situation (6) by allowing the data set (1) be prefiltered.

In any case we shall call $\varepsilon(t)$ the *Model Residuals* ("model leftovers").

3   Some Statistics Around the Residuals

Typical model validation tests amount to computing the model residuals and giving some statistics about them. Note that this as such has nothing to do with probability theory. (It is another matter that *statistical model validation* often is complemented with probability theory and model assumptions to make probabilistic statements based on the residual statistics. See, e.g., [2]. We shall not do that in this contribution.)

The following statistics for the model residuals are often used:

- The maximal absolute value of the residuals

$$M_N^\varepsilon = \max_{1 \le t \le N} |\varepsilon(t)| \tag{7}$$

- Mean, Variance and Mean Square of the residuals

$$m_N^\varepsilon = \frac{1}{N} \sum_{t=1}^{N} \varepsilon(t) \tag{8}$$

$$V_N^\varepsilon = \frac{1}{N} \sum_{t=1}^{N} (\varepsilon(t) - m_N^\varepsilon)^2 \tag{9}$$

$$S_N^\varepsilon = \frac{1}{N} \sum_{t=1}^{N} \varepsilon(t)^2 = (m_N^\varepsilon)^2 + V_N^\varepsilon \tag{10}$$

- Correlation between residuals and past inputs.
  Let

$$\varphi(t) = [u(t-1), u(t-1), \ldots, u(t-M)]^T \tag{11}$$

  and

$$R_N = \frac{1}{N} \sum_{t=1}^{N} \varphi(t)\varphi(t)^T \tag{12}$$

Now form the following scalar measure of the correlation between past inputs (i.e. the vector φ) and the residuals:

$$\xi_N^M = \frac{1}{N} \| \sum_{t=1}^{N} \varphi(t)\varepsilon(t) \|_{R_N^{-1}} \tag{13}$$

Now, if we were prepared·to introduce assumptions about the true system (the measured data Z^N), we could used the above statistical measures to make statements about the relationship between the model and the true system, typically using a probabilistic framework.

Using Induction

If we do not introduce any explicit assumptions about the true system, what is then the value of the statistics (7)-(13)? Well, we are essentially left only with *induction*. That is to say, we take the measures as indications of how the model will behave also in the future: "Here is a model. On past data it has never produced a model error larger than 0.5. This indicates that in future data and future applications the error will also be below that value." This type of induction has a strong intuitive appeal.

In essence, this is the step that motivates the "unknown-but-bounded" approach. Then a model or a set of models is sought that allows the preceeding statement with the smallest possible bound, or perhaps a physically reasonable bound.

Note, however, that the induction step is not at all tied to the unknown-but-bounded approach. Suppose we instead select the measure S_N^ε as our primary statistics for describing the model error size. Then the Least Squares (Maximum Likelihood/Prediction Error) identification method emerges as a way to come up with a model that allows the "strongest" possible statement about past behavior.

How reliable is the induction step? It is clear that some sort of invariance assumption is behind all induction. Here the statistics (13) plays a major role.

4   A Fundamental Split of the Residuals

It is very useful to consider two sources for the model residual ε: One source that originates from the input $u(t)$ and one that doesn't. With the (bold) assumption that these two sources are additive and the one that originates from the input is linear, we could write

$$\varepsilon(t) = \Delta(q)u(t) + w(t) \tag{14}$$

Note that the distinction between the contributions to ε is fundamental and has nothing to to with any probabilistic framework. We have not said anything about $w(t)$, except that it would not change, if we changed the input $u(t)$. We refer to (14) as the separation of the model residuals into *Model Error* and *Disturbances*.

The division (14) shows one weakness with induction for measures like M_N^ε and S_N^ε going from one data set to another. The implicit invariance assumption would require the input to be the same (or at least similar) in the two sets, unless we would have indications that Δ is of insignificant size. The purpose of the statistics ξ_N^M in (13) is exactly to assess the size of Δ. We shall see this clearly in Section 5. (One might add that more sophisticated statistics will be required to assess more complicated contributions from u to ε.)

In any case, it is clear that the induction about the size of the model residuals from one data set to another is much more reasonable if the statistics ξ_N^M has given a small value ("small" must be evaluated in comparison with S_N^ε in (10)).

5   Model Validation Tests

We are now in the situation that we are given a *nominal model* $\hat{G}$ along with a validation data set Z^N. We would like to devise a test by which we may *falsify* the model using the data, that is to say that it is not possible, reasonable or acceptable to assume that the validation data have been generated the nominal model. If the model is not falsified, we say that the model has *passed the model validation test*.

Now what tests are feasible? Let us list some typical choices, based on the discussion above. In all cases we first compute the residuals ε from the nominal model as in (5).

1. Test if $M_N^\varepsilon < \alpha$ (cf (7)) This corresponds to an assumption that the output noise is amplitude limited, and a model is valid if it does not break this assumption.
2. Test if $S_N^\varepsilon < \alpha$ (cf (10)) This test is perhaps not common, but is based on an assumption that the variance of the output noise is known, and if the residuals show a significantly larger value, the model is rejected
3. Test if $\xi_N^M < \alpha$ (cf(13)). This is the standard, "classical" residual analysis test, see e.g. [2].
4. Test if $\exists \Delta, \|\Delta\|_\infty < \alpha_1$ such that $\max_t |\varepsilon(t) - \Delta u(t)| < \alpha_2$ (cf (14)). This is, in simplified summary, the model validation test proposed in the "identification-for-robust-control" community, see, e.g., [3], [10], [11].
5. Estimate Δ in (14), and let $\hat{G}$ be unfalsified if the estimate $\hat{\Delta}$ is not significantly different from zero.

In all these cases, one might ask where the threshold α comes from. In a sense this has to rely upon prior information about the noise source, and we shall later discuss this issue. Only test number 3 is "self-contained", in the sense that it corresponds to a hypothesis test that ε is white noise, and then the hypothesis to be tested also comes with a natural estimate of the size of the noise.

Let us also comment on test number 5. An estimate $\hat{\Delta}$ can be viewed as a "model error model", cf [4], [5], but there is a very intimate relationship to test number 3:

If Δ is parameterized as a FIR model, its impulse response coefficients are estimated as

$$\hat{\eta}_N = R_N^{-1} \frac{1}{N} \sum_{t=1}^{N} \varphi(t)\varepsilon(t) \tag{15}$$

with covariance matrix $\hat{\lambda} R_N^{-1}$, where $\hat{\lambda}$ is an estimate of the variance of w. This means that a standard χ^2 test whether the true Δ is zero has the form

$$\hat{\Delta}^T (\hat{\lambda} R_N^{-1})^{-1} \hat{\Delta} < \alpha' \tag{16}$$

or (cf (13))

$$[\sum_{t=1}^{N} \varphi(t)\varepsilon(t)]^T R_N^{-1} \frac{1}{\hat{\lambda}} R_N R_N^{-1} [\sum_{t=1}^{N} \varphi(t)\varepsilon(t)] = \frac{N}{\hat{\lambda}} \xi_N^M < \alpha' \tag{17}$$

That is, test number 5 is equivalent to test number 3 for FIR model error models Δ.

6   Model Validation as Set Membership Identification

Each of the five model validation test can also be seen as *set membership identification methods* in the sense that we may ask, for the given data set Z^N, which models within a certain class would pass the test. This set of "unfalsified models" would be the result of the validation process, and could be delivered to the user. The interpretation would be that any model in this set could have generated the data, and that thus a control design must give reasonable behavior for all models in this set. Let us now further discuss what sets are defined in the different cases.

To more clearly display the basic ideas we shall here work with models of FIR structure, i.e. we ask which models of the kind

$$\hat{G}(q)u(t) = \sum_{t=1}^{N} g_k u(t-k) = \theta^T \varphi(t) \tag{18}$$

will pass the test. φ is defined by (11). The validation measures above will then be given the argument θ as in $\varepsilon(t,\theta)$ and $M_N^\varepsilon(\theta)$ to emphasize the dependence. See also [7].

6.1   Limited residual amplitude

The first test $S_N^\varepsilon < \alpha$ gives the standard set membership approach to system identification. All models that produce an output error less than α are computed, and for linear regression model structures, this problem can be solved by linear programming or bounding ellipsoids. See among many references, e.g. [1], [9], [8]

6.2   Limited MSE of the residuals

Suppose now we use test number 2. Let us define the LS-estimate of θ for the validation data as $\hat{\theta}_N$. Then simple manipulations give that

$$\varepsilon(t,\theta) = -\varphi^T(t)(\theta - \hat{\theta}_N) + \varepsilon(t,\hat{\theta}_N)$$

and hence

$$\frac{1}{N}\sum_{t=1}^{N}\varepsilon^2(t,\theta) = \frac{1}{N}\sum_{t=1}^{N}\varepsilon^2(t,\hat{\theta}_N) - \frac{2}{N}\sum_{t=1}^{N}\varepsilon(t,\hat{\theta}_N)\varphi^T(t)(\theta - \hat{\theta}_N)$$

$$+ (\theta - \hat{\theta}_N)^T\frac{1}{N}\sum_{t=1}^{N}\varphi(t)\varphi^T(t)\ (\theta - \hat{\theta}_N)$$

$$= \frac{1}{N}\sum_{t=1}^{N}\varepsilon^2(t,\hat{\theta}_N) + (\theta - \hat{\theta}_N)^T R_N(\theta - \hat{\theta}_N)$$

where the second equality follows from the fact that

$$\frac{1}{N}\sum_{t=1}^{N}\varepsilon(t,\hat{\theta}_N)\varphi(t) = 0 \tag{19}$$

and R_N is defined as in (12).

This shows that the validation test will pass for exactly those models θ for which

$$(\theta - \hat{\theta}_N)^T R_N(\theta - \hat{\theta}_N) \le \alpha - S_N^\varepsilon(\hat{\theta}_N) \tag{20}$$

Note the connection between this result and traditional *confidence ellipsoids*. In a probabilistic setting, the covariance matrix of the LS estimate $\hat{\theta}_N$ is proportional to R_N (see e.g. [6]). This means that (20) describes those models θ that are within a standard confidence area from the LSE. The level of confidence depends on α.

6.3   Uncorrelated residuals and inputs

Suppose now we use test number 3. We have

$$\frac{1}{N}\sum_{t=1}^{N}\varepsilon(t,\theta)\varphi(t) = \frac{1}{N}\sum_{t=1}^{N}(\varepsilon(t,\theta) - \varepsilon(t,\hat{\theta}_N))\varphi(t)$$

$$= -\frac{1}{N}\sum_{t=1}^{N}\varphi(t)\varphi^T(t)(\theta - \hat{\theta}_N)$$

$$= -R_N(\theta - \hat{\theta}_N) \tag{21}$$

where the first step follows from(19). We then find that

$$\xi_N^M(\theta) = (\theta - \hat{\theta}_N)^T R_N R_N^{-1} R_N (\theta - \hat{\theta}_N)$$
$$= (\theta - \hat{\theta}_N)^T R_N(\theta)(\theta - \hat{\theta}_N) \tag{22}$$

Inserting this in $\xi_N^M < \alpha$ gives the the set of non-falsified models is given by

$$(\theta - \hat{\theta}_N)^T R_N(\theta - \hat{\theta}_N) \le \alpha \tag{23}$$

Here, again $\hat{\theta}_N$ is the LSE for the validation data.

From these results we conclude that, for FIR models, the tests 2 and 3 are very closely related criteria. Furthermore, the results provide an alternative interpretation of probabilistic confidence regions. They are regions in the parameter space where, simultaneously, the *sample* cross correlation, ξ_N^M, and the Mean Square of the Model Residuals, S_N^ε, are small.

6.4   Control oriented model validation

Model validation test 4 has been suggested for control oriented model validation. In this context is has also been customary to compute the set of unfalsified models, parameterized by α_1 and α_2. This is quite a formidable computational task, but results in a curve in the α_1-α_2 plane, below which the set of unfalsified models is empty. See figure 1. The shaded area corresponds to "possible" model descriptions, but it is normally interesting to consider just the models on the boundary.

7   Discussion

In a control oriented perspective, the reason for model validation is to find out if the nominal model can be used for reliable, robust control design. From this perspective, the set membership aspect of model validation can be interpreted so that all models that pass the test should also give an acceptable closed loop system together with the constructed regulator. This indeed is closely related to the classical use of confidence intervals and uncertainty in connection with decision making.

A crucial question for this to hold is however that the validation process in some sense is successful in distinguishing the error model Δ from the "noise" w. As a trivial example, let $\varepsilon(t) = u(t)$. If the input in the validation data happened to be less than 0.7, then, using a "deterministic setting" as in the tests 1) and 4) a model $\Delta = 0$, $|w(t)| < 0.7$ would not be falsified. Such a model clearly would not work for control design, however. It is really not up to the designer to pick any model in the shaded area for this reason: there is more information in the data. The notion of "independence" (between u and w) is unavoidable, but not so easy to deal with if a probabilistic setting is rejected.

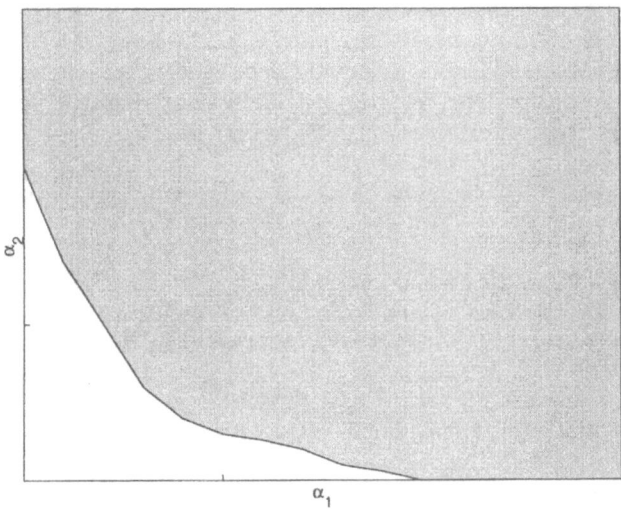

Fig. 1. Shaded area: Models that pass the test that they can explain data with a model error less than α_1 and an additive disturbance less than α_2.

One useful possibility to get over this dilemma, in a non-probabilistic setting, is to use periodic inputs. That allow a distinction of the two terms in (14): the first one $\Delta(q)u(t)$ is periodic with the same period as the input, while the disturbance w is not. This allows an independent, and non-parametric estimate of the size of the model error $\|\delta\|$ and of the amplitude or MSE of the disturbance w. Note that the periodicity is just a way to pinpoint "independence": u and w are independent if they are not both periodic. Note also that in the pessimistic "worst-case" identification setup this would not help: the worst case disturbance is likely to be one that is periodic with the same period as the input.

With this we are back at yet another interpretation of the basic test 3): an interpretation in the frequency domain of this measure is that we are checking if the different harmonic components of u and w are systematically in phase with each other.

References

1. J.R. Deller. Set membership identification in digital signal processing. *IEEE ASSP Magazine*, 4:4–20, 1989.
2. N.R. Draper and H. Smith. *Applied Regression Analysis, 2nd ed.* Wiley, New York, 1981.
3. R. Kosut, M. K. Lau, and S. P. Boyd. Set-membership identification of systems with parametric and nonparametric uncertainty. *IEEE Trans. Automatic Control*, AC-37:929–941, 1992.

16   Lennart Ljung

4. L. Ljung. Identification, model validation and control. In *Plenary Presentation at the 36th IEEE Conference on Decision and Control*, San Diego, Dec 1997.
5. L. Ljung. Model validation and model error models. In *Symposium, in Honor of Karl Johan Åström, To appear*. Lund, Sweden, August 1999.
6. L. Ljung. *System Identification - Theory for the User*. Prentice-Hall, Upper Saddle River, N.J., 2nd edition, 1999.
7. L. Ljung and H. Hjalmarsson. System identification through the eyes of model validation. In *Proc. Third European Control Conference*, volume 3, Rome, Italy, Sep 1995.
8. P. M. Mäkilä, J. R. Partington, and T.K. Gustafsson. Worst-case control-relevant identification. *Automatica*, 31(12):1799–1819, 1995.
9. B. Ninness and G. C. Goodwin. Estimation of model quality. *Automatica*, 31(12):1771–1797, 1995.
10. R. Smith and J. C. Doyle. Model validation: a connection between robust control and identification. *IEEE Trans. Automatic Control*, AC-37:942–952, 1992.
11. R. Smith and G.E. Dullerud. Continuous-time control model validation using finite experimental data. *IEEE Trans. Automatic Control*, AC-41:1094–1105, 1996.

SM Identification of Model Sets for Robust Control Design from Data

Mario Milanese and Michele Taragna

Dipartimento di Automatica e Informatica, Politecnico di Torino,
Corso Duca degli Abruzzi 24, I-10129 Torino, Italy
E-mail: milanese@polito.it, taragna@polito.it

Abstract. The problem of identifying complex linear systems from noise corrupted data is investigated, considering that only approximated models can be estimated and the effects of unmodeled dynamics have to be accounted for. The paper presents a unified view of the Set Membership Identification Theory (SMIT), as recently evolved by the authors and coworkers, aiming to deliver not a single model of the system to be identified, but a set of models, indicated as (uncertainty) model set, describing the inherent uncertainty about the system under consideration and to be used for analysis and design purposes. Optimality and convergence results of algorithms for identifying model sets are reported, related to identification problems with different settings of experimental conditions, noise assumptions and norms measuring the system approximation.

1 Set Membership Identification Theory: general formulation

The typical problem a control designer has to face in most practical situations can be briefly described as follows: a control law has to be designed, able to drive a plant to reach, if possible, given performance specifications. The plant typically is not known, but some prior information on it is available and it is possible to perform some kind of input-output measurements.

The classical approach consists in building a mathematical model of the plant, on the basis of available information on it (priors and measurements) and then designing a control that meets the desired performance specifications for the identified model. However, in this way it is not taken into account that any identified model is only an approximation of the actual plant. Indeed, the performances that can be actually achieved on the plant

---

[1] This research was supported in part by funds of Ministero dell'Università e della Ricerca Scientifica e Tecnologica, Italy, and of Japan Society for the Promotion of Science.

may be very poor, according to the size of the modeling error, and even the closed loop stability may be missed. In order to face these problems, robustness had become in past years a central issue in system and control. This shifted the attention of researchers from the study of a single model, to the investigation of set of models. Such set, often indicated as uncertainty model set, has to be suitably constructed to describe the inherent uncertainty about the system under consideration and to be used for analysis and design purposes. Model set represents the lack of sufficient information to uniquely identify the system. In this sense, model set is a natural way of representing the real world.

To be specific, let us consider a dynamic system S^0 to be controlled on the basis of some prior knowledge on it, expressed by assuming that S^0 belongs to some subset K of the set S of dynamical systems ($S^0 \in K \subseteq S$), and on N noise corrupted measurements:

$$y^N = F_N(S^o) + e^N$$

where y^N is the N-dimensional vector of measurements, F_N is the information operator indicating how the measurements depend on S^o, according to the used experimental conditions, and e^N is the noise.

Different theories have been developed, according to different assumptions on K and on noise.

Classical statistical identification theory gives deep and extensive results for the case that:

- $K = M_n(p)$, $p \in \Re^n$: set of parametric models
- e: stochastic noise with known p.d.f., filtered by a parametric noise model.

This way, the identification problem is reduced to a statistical parameter estimation problem:

$$y^N = F_N(M_n(p)) + e^N$$

Since it is assumed that $S^0 = M_n(p^0)$ for some "true" parameter p^0, only parametric uncertainty is considered, and a model set $\mathcal{M} = \{M(p) : p \in \Pi\}$ can be obtained, where Π represents the uncertainty in parameter estimation. Indeed, the assumption $S^0 = M_n(p^0)$ is in most cases not satisfied and then it results:

$$y^N = F_N(M_n(p)) + e^N + \text{unmodeled dynamics}$$

Then the problem arises of considering the case that S^o does not belongs to the model set used for the identification, so that the effects of unmodeled dynamics have to be accounted for. Some results are available in statistical identification and learning literature, mainly related to asymptotic analysis, see e.g. [1]-[3]. Indeed, evaluation of the identification accuracy with finite sample is of great importance, for example in providing suitable model sets to be used by robust control design methods.

Set Membership Identification Theory (SMIT) allows one to deal with these problems, see e.g. [4]-[17] and the references therein. SMIT develops on the following assumptions:

- K : compact subset of dynamic systems
- $e^N \in \mathcal{B}^e$, where $\mathcal{B}^e$ is a bounded set

and aims to derive, on the basis of these assumptions and of the finite number N of noise corrupted measurements y^N, a set of models $\mathcal{M}$ guaranteeing to contain S^0.

Definition 1: A set of models $\mathcal{M}$ is called a *model set* for S^0 if:

$$S^0 \in \mathcal{M}$$

■

A key role in SMIT is played by the Feasible Systems Set, defined as follows.

Definition 2: *Feasible Systems Set*

$$FSS = \left\{ S \in K : y^N = F_N(S) + \tilde{e}^N, \ \tilde{e}^N \in \mathcal{B}^e \right\}$$

■

This is the set of all systems satisfying prior information and not invalidated by measured data, and is also indicated in the literature as the unfalsified systems set. The relevance of FSS in SMIT is due to the following property.

Result 1

i) FSS is a model set for S^0
ii) If $\mathcal{M}$ is a model set for S^0, then $FSS \subseteq \mathcal{M}$

■

Thus, FSS is the "best" model set for S^0 that can be obtained on the basis of the given system and noise assumptions and of available measurements. However, in most cases FSS is not represented in a suitable form to be used by robust control design techniques, and model sets with such a property have to be looked for, e.g. in LFT description. In this paper, additive model sets are considered, of the form $\mathcal{M} = \{M + \Delta : \|\Delta\|_S \leq E(\mathcal{M})\}$, where $\| \cdot \|_S$ is a norm in the system space S used to measure the model approximation and $E(\mathcal{M}) = \sup_{S \in FSS} \|S - M\|_S$.

An identification algorithm is an operator ϕ mapping the available information, i.e. the triple $(K, \mathcal{B}^e, y^N)$, into a model $\hat{M} \in S$: $\phi(K, \mathcal{B}^e, y^N) = \hat{M}$. For notational simplicity, the dependence on y^N only will be evidenced.

A given algorithm ϕ determines a model set for S^0, as given by the following result.

Result 2

For given algorithm $\phi(y^N) = \hat{M}$, let $\hat{\mathcal{M}}$ be defined as:

$$\hat{\mathcal{M}} = \{\hat{M} + \Delta : \|\Delta\|_S \leq E(\hat{\mathcal{M}})\} \tag{1}$$

Then, $\hat{\mathcal{M}}$ is a model set for S^0

■

The identification error of ϕ and $\hat{\mathcal{M}}$ are defined as follows.

Definition 3: *Algorithm and model set identification error*
$$E_y(\phi) = E(\hat{\mathcal{M}}) = \sup_{S \in FSS} \|S - \phi(y^N)\|_S$$

■

Definition 4: *Optimal algorithm and model set, radius of information*
An algorithm ϕ^* is called optimal if:
$$\inf_\phi E_y(\phi) = E_y(\phi^*) = r_y$$
The corresponding model set $\mathcal{M}^*$ is called optimal model set and r_y is called radius of information.

■

Note that if an algorithm ϕ^* is optimal, the corresponding model set $\mathcal{M}^*$ has identification error equal to the radius of information r_y, which is the minimal identification error among all model sets for S^0. A basic result in IBC [18] relates optimal algorithms and the Chebicheff center M^c of FSS (defined as $\sup_{S \in FSS} \|S - M^c\|_S = \inf_{M \in S} \sup_{S \in FSS} \|S - \phi^c(y^N)\|_S$). Thus, the following result holds.

Result 3
Let M^c be a center of FSS. Then the algorithm $\phi^c(y^N) = M^c$ is an optimal algorithm and the corresponding model set $\mathcal{M}^c$ is an optimal model set.

■

Algorithm ϕ^c is called *central algorithm*. In many cases, methods for computing central algorithms are not known or, if known, are computationally complex. Moreover, the central estimate $M^c = \phi^c(y^N)$ may be not suitable for control design (e.g. may have high order transfer function). This motivates the interest in deriving suboptimal algorithms, giving suboptimal but simpler model sets, at the expense of some degradation in the identification accuracy. The following definition is introduced to give a measure of such a degradation.

Definition 5: *Model set suboptimality level α*
A model set $\hat{\mathcal{M}}$ is called α-optimal if:
$$E(\hat{\mathcal{M}}) \le \alpha \cdot r_y$$

■

Note that $\alpha \ge 1$, since $\inf_{\mathcal{M}} E(\mathcal{M}) = r_y$.
Widely investigated suboptimal algorithms are interpolatory algorithms, where $\phi^I(y^N) = M^I \in FSS$. The corresponding model set $\mathcal{M}^I$ is 2-optimal for any norm $\|\cdot\|_S$, see e.g. [18].

2 Specific settings

In the next sections, results will be reported on the derivation of optimal and suboptimal algorithms, on the evaluation of their finite sample identification errors and on their convergence properties for some specific settings, defined by the following choices of K (prior information on S^o), $\mathcal{B}^e$ (prior information on noise), $F_N(S)$ (experimental information) and $\|\cdot\|_S$ (measure of the identification error).

- *Prior information on S^o*
  The following sets K will be considered:
  - $K_\infty^m = \{S \in \mathcal{S} : |h_j^S| \leq L\rho^j, \ \forall j \geq m\}$
  - $K_2^m = \{S \in \mathcal{S} : \sum_{j=m}^\infty |h_j^S|^2 \rho^{-2j} \leq L^2\}$
  where $L \geq 0$ and $0 < \rho < 1$, $\mathcal{S}$ is the set of causal, LTI, SISO, BIBO stable, discrete-time systems and $h^S = \{h_0^S, h_1^S, \ldots\}$ is the impulse response of $S \in \mathcal{S}$.

- *Prior information on noise*
  The following sets $\mathcal{B}^e$ will be considered:
  - Power bounded noise:
    $$\mathcal{B}_2^e = \left\{e^N \in \Re^N : \frac{1}{\sqrt{N}}\|e^N\|_2 \leq \varepsilon\right\}$$
  - Pointwise bounded noise:
    $$\mathcal{B}_\infty^e = \{e^N \in \Re^N : \|e^N\|_\infty \leq \varepsilon\}$$
  - Boundary visiting noise: e^N is a pointwise bounded random variable such that any neighborhood of the boundary is visited with finite probability.

- *Experimental information*
  The following experimental conditions will be considered:
  - Time domain measurements of system output y with known input u, system initially at rest:
    $$y_\ell = \sum_{j=0}^\ell h_j^{S^o} u_{\ell-j} + e_\ell, \quad \ell = 0, \ldots, N-1$$
    In this case the information operator is given by:
    $$F_N\left(S^0\right) = F_N h^{S^o} \tag{2}$$
    where F_N is the semi-infinite matrix defined as:
    $$F_N = \begin{bmatrix} U_N & 0_{N \times \infty} \end{bmatrix} \in \Re^{N \times \infty} \tag{3}$$
    $$U_N = \begin{bmatrix} u_0 & 0 & \cdots & 0 \\ u_1 & u_0 & \cdots & 0 \\ \vdots & \vdots & \ddots & \vdots \\ u_{N-1} & u_{N-2} & \cdots & u_0 \end{bmatrix} \in \Re^{N \times N}$$
  - Measurements of real and imaginary part of the frequency response $S^0(\omega_k)$, $k = 1, \ldots, N/2$:
    $$y_{2k-2} = \Re e\left(S^0\left(\omega_k\right)\right) + e_{2k-2},$$
    $$y_{2k-1} = \Im m\left(S^0\left(\omega_k\right)\right) + e_{2k-1},$$
    In this case the information operator is given by:
    $$F_N\left(S^0\right) = F_N h^{S^o}$$
    where F_N is the semi-infinite matrix defined as:
    $$F_N = \begin{bmatrix} \Omega(\omega_1) & \cdots & \Omega(\omega_{N/2}) \end{bmatrix}^T \in \Re^{N \times \infty} \tag{4}$$
    $$\Omega(\omega) = \begin{bmatrix} \Re e\left(\Psi\left(\omega\right)\right) & \Im m\left(\Psi\left(\omega\right)\right) \end{bmatrix} \in \Re^{\infty \times 2}$$
    $$\Psi\left(\omega\right) = \begin{bmatrix} 1 & e^{-j\omega} & e^{-j2\omega} & \cdots \end{bmatrix}^T \in \mathcal{C}^\infty$$

Note that the case that both time domain and frequency domain measurements are available can be considered as well. In such a case, the information operator can be obtained by stacking together the information operators (3) and (4).

- *Norms $\|\cdot\|_S$ measuring the identification error*
  The following cases will be considered:
  - H_∞ identification:
    $$\|S\|_\infty^W = \sup_\omega |W^{-1}(\omega) S(\omega)|$$
    $W(\omega)$: given weighting function
  - H_2 identification:
    $$\|S\|_2 = \sqrt{\int_0^\infty |S(\omega)|^2 \, d\omega} = \sqrt{\sum_{j=0}^\infty |h_j^S|^2}$$

The following notations are used throughout the paper:

$(\,\cdot\,)^T$ Transpose of a matrix; adjoint for complex matrix;

$\bar{\sigma}(\Sigma)$ Maximal singular value of matrix Σ;

u^N Column vector $[u_0 \ldots u_{N-1}]^T \in \Re^N$;

T_N Truncation operator mapping one$-$sided sequences h into $h^N \in \Re^N$;

$\Omega_N(\omega) = T_N \Omega(\omega)$

3   Time data, power bounded noise

This setting considers:

- Time domain measurements
- Power bounded noise: $e^N \in \mathcal{B}_2^e$
- Prior information on S^o: $K = K_\infty^N$ or $K = K_2^N$
  These types of priors on S^o are indicated as *residual*, since provide *information only on the tail* of h^{S^o}. The *FSS* is given by:
  $$FSS = \left\{ S \in K : \tfrac{1}{\sqrt{N}} \|y^N - F_N h^S\|_2 \le \varepsilon \right\}$$

where F_N is given in (3). It is clear that a residual K is the "minimal" type of prior information on S^o necessary for *FSS* to be bounded, since measurements give information only on the first N samples of h^{S^o}.

In this case, the optimal model set can be easily computed as $\mathcal{M}^c = \{M^c + \Delta : \|\Delta\|_S \le r_y\}$, where the central estimate M^c and the corresponding identification error r_y are given by the next result.

Result 4 [19]-[21]

i) The optimal (central) estimate $M^c = \phi^c(y^N)$ is the FIR_N model with impulse response:
$$h^{M^c} = \{h_0^{M^c}, \ldots, h_{N-1}^{M^c}, 0, 0, \ldots\}$$
$$T_N h^{M^c} = U_N^{-1} y^N$$

ii) If $\|\cdot\|_{\mathcal{S}} = \|\cdot\|_2$ and $K = K_2^N$, then:
$$E(\mathcal{M}^c) = r_y = \sqrt{N\varepsilon^2\bar{\sigma}^2\left(U_N^{-1}\right) + L^2\rho^{2N}}$$

iii) If $\|\cdot\|_{\mathcal{S}} = \|\cdot\|_\infty^W$ and $K = K_\infty^N$, then:
$$\underline{r}_y = \sup_\omega \left| W^{-1}(\omega)\left[\sqrt{N}\varepsilon\bar{\sigma}\left(\Sigma(\omega)\right) - \frac{L\rho^N}{1-\rho}\right]\right| \leq$$
$$\leq \overset{?}{E}(\mathcal{M}^c) = r_y \leq$$
$$\leq \sup_\omega \left| W^{-1}(\omega)\left[\sqrt{N}\varepsilon\bar{\sigma}\left(\Sigma(\omega)\right) + \frac{L\rho^N}{1-\rho}\right]\right| = \bar{r}_y$$

where $\Sigma(\omega) = \Omega_N(\omega)U_N^{-1}$ ∎

Note that for obtaining reasonable identification errors, the number of samples N has to be chosen sufficiently large with respect to the impulse response decay rate of the system S^0, so that the tail contribution $\frac{L\rho^N}{1-\rho}$ is negligible. In such a case, the lower and upper bounds of the radius of information provided by the previous result for the H_∞ case are quite tight.

The optimal estimate M^c has transfer function of order N and this may be undesirable, since N may be large. For example, robust control design based on the corresponding optimal model set $\mathcal{M}^c$ may be computationally cumbersome and/or may lead to high order controllers. This motivates the interest in deriving lower order model sets.

Let $\hat{M}_n$ be a model of order n derived from some order reduction method (e.g. balanced realization of M^c and SVD reduction) and $\hat{\mathcal{M}}_n$ be the corresponding model set.

Result 5 [19,20]

i) Model set $\hat{\mathcal{M}}_n$ is α-optimal with:
$$\alpha \leq 1 + \frac{\|M^c - \hat{M}_n\|_2}{r_y}, \text{ if } \|\cdot\|_{\mathcal{S}} = \|\cdot\|_2, K = K_2^N$$
$$\alpha \leq 1 + \frac{\|M^c - \hat{M}_n\|_\infty^W}{\underline{r}_y}, \text{ if } \|\cdot\|_{\mathcal{S}} = \|\cdot\|_\infty^W, K = K_\infty^N$$

ii) The identification error of $\hat{\mathcal{M}}_n$ is bounded by:
$$E(\hat{\mathcal{M}}_n) \leq r_y + \|M^c - \hat{M}_n\|_2, \text{ if } \|\cdot\|_{\mathcal{S}} = \|\cdot\|_2, K = K_2^N$$
$$E(\hat{M}_n) \leq \bar{r}_y + \|M^c - \hat{M}_n\|_\infty^W, \text{ if } \|\cdot\|_{\mathcal{S}} = \|\cdot\|_\infty^W, K = K_\infty^N$$

where $r_y, \underline{r}_y$ and $\bar{r}_y$ are given by Result 4 ∎

Note that as $n \to N$, $\|M^c - \hat{M}_n\|_{\mathcal{S}} \to 0$ and then $\alpha \to 1$. Typically, yet for $n \ll N$, $\|M^c - \hat{M}_n\|_{\mathcal{S}} \ll r_y$ and then $\alpha \approx 1$.

Another approach for obtaining low order models is to consider algorithms ϕ_n, estimating models in a linearly parametrized set of models $M_n(p) = \sum_{i=1}^n p_i B_i(z)$, where the $B_i(z) \in \mathcal{S}$ are given (low order) transfer functions, such as Laguerre, Kautz, generalized orthonormal basis etc., see e.g. [21]-[23]. Note that $M_n(p)$ is not a model set for S^0, since in general $S^0 \notin M_n(p)$. Indeed, $M_n(p)$ is an n-dimensional linear subspace of S. Let us define:

$P_n = \begin{bmatrix} b^{(1)} & \ldots & b^{(n)} \end{bmatrix} \cdot \begin{bmatrix} b^{(1)} & \ldots & b^{(n)} \end{bmatrix}^T$;

$b^{(1)}, \ldots, b^{(n)}$: orthonormal base in subspace $T_N M_n(p)$;

$\Pi = P_n U_N^{-1}, \quad \bar{\Pi} = (I - P_n) U_N^{-1}$.

Result 6 [19]

Let $\| \cdot \|_S = \| \cdot \|_2$ and $K = K_2^N$.

Consider the algorithm $\phi_n^p(y^N) = M_n(\hat{p})$, where $\hat{p} = P_n U_N^{-1} y^N$. Let $\hat{\mathcal{M}}_n^p$ be the corresponding model set. Then:

i) the identification error of $\hat{\mathcal{M}}_n^p$ is bounded by:

$$E(\hat{\mathcal{M}}_n^p) \le \sqrt{[\sqrt{N}\varepsilon\bar{\sigma}(U_N^{-1}) + \|\bar{\Pi}y^N\|_2]^2 + L^2\rho^{2N}}$$

ii) $\hat{\mathcal{M}}_n^p$ is α-optimal with:

$$\alpha \le \frac{\sqrt{1 + \|\bar{\Pi}y^N\|_2^2}}{\sqrt{N}\varepsilon\bar{\sigma}(U_N^{-1})}$$

∎

4 Frequency data, pointwise bounded noise

This setting considers:

- Frequency domain measurements
- Pointwise bounded noise: $e^N \in \mathcal{B}_\infty^e$
- Prior information on S^o: $K = K_\infty^0$

The *FSS* is given by:

$$FSS = \left\{ S \in K_\infty^0 : \|y^N - F_N h^S\|_\infty \le \varepsilon \right\}$$

where F_N is given in (4). This *FSS* is an ∞-dimensional polytope. Methods for finding its Chebicheff center are unknown. An interpolatory estimate can be easily found as the FIR model $\hat{M}_\mu$ of order μ with impulse response:

$$h^{\hat{M}_\mu} = \arg \min_{h^\mu \in T_\mu K} \|y^N - F_{N\mu} h^\mu\|_\infty \tag{5}$$

where $F_{N\mu}$ is the matrix given by the first μ columns of F_N and μ is such that $\|y^N - F_{N\mu} h^{\hat{M}_\mu}\|_\infty \le \varepsilon$. Solution of (5) can be obtained by LP, see e.g. [24].

The model $\hat{M}_\mu$ has transfer function of order μ. Let $\hat{M}_n$ be a model of order n derived from some order reduction method (e.g. balanced realization of $\hat{M}_\mu$ and SVD reduction, or best H_∞ approximation of $\hat{M}_\mu$ within parametrized model set $M_n(p)$) and $\hat{\mathcal{M}}_n$ be the corresponding model set. The following result allows one to evaluate the H_∞ identification error and the suboptimality level of such a model set. For the sake of simplicity, the case is considered that $\hat{M}_n \in K_\infty^0$, as typically comes out. Only very slight modifications are needed if this assumption is dropped out.

Result 7 [24,25]

Let $\| \cdot \|_S = \| \cdot \|_\infty$ and $\hat{M}_n \in K_\infty^0$. Then, for given positive integers ν and ξ:

i) the identification error of $\hat{\mathcal{M}}_n$ is bounded by:

$$\sup_\omega W_\nu(\omega) - \varepsilon_\nu \le E(\hat{\mathcal{M}}_n) \le \sup_\omega W_\nu(\omega) + \varepsilon_\nu = \bar{E}$$

where:

$$W_\nu(\omega) = \max_{FIR_\nu \in FES_\nu} |FIR_\nu(\omega)|$$

$$FES_\nu = \left\{ h^\nu \in T_\nu 2K_\infty^0 : \|y^N - F_N h^{\hat{M}} - F_{N\nu} h^\nu\|_\infty \le \varepsilon + \varepsilon_\nu \right\}$$

$$\varepsilon_\nu = \frac{2L\rho^\nu}{1-\rho}$$

ii) model set $\hat{\mathcal{M}}_n$ is α-optimal with:

$$\alpha \le \overline{E}/\underline{r_y}$$

where:

$$\underline{r_y} = \sup_\omega \frac{\overline{FIR_\xi}(\omega) - \underline{FIR_\xi}(\omega)}{2} - \varepsilon_\xi$$

$$\underline{FIR_\xi}(\omega) = \min_{FIR_\xi \in T_\xi FSS} |FIR_\xi(\omega)|$$

$$\overline{FIR_\xi}(\omega) = \max_{FIR_\xi \in T_\xi FSS} |FIR_\xi(\omega)|$$

$$\varepsilon_\xi = \frac{2L\rho^\xi}{1-\rho}$$

∎

Computation of $\underline{FIR_\xi}(\omega)$ requires to evaluate the minimal euclidean distance of a polytope from the origin. This is a convex problem which can be solved by standard techniques. Computations of $W_\nu(\omega)$ and $\overline{FIR_\xi}(\omega)$ require to evaluate the maximal euclidean distance of a polytope from the origin, which is not a convex problems. An efficient method for solving such a problem can be found in [26].

$W_\nu(\omega)$ can be actually evaluated only on a finite grid of frequencies. Since $FIR_\nu \in FES_\nu$, the variation rate of $|FIR_\nu(\omega)|$ is bounded by $\max_{0 \le \omega \le \pi} \frac{d}{d\omega}|FIR_\nu(\omega)| \le \frac{2L\rho}{(1-\rho)^2}$. This information can be used to choose how coarse the gridding has to be for the intersample variation to be negligible. Moreover, the value of ν can be chosen to let the term ε_ν be negligible with respect to $W_\nu(\omega)$. Thus the identification error of the derived model set can be estimated within any desired accuracy.

5   Time data, boundary visiting noise

This setting considers:

- Time domain measurements
- Pointwise bounded noise: $e^N \in B_\infty^e$
- Prior information on S^o: $K = K_\infty^0$

The *FSS* is given by:
$$FSS = \{S \in K_\infty^0 : \|y^N - F_N h^S\|_\infty \le \varepsilon\} \tag{6}$$
where F_N is given in (3). This *FSS* is an ∞-dimensional polytope. The Chebicheff center of *FSS* can be proved to be a FIR model of order N, [19]. However, methods for its computation either are unknown or, when known (e.g. if $\| \cdot \|_S = \| \cdot \|_2$), are computationally "hard". Similarly to the previous setting, where frequency domain data are considered, an interpolatory estimate can be easily found as the FIR model $\hat{M}_\mu$ of order μ with impulse response given by (5), using for $F_{N\mu}$ the matrix of first μ columns of F_N given in (3) and $\mu \le N$ is such that $\|y^N - U_{N\mu} h^{\hat{M}_\mu}\|_\infty \le \varepsilon$. Such a value of μ can be found if (and only if) *FSS* is not empty. This last fact, interesting on its own, being equivalent to the validation of prior assumptions on S^0 and on the noise using the measured data, can be checked by solving a LP problem, as given by the following result.

Result 8
FSS is not empty if and only if $\varepsilon_p \le \varepsilon$, where:
$$\varepsilon_p = \min_{h^N \in T_N K} \|y^N - F_{NN} h^N\|_\infty \tag{7}$$
∎

As discussed in the previous section, the model set corresponding to $\hat{M}_\mu$ may be too complex to be used for robust control design and simpler model set $\hat{\mathcal{M}}_n$ can be obtained from some order reduction method. H_∞ identification error and the suboptimality level of such a model set can be obtained from Result 7, using for $F_{N\mu}$ the matrix of first μ columns of F_N given in (3). The only other difference is that $\nu \le N$ has to be used. However, for obtaining reasonable identification errors, the number of samples N has to be chosen sufficiently large with respect to the impulse response decay rate so that the tail contribution $\frac{L\rho^N}{1-\rho}$ is negligible. In such a case, ν can be chosen so that the lower and upper bounds of the model set identification error provided by Result 7 are quite tight.

Some interesting convergence properties can be proved under the further assumptions that:
- the noise e^N is boundary visiting
- the input u is persistently performing

For definitions of these concepts, see [27,28]. Indeed, these assumptions are not very severe. For example, independent random variables, whose p.d.f. has finite support $[-\varepsilon, \varepsilon]$ and is non-zero at the boundary, are boundary visiting. On the other hand, inputs typically used in identification such as Galois and random binary inputs are persistently performing, [28].

Consider the model set $\hat{\mathcal{M}}_n^p$, obtained by estimating model $M_n(\hat{p})$ in a linearly parametrized model set $M_n(p)$ of the form (3) by solving the convex problem:
$$\hat{p} = \arg\inf_p \|\hat{M}_\mu(z) - M_n(p)\|_\infty \tag{8}$$

The following convergence results hold for model sets $\hat{\mathcal{M}}_\mu$ and $\mathcal{M}_n(\hat{p})$.

Result 9 [27]
   If:

- e^N is boundary visiting
- u is persistently performing
- $\|\cdot\|_{\mathcal{S}} = \|\cdot\|_\infty$

   then, w.p.1:

i) $\lim\limits_{\mu\to\infty}\lim\limits_{N\to\infty} E(\hat{\mathcal{M}}_\mu) = 0$

ii) $\lim\limits_{\mu\to\infty}\lim\limits_{N\to\infty} E(\mathcal{M}_n(\hat{p})) = \|S^0 - M_n(p_0)\|_\infty$
   where $M_n(p_0)$ is an optimal approximation of S^0:
$$p_0 = \arg\inf_p \left\| S^0 - M_n(p) \right\|_\infty$$

iii) if $M_n(p_0)$ is unique, then:
$$\lim\limits_{\mu\to\infty}\lim\limits_{N\to\infty} \|M_n(\hat{p}) - M_n(p_0)\|_\infty = 0$$
                                                                                    ∎

Note that these convergence results hold for finite ε, while convergence results in previous SMIT literature are derived for noise bound $\varepsilon \to 0$, see e.g. [10]-[12] and the surveys [7]-[9].
   Point i) of Result 9 implies that if the measurement noise is boundary visiting and the input is persistently performing, then a FIR model can be estimated whose modeling error can be made w.p.1 asymptotically as small as desired by allowing the FIR order to appropriately increase. A similar result holds in the stochastic setting, as proved in [29]. However, such a convergence property does not hold in a SM setting if FIR estimates are obtained by least-squares, see e.g. [30]-[32].
   Points ii)-iii) of Result 9 imply that if models within a given parametric class $M_n(p)$ are looked for, then the estimated model $M_n(\hat{p})$ w.p.1 asymptotically converges to the best H_∞ approximation of S^o. A similar result holds in the stochastic setting for prediction-error approach, see e.g. [1], Th. 8.2.

6   Model sets for H_∞ robust control design

In the case that the H_∞ norm is used in the system space $\mathcal{S}$ to measure the model approximation, some more advanced results can be obtained, by taking advantage of specific features of such a norm. In this section results are presented related to the following problems:

a) looking for frequency shaped model sets;
b) tuning model sets to the closed loop measure of performance that is underlying the control design.

6.1   Frequency shaped model sets

The motivations for considering the first problem are as follows. So far, model sets of the form (1) have been considered, where uncertainty Δ has *constant bound in frequency*:

$$|\Delta(\omega)| \leq E(\hat{\mathcal{M}}), \ \forall \omega$$

A frequency shaping bound $W^\Delta(\omega)$ is more convenient for H_∞ control design. Let us consider the following model set:

$$\hat{\mathcal{M}}^W = \{\hat{M} + \Delta : |\Delta(\omega)| \leq W(\omega), \ \forall \omega\} \tag{9}$$

where:

$$W(\omega) = \sup_{\Delta \in FES} |\Delta(\omega)|$$
$$FES = \left\{\Delta \in 2K : y^N = F_N \, h^{\hat{M}} + F_N \, h^\Delta + \tilde{e}^N, \ \ \tilde{e}^N \in \mathcal{B}^e\right\}$$

Then we have the following inclusion result.

Result 10
Let $\hat{M} \in K$. Then:

i) $\hat{\mathcal{M}}^W$ is a model set for S^0
ii) $E(\hat{\mathcal{M}}^W) = E(\hat{\mathcal{M}})$
iii) $\hat{\mathcal{M}}^W \subseteq \hat{\mathcal{M}}$

∎

Thus, though the two model sets have the same identification error, designing H_∞ robust control for $\hat{\mathcal{M}}^W$ instead of $\hat{\mathcal{M}}$ can give better guaranteed performances.

We report now two results on the computation of $W(\omega)$ for the cases of power bounded and pointwise bounded noise.

Result 11 [21]
Let us consider:

- Time domain measurements
- Power bounded noise: $e^N \in \mathcal{B}_2^e$
- Prior information on S^o: $K = K_2^N$

For given $\omega \in [0, \pi]$, if $\lambda_1 \neq \lambda_2$ and $T_\Sigma^{-1}\hat{x}(\omega) \neq 0$, then:

$$W(\omega) \leq \max_i |\gamma_i| \sqrt{\hat{x}(\omega)^T \Sigma(\omega)(I + \gamma_i \Sigma(\omega))^{-2} \Sigma(\omega)\hat{x}(\omega)} + \frac{2L\rho^N}{1-\rho}$$

where:

$$\hat{x}(\omega) = \Omega_N(\omega)U_N^{-1}(y^N - F_N \, h^{\hat{M}})$$
$$\Sigma(\omega) = \left[\Omega_N(\omega)(U_N^T U_N)^{-1}\Omega_N^T(\omega)\right]^{-1}$$
$$c = N\varepsilon^2$$
$$T_\Sigma = \text{modal matrix of } \Sigma$$
$$\lambda_1, \lambda_2 = \text{eigenvalues of } \Sigma$$

and γ_i are the real zeros of the polynomial:

$$p\left(\gamma\right) = \gamma^4 c \det_\Sigma^2 + 2\gamma^3 c \det_\Sigma \operatorname{tr}_\Sigma + \gamma^2 [c(2\det_\Sigma + \operatorname{tr}_\Sigma^2) - \det_\Sigma \hat{x}^T \operatorname{Adj}_\Sigma \hat{x}] +$$
$$+ 2\gamma(c\operatorname{tr}_\Sigma - \det_\Sigma \hat{x}^T \hat{x}) + c - \hat{x}^T \Sigma \hat{x} \qquad \blacksquare$$

Result 12 [24,25]

Let us consider:

- Time or frequency domain measurements
- Pointwise bounded noise: $e^N \in \mathcal{B}_\infty^e$
- Prior information on S^o: $K = K_\infty^0$

Then, for given $\omega \in [0, \pi]$ and positive integer ν ($\nu \leq N$ for time domain measurements):

$$W_\nu\left(\omega\right) - \varepsilon_\nu \leq W\left(\omega\right) \leq \sup_\omega W_\nu\left(\omega\right) + \varepsilon_\nu$$

where $W_\nu\left(\omega\right)$ and ε_ν are given in Result 7 $\qquad \blacksquare$

Note that, in the above results, the appropriate expression (3) or (4) has to be considered for the information operator F_N, depending on whether time or frequency domain measurements are used.

6.2   Tuning model sets to control design

Let us now consider the problem of tuning the model set to the closed loop measure of performance that is underlying the control design.

A performance function of a closed loop configuration composed of plant S^0 and controller C can be formalized as an element $J\left(S^0, C\right)$ in some normed (Banach) space. The control performance cost is then measured by the norm $\left\|J\left(S^0, C\right)\right\|$, and control design methods aim to provide a controller that minimizes this cost. For given model M and controller C, it holds that:

$$\left| \left\|J\left(M, C\right)\right\| - \left\|J\left(S^0, C\right) - J\left(M, C\right)\right\| \right| \leq \left\|J\left(S^0, C\right)\right\| \leq$$
$$\leq \left\|J\left(M, C\right)\right\| + \left\|J\left(S^0, C\right) - J\left(M, C\right)\right\|$$

Minimization of the R.H.S. upper bound of the performance cost should be of interest. However, this simultaneous optimization over both M and C appears to be intractable. This has led to the introduction of several iterative schemes making use of separate stages of identification and control design, see e.g. [33]-[35] and the references therein. In the identification stage of the i-th iteration, a new model M_i is obtained by minimizing the performance degradation $\left\|J\left(S^0, C_{i-1}\right) - J\left(M, C_{i-1}\right)\right\|_\infty$, due to the fact that the controller C_{i-1} obtained in the previous iteration has been designed from M rather than from S^0. These considerations on the interaction between identification and control suggest, for given C, to use as measure of the model approximation $\|S - M\|_S = \left\| J\left(S^0, C\right) - J\left(M, C\right) \right\|_\infty$. If H_∞ performance are used, $\| \cdot \|_S$ turns out to be equivalent to a weighted H_∞ norm of the dual Youla parametrization of the plant and of the model.

To be more specific, consider a cascade unitary feedback control configuration, with C such that the closed loop is stable. Let C have right coprime factorization (rcf) $C = N_c D_c^{-1}$, P_x be any auxiliary system stabilized by C with rcf $P_x = N_x D_x^{-1}$ and define:

$$x = (D_x + C N_x)^{-1} (u + Cy) \tag{10}$$

$$v = (D_c + P_x N_c)^{-1} (y - P_x u) \tag{11}$$

$$S = D_c^{-1} \left(I + S^0 C\right)^{-1}$$

$$R^0 = D_c^{-1} \left(I + S^0 C\right)^{-1} \left(S^0 - P_x\right) D_x$$

Then it results, [33]:

$$v = R^0 x + S e$$

The next propositions show that identifying a model $\hat{M}$ of S^0 minimizing $\left\| J\left(S^0, C\right) - J\left(M, C\right) \right\|_\infty$ is equivalent to finding a model $\hat{R}$ of R^0, using input-output data x and v, obtained through (10) and (11) from input-output measurements on S^0, and minimizing a suitably weighted H_∞ norm, depending on the chosen performance measure. Two typical choices of H_∞ optimization are considered:

- *Mixed sensitivity optimization.* The mixed sensitivity design is reflected by the choice:

$$\left\| J\left(S^0, C\right) \right\|_\infty = \left\| \begin{bmatrix} V_1 \left(I + S^0 C\right)^{-1} \\ V_2 S^0 C \left(I + S^0 C\right)^{-1} \end{bmatrix} \right\|_\infty$$

with $V_1, V_2 \in \mathcal{R}H_\infty$.

- *H_∞ design based on robustness optimization.* This control design scheme proposed in [36] is reflected by the choice:

$$\left\| J\left(S^0, C\right) \right\|_\infty = \left\| \begin{bmatrix} S^0 \\ I \end{bmatrix} \left(I + C S^0\right)^{-1} \begin{bmatrix} C & I \end{bmatrix} \right\|_\infty$$

Result 13 [20]

For given M, let $R = D_c^{-1} (I + MC)^{-1} (M - P_x) D_x$. Then:

$$\left\| J\left(S^0, C\right) - J\left(M, C\right) \right\|_\infty = \left\| R^0 - R \right\|_\infty^{W_C} = \sup_\omega W_C^{-1}(\omega) \left| R^0(\omega) - R(\omega) \right|$$

where:

- for the mixed sensitivity optimization:

$$W_C^{-1} = \frac{\sqrt{\left|V_1(\omega)\right|^2 + \left|V_2(\omega)\right|^2} N_c(\omega)}{D_x(\omega) \left[1 + C(\omega) P_x(\omega)\right]}$$

- for the H_∞ design based on robustness optimization:

$$W_C^{-1} = \frac{D_c(\omega) \left[1 + \left|C(\omega)\right|^2\right]}{D_x(\omega) \left[1 + C(\omega) P_x(\omega)\right]}$$

■

Then, the problem of finding a model set for S^0, tuned to control performance, is reduced to the problem of identifying a model set for R^0, using the samples of its input and output signals x and v, obtained through (10) and (11) from measured samples of input and output of S^0, and assuming the above given weighted H_∞ norm for measuring the modeling error. Thus, previous theory can be used for deriving a model set $\hat{\mathcal{R}}^{W_{\hat{R}}} = \{\hat{R} + \Delta : |\Delta(\omega)| \leq W_{\hat{R}}(\omega), \forall\omega\}$ for R_0. Since the correspondence between systems and their dual Youla parametrization is one-to-one, once a model set for R^0 has been obtained, the corresponding model set for S^0 can be derived, as shown by the next result, which considers the case of stable controllers.

Result 14 [20]

Let C be stable and choose $N_x = 0$, $D_x = 1$, $N_c = C$ and $D_c = 1$.

Let $\hat{\mathcal{R}}^{W_{\hat{R}}} = \{\hat{R} + \Delta : |\Delta(\omega)| \leq W_{\hat{R}}(\omega), \forall\omega\}$ be a model set for R_0.

Then, $\hat{\mathcal{M}}^{W_{\hat{M}}} = \{\hat{M} + \Delta : |\Delta(\omega)| \leq W_{\hat{M}}(\omega), \forall\omega\}$ is a model set for S^0, where:

$$\hat{M} = \frac{1}{C}\left(\frac{(1-C\hat{R})^T}{|1-C\hat{R}|^2 - |C|^2 W_{\hat{R}}^2} - 1\right)$$

$$W_{\hat{M}} = \frac{W_{\hat{R}}}{|1-C\hat{R}|^2 - |C|^2 W_{\hat{R}}^2}$$

∎

7   Conclusions

In the paper, a unified view of Set Membership Identification Theory is presented. Some results developed by the authors and their coworkers are reported, related to specific settings, in particular considering open loop measurements and evaluating "hard" (guaranteed) model sets.

Results related to closed loop measurements can be found in [20] and results related to identification of "soft" (in probability) model sets can be found in [37]. A survey on results related to ℓ_1 identification can be found in [38]. The use of SMIT for robust prediction of time series is reported in [39,40]. The use of SMIT for robust control design has been investigated in [41]-[43].

For the sake of space limitation, the relations of the presented approach and results with the existing literature on related topics, as documented e.g. in [4]-[17] and in many papers of the present volume, have not been discussed. Such discussions can be found in the papers reporting each specific result here cited.

References

1. Ljung, L. (1987) System identification: theory for the user. Prentice Hall, Englewood Cliffs, NJ
2. Vidyasagar, M. (1996) A Theory of Learning and Generalization with Application to Neural Networks and Control Systems. Springer-Verlag
3. Ljung, L., Guo, L. (1996) Estimating the total error from standard model validation tests. Proc. of the 13th IFAC World Congress, vol. I, 133–138
4. Milanese, M., Tempo, R., Vicino, A., eds. (1989) Robustness in Identification and Control. Plenum Press, New York
5. Smith, R.S., Dahleh, M., eds. (1994) The Modeling of Uncertainty in Control Systems. Springer-Verlag, London
6. Milanese, M., Norton, J., Piet-Lahanier, H., Walter, É., eds. (1996) Bounding Approaches to System Identification. Plenum Press, New York
7. Milanese, M., Vicino, A. (1993) Information-based complexity and nonparametric worst-case system identification. Journal of Complexity 9, 427–446
8. Ninness, B., Goodwin, G.C. (1995) Estimation of model quality. Automatica 31, 1771–1797
9. Mäkilä, P.M., Partington, J.R., Gustafsson, T.K. (1995) Worst-case control-relevant identification. Automatica 31, 1799–1819
10. Helmicki, A.J., Jacobson, C.A., Nett, C.N. (1991) Control oriented system identification: a worst-case/deterministic approach in H_∞. IEEE Trans. Automat. Control 36, 1163–1176
11. Partington, J.R. (1991) Robust identification and interpolation in H_∞. International Journal of Control 54, 1281–1290
12. Gu, G., Khargonekar, P.P. (1992) A class of algorithms for identification in H_∞. Automatica 28, 299–312
13. Tse, D.N.C., Dahleh, M.A., Tsitsiklis, J.N. (1993) Optimal identification under bounded disturbances. IEEE Trans. Automat. Control 38
14. Zames, G., Lin, L., Wang, L.Y. (1994) Fast identification n-widths and uncertainty principles for LTI and slowly varying systems. IEEE Trans. Automat. Control 39
15. de Vries, D.K., Van den Hof, P.M.J. (1995) Quantification of uncertainty in transfer function estimation: a mixed probabilistic–worst-case approach. Automatica 31, 543–557
16. Zhou, T., Kimura, H. (1995) Structure of model uncertainty for a weakly corrupted plant. IEEE Trans. Automat. Control 40, 639–655
17. Venkatesh, S.R., Dahleh, M.A. (1997) Identification in the presence of unmodeled dynamics and noise. IEEE Trans. Automat. Control 42, 1620–1635
18. Traub, J.F., Wasilkowski, G.W., Woźniakowski, H. (1988) Information-Based Complexity. Academic Press, New York
19. Giarré, L., Kacewicz, B.Z., Milanese, M. (1997) Model quality evaluation in set membership identification. Automatica 33, 1133–1139
20. Milanese, M., Taragna, M., Van den Hof, P.M.J. (1997), Closed-loop identification of uncertainty models for robust control design: a set membership approach. Proc. of the 36th IEEE Conference on Decision and Control
21. Giarré, L., Milanese, M., Taragna, M. (1997) H_∞ identification and model quality evaluation. IEEE Trans. Automat. Control 42, 188–199

22. Wahlberg, B., Mäkilä, P.M. (1996) On approximation of stable linear dynamical systems using orthonormal functions. Automatica **32**, 693–708

23. Van den Hof, P.M.J., Heuberger, P.S.C., Bokor, J. (1995) System identification with generalized orthonormal basis functions. Automatica **31**, 1821–1834

24. Giarré, L., Milanese, M. (1996) SM identification of approximating models for H_∞ robust control. Proc. of the 35th IEEE Conference on Decision and Control, 4184–4189. Also in Int. J. of Nonlinear and Robust control (1999)

25. Milanese, M., Taragna, M. (1999) Suboptimality of approximated models in H_∞ identification. Technical Report MiTa99, Politecnico di Torino

26. Taragna, M. (1998) Uncertainty model identification for H_∞ robust control. Proc. of the 37th IEEE Conference on Decision and Control

27. Milanese, M., Taragna, M. (1997) Convergence properties in H_∞ identification with approximated models. Proc. of IFAC SYSID'97, 105–110

28. Milanese, M., Taragna, M. (1998) Inputs for convergent SM identification with approximated models. Proc. of the 37th IEEE Conference on Decision and Control

29. Ljung, L., Yuan, Z.D. (1985) Asymptotic properties of black box identification of transfer functions IEEE Trans. Automat. Control **37**, 514–530

30. Hjalmarsson, H., Ljung, L. (1994) A unifying view of disturbances in identification. Proc. of IFAC SYSID'94, vol. 2, 73–78

31. Akçay, H., Hjalmarsson, H. (1994) The least-squares identification of FIR systems subject to worst-case noise. Systems & Control Letters **23**, 329–338

32. Partington, J.R., Mäkilä, P.M. (1995) Worst-case analysis of the least-squares method and related identification methods. Systems & Control Letters **24**, 193–200

33. Van den Hof, P.M.J., Schrama, R.J.P. (1995) Identification and control – closed-loop issues. Automatica **31**, 1751–1770

34. Zang, Z., Bitmead, R.R., Gevers, M. (1995) Iterative weighted least-squares identification and weighted LQG-control. Automatica **31**, 1577–1594

35. Bitmead, R.R., Gevers, M., Partanen, A.G. (1997) Introducing caution in iterative controller design. Proc. of IFAC SYSID '97, 1701–1706

36. McFarlane, D., Glover, K. (1990) Robust Controller Design Using Normalized Coprime Factor Plant Descriptions. Springer, Berlin

37. Milanese, M., Taragna, M. (1996) H_∞ identification of "soft" uncertainty models. Proc. of the 35th IEEE Conference on Decision and Control, 2418–2423. Also in Systems & Control Letters (1999)

38. Milanese, M. (1996) Worst case ℓ_1 identification. In Milanese, M., Norton, J., Piet-Lahanier, H., Walter, É., eds., Bounding Approaches to System Identification. Plenum Press, New York

39. Milanese, M., Tempo, R. (1985) Optimal algorithms theory for estimation and prediction. IEEE Trans. Automat. Control **30**, 730–738

40. Genesio, R., Milanese, M., Tempo, R., Vicino, A. (1987) Optimal error and GMDH predictors: a comparison with some statistical techniques. International Journal of Forecasting **3**, 313–328

41. Canale, M., Malan, S., Milanese, S., Taragna, M. (1996) Model structure selection in identification for control. Proc. of the 13th IFAC World Congress, vol. I, 109–114

42. Fiorio, G., Malan, S., Milanese, M., Taragna, M. (1997) Robust design of low order controllers via uncertainty model identification. Proc. of the 2nd IFAC Symposium on Robust Control Design, 49–54

43. Canale, M., Malan, S.A., Milanese, M. (1998) Model quality evaluation in identification for H_∞ control. IEEE Trans. Automat. Control **43**, 125–132

Robust Identification and the Rejection of Outliers

Jonathan R. Partington[1] and Pertti M. Mäkilä[2]

[1] School of Mathematics, University of Leeds, Leeds LS2 9JT, UK
[2] Automation and Control Institute, Tampere University of Technology, P.O. Box 692, FIN-33101 Tampere, FINLAND

Abstract. We present some of the ideas behind the theory of worst-case identification of discrete-time systems that are approximately linear and time-invariant. The problem of outliers is discussed, and negative results are presented showing the problems that arise in H_∞ and ℓ_1 identification if the disturbances are power-bounded rather than uniformly bounded. It is possible to give a partial remedy to this situation, and it is shown that certain classes of disturbances, not uniformly small but small in various Orlicz and Lorentz norms, can be filtered out by sufficiently robust techniques.

1 Introduction

In an ideal world, we could model many processes exactly as linear time-invariant BIBO-stable systems, given by an input/output relationship in convolution form $h * u$, or

$$y(t) = \sum_{k=0}^{\infty} h(k)u(t-k), \qquad t = 0, 1, 2, \ldots, \tag{1}$$

where, conventionally, $(u(t))_{t=-\infty}^{\infty}$ is the sequence of inputs to our system, $(y(t))_{t=0}^{\infty}$ the sequence of outputs, and $(h(k))_{k=0}^{\infty}$ the system's impulse response. The standing assumptions will be that $(u(t))$ and $(y(t))$ are bounded real sequences, and that $h \in \ell_1$, i.e., $\sum_{k=0}^{\infty} |h(k)| < \infty$. Indeed, we shall generally normalize the problem in order to assume that $-1 \le u(t) \le 1$ for all t. It is also often convenient to suppose that $u(t) = 0$ for $t < 0$. An exact identification experiment would consist in supplying input values $u(k)$, measuring output values $y(k)$, and using this information to obtain h. In this highly Utopian situation, the entire process would be an exercise in elementary linear algebra.

In practice we would only be able to work with finitely many measurements, and so it would be convenient to find some finite parametrization of h: in other words, we should seek a model for h in some finite-dimensional subspace of the Banach space ℓ_1. One obvious possibility is to use a Finite Impulse Response (FIR) model for h, but there are other well-established model sets available, such as, for example, the use of Laguerre or Kautz models (see, for example, [15,20]).

The above was a pleasant dream, but now comes the awakening. There are several obvious reasons why (1) does not correspond to real life. Let us list a couple.

- Systems are hardly ever exactly linear and time-invariant.
- We expect measurement errors and other disturbances, not included in the input/output model presented above.

The first of these is a killer: if there is no genuine linear time-invariant relationship between u and y, then at first sight it makes little sense to attempt to identify h, unless we decide in what sense (1) is to be regarded as a good approximation to reality. The second objection is commonly treated in the classical (stochastic) identification literature (see, e.g., [7]) by assuming some statistical properties of the disturbance (and the term 'noise' is often employed in this case).

Let us therefore replace (1) by a more appropriate input/output relationship, as follows.

$$y(t) = \sum_{k=0}^{\infty} h(k)(u(t-k) + w(t-k)) + \Delta G(P_t u, P_t w) + e(t), \qquad (2)$$

$$t = 0, 1, 2, \ldots,$$

where w is an additive 'input error' term which is independent of u, $P_t u = (u_0, \ldots, u_t)$, $P_t w = (w_0, \ldots, w_t)$, ΔG models any distortions due to nonlinear and similar effects, and e is an 'output error' term. The use of P_t reflects the fact that, for physical reasons, we are not prepared to abandon the idea of causality—the fact that lotteries are considered successful by the organizers rather than most of the players suggests that this is a reasonable assumption.

In fact, (2) is unnecessarily complex for our purposes, since we are only interested in identifying the linear term h, and we shall rewrite it in the following form.

$$y(t) = \sum_{k=0}^{t} h(k)u(t-k) + v(t), \qquad t = 0, 1, 2, \ldots. \qquad (3)$$

Here the *disturbance term* v includes terms attributable to nonlinearities, noise, the effects of inputs $u(t)$ for $t < 0$, and so forth. The latter can in fact be assumed to be small, by taking $u(t) = 0$ for $-N \le t \le -1$, where N is sufficiently large, since in this case the additional error in $y(t)$ seen as the ghost of a long-departed input is at most

$$\left| \sum_{k=t+N+1}^{\infty} h(k)u(t-k) \right| \le \sum_{k=t+N+1}^{\infty} |h(k)|,$$

which goes to zero as $N \to \infty$.

This article gives a tutorial account of our recent attempts to understand what is meant by an acceptable disturbance, at least from the point of view of worst-case identification [9,10], together with some recent simplifications and observations that have come to light.

2   Worst-case errors

One philosophy which has attracted some interest recently is that of worst-case identification or set-membership identification, for which we refer the reader to the survey articles [11,12] as well as to the books [15,19]. The terms 'robust identification' [8] and 'control oriented identification' [4] have also been used, but these are possibly more controversial terms.

The idea here is that we may make some *a priori* assumptions concerning the sets in which h and v lie, and build such information into an identification algorithm. To some extent, is possible to treat statistical properties of disturbances in this framework, although we may then have to make assumptions based on events that happen with high probability.

A brief example will explain some of the possibilities. Suppose that h may be assumed to be lie in some given set H of systems—for example, it might be that we can specify constants $M > 0$ and $\rho > 1$ such that $|h(k)| \le M\rho^{-k}$ for all k (a measure of exponential stability). Suppose also that v may be assumed to lie in some set V—for example, there might be a fixed constant $\epsilon > 0$ such that $|v(k)| \le \epsilon$ for all k. (We shall have much more to say about this latter assumption later.) Then we could quantify the worst-case error of an identification algorithm A, producing a model $\tilde{h}_N = A(P_N y, P_N u)$ from a set of problem data $u(0), \ldots, u(N)$ and $y(0), \ldots, y(N)$, by writing

$$E_{N,\epsilon} = \sup_{h \in H, v \in V} \|h - \tilde{h}_N\|, \tag{4}$$

where we take the largest possible modelling error over all possible experiments which could have produced the given data.

There are several norms of system-theoretic interest that we can use here, of which we mention the ℓ_1 norm, $\|h\|_1 := \sum_{k=0}^{\infty} |h(k)|$, the ℓ_2 norm, $\|h\|_2 := \left(\sum_{k=0}^{\infty} |h(k)|^2 \right)^{1/2}$, and the H_∞ norm, $\|\hat{h}\|_\infty := \sup_{|z|<1} \left| \sum_{k=0}^{\infty} h(k)z^k \right|$. These are related by the well-known inequalities

$$\|h\|_2 \le \|\hat{h}\|_\infty \le \|h\|_1. \tag{5}$$

Note that if $(h(k))_{k=0}^{\infty}$ is any bounded sequence, then by convention we write

$$\hat{h}(z) = \sum_{k=0}^{\infty} h(k)z^k, \qquad |z| < 1, \tag{6}$$

and then $\hat{h}$ is analytic in the open unit disc in the complex plane.

In fact we shall not be concerned here with assuming any *a priori* information about h and v, but we shall seek 'black-box' methods of identifying h from input/output data. There will in general be some error estimate depending on h and v, and we hope to ensure that the error is small if N is sufficiently large and the disturbance satisfies some suitable conditions (for example, if it is small in some sense).

The most natural way to quantify the size of a disturbance sequence $v(0), \ldots, v(N)$ is by means of a norm on $\mathbf{R}^{N+1}$.

Possible choices include the classical ℓ_p norms,

$$\|(v(0), \ldots, v(N))\|_p := \left(\sum_{k=0}^{N} |v(k)|^p \right)^{1/p}, \qquad 1 \leq p < \infty, \tag{7}$$

with $\|(v(0), \ldots, v(N))\|_{\infty} = \max_{0 \leq k \leq N} |v(k)|$. However, it is also of interest to talk about mean p-power norm, and we write

$$\|(v(0), \ldots, v(N))\|_p' := \left(\frac{1}{N+1} \sum_{k=0}^{N} |v(k)|^p \right)^{1/p}, \qquad 1 \leq p < \infty. \tag{8}$$

For example, in the case $p = 2$, the norm $\| \cdot \|_2'$ is associated with the notion of mean power [21].

We summarise briefly what is known when we use the norm $\|v\|_{\infty}$ to quantify the size of the disturbances. We define the worst-case error of an identification algorithm based on a bounded input sequence u by

$$e_{N,\epsilon}(h) = \sup_{\|P_N v\|_{\infty} \leq \epsilon} \|h - \widetilde{h}_N\|. \tag{9}$$

Then the worst-case convergence condition is as follows:

$$e_{N,\epsilon}(h) \to 0 \qquad \text{as } N \to \infty \text{ and } \epsilon \to 0, \tag{10}$$

and this is to hold for all possible systems $h \in \ell_1$.

Not all inputs u permit a robustly convergent linear identification algorithm. Clearly the zero input will tell us nothing about h, but even a reasonable input such as a pulse fails to satisfy the given condition: for now $y = h + v$ and two distinct impulse responses h and h' can produce outputs y, y' that are close (i.e., $|y(k) - y'(k)|$ is small for all k), even though $\|h - h'\|$ is large; for the latter is to be measured in an ℓ_1, ℓ_2 or H_{∞} sense.

A necessary and sufficient condition on the input sequence u for there to exist an identification procedure satisfying (10) is as follows [13,16].

- There must exist a constant $\delta > 0$ such that

$$\|h * u\|_\infty \geq \delta\|h\| \qquad \text{for all } h \in \ell_1. \tag{11}$$

That is, we should be able to extract the signal from the disturbances.

This condition is difficult to satisfy when we use the ℓ_1 norm to measure identification errors: the *sample complexity*, the number of measurements necessary to achieve (11) for all FIR models of degree at most N, is exponential in N [1,5,18]. Fortunately, the situation is much better in other norms: for the rather friendlier H_∞ norm it is $O(N^2)$ [3], and for the truly benevolent ℓ_2 norm it is $O(N)$ [14]. All the above results were formulated for FIR models, but a recent extension to more general rational models can be found in [2].

To attain these bounds, particular inputs u are required—in the ℓ_1 case rich binary sequences (Galois sequences) are employed, in the H_∞ case sequences based on roots of unity, and in the ℓ_2 case sequences based on polynomials $p(z)$ which are large in modulus throughout the circle $|z| = 1$.

3   Rejection of outliers

To use the norm $\|v\|_\infty$ as a measure of the size of a disturbance is regarded as a rather conservative approach, since a single outlier (a measurement for which the corresponding disturbance is exceptionally large) will make the norm of v large, leading to possible doom for an identification algorithm which was designed to perform well only when the disturbances are small. We therefore ask whether worst-case convergence will be possible if another norm is used, in particular one which is less sensitive to outliers. The answer was given in [9].

Theorem 1. *Suppose that we have a sequence of seminorms $\| \cdot \|_N$ on $\mathbf{R}^{N+1}$, each satisfying*

$$\|(v_0, \ldots, v_N)\|_N \leq \max_{0 \leq k \leq N} |v_k| \qquad \text{for all } (v_0, \ldots, v_N) \in \mathbf{R}^{N+1}.$$

Then a necessary and sufficient condition for the existence of a worst-case convergent identification algorithm, that is, one satisfying $e_{N,\epsilon}(g) \to 0$ as $N \to \infty$ and $\epsilon \to 0$, for every $g \in \ell_1$, where

$$e_{N,\epsilon}(g) = \sup_{\|P_N v\|_N \leq \epsilon} \|g - \hat{g}_N\|,$$

is that there is a constant $\delta > 0$ such that

$$\liminf_{n \to \infty} \|P_n(h * u)\|_n \geq \delta\|h\| \qquad \text{for all } h \in \ell_1. \tag{12}$$

This generalizes the condition (11) given above. The theorem can also be placed in the more abstract context of recovering elements x of a normed space from values $(f_n(x))$ where (f_n) are linear functionals, and thus applies to a wide variety of other estimation problems. However, we shall restrict attention to the standard identification set-up.

In the case in which the condition is satisfied, we can give a very explicit algorithm for identification. It begins by choosing a suitable model set (such as FIR models, Laguerre models or Kautz models), and making a decision about what order our candidate model should have. This corresponds to a linear parametrization of the space of models, where we have specified how many parameters will be deemed appropriate.

Finally, we perform a minimization of the form

$$\inf_{g \in S} \|g * u - y\|_N, \tag{13}$$

using the appropriate choice of seminorm $\| \cdot \|_N$ to produce a candidate model g. Here y is our output data given by (3), and S is the set of candidate models.

For example, if the norm used were an ℓ_∞ (supremum) norm, then we should try a Chebyshev fit to the data; were it the norm $\| \cdot \|'_2$, then a least-squares fit would be called for. Looking ahead, we shall be attempting something more subtle—a Chebyshev fit, subject to allowing ourselves to reduce the impact of a certain number of outliers.

The situation may be clarified somewhat by considering another example, one of the most basic of all estimation problems. Suppose we wish to fit a straight line $y = mx + c$ to a set of data points $(x_1, y_1), \ldots, (x_N, y_N)$. Then we know that in many situations the 'best' fit is the least-squares fit, minimizing $\sum_{k=1}^{N}(y_k - (mx_k + c))^2$, or in vector notation $\min \|y - (mx + c)\|_2$. Under some circumstances, however, we might prefer the Chebyshev fit, choosing m and c to minimize $\|y - (mx + c)\|_\infty$. The problem here is that one outlier ruins the fit. However, there are many other norms of the form $\|y - (mx + c)\|$ that we might try to minimize, which are more robust against outliers, and which may be more appropriate if the errors in the data are the result of non-random factors (for example, it is conceivable that the real relation between x and y is actually $y = x + x^2/10$).

Returning to the question of identification, let us see why we might need to involve ourselves in such complications. We begin by stating without proof a negative result for ℓ_1 identification. The proof, given in [9], proceeds by testing condition (12) against the set of impulse responses of the form $(\pm 1, \pm 1, \ldots, \pm 1)$.

Theorem 2. *Let $1 \leq p < \infty$. Then, with the notation and conventions above, there is no worst-case convergent algorithm for identification where the error is measured by the ℓ_1 norm, in the presence of disturbances whose size is measured by the norm $\|v\|'_p$.*

For the H_∞ situation, it is possible to choose inputs to change the problem to a frequency-domain experiment, as suggested by Harrison et al. [3]. To do this, we choose an integer m and concatenate inputs of the form

$$c_k := (\cos(mk\lambda), \cos((m-1)k\lambda), \ldots, \cos(k\lambda), 1), \qquad (14)$$

and

$$s_k := (\sin(mk\lambda), \sin((m-1)k\lambda), \ldots, \sin(k\lambda), 1), \qquad (15)$$

where $0 \le k \le m$ and $\lambda = 2\pi/(m+1)$. If the true system h is a FIR model of degree at most m, then the linear combination $h * c_k + ih * s_k$ includes an output equal to $\hat{h}(e^{ik\lambda})$. For more details, we refer to [10], where it is also shown that the corresponding frequency-domain experiment does not show worst-case H_∞-norm convergence in the presence of disturbances whose size is measured by the norm $\|v\|_p'$ if $1 \le p < \infty$.

However the following result, which we give only for the most important case, $p = 2$ (implying an analogous result for all $1 \le p \le 2$), shows that no input sequence will enable us to avoid this problem.

Theorem 3. *With the notation and conventions above, there is no worst-case convergent algorithm for identification where the error is measured by the H_∞ norm, in the presence of disturbances whose size is measured by the norm $\|v\|_2'$.*

Proof. We fix $m \ge 1$ and consider the FIR sequences c_k and s_k defined above. Note that $(\hat{c}_k + i\hat{s}_k)(z) = e^{imk\lambda} + e^{i(m-1)k\lambda}z + \ldots + e^{ik\lambda}z^{m-1} + z^m$, so that

$$\|\hat{c}_k\|_\infty + \|\hat{s}_k\|_\infty \ge m + 1. \qquad (16)$$

Moreover, we have the following identity, valid for $|z| = 1$:

$$\sum_{k=0}^{m} |\hat{c}_k(z) + i\hat{s}_k(z)|^2 = \sum_{k=0}^{m} \left(\sum_{r=0}^{m} e^{i(m-r)k\lambda} z^r \right) \left(\sum_{r=0}^{m} e^{-i(m-r)k\lambda} z^{-r} \right)$$

$$= \sum_{k=0}^{m} \sum_{j=-m}^{m} (m+1-|j|) z^j e^{-ijk\lambda}$$

$$= (m+1)^2. \qquad (17)$$

The same is clearly true for $\hat{c}_k - i\hat{s}_k$, and by adding we obtain

$$\sum_{k=0}^{m} |\hat{c}_k(z)|^2 + |\hat{s}_k(z)|^2 = (m+1)^2. \qquad (18)$$

Let $u = (u_0, \ldots, u_n)$ be an arbitrary finite input sequence with $|u_k| \le 1$ for all k. We consider the expression

$$I := \sum_{k=0}^{m} \frac{1}{n+1} \frac{1}{2\pi} \int_{-\pi}^{\pi} (|\hat{c}_k(e^{it})|^2 + |\hat{s}_k(e^{it})|^2) |\hat{u}(e^{it})|^2 \, dt. \qquad (19)$$

We have

$$I = \frac{(m+1)^2}{n+1} \|u\|_2^2 \le (m+1)^2. \tag{20}$$

Thus there is an index k for which

$$(\|c_k * u\|')_2^2 + (\|s_k * u\|')_2^2 \le (m+1) \tag{21}$$

and hence, by the Cauchy–Schwarz inequality,

$$\|c_k * u\|_2' + \|s_k * u\|_2' \le \sqrt{2(m+1)}. \tag{22}$$

Now consider an infinite input u. Then condition (12), together with (16), implies that

$$\liminf_{n\to\infty} \|c_k * P_n u\|_2' + \|s_k * P_n u\|_2' \ge \delta(m+1). \tag{23}$$

By (22), this is only possible for each choice of k if

$$\delta \le \left(\frac{2}{(m+1)}\right)^{1/2}. \tag{24}$$

But m was arbitrary and δ was fixed, and hence we obtain a contradiction. This completes the proof.

A closely-related result can be found in [17], where connexions with the least-squares method are analysed; however, Theorem 3 does not seem to follow directly from the results of [17].

It might therefore be thought that worst-case convergence could not be obtained in any situation in which the size of the norm was measured by a less conservative bound than the supremum norm. This is not the case, and the reason that this is not the case is that one can make use of redundancies in the information. Roughly speaking, an input u must be rich enough that for any non-zero h it will generate one sufficiently large output $(h * u)(k)$; however, in practice it will generate several, and this information can be made use of.

Let us recall the definitions of two types of norm on $\mathbf{R}^N$, for which we refer to [6] for more information.

Let ϕ be a convex non-decreasing function on $[0, \infty)$ such that $\phi(0) = 0$ and $\phi(t) \to \infty$ as $t \to \infty$. These are called *Orlicz functions*, and the sort of functions we have in mind are $\phi(t) = t^p$ for fixed $p \ge 1$, and the function $\phi(t) = te^{-1/t}$. (In [9] we suggested the function

$$\phi(t) = \begin{cases} \exp(2 - 2/t) & \text{for } 0 < t \le 1, \\ 2t - 1 & \text{for } t \ge 1, \end{cases} \tag{25}$$

but $te^{-1/t}$ is a simpler alternative.)

Define the *Orlicz norm*

$$\|(v_1,\ldots,v_N)\|_\phi = \inf\{\rho > 0 : \sum_{k=1}^N \phi(|v_k|/\rho) \le 1\}, \tag{26}$$

and then renormalize it by defining

$$\|(v_1,\ldots,v_N)\|^{(\phi)} = \|(v_1,\ldots,v_N)\|_\phi/\|(1,\ldots,1)\|_\phi. \tag{27}$$

For the special case $\phi(t) = t^p$, we obtain the norms $\|(v_1,\ldots,v_N)\|_\phi = \|(v_1,\ldots,v_N)\|_p$ and $\|(v_1,\ldots,v_N)\|^{(\phi)} = \|(v_1,\ldots,v_N)\|'_p$.

An alternative class of norms is the family of *Lorentz norms*, for which we take a sequence $w = (w_k)_{k=1}^N$ of positive numbers and define

$$\|(v_1,\ldots,v_N)\|_w = \sup_{\pi \in \Sigma_N} \sum_{k=1}^N |a_{\pi(k)}| w_k, \tag{28}$$

where Σ_N denotes the set of all permutations of $\{1,\ldots,N\}$. Again we normalize the norm by

$$\|(v_1,\ldots,v_N)\|^{(w)} = \|(v_1,\ldots,v_N)\|_w/\|(1,\ldots,1)\|_w. \tag{29}$$

In fact one can construct slightly more general versions of the Lorentz norms, but we shall not require them. The choice $w = (1,0,\ldots,0)$ makes $\|v\|^{(w)} = \|v\|_\infty$. We shall principally be interested in the choice $w = (1,\ldots,1,0,\ldots,0)$, with the first K terms equal to 1. The resulting Lorentz norm is then

$$\|(v_1,\ldots,v_N)\|^{(w)} = \frac{1}{K} \sup_{S \subseteq \{1,\ldots,N\}, |S|=K} \sum_{k \in S} |v_k|, \tag{30}$$

where $|S|$ denotes the number of elements of S.

These norms are in general rather smaller than the supremum norm, and less sensitive to outliers. For example, take $N = 25$ and $v(k) = 1$ for $1 \le k \le 24$, but $v(25) = 10$. If we take $K = 5$ and compute the Lorentz norm of v, it is simply one fifth of the sum of the 5 largest elements, i.e., 14/5 or 2.8. This compares with the supremum norm of 10 and a mean square norm $\|v\|'_2 = \sqrt{124/25}$, about 2.22. The Orlicz norm is rather harder to compute (the value turns out to be about 2.42), but this norm is differentiable, and it can be used as the basis of a convex optimization algorithm.

The following result helps us distinguish between outliers and consistently large elements of a vector.

Proposition 1. *Suppose that $N \ge 3$, that $1 \le K \le N$, and that K components of a vector $v = (v_1,\ldots,v_N)$ are all at least 1 in magnitude. Let $\phi(t) = te^{-1/t}$ and $w = (1,\ldots,1,0,\ldots,0)$, with K coordinates equal to 1. Then*

$$\|v\|^{(\phi)} \ge \frac{\log K}{2\log N} \quad \text{and} \quad \|v\|^{(w)} \ge 1. \tag{31}$$

Proof.   Consider the root of the equation $xe^{-1/x} = 1/a$, where $a > e$. Writing $u = 1/x$, we have $e^{-u}/u = 1/a$; hence $u = \log a$ is larger than the root, since $e^{-\log a}/\log a < 1/a$; also $u = (\log a)/2$ is smaller than the root, since $2e^{-\log a/2}/\log a > 1/a$. Hence $\|v\|_\phi \geq \log(K)/2$ and $\|(1,\ldots,1)\|_\phi \leq \log N$, which gives the first inequality. The second inequality is clear from the definition.

Let us describe two typical experiments—a 'long' one for ℓ_1 identification [9] and a 'short' one for H_∞ identification [10].

The long experiment can be performed by convolution with a suitable input sequence of ± 1 values, which we interpret by saying that for a FIR system h of length $n + 1$ we have available 2^{n+1} corrupted measurements of the form $h_\epsilon = \epsilon_0 h(0) + \ldots + \epsilon_n h(n)$, where each $\epsilon_j = \pm 1$. If we take $N = 2^{n+1}$ and $K \sim 2^{(n+1)/4}$ (or at least the nearest integer below this), we find that at least K of the expressions h_ϵ will be of magnitude $\|h\|_1/2$ or greater, and hence the ϕ and w norms of the sequence (h_ϵ) are greater than or equal to $\delta\|h\|_1$ for some absolute constant $\delta > 0$.

As we have already seen, the short experiment uses the inputs c_k and s_k defined above, and produces a set of output values of which $2N$ are closely linked to the N values of a function $\hat{h}$ at the Nth roots of unity (here $N = m + 1$ in the notation used previously). If we look at FIR models of degree $n + 1$, and take $N \sim n^2$, then it can be shown that $K \sim n$ is a suitable choice.

In words, what is happening is as follows: in the long experiment we take $N = 2^{n+1}$ measurements, and we allow about $N^{1/4}$ of them to be unreliable. In the short experiment we take $N = n^2$ measurements, and we allow about $N^{1/2}$ of them to be unreliable. These exponents can be adjusted up or down, according to how reliable we think the information is—in informal terms, if there are more outliers, then we build in more redundancy.

4   Issues raised

We feel that there is still plenty of scope for further research. Here are a few related comments and questions.

- It remains to construct some appropriate identification algorithms. A convex optimization approach works, but there will surely be better alternatives.
- It would be of interest to consider further the ideas behind set membership identification, where the disturbances are now supposed to lie in a set given by Orlicz or Lorentz norm constraints.
- New questions of sample complexity arise—how much data is needed to identify all models of given complexity when certain assumptions are made about the allowable disturbances?

- Questions of optimal experiment design are raised. What inputs should we take?
- Various links with stochastic identification may be explored—how can **random** disturbances be included in this framework?
- For what mildly nonlinear systems can the linear part be identified by these means?

Acknowledgement

The authors are grateful to K.J. Harrison and J.A. Ward for their comments on the manuscript.

References

1. Dahleh, M.A., Theodosopoulos, T.,Tsitsiklis, J.N. (1993). The sample complexity of worst-case identification of F.I.R. linear systems. Systems Control Lett. **20**, 157–166.
2. Harrison, K.J., Partington, J.R., Ward, J.A. (1997). Complexity of identification of linear systems with rational transfer functions. Mathematics of Control, Signals and Systems, accepted.
3. Harrison, K.J., Ward, J.A., Gamble, D.K. (1996). Sample complexity of worst-case H^∞-identification. Systems Control Lett. **27**, 255–260
4. Helmicki, A.J., Jacobson, C.A., Nett, C.N. (1991). Control oriented system identification: a worst-case/ deterministic approach in H^∞. IEEE Trans. Automat. Control **36**, 1163–1176
5. Kacewicz, B.Z., Milanese, M. (1995). Optimality properties in finite sample ℓ_1 identification with bounded noise. Int. J. Adapt. Control Signal Process. **9**, 87–96.
6. Lindenstrauss, J., Tzafriri, L. (1977). Classical Banach spaces I: sequence spaces. Springer-Verlag, Berlin
7. Ljung, L. (1987). System Identification. Prentice-Hall, Englewood Cliffs, NJ
8. Mäkilä, P.M. (1991). Robust identification and Galois sequences. Int. J. Control **54**, 1189–1200.
9. Mäkilä, P.M., Partington, J.R. (1998a). On robustness in system identification. Automatica, to appear
10. Mäkilä, P.M., Partington, J.R. (1998b). Robustness in H_∞ Identification. Submitted
11. Mäkilä, P.M., Partington, J.R., Gustafsson, T.K. (1995). Worst-case control-relevant identification. Automatica **31**, 1799–1819
12. Milanese, M., Vicino, A. (1993). Information based complexity and nonparametric worst-case system identification. J. Complexity **9**, 427–446
13. Partington, J.R. (1994). Interpolation in normed spaces from the values of linear functionals. Bull. London Math. Soc. **26**, 165–170
14. Partington, J.R. (1994). Worst-case identification in ℓ_2: linear and nonlinear algorithms. Systems Control Lett. **22**, 93–98.
15. Partington, J.R. (1997). Interpolation, Identification and Sampling. Oxford University Press

16. Partington, J.R., Mäkilä, P.M. (1994). Worst-case analysis of identification—BIBO robustness for closed-loop data. IEEE Trans. Automat. Control **39**, 2171–2176.
17. Partington, J.R., Mäkilä, P.M. (1995). Worst-case analysis of the least-squares method and related identification methods. Systems Control Lett. **24**, 193–200.
18. Poolla, K., Tikku, A. (1994). On the time-complexity of worst-case system-identification. IEEE Trans. Automat. Control **39**, 944–950.
19. Smith, R.S., Dahleh, M. (Ed.) (1994). The modeling of uncertainty in control systems. Lecture Notes in Control and Information Sciences 192, Springer-Verlag.
20. Wahlberg, B., Mäkilä, P.M. (1996). On approximation of stable linear dynamical systems using Laguerre and Kautz functions. Automatica **32**, 693–708,
21. Wiener, N. (1933). The Fourier integral. Cambridge University Press.

Semi-parametric Methods for System Identification *

Michiel Krüger, Eric Wemhoff, Andrew Packard, and Kameshwar Poolla[†]

Department of Mechanical Engineering, University of California, Berkeley CA 94720, USA

Abstract. This paper is concerned with identification problems in general, structured interconnected models. The elements to be identified are of two types: parametric components and non-parametric components. The non-parametric components do not have a natural parameterization that is known or suggested from an analytical understanding of the underlying process. These include static nonlinear maps and noise models.

We suggest a novel procedure for identifying such interconnected models. This procedure attempts to combine the best features of traditional parameter estimation and non-parametric system identification. The essential ingredient of our technique involves minimizing a cost function that captures the "smoothness" of the estimated nonlinear maps while respecting the available input-output data and the noise model. This technique avoids bias problems that arise when imposing artificial parameterizations on the nonlinearities. Computationally, our procedure reduces to iterative least squares problems together with Kalman smoothing.

Our procedure naturally suggests a scheme for control-oriented modeling. The problem here is to construct uncertainty models from data. The essential difficulty is to resolve residual signals into noise and unmodeled dynamics. We are able to do this using an iterative scheme. Here, we repeatedly adjust the noise component of the residual until its power-spectral density resembles that supplied in the noise model. Our computations again involve Kalman smoothing. This scheme is intimately related to our procedure for semi-parametric system identification.

Finally, we offer illustrative examples that reveal the promise of the techniques developed in this paper.

1 Introduction

Nonlinear system identification is in its infancy. Much of the available literature treats nonlinear system identification in extreme generality, for example

---

* Supported in part by an NSF Graduate Fellowship, by NASA-Dryden under Contract NAG4-124, and by NSF under Grant ECS 95-09539.

[†] Corresponding author. Department of Mechanical Engineering, University of California, Berkeley, CA 94720, Tel. (510) 642-4642, Email: poolla@jagger.me.berkeley.edu

using Volterra kernel expansions, neural networks, or radial basis function expansions [1,3,17,28]. These studies offer asymptotic analyses, and local convergence results. While there is occasional work that incorporates *a priori* structural information about the system to be identified [31,22,30], this is the not commonly the case. It is our conviction that a completely general theory of nonlinear system identification will have little material impact on the many practical problems that demand attention. We believe that it is more fruitful to study specific classes of nonlinear system identification problems, to devise appropriate algorithms for these, and to analyze the behavior of these algorithms. This experience can be collated together with experimental studies into a broader theory.

There is available limited past work on identification of structured nonlinear systems. These include studies of the identification of Hammerstein and Weiner system [2,22,26]. However, many of the simplest problems here remain open.

This paper is concerned with identification problems of interconnected structured nonlinear systems. In particular, we are concerned with problems in which the nonlinear elements to be identified are *non-parametric*. By this we mean that these elements do not have a natural parameterization that is known or suggested from an analytical understanding of the underlying process. Such problems are particularly common in process control applications, or in nonlinear model reduction problems where an approximation is sought for a subsystem containing complex nonlinear dynamics. The other *non-parametric* components in an interconnected model are the noise models.

Non-parametric elements in an interconnected model are often handled by coercing them to assume artificial parameterizations, and then to invoke a parameter estimation procedure. For example, static nonlinearities may be parameterized by radial basis functions, polynomial expansions, or (finite) Fourier series [14,20]. Noises in an interconnected model are also treated in this fashion. The noise processes are represented as white noises driving an unknown coloring filter. A convenient order is chosen for this filter, and the filter model is determined by conventional parameter estimation methods (see for example the classical text [21]).

We submit that this technique of coercing arbitrary "parameterizations" on the inherently non-parametric elements in an interconnected model is at best inefficient, and at worst, can yield substantially biased model estimates.

We therefore suggest an alternate framework for structured nonlinear system identification. In our framework, which we term *semi-parametric system identification*, we combine parametric and non-parametric identification techniques. This permits us to identify various components in a complex interconnected model using the "most" appropriate methodology for each component. This marriage of parametric and non-parametric identification methodologies spawns several fundamental and challenging problems that are at the core of this paper.

The remainder of this paper is organized as follows. In Section 2 we develop our approach to Semi-parametric System Identification. Following this, we treat Control-oriented Modeling in Section 3. Section 4 contains several examples that reveal the early promise of our methods. Finally, we draw some conclusions in Section 5.

2   Semi-parametric Modeling

2.1   Problem Formulation

We are concerned with identification problems of interconnected structured nonlinear systems. As is well known [32], we may rearrange a general interconnected system to the linear fractional transformation (LFT) form shown in Figure 1. Here, u is the applied input signal, y is the measured output signal, and e is white noise. Also, $L(\theta)$ is a linear time-invariant system that may contain unknown parameters θ, and all the nonlinearities in the interconnection are gathered into the *static* block $\mathcal{N}$. Some or all of the components of $\mathcal{N}$ may require identification. Note also that the nonlinear block $\mathcal{N}$ will often have a block-diagonal structure, owing to the fact that nonlinearities appear in specific components in our interconnection.

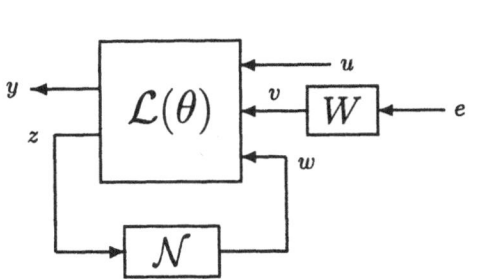

1	initialize $\mathcal{N}$, W
	while stopcheck
2	estimate θ
3	estimate w and e
	compute z
4	update W
	update $\mathcal{N}$
	end

Fig. 1. Interconnected Nonlinear System   **Fig. 2.** Solution Strategy

This is admittedly a extremely general situation. Indeed, *every* nonlinear system (and associated identification problems) can be rearranged to this LFT framework. The utility of this framework is that it allows us to explore specific classes of structured nonlinear system identification problems using a single paradigm.

2.2   Solution Strategy

We suggest a combined parametric/non-parametric approach that retains the advantages of each method. The basic structure of our approach is shown below. We begin with an input-output data record (u, y).

Step 1 is delicate. One choice is to use data from small signals experiments so as not to excite the nonlinear dynamics. We may then use classical linear time-invariant system identification methods [21] to make initial estimates of θ and W. Initial estimates of $\mathcal{N}$ are harder, and probably require *a priori* problem specific information. The estimation of θ in Step 2 may be conducted using classical parameter estimation techniques such as (linearized) maximum likelihood estimation [5,21]. We should remark that this step is quite delicate, and often results in convergence to false local minima of the likelihood function.

In Step 4, the noise model W can be updated by computing the correlation of the estimated noise e, and realizing its power spectral density. This is a classical problem in signal processing [23]. The nonlinearities $\mathcal{N}$ may be revised graphically using a scatter-plot of $\hat{z}$ versus $\hat{w}$, by using a smoothed table look-up curve fit, or by standard spline interpolation. These methods respect the "non-parametric" nature of the nonlinearities $\mathcal{N}$ and the noise model W.

The critical component of our solution strategy is Step 3. At this point, an estimate of the forward linear system $\mathcal{L}$ is available. Using the input-output data (u, y) we seek to estimate the signals (e, w, z). This is guided by two requirements. First, the graphs from the (vector-valued) signals z to w should appear to be *static* nonlinear maps, and be consistent with the *structure* of the block $\mathcal{N}$. Second, we should insist that the estimated signal e be *consistent* with any *a priori* statistical information we may have regarding the noise. Amalgamating these requirements into a computationally attractive optimization problem is at the heart of this paper, and this development is conducted below.

2.3   Cost-function formulation

To better illustrate our ideas, let us consider the simplified case shown in Figure 3 below. Here, the forward linear system $\mathcal{L}$ is known. The noise model W being known is incorporated in $\mathcal{L}$, leaving e to be a white noise sequence. The only unknown elements are the static nonlinearities, which are collected in the *structured* block labeled $\mathcal{N}$. Equivalently, this situation arises in the general problem at Step 3 of our proposed solution strategy.

Partition $\mathcal{L}$ conformably as

$$\mathcal{L} = \begin{bmatrix} \mathcal{L}_{yu} & \mathcal{L}_{ye} & \mathcal{L}_{yw} \\ \mathcal{L}_{zu} & \mathcal{L}_{ze} & \mathcal{L}_{zw} \end{bmatrix}.$$

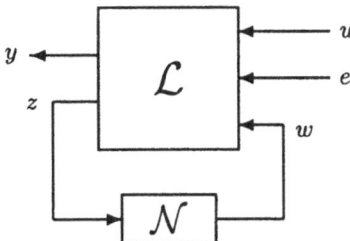

Fig. 3. Simplified Nonlinear System

We can write the set of all signals (w, e) consistent with the input-output data (u, y) as

$$\begin{bmatrix} w \\ e \end{bmatrix} = \begin{bmatrix} w^\circ \\ e^\circ \end{bmatrix} + \begin{bmatrix} B_w \\ B_e \end{bmatrix} f \tag{1}$$

where f is a free signal. Here B is a basis for the null space of $\begin{bmatrix} \mathcal{L}_{ye} & \mathcal{L}_{yw} \end{bmatrix}$, and (w°, e°) is a particular solution. Note also that the resulting signal z can be readily parameterized by f as

$$z = z^\circ + (\mathcal{L}_{zw} B_w + \mathcal{L}_{ze} B_e) f \tag{2}$$

The problem is then to select the free signal f. There are two competing criteria that must be considered in this selection. We should require that the graphs from the (vector-valued) signals z to w appear to be static nonlinear maps, and are consistent with the structure of the block $\mathcal{N}$. In addition, we should insist that this choice of f results in a signal e which could likely be a sample path produced from a noise process, consistent with any *a priori* statistical information we may have.

We now introduce a cost criterion that allows us to navigate this tradeoff. To this end, let us restrict our attention to the case where the nonlinear block $\mathcal{N}$ has block diagonal structure as

$$\mathcal{N} = \begin{bmatrix} \mathcal{N}_1 & \cdots & 0 \\ \vdots & \ddots & \vdots \\ 0 & \cdots & \mathcal{N}_r \end{bmatrix},$$

with each of the component nonlinearities $\mathcal{N}_i$ being *single-input single-output*. Partition the signals w and z conformably with the structure of $\mathcal{N}$ as

$$w = \begin{bmatrix} w_1 \\ \vdots \\ w_r \end{bmatrix}, \quad w = \begin{bmatrix} z_1 \\ \vdots \\ z_r \end{bmatrix}$$

For a *scalar* real-valued sequence q, let S_q denote the permutation operator (matrix !) that sorts the values of q in ascending order. Also, let ∇ be the difference operator with action

$$(\nabla(u))_t = u_{t+1} - u_t$$

Suppose we are given candidate signals (z, w) that are allegedly generated by $\mathcal{N}$. We wish to develop a measure of "smoothness" of the static maps obtained by interpolating these signals. Observe that $\mathcal{N}_i(z_i) = w_i$. Since $\mathcal{N}_i$ is static, it follows that $\mathcal{N}_i S_{z_i}(z_i) = S_{z_i} w_i$. This is because the operator S_{z_i} only conducts a permutation. We suggest that the cost function

$$J(z, w) = \sum_1^r \left(\|\nabla S_{z_i} z_i\|^2 + \|\nabla S_{z_i} w_i\|^2 \right) = \|\nabla S_z z\|^2 + \|\nabla S_z w\|^2 \qquad (3)$$

measures the "smoothness" of the relations from z to w. In (3) we sum the squared total-variation of the graphs of the component nonlinearities determined by interpolating the data (w, z).

In selecting the free signal f, we should reasonably expect the resulting signals w and z be interpolable by a "smooth" nonlinear map, or that $J(z, w)$ be small. In addition, we should demand that the resulting noise trajectory e retain the character of white noise. To the first approximation, if e is presumed Gaussian, we may ask that $\|e\|$ be minimized. This corresponds to selecting the most-likely noise sample path consistent with the input-output data. Combining both criteria suggests the cost function

$$J(f) = \gamma \|e\|^2 + \|\nabla S_z z\|^2 + \|\nabla S_z w\|^2 \qquad (4)$$

where $J(z, w)$ is as in (3). Note that the functional dependence of e, z, w on f is affine (see equations (1) and (2)). The parameter γ allow us to balance "noise" and "smoothness" in identifying the nonlinear maps of $\mathcal{N}$.

The optimization problem to be solved for Step 3 of our solution strategy is then

$$\min_f J(f) = \min_f \left(\gamma \|e\|^2 + \|\nabla S_z z\|^2 + \|\nabla S_z w\|^2 \right) \qquad (5)$$

We should stress that minimizing the functional $J(f)$ does not capture all the aspects we might demand of a solution to Step 3 of our solution strategy. Indeed, *whiteness* of the noise signal e, requiring that e and u be uncorrelated (as in open-loop identification) are not imposed though the optimization problem (5).

The problem (5) is, in general, a nonlinear programming problem. In the special case where the signal z is measured (i.e. $[\mathcal{L}_{ze} \mathcal{L}_{zw}]B = 0$), the optimization problem (5) simplifies considerably. In this case, the free signal f does not affect the inputs z to $\mathcal{N}$. Then, the cost criterion simplifies to

$$J(f) = \gamma \|e\|^2 + \sum_1^r \|\nabla S_{z_i} w_i\|^2$$

Since (e, w) are affine in f, and the permutations S_* are independent of f, this is a *least squares* problem !

2.4 Computational Issues

Solution of the central optimization problem (5) is nontrivial. The source of the difficulty is that the permutation operators S_z are themselves dependent on f. However, if S is fixed, the problem reduces to an instance of least squares. This immediately suggests a bootstrapping algorithm to solve (5).

To make our presentation lucid, we abuse notation and make explicit the dependence of S_z on f by writing it as $S(f)$. The problem of interest is then

$$\min_f \left(\gamma \|e\|^2 + \|\nabla S(f)z\|^2 + \|\nabla S(f)w\|^2 \right) \qquad (6)$$

The natural bootstrapping algorithm is as follows:

0	initialize $f^{(k)} = 0$
	while stopcheck
	$\quad$ compute $S = S(f^{(k)})$
1	$\quad$ compute $f^{(k+1)} =$
	$\quad\quad \arg\min_f \left(\gamma \|e\|^2 + \|\nabla Sz\|^2 + \|\nabla Sw\|^2 \right)$
	$\quad$ update $k \leftarrow k + 1$
	end

Note that Step 1 above is simply least squares. We would like to remark that in the event z is measured, this bootstrapping procedure becomes unnecssary as the problem (5) reduces to just this step. Our initial experience indicates the promise of this technique. While it is unreasonable to expect global convergence, it appears straightforward to establish a local convergence result here using the methods of [4,11].

The minimization problem of Step 1, which we capriciously refer to as "simply" least squares, requires serious attention. In the event the data record is of modest length (fewer than 200 samples), we may explicitly form the matrices ∇, S and the toeplitz matrices B_w, B_e that appear in e, w (see 1) and then use a standard least squares solver. These toeplitz matrices may be efficiently formed by determining determining state-space realization of the complimentary inner factors of $[\mathcal{L}_{ye} \mathcal{L}_{ze}]$ (see [10] for details). Such spectral factorization methods are numerically necessary because $\mathcal{L}$ may be unstable and/or non-minimum phase. This is the method employed in the second example of the next section.

More ofter, however, we are faced with having a sizeable input-output data record. In this case, storage requirements render the direct method above impractical. We therefore suggest an iterative least squares procedure [12]. For the least squares problem $\min_x \|Ax - b\|$ this iteration proceeds as

$$x^{(k+1)} = x^{(k)} - \frac{\|A^r\|^2}{\|AA^*r\|^2}r$$

where $r = Ax^{(k)} - b$. We therefore have to repeatedly compute the action of the adjoint operator A^*. In our situation A is a product of various Toeplitz operators, various permutation operators S_z, and the difference operator ∇. The adjoints of the Toeplitz operators involve anti-causal filtering. Adjoints of permutation operators are themselves permutations and can be rapidly found with a sorting procedure. The adjoint of the difference is again a (backward) difference operator. This iterative least squares procedure is employed in the first example of the next section.

These methods for solving Step 1 above can be slow. An interesting open problem is to exploit the special structure of our least squares problems (i.e. Toeplitz, difference, and permutation operator products) to devise a fast recursive least squares solver. The impediment here appears to be the permutation operators, without which the problem reduces to standard Kalman Smoothing [15].

3   Control-oriented modeling

Much of control theory is predicated upon mathematical models for physical systems. These models describe how inputs, states, and outputs are related. It is well recognized that one rarely, if ever, has an exact and complete mathematical model of a physical system. Thus, along with a mathematical model, one should also explicitly describe the uncertainty which represents the possible mismatch between the model and the physical system. This uncertainty is both in the signals and the systems. Thus, a complete model must include a nominal model together with descriptions of the signal and the system uncertainty. We will refer to such models as *uncertainty models*. Such uncertainty models are the starting point for modern robust control methods such as $\mathcal{H}_\infty$ or ℓ_1 optimal control.

An uncertainty model has four components: a nominal model M, system uncertainty Δ, inputs and outputs u, y and uncertain signals e. A very general class of uncertainty models is depicted in Fig. 4. Here e is unit-variance white noise (any color in the noise may be incorporated into the model M), and the operator Δ represents modeling uncertainty. The operator Δ is not known, but is norm-bounded by the known constant γ. The objective of control-oriented modeling is to arrive at such uncertainty models using *a priori* information together with input-output data.

To this end, suppose we postulate the *structure* of an uncertainty model as shown in Fig. 4. For instance, this could consist of nominal models of various components in a complicated interconnected system (determined using classical parameter estimation of system identification methods) together with standard multiplicative uncertainty models for each component. Suppose, in addition, we have available an input output data record $\{u_t, y_t\}_{t=0}^{L-1}$. We wish to determine the smallest norm bound γ so that the uncertainty model could have "reasonably" generated the given input-output data.

The essential problem here is as follows. We can readily compute the residual signal using the nominal model and the given input-output data. The difficulty is to decompose this residual into a component due to the noise e and a component due to the undermodeling Δ. The only information that assists us in making this decomposition is that the noise e is white, and uncorrelated with the input signal u. Unfortunately, we have available only a finite sample data record on which to impose these ensemble average restrictions. Once the decomposition is complete, the signals w and z may be computed and we can use standard deterministic model validation methods to compute the minimal norm bound γ.

Earlier treatments of control-oriented modeling have relied on treating the noise e as *deterministic*. While this may be appropriate in some circumstances, it leads to overestimating the amount of unmodeled dynamics if the "true" noise is random. As suggested by [16,13], regarding the noise signals as *deterministic* results in a worst-case assignment of their sample paths, resulting in their being highly correlated with the input signal. In addition, hard bounded noise models preclude special treatment of outliers in the data, again contributing to under-estimating the amount of undermodeling. This is due to the fact that accounting for occasional but unlikely large values of the disturbance signal, the noise norm bound must be unnecessary large.

In order to address this conservatism in assessing the amount of unmodeled dynamics, it seems far more appropriate [6,7] to use a norm-bounded modeling uncertainty description together with a stochastic noise model. We suggest a particular realization of this approach which is described below.

Consider the uncertainty model of Fig. 4. The set of signals (e, v) consistent with this data record forms an (affine) subspace. It then follows that (assuming normally distributed noises) of these signals, the particular choice e^{opt} with smallest $\|e\|$ is most likely. This choice, however, may not well correspond to a sample path of unit-variance white noise. Based on the above discussion, the problem we wish to solve is

$$\min_e \left\{ \|e\| \text{ subject to } \begin{bmatrix} y \\ z \end{bmatrix} = M \begin{bmatrix} u \\ e \\ v \end{bmatrix} \text{ and } e \text{ is white} \right\} \qquad (7)$$

Note that the first constraint above requires that the estimates signals (e, v, z) be consistent with the input-output data. It is straightforward to solve this problem *without* the requirement that e be white using standard Kalman smoothing. This suggests an iterative scheme to solve the optimization problem (7) as outlined in Fig. 5 (compare with our scheme for semi-parametric modeling Fig. 2). Essentially, we update a whitening filter F to force the estimated signal e to appear white. Numerical experiments reveal that a Newton update $W = WF^{\frac{1}{4}}$ performs quite well.

The details of this procedure can be found in [19].

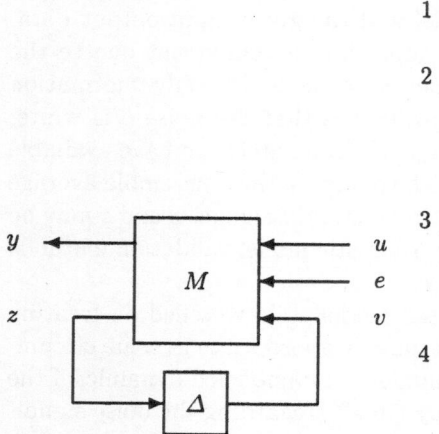

1 initialize $W = I$
while stopcheck
2   estimate e, v
    by Kalman smoothing :
    $$\min_{e,v} \lambda^2 \|e\|^2 + \|v\|^2$$
    subject to i-o data
3   find whitening filter F
    using estimated e
    set $W = WF^{\frac{1}{4}}$
end
4 compute z, estimate $\|\Delta\|$
using deterministic validation
mmethods [25,29]

Fig. 4. Uncertainty Model   **Fig. 5.** Control-oriented modeling scheme

4 Examples

Consider the model of Figure 6. Here, the nonlinearities $\mathcal{N}_1$ and $\mathcal{N}_2$ are to be identified. The (possibly unstable) linear systems L_1 and L_2 are known.

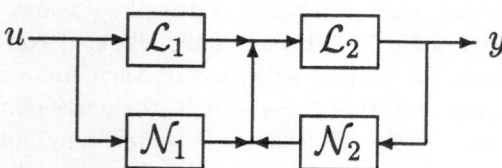

Fig. 6. Example 1

In this situation, it is easy to verify that we essentially have access to the signal $q = (\mathcal{N}_1 u + \mathcal{N}_2 y)$. This follows because we can write $q = \mathcal{L}_2^{-1} y - \mathcal{L}_1 u$. The "inversion" here is to be conducted using Kalman smoothing methods as $\mathcal{L}_2$ need not be minimum-phase. From the signal q we are to infer the individual nonlinear maps. Plotted in the panels above are the results of

our technique as applied to this example. The solid lines are the graphs of the true nonlinearities. The dots are the estimates based on 2000 samples

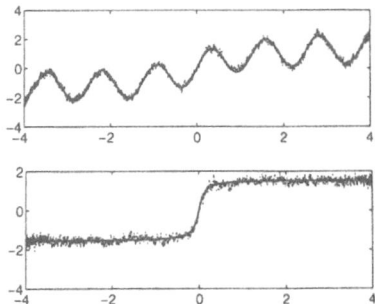

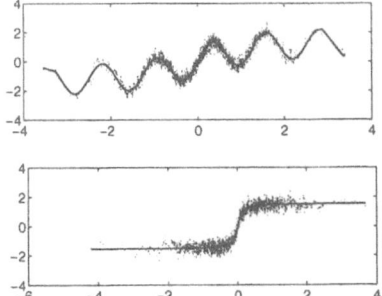

Fig. 7. Example 1: White-noise input   **Fig. 8.** Example 1: Multi-tone input

of input-output data using a white-noise input and a bandpass multi-tone input. In both cases, the data was corrupted by additive output noise at a signal-to-noise ratio of $15db$. The noise is not, however, incorporated in the modeled structure. Here, we used the iterative least squares procedure (see Section 4), for which we found 200 iterations to be adequate. The size of our input-output data record compelled us to use the iterative method.

In our second example, we have a single static nonlinearity in feedback around a known, randomly generated linear system $\mathcal{L}$ with 2 states, 2 exogenous inputs u, one noise input e, and a single measured output y. The model structure is shown in Figure 9 below. Other details about this example may be found in the thesis [8].

With the noise being incorporated explicitly in our model structure we can use the methods of Section 3 to estimate both $\mathcal{N}$ and the signal e. A 150 sample data record was generated with no undermodeling. We did not employ iterative least squares methods, but instead chose to form the permutation matrices and invoke the MATLAB solver. This was possible given the modest size of our input-output data record. Plotted below are the true nonlinearity (solid line) and its estimate (dots). In the second panel, we show 100 samples of the "true" (solid) and estimated (dotted) noise sequence.

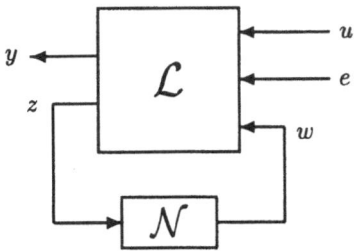

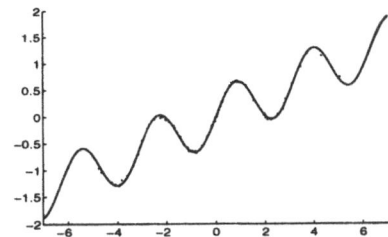

Fig. 9. Example 2: Model Structure   **Fig. 10.** Example 2: Estimates of the nonlinearity

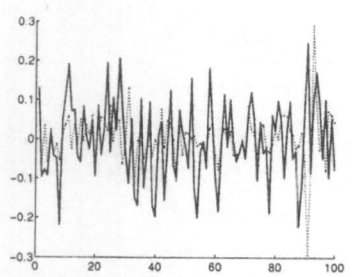

Fig. 11. Example 2: Estimates of the noise

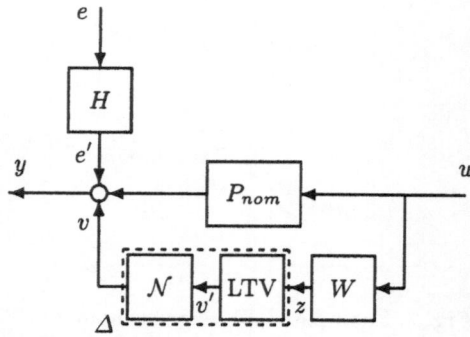

Fig. 12. Data-generating system

A simple numerical example will be used to show the ability of the proposed iterative approach to recover the noise and unmodeled dynamics residual signals e and v. Once these signals are estimated, standard validation techniques [25,29] can be used to estimate the size of the uncertainty. The model in this example consists of a second order low-pass filter P_{nom} with a lightly damped resonance peak at $z = 0.9893 \pm 0.0396i$,

$$x_{k+1} = \begin{bmatrix} 0.9797 & -0.0407 \\ 0.0407 & 0.9988 \end{bmatrix} x_k + \begin{bmatrix} -0.4469 \\ 0.0445 \end{bmatrix} u_k$$

$$y_k = \begin{bmatrix} -0.4469 & -0.0445 \end{bmatrix} x_k + 0.0995 u_k$$

disturbed by a filtered white noise sequence and a weighted additive uncertainty, which consists of an LTV system and a static nonlinearity $\mathcal{N}$, see Fig. 12. The weighting for the uncertainty is given by

$$W(z) = \frac{0.4786 - 0.4738z^{-1}}{1 - 0.9048z^{-1}}$$

and the white noise disturbance signal e is filtered by

$$H(z) = \frac{0.0153 - 0.0141z^{-1}}{1 - 0.9608z^{-1}}.$$

The static nonlinearity is defined as

$$\mathcal{N} : v_k = \frac{\mathrm{atan}(100v'_k)}{50}.$$

The LTV system consisted of a slowly varying gain between -1 and 1. A low frequency input signal u and a unit variance white noise signal e were used to generate data for the validation procedure; 4096 uniformly spaced samples were used. The signals v and $e' = H(z)e$ were chosen such that $\|e'\|^2 \approx \|v\|^2$.

For each estimated noise signal $\hat{e}$ the spectral density was computed at 128 uniformly spaced frequency points between 0 and π. A fourth order, stable, minimum-phase (to avoid unstable behavior of M_W) system $F(e^{i\omega})$ was then estimated such that $|F(e^{i\omega})|^2 \approx \Phi_{\hat{e}}(\omega)$, where $\Phi_{\hat{e}}(\omega)$ denotes the spectrum of e. After 10 iteration steps the spectrum of e was (almost) flat. Fig. 13 shows the estimated noise and uncertainty signals, resp. $\hat{e}$ and $\hat{v}$

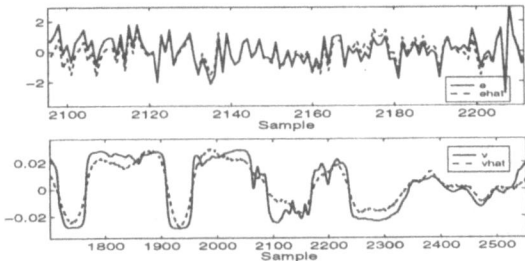

Fig. 13. Estimated signals $\hat{e}$ and $\hat{v}$ and real signals e and v.

together with the real signals e and v. Both $\hat{e}$ and $\hat{v}$ are good approximations of the real signals e and v. Besides, the cross-correlation between $\hat{e}$ and u is

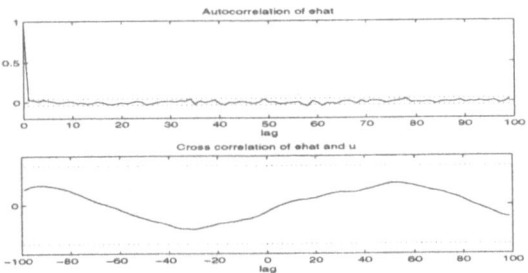

Fig. 14. Auto- and cross-correlation of resp. $\hat{e}$ and $\hat{e},u$ together with 99% confidence bounds.

within the 99% confidence bounds, see Fig. 14. Hence, we can conclude that the iterative procedure has successfully approximated the real signals e and v such that the $(\hat{e}, \hat{v})$-pair is feasible without violating the assumptions on the estimated noise signal $\hat{e}$.

5   Conclusions

We have presented a novel method for identification of static nonlinearities embedded in a general interconnected system with known structure and linear components. The appeal of this method is twofold. First, it combines a smoothness criteria (to reflect our desire that the estimated nonlinearity be static) with a noise penalty term. Second, we have offered computationally tractable methods exist for optimizing this cost functional. We have also described how our method fits into a larger framework for identification of general interconnected nonlinear systems.

In the future, we need to develop and implement more sophisticated computational algorithms for these methods, as well as address convergence issues. Further, a consideration of identifiability is essential. Finally, we wish to further address the situation when the linear system is unknown, as described in Section 2.2.

References

1. S. A. Billings, S. Y. Fakhouri, "Identification of a Class of Non-linear Systems Using Correlation Analysis," *Proc. IEE*, v.125, pp. 691-7, 1978.
2. M. Boutayeb, M. Darouach, H. Rafaralahy, G. Krzakala, "A New Technique for Identification of MISO Hammerstein Model," *Proc. American Control Conf.*, San Francisco, CA, v.2, pp. 1991-2, 1993.
3. C. Chen, S. Fassois, "Maximum likelihood identification of stochastic Wiener-Hammerstein-type non-linear systems," *Mech. Syst. Sig. Proc.*, vol. 6, no. 2, pp. 135-53, 1992.
4. J. E. Dennis, Jr. and R. B. Schnabel, *Numerical methods for unconstrained optimization and nonlinear equations*, Prentice-Hall, 1983.
5. R. Deutsch, *Estimation Theory*, Prentice-Hall, 1965.
6. D. K. De Vries, *Identification of Model Uncertainty for Control Design*, Ph.D. thesis, Delft University of Technology, Delft, the Netherlands.
7. D. K. De Vries and P. M. J. Van den Hof, "Quantification of uncertainty in transfer function estimation: a mixed probabilistic - worst-case approach," *Automatica*, vol. 31, no. 4, pp. 543–557, 1995.
8. E. Wemhoff, "A Nonparametric Method for the Identification of a Static Non-linearity in a Structured Nonlinear System," M.S. Thesis, Department of Mechanical Engineering, University of California, Berkeley, August 1998.
9. A. H. Falkner, "Identification of the system comprising parallel Hammerstein branches," *Int. J. Syst. Sci.*, vol. 22, no. 11, pp. 2079-87, 1991.
10. B. Francis, *A course in H_∞ control theory*, Springer-Verlag, 1987.
11. P. E. Gill, W. Murray, and M. H. Wright, *Practical optimization*, Academic Press, 1981.
12. G. Golub and C. F. Van Loan, *Matrix Computations*, 2nd edition, The Johns Hopkins University Press, 1989.
13. G. C. Goodwin, M. Gevers and B. Ninness, "Quantifying the error in estimated transfer functions with application to model order selection," *IEEE Trans. Autom. Control*, vol. 37, no. 7, pp. 913–928, 1992.
14. W. Greblicki, "Non-parametric orthogonal series identification of Hammerstein systems," *Int. J. Syst. Sci.*, vol. 20, no. 12, pp. 2355-67, 1989.
15. M. S. Grewal, *Kalman filtering : theory and practice*, Prentice-Hall, 1993.
16. H. Hjalmarsson, *Aspects on Incomplete Modeling in System Identification*, Ph.D. thesis, Electrical Engineering, Linköping University, Linköping, Sweden, 1993.

17. T. A. Johansen, "Identification of Non-linear Systems using Empirical Data and Prior Knowledge–An Optimization Approach," *Automatica*, vol.32, no.3, pp. 337–56, 1997.
18. A. Juditsky, H. Hjalmarsson, A. Benveniste, B. Delyon, L. Ljung, J. Sjöberg, and Q. Zhang, "Nonlinear Black-box Models in System Identification: Mathematical Foundations," *Automatica*, vol.31, no.12, pp. 1752–1750, 1995.
19. M. V. P. Krüger and K. Poolla, "Validation of uncertainty models in the presence of noise," *Sel. Topics in Identification, Modeling and Control*, vol. 11, Delft University Press, pp. 1–8, 1998.
20. A. Kryzak, "Identification of discrete Hammerstein systems by the Fourier series regression estimate," *Int. J. Syst. Sci.*, vol. 20, no. 9, pp. 1729-44, 1989.
21. L. Ljung, *System Identification, Theory for the User*, Prentice-Hall, Inc., Englewood Cliffs, New Jersey, 1987.
22. K. S. Narendra, P. G. Gallman, "An Iterative Method for the Identification of Nonlinear Systems Using the Hammerstein Model," *IEEE Trans. Autom. Control*, vol. 11, no. 7, pp. 546-50, 1966.
23. A. V. Oppenheim and R. W. Schafer, *Discrete-time signal processing*, Prentice-Hall, 1989.
24. A. Poncet and G. S. Moschytz, "Selecting inputs and measuring nonlinearity in system identification," *Proc. Int. Workshop on Neural Networks for Identification, Control, Robotics, and Signal/Image Processing*, Venice, pp. 2–10, 1996.
25. K. Poolla, P. Khargonekar, A. Tikku, J. Krause and K. Nagpal, "A time-domain approach to model validation," *IEEE Trans. Autom. Control*, vol. 39, no.5, pp. 951–959, 1994.
26. M. Pawlak, "On the series expansion approach to the identification of Hammerstein systems," *IEEE Trans. Auto. Contr.*, vol. 36, no. 6, pp. 763-7, 1991.
27. R. Sen, P. Guhathakurta, "On the solution of nonlinear Hammerstein integral equation in $L_2(0,1)$," *Trans. Soc. Comp. Simu.*, vol. 8, no. 2, pp. 75-86, 1991.
28. J. Sjöberg, Q. Zhang, L. Ljung, A. Benveniste, B. Delyon, P. Y. Glorennec, H. Hjalmarsson, and A. Juditsdy, "Nonlinear Black-box Modeling in System Identification: a Unified Overview," *Automatica*, vol.31, no.12 , pp. 1691–1724, 1995.
29. R. Smith, G. Dullerud, S. Rangan and K. Poolla, "Model validation for dynamically uncertain systems," *Mathematical Modelling of Systems,* vol. 3, no. 1, pp. 43–58, 1997.
30. P. Stoica, "On the Convergence of an Iterative Algorithm Used for Hammerstein System Identification," *IEEE Trans. Autom. Control*, vol. 26, no. 4, pp. 967-69, 1981.
31. G. Vandersteen and J. Schoukens, "Measurement and Identification of Nonlinear Systems consisting out of Linear Dynamic Blocks and One Static Nonlinearity," *IEEE Instrumentation and Measurement Technology Conference*, vol.2, pp. 853–8, 1997.
32. G. Wolodkin, S. Rangan, and K. Poolla, "An LFT approach to parameter estimation," *Proc. American Control Conf.*, Albuquerque, NM, pp. 2088-2092, 1997.

Modal Robust State Estimator with Deterministic Specification of Uncertainty

John Norton

School of Electronic and Electrical Engineering, University of Birmingham
Edgbaston, Birmingham B15 2TT, UK

Abstract. The subject of the paper is updating of hard bounds on state for a system with a linear, time-invariant model with additive process and observation noise and a deterministic specification of uncertainty in the parameters of the model and the noise variables. The instantaneous values of the uncertain parameters and the noise are individually confined within specified bounds. The basic operations involved in the time and observation updates are examined and the approximations needed to deal with non-linearity, non-convexity and excessive complication of the bounds are considered.

1 Introduction

In the long history of discrete-time state estimation, uncertainties in the state-transition equation and observation equation have most often been treated as random variables adding to the process and observation noise and contributing to their covariances [1,2]. A common alternative is to deal with structured uncertainty by augmentation of the state-variable model. For example, an unknown bias in an instrument, an unknown constant parameter or an unknown constant disturbance might be modelled as a state variable with no forcing and *new value=old value*, and unknown time-varying parameters as state variables executing random walks or evolving with slow poles [3-5]. A drifting mean or autocorrelation structure in the noise can be handled similarly. As the characteristics of the uncertainties are themselves often not well known, the extra state variables in such an auxiliary model may be accompanied by extra unknown parameters. As well as increasing the computing load, such augmentation carries the risk of overmodelling, reducing statistical efficiency and possibly causing ill conditioning. A further difficulty is that treating model uncertainty as noise or modelling it through auxiliary state variables may obscure the source of uncertainty and complicate its description. As an example, consider an uncertain time constant in a linear, time-invariant system. This is a simple parameter uncertainty if the model is in modal form but a more complicated parametric uncertainty if not. If its effects are treated as pseudo-noise, the noise sequence is autocorrelated, input-dependent and quite possibly non-stationary and non-zero-mean. For these reasons, retention of the original parametric uncertainty, in a model structure chosen to make it explicit, is a better option.

Parametric uncertainty may be regarded as probabilistic or deterministic, specifying the uncertainty respectively through a parameter-error covariance or through bounds on some norm of the parameter error. The former superficially suits the Kalman filter formulation but involves products of random variables (an uncertain parameter multiplying unknown forcing or an uncertain state variable), rendering a linear model non-linear in the unknowns. To avoid this difficulty, a conventional probabilistic noise specification might be combined with a deterministic specification of parameter uncertainty, using bounds on the parameter values to define worst cases for updating the state estimate and state-error covariance. So-called sensitivity analysis is then carried out, testing the estimator empirically over a range of cases. A more systematic treatment of deterministic uncertainty in the model is desirable, and it would also be of interest to permit the system to vary with time within the model bounds. For these reasons, a completely deterministic problem formulation is considered here; in addition to the uncertain parameters being specified only as between given bounds, the process and observation noise are characterised only by bounds on the instantaneous values of their constituent variables (ℓ_∞ bounds on the vector of successive values of each scalar noise variable but "box" bounds on the noise vector at any one instant). Independent bounds on instantaneous values allow the physically separate sources of forcing and separate origins of observation noise to be distinguished, and avoid any assumption about the serial behaviour of the noises. To retain contact with the uncertainties in the underlying continuous-time system, the model will be assumed to be in decoupled, modal form, with uncertain poles and modal gains. The problem is then to compute, on receipt of each new observation, the feasible set of all state values compatible with the observations so far, the uncertain model and the bounds on process and observation noise.

The aim here is to identify the basic computational operations and suggest approximations where necessary. One starts from a fairly mature technology for state bounding with models which are not uncertain. With instantaneously bounded process and observation noise in a linear model, linear (hyperplane) initial bounds on state define a polytope feasible set, which remains a polytope as the bounds evolve according to the dynamics and as new linear bounds are introduced by the observations. However, the polytope typically becomes complicated as more observations are processed. There are established algorithms [6,7] for exact bounding while the computing load is acceptable. When it is not, the feasible set must be approximated. For most purposes an outer approximation is required, including all feasible

values but exaggerating the uncertainty. The most economical approximation schemes employ ellipsoids [8,9] or parallelotopes [10] with complexity
independent of the number of observations. These approximating sets are
symmetrical and convex. Symmetry implies looseness whenever the exact
bounds are strongly asymmetrical. Cumulatively, the errors due to outerbounding approximations may make the bounds too loose to be useful.
Convexity is not a serious restriction in a completely linear problem, but
in state-estimation problems with linear models containing some unknown
parameters, the non-convexity incurred by the presence of products of unknowns prevents tight approximation.

With these factors in mind, ellipsoidal or parallelotopic approximation
will not be considered, but the potential for piecewise linear approximation
will be considered.

2   Problem formulation

The LTI model is

$$x_{k+1} = \Lambda x_k + \Gamma w_k \ , \quad \Lambda \equiv \mathrm{diag}(\lambda_i) \tag{1}$$
$$y_{k+1} = H x_{k+1} + v_{k+1} \tag{2}$$

where k denotes sample instant, state $x_k \in \mathbb{R}^n$, unknown forcing $w_k \in \mathbb{R}^m$
and observation $y_k \in \mathbb{R}^r$. The poles λ_i, $i = 1, 2, \ldots, n$ are for simplicity
assumed distinct and non-zero. Each non-zero element of Λ, Γ and H is in
a specified range (with coincident ends if it is known exactly):

$$\Lambda \in \mathcal{L} \equiv \{\Lambda|\ \check{\lambda}_i \le \lambda_i \le \hat{\lambda}_i,\ i = 1, \ldots, n\}$$

$$\Gamma \in \mathcal{G} \equiv \{\Gamma|\ \check{\gamma}_{ij} \le \gamma_{ij} \le \hat{\gamma}_{ij},\ i = 1, \ldots, n,\ j = 1, \ldots, m\} \tag{3}$$

$$H \in \mathcal{H} \equiv \{H|\ \check{h}_{li} \le h_{li} \le \hat{h}_{li},\ l = 1, \ldots, r,\ i = 1, \ldots, n\}$$

and Γ is of full column rank, since otherwise the number of variables making
up w could be reduced. Each element of w_k, v_{k+1} is in a specified constant
range:

$$w_k \in \mathcal{W} \equiv \{w_k|\ \check{w}_j \le w_{kj} \le \hat{w}_j,\ j = 1, \ldots, m\} \tag{4}$$
$$v_{k+1} \in \mathcal{V} \equiv \{v_{k+1}|\ \check{v}_l \le v_{kl} \le \hat{v}_l,\ l = 1, \ldots, r\} \tag{5}$$

Note that the noise sequences $\{w\}$ and $\{v\}$ need not be white or zero-mean.

In a practical context, each real λ and the real part of each complex
λ is typically known in sign and approximate size from the corresponding
time constant. The approximate angle of each complex-conjugate pair of
λ's is typically known from the ringing frequency. Each modal gain is approximately known and normalisation of x can be chosen to make any n
of the non-zero h_{li} or γ_{ij} unity, simplifying the problem if the system is

single-input or single-output. Further, the sign of each h_{li} and γ_{ij} (but not w_{kj} or v_{kl}) is usually known.

The process at each of the sample instants $k = 0, 1, 2, \ldots$, is to update the feasible set $\mathcal{X}$ of state x by a time update and an observation update.

Time update: form

$$\mathcal{X}'_{k+1} \equiv \{x'_{k+1} | \; x'_{k+1} = \Lambda x_k, \; x_k \in \mathcal{X}_k, \; \Lambda \in \mathcal{L}\} \tag{6}$$
$$\mathcal{X}''_{k+1} \equiv \{x''_{k+1} | \; x''_{k+1} = x'_{k+1} + \Gamma w_k, \; x'_{k+1} \in \mathcal{X}'_{k+1}, \Gamma \in \mathcal{G}, w_k \in \mathcal{W}\} \tag{7}$$

Observation update: form

$$\mathcal{X}_{k+1} \equiv \{x_{k+1} | \; y_{k+1} = H x_{k+1} + v_{k+1}, \; x_{k+1} \in \mathcal{X}''_{k+1}, H \in \mathcal{H}, v_{k+1} \in \mathcal{V}\} \tag{8}$$

3   Basic operations involved in updates

3.1   Time update

Consider first the effect of the state transition $x'_{k+1} = \Lambda x_k$.

For **all poles real**: at any one value of Λ, $\mathcal{X}_k \to \mathcal{X}'_{k+1}$ merely rescales each coordinate of $\mathcal{X}_k$, so there is no change in which bounds are active in defining the feasible set and no rotation of the set. In particular, if $\mathcal{X}_k$ is an axis-aligned box, so is $\mathcal{X}'_{k+1}$. Within any orthant, the behaviour of the new bound due to any hyperplane bound of $\mathcal{X}_k$ as Λ varies over its range can be seen by considering its intercepts on the coordinate axes. It remains a hyperplane and if its axis intercepts are all on the half axes defining the orthant, all points on it reach their furthest from the origin when the intercepts of the transformed hyperplane on those half axes are all at their furthest from the origin. This is when the λ_i, $i = 1, \ldots, n$ are at their appropriate extremes (upper if, as usual, all feasible values of each λ_i are positive). If the transformed hyperplane intersects some half axes *not* defining the orthant, its intercepts there must be at their smallest to maximise the distance of the hyperplane from the origin in the orthant. The opposite extremes of the λ_i determine the closest approach of the transformed hyperplane to the origin. Fig. 1 shows an example with n=2.

Within an orthant, the hyperplane bound of $\mathcal{X}_k$ thus transforms, as Λ varies, to all hyperplanes between an inner and an outer hyperplane, not generally parallel, one of which will dominate as a bound on $\mathcal{X}'_{k+1}$. The dominant bounds due to all the hyperplane bounds of $\mathcal{X}_k$ active in that orthant thus form the bounds of $\mathcal{X}'_{k+1}$ there. The feasible set $\mathcal{X}'_{k+1}$ is generally non-convex overall, even if $\mathcal{X}_k$ is a convex polytope, but convex in any one orthant. In practice, $\mathcal{X}'_{k+1}$ may well be confined to one orthant and thus convex.

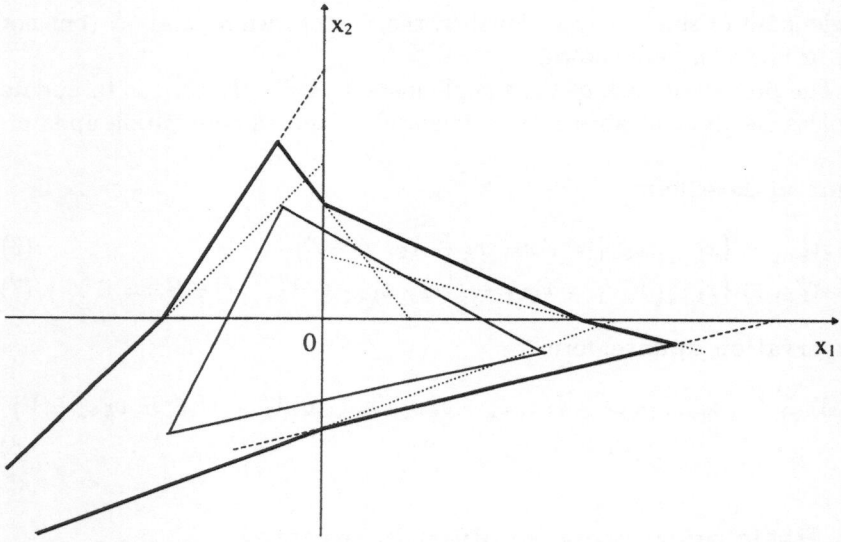

Fig. 1. Effects of homogeneous state transition on bounds of $\mathcal{X}_k$ (fine lines). Bold lines show bounds of $\mathcal{X}'_{k+1}$. Dashed lines are bounds deriving from both λ's being at upper extremes, dotted lines from one λ at upper and one at lower extreme.

For a **complex-conjugate pole pair** with state transition

$$x'^c_{k+1} = \begin{bmatrix} p & -q \\ q & p \end{bmatrix} x^c_k \tag{9}$$

we have prior knowledge

$$\check{\rho}^2 \le p^2 + q^2 \le \hat{\rho}^2 \tag{10}$$

$$0 < \check{\theta} \le \theta = \tan^{-1}(q/p) \le \hat{\theta} < \pi \tag{11}$$

from bounds on the damping time constant and ringing frequency respectively. Since $\|x'^c_{k+1}\|^2 = \rho^2\|x^c_k\|^2$, the extreme of $\|x'^c_{k+1}\|$ is *extreme $\rho \times$ extreme $\|x^c_k\|$*. Hence while θ is within its range, any hyperplane (line) bound in the x^c subspace of $\mathcal{X}_k$ becomes a circular arc in that subspace of $\mathcal{X}'_{k+1}$. The lower bound on x'^c_{k+1} is thus non-convex, and both the lower and upper bounds must be replaced by linear approximations for computational convenience. At either end of the range of θ, a linear bound on x^c_k yields a linear bound on x'^c_{k+1}.

Next consider the effects of forcing. It transforms $\mathcal{X}'_{k+1}$ into $\mathcal{X}''_{k+1}$ by a vector-summing operation, denoted by $\oplus$:

$$\mathcal{X}''_{k+1} \equiv \mathcal{X}'_{k+1} \oplus \mathcal{F}_k \quad \text{where} \quad \mathcal{F}_k \equiv \{\Gamma w_k |\ w_k \in \mathcal{W},\ \Gamma \in \mathcal{G}\} \tag{12}$$

If the system is fully controllable (every mode is excited by w_k) and $m < n$, at least one forcing variable w_{kj} influences two or more state variables x_{ki}, so it does not merely add to uncertainty in one state variable. Uncertainty in γ_1 and γ_2 gives a box with size proportional to w_{kj}. Uncertainty in w_{kj} gives a range of boxes; if w_{kj} ranges between equal positive and negative values, the envelope of the boxes consists of two sections which are antisymmetric about the origin. Such a case is shown in Fig. 2 for $n = 2$, $m = 1$.

Several important points emerge: $\mathcal{F}_k$ is much smaller than the box formed by the overall bounds on each $\gamma_{ij}w_{kj}$; the bounds are piecewise linear; $\mathcal{F}_k$ is convex in each orthant but non-convex overall; finally and crucially, vector addition of $\mathcal{F}_k$ to $\mathcal{X}'_{k+1}$ will double the number of convex components of $\mathcal{X}$ as each scalar forcing variable is applied, unless $\mathcal{F}_k$ is approximated by a convex set.

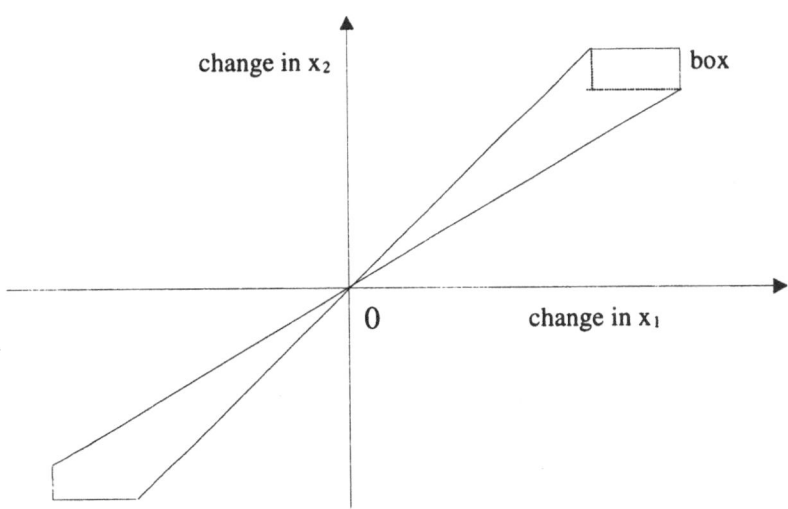

change in x_2

box

0

change in x_1

Fig. 2. Feasible set $\mathcal{F}_k$ of forcing for case $n = 2$, $m = 1$.

The tightest convex approximation is the convex hull co($\mathcal{F}_k$), as in Fig. 3.

The approximation increases the volume by a factor $\approx n$ but the set remains much smaller than that of the overall box, also shown in Fig. 3. The convex hull may be regarded as *box* $\oplus$ *lower vector* $\oplus$ *upper vector*, where the vectors are respectively $\check{w}_j\bar{\gamma}$ and $\hat{w}_j\bar{\gamma}$, in opposite directions as the bounds on w_j have opposite signs. The vector $\bar{\gamma}$ is composed of the mid-points of the γ_{ij} ranges and the size of the box is determined by the ranges of the γ_{ij}.

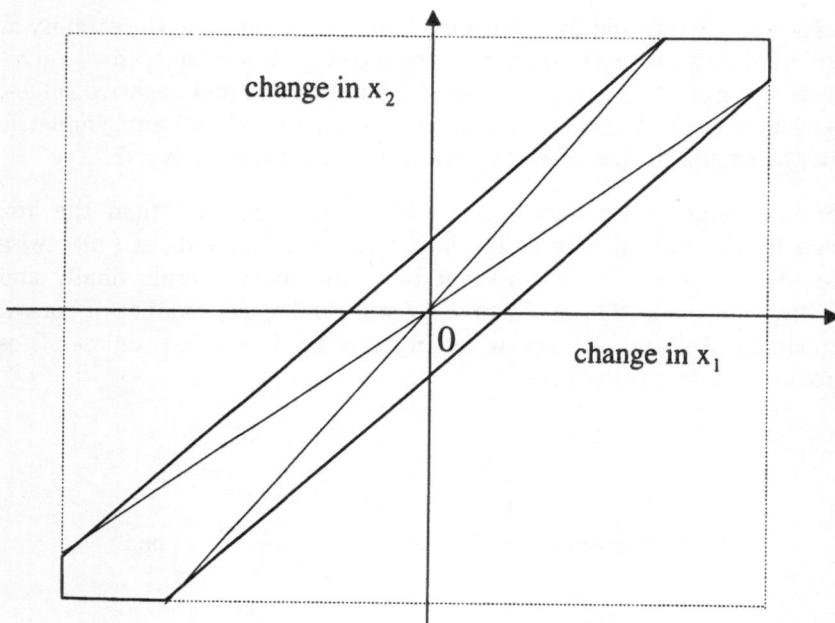

Fig. 3. Approximation of $\mathcal{F}_k$ by $\text{co}(\mathcal{F}_k)$ (heavy lines). The dashed lines are the overall bounds on $\gamma_{11}w_{k1}$ and $\gamma_{21}w_{k1}$.

3.2 Observation update

The operation transforming $\mathcal{X}''_{k+1}$ to $\mathcal{X}_{k+1}$ is intersection with the set

$$\mathcal{O}_{k+1} \equiv \{x_{k+1} | \, y_{k+1} - Hx_{k+1} \in \mathcal{V}, \, H \in \mathcal{H}\} \tag{13}$$

In each orthant, $\mathcal{O}_{k+1}$ is bounded by r pairs of non-parallel hyperplanes

$$y_{k+1,l} - \check{v}_{k+1,l} \leq \inf_{H \in \mathcal{H}}(h_l^T x_{k+1}); \quad \sup_{H \in \mathcal{H}}(h_l^T x_{k+1}) \leq y_{k+1,l} - \hat{v}_{k+1,l} \tag{14}$$

so a standard polytope-updating algorithm [6,7] can be used, $2r$ times per update. As the hyperplanes in a pair are not parallel, $\mathcal{X}_{k+1}$ is asymmetrical. Within each orthant, the normal to each hyperplane (made up of the $\check{h}_{li}$'s and $\hat{h}_{li}$'s according to the signs of the corresponding elements of x_{k+1}) is fixed. Fig. 4 illustrates $\mathcal{O}_{k+1}$ due to a single observation for $r = 1$, $n = 2$.

The next section discusses the implications of these basic updating operations for economical and relatively tight approximation of the bounds of the feasible set.

4   Implications of basic operations for updating approximate feasible set

Consider first the effects of the model dynamics on the feasible set $\mathcal{X}$: evolution according to $x'_{k+1} = \Lambda x_k$ then addition of the forcing Γw_k. The

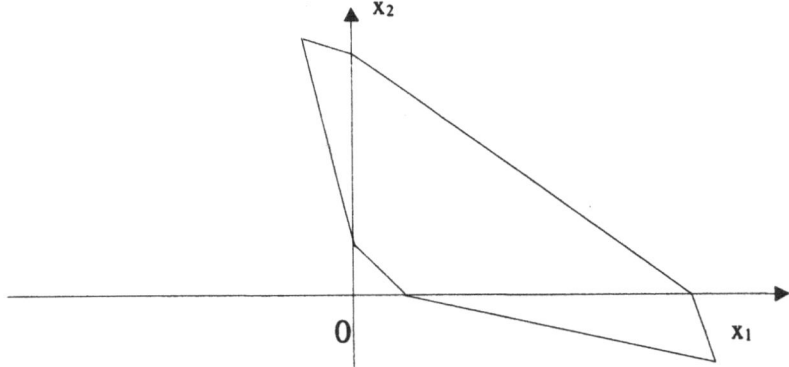

Fig. 4. Feasible set $\mathcal{O}_{k+1}$ due to single observation, $r = 1$, $n = 2$.

complex-pole case has non-linear bounds which immediately require approximation, as noted in Section 3.1. For entirely real poles, the bounds are all linear but the question is how far they must be approximated to produce a simple representation of the feasible set, suitable for updating. Recall that evolution rescales $\mathcal{X}$ in each coordinate direction, and that with the approximation suggested in Section 3.1, forcing adds to $\mathcal{X}'_{k+1}$ the vector sum of a fixed axis-aligned box $\mathcal{B}_j$, say, and the fixed, collinear, opposite-signed vectors $f_j^- \equiv \check{w}_j\overline{\gamma}$ and $f_j^+ \equiv \hat{w}_j\overline{\gamma}$. The effect of the time-update stages can be seen by looking separately at the box and the vector components of the vector sum. If initially the feasible set is an axis-aligned box $\mathcal{A}_0$ (the vector sum of n vectors in the axis directions), $\mathcal{B}_j$ is vector-added at each subsequent forcing, making the box larger but not otherwise altering it. The contributions from f_j^- and f_j^+ entering at each time update are then repeatedly rescaled (shrunk, if the system is stable, and altered in direction) by multiplication by Λ. After k time updates without yet considering the observations, the feasible set consists of the vector sum of an axis-aligned box

$$\mathcal{A}_k \equiv \Lambda^k \mathcal{A}_0 \oplus \Lambda^{k-1}\mathcal{B} \oplus \Lambda^{k-2}\mathcal{B} \oplus \ldots \oplus \mathcal{B} \tag{15}$$

and a vector sum

$$\mathcal{V}_k \equiv \Lambda^k(f_j^- \oplus f_j^+) \oplus \Lambda^{k-1}(f_j^- \oplus f_j^+) \oplus \ldots \oplus (f_j^- \oplus f_j^+). \tag{16}$$

$\mathcal{A}_k$ is very easily computed but $\mathcal{V}_k$ poses a problem. It is the vector sum of k generally non-collinear pairs of collinear vectors and thus has a large number of faces for realistic k. For example, the vector sum of k vectors in 3 dimensions has $k(k-1)$ plane faces in general.

The bounds introduced by the observations will reduce the feasible set, possibly simplifying it by removing more faces than they introduce, but as the observation-induced bounds are asymmetrical, they will destroy the symmetry of $\mathcal{A}_k \oplus \mathcal{V}_k$. It would not help if these pairs of non-parallel hyperplanes were approximated by pairs of parallel ones, as their centre hyperplanes do not pass through the centre of $\mathcal{A}_k \oplus \mathcal{V}_k$.

As k increases, $\mathcal{A}_k$ contracts (if the system is stable) but $\mathcal{V}_k$ expands, and unless the range of possible forcing is small, $\mathcal{A}_k \oplus \mathcal{V}_k$ expands. If so, informative observations cut deeply into $\mathcal{A}_k \oplus \mathcal{V}_k$ and make many of its bounds inactive. That is to say, bounds resulting from the state equation eventually become obsolete, so it may well suffice to keep only a limited number of recent such bounds in play.

5   Summary and conclusions

Updating of the feasible state set of an LTI system with an uncertain model has been considered. The model is modal and has individually, instantaneously bounded uncertain elements in the transition, excitation and observation matrices, and forcing and observation noise bounded in the same way. To summarise the effects of updating, the homogeneous part of uncertain real-pole modal dynamics yields linear bounds from linear bounds, axis-aligned box bounds from axis-aligned box bounds. For uncertain complex-pole modal dynamics it yields linear and circular-arc bounds, of which the lower bound on $\|x^c\|$ includes a non-convex section which must be approximated by a linear bound. Each scalar forcing variable with uncertain gain vector-adds the non-convex union of two convex sets to the feasible state set, so the union must be approximated, the convex hull being the tightest linear approximation. An observation with uncertain gain yields piecewise linear bounds: hyperplanes with known, constant normals within each orthant. The time-evolved bounds from the state equation are symmetrical and give a feasible set (ignoring the observations) with a simple structure but perhaps many faces. The observation-induced bounds are asymmetrical and in most practical situations soon supersede the state-equation bounds.

These features suggest that for simplicity and reasonable efficiency, one should employ for the time updates:

- a modal form for the model
- expression of the feasible set partly as a sum of vectors which include a set along the coordinate axes
- approximation of the uncertain forcing set by its convex hull
- separate handling of the orthants if the feasible set spans more than one
- approximation of the circular-arc bounds by linear bounds if there are complex poles.

However, asymmetry imposed by the observation-derived bounds and eventual obsolescence of the state-equation bounds prevent a completely symmetrical set of bounds from being tight, so it is suggested that the feasible set be approximated, immediately after each update, by reviewing the currently active bounds and deleting some on a systematic but easily computed basis. Those farthest from the Chebyshev centre of the feasible set would be good candidates; the centre is expensive to compute from scratch unless the number of active bounds is modest, but the standard polytope-updating algorithms keep a list of vertices which allows the centre to be found cheaply. Another possibility is to note how much of the existing feasible set is cut off by each new bound (e.g. in the direction of its normal) and delete those bounds which cut off least. This scheme represents an intermediate course between exact (but too complicated) polytope bounding and the cheap but symmetrical and therefore often too loose bounding of the parallelotope technique [10].

It is disappointing that no benefit in simple updating seems to result from the special form of the model, because of the asymmetry in the observation-induced bounds. As noted above, even when there is no uncertainty in the observation equation, the symmetry of $\mathcal{A}_k \oplus \mathcal{V}_k$ is generally lost because its centre is not on the central hyperplane of the parallel pair imposed by the observation. On the positive side, all the operations required are covered by existing polytope-updating algorithms, together with a pruning procedure to control the number of active bounds defining the feasible set.

References

1. Jazwinski, A. H. (1970) Stochastic Processes and Filtering Theory. Academic Press, New York.
2. Maybeck, P. S. (1982) Stochastic Models, Estimation and Control, Vol. 2. Academic Press, New York.
3. Mayne, D. Q. (1963) Optimal non-stationary estimation of the parameters of a linear system with Gaussian inputs. J. Electron. Control 14, 101-112.
4. Norton, J. P. (1975) Optimal smoothing in the identification of linear time-varying systems, Proc. IEE 122, 663-668.
5. Young, P. C. (1984) Recursive Estimation and Time Series Analysis. Springer-Verlag, Berlin.
6. Mo, S.H., Norton, J. P. (1990) Fast and robust algorithm to compute exact polytope parameter bounds. Math. and Comput. in Simulation 32, 481-493.
7. Walter, E., Piet-Lahanier, H. (1989) Exact recursive polyhedral descriptions of the feasible parameter set for bounded-error models. IEEE Trans. on Autom. Control 34, 911-915.
8. Schweppe, F. C. (1968) Recursive state estimation: unknown but bounded errors and system inputs. IEEE Trans. on Autom. Control 13, 22-29.
9. Maksarov, D. G., Norton, J. P. (1996) State bounding with ellipsoidal set description of the uncertainty. Int. J. of Control 65, 847-866.
10. Chisci, L., Garulli, A., Zappa, G. (1996) Recursive state bounding by parallelotopes. Automatica 32, 1049-1056.

The Role of Experimental Conditions in Model Validation for Control*

Michel Gevers[1], Xavier Bombois[1], Benoît Codrons[1],
Franky De Bruyne[2], and Gérard Scorletti[3]

[1] Centre for Systems Engineering and Applied Mechanics (CESAME)
   Université Catholique de Louvain, B-1348 Louvain-la-Neuve, Belgium
[2] Department of Systems Engineering, Research School of Information Sciences
   and Engineering, The Australian National University, Canberra ACT 0200,
   Australia
[3] LAP ISMRA, 6 boulevard du Maréchal Juin, F-1405 Caen Cedex, France

Abstract. Within a stochastic noise framework, the validation of a model yields an ellipsoidal parameter uncertainty set, from which a corresponding uncertainty set can be constructed in the space of transfer functions. We display the role of the experimental conditions used for validation on the shape of this validated set, and we connect a measure of the size of this set to the stability margin of a controller designed from the nominal model. This allows one to check stability robustness for the validated model set and to propose guidelines for validation design.

1 Introduction

Model validation is the exercise that consists in assessing whether a model of some underlying system is *good enough*. Such quality control step cannot be decoupled from the purpose for which the model is to be used. And just as the research on system identification has, in the last 10 years, focused on issues of *design* in order to obtain a model that suited the objective, so must the validation experiment similarly be designed in such a way that the model is guaranteed to deliver what the model is supposed to deliver. Thus, one must think in terms of "goal-oriented validation".

In this chapter we focus on the situation where a model is to be validated with the purpose of designing a controller for the underlying system. This is called *model validation for control*.

The assessment of the quality of a model can take a variety of forms, such as a frequency-domain bound on the error between the system and the model transfer functions, or a worst-case bound on such error over all

---

* The authors acknowledge the Belgian Programme on Inter-university Poles of Attraction, initiated by the Belgian State, Prime Minister's Office for Science, Technology and Culture. The scientific responsibility rests with its authors.

frequencies, or the certification of a region in the complex plane in which the system and the model are guaranteed to lie (set membership validation). Depending on the application some of these quality statements can be more useful than others.

Spurred by the strong reliance of robust control theory on specific uncertainty descriptions, the research on model uncertainty estimation and on model validation gathered momentum in the 1990s. Two directions have been pursued.

1. The first consists in estimating uncertainty regions around estimated models. In the stochastic framework, estimates of the total mean square transfer function error were obtained by adopting, for the bias error, a parametrized probability distribution and by estimating the parameters of this distribution from the data, just as is done for the noise error [4]. In the "hard-bound" framework, uncertainty models have been derived under a variety of hard-bound assumptions on the error model and on the noise: see e.g. [3], [5].

2. The second direction consists in reducing a prior set of admissible models by invalidating models on the basis of observed data and prior hard-bound assumptions: see e.g. [11], [9]. The concept of model invalidation, on the basis of an observed incompatibility between a model, prior assumptions and data, was extended to *controller invalidation* in [10].

The validation theory presented in this chapter is inspired by recent validation results of Ljung and collaborators [6], [8], [7] that are based on signal statistics, with essentially no prior assumptions other than some unavoidable *invariance assumption*. To paraphrase Swedish literature [7], one would like to approach the model validation problem 'as naked as possible' and strip off common covers such as *prior assumptions, probabilistic framework, worst case model properties*. What we are then left with are experimental data that we can collect on the true system and compare with simulated data generated by the model, statistics that we can compute from these data, and some invariance assumption that states that the future statistics will not be different from those observed so far.

The key idea of the method proposed by Ljung for the validation of a model $\hat{G}$ is that the residuals ϵ, obtained by substracting simulated outputs from measured outputs, contain information about the model error $G_0 - \hat{G}$. The identification of an unbiased model for the dynamics connecting the input signal u to the residuals ϵ delivers an estimate of the model error $G_0 - \hat{G}$ and a covariance for this estimate.

Our departure from the validation results of Ljung and collaborators, and the new contributions of this chapter, are contained in the following sequence of new ideas and observations whose presentation will form the essence of this chapter.

1. The validation results of Ljung and Guo [8] allow one to define an uncertainty region $\mathcal{D}$ in the frequency domain, that contains G_0, and also $\hat{G}$ if the model is validated. Our first observation is that different experimental conditions for the collection of validation data, will produce different uncertainty regions $\mathcal{D}_i$, some of which may result in a successful validation and some of which may not. Thus we shall elaborate on **the role of experimental conditions in the validation of a model** and on the concept of **validation design.**

2. We then observe that a model $\hat{G}$ may be validated under closed loop experimental conditions. By collecting data on the closed loop system (G_0, C) with some controller C, one can apply the validation procedure to the closed loop transfer function model $\hat{T} = \frac{\hat{G}C}{1+\hat{G}C}$ of the true closed loop system $T_0 = \frac{G_0 C}{1+G_0 C}$. This defines a closed loop uncertainty set $\mathcal{D}(\hat{T})$, from which the corresponding open loop set $\mathcal{D}(\hat{G})$ can be computed. Thus, we have introduced the concept of **validation in closed loop.**

3. Since each validation experiment leads to a different set of validated models $\mathcal{D}_i(\hat{G})$ that contains G_0, some of these validated regions may be more useful than others, depending on the intended use of the model. This suggests that one should design the validation experiment so that the uncertainty regions are tuned towards the intended use of the model. This leads to **goal-oriented model validation** and to *tuned uncertainty regions.*

4. Vinnicombe [12] has shown that a controller C that stabilizes $\hat{G}$ with a *generalized stability margin* denoted $b_{\hat{G},C}$ stabilizes all plants G for which $\delta_\nu(\hat{G}, G) < b_{\hat{G},C}$, where $\delta_\nu(\hat{G}, G)$ is a metric that measures the distance between $\hat{G}$ and G. Details will be given later in the chapter. We shall introduce the concept of *worst case gap* $\delta_{WC}(\hat{G}, \mathcal{D})$ between a model $\hat{G}$ and all plants in a validated set $\mathcal{D}(\hat{G})$. This leads us to introduce the idea of **model validation for control**: a validation experiment that delivers a validated model set with a smaller worst case gap than another one allows for a larger class of robustly stabilizing controllers.

The validated uncertainty regions constructed in this chapter are based on ellipsoidal confidence regions obtained in parameter space from covariance estimates. Thus, all statements about a system belonging to an uncertainty set are understood to be probabilistic; note, however, that the probability level is left to the user to decide.

2   Model and controller validation concepts

We consider that the input-output data that are used to validate a model are generated from a *"true system"*:

$$y(t) = G_0(q)u(t) + v(t), \tag{1}$$

where $G_0(q)$ is a linear time-invariant causal operator. We make no special assumptions about the input signal $u(t)$ and the noise $v(t)$. We consider that somebody has delivered to us a model $\hat{G}(q)$ for $G_0(q)$, and our task is to validate that model. We are allowed to perform experiments on the true system by applying N input data $u(t)$ to it and by observing the corresponding N output data $y(t)$. Given this framework, the following particular *validation questions* will be addressed.

Model validation question. On the basis of the data I collect, can I define an uncertainty set $\mathcal{D}$ in which G_0 is guaranteed to lie, at a certain probability level? If $\hat{G} \in \mathcal{D}$, then $\hat{G}$ will be called validated.

Controller validation question. On the basis of the data I collect, can I guarantee that a given controller $C(q)$, typically computed from $\hat{G}(q)$, stabilizes not just $\hat{G}$ but also the true $G_0(q)$? If the answer is positive, the controller is said to be *unfalsified* by the data; in the converse case, it is said to be *falsified* or *invalidated*.

Our results provide a contribution to both of these validation questions, **in a stochastic framework.** Our validation procedure will lead to the validation of sets of transfer functions; it could appropriately be called *set membership validation*. We insist that we do not a posteriori validate an a priori given uncertainty set, but rather the validation of a nominal model $\hat{G}$ under specific experimental conditions determines a validated uncertainty set.

3   The model validation procedure

Consider the true system (1) and a model $\hat{G}$ that requires validation. If we apply some input sequence $U^N = \{u(t), t = 1, \ldots, N\}$ to the system, it generates the noisy output sequence $Y^N = \{y(t), t = 1, \ldots, N\}$ using (1). The corresponding simulated outputs are given by

$$\hat{y}(t) = \hat{G}(q)u(t). \tag{2}$$

Consider now the *model residuals* $\epsilon(t)$ defined as the difference between measured and simulated outputs:

$$\epsilon(t) = y(t) - \hat{y}(t) = y(t) - \hat{G}(q)u(t) \tag{3}$$

Inserting the system equation (1) these residuals can then be written as

$$\epsilon(t) = [G_0(q) - \hat{G}(q)]u(t) + v(t) = \partial G(q)u(t) + v(t). \tag{4}$$

The transfer function ∂G is called the *model error* in [6]. Using the assumption of a linear true system and the independence between $v(t)$ and $u(t)$,[1]

---

[1] By this we mean that $v(t)$ would not change if we were to change the input signal $u(t)$.

we have thus decomposed the residual error $\epsilon(t)$ into the sum of two independent sources: one, $\partial G(q)u(t)$ that is due to a *model error*, and one, called *disturbance*, that is not due to a model error. The distinction between these two sources of signal error is very fundamental, and has nothing to do with a probabilistic framework. It is at the heart of all validation theories.

Observations

- The essential difference between the two sources of residual error $\epsilon(t)$ is that one can be manipulated by the user by experimenting with $u(t)$ while the other is totally outside the range of experimentation.
- Without any assumption on the disturbance $v(t)$, any observed error $\epsilon(t)$, however large, can always be attributed to the occurrence of a very large disturbance $v(t)$. Thus, one cannot invalidate a model on the basis of an observed data unless some bounded noise assumption is made.
- If an *invariance assumption* is made on the mechanism that generates the disturbance $v(t)$, then one can evaluate whether ∂G is significantly different from zero by estimating an unbiased model for ∂G from $[\epsilon \ u]$ data.

This last observation is at the heart of the validation procedure proposed by Ljung [6] that we adopt here, with some modifications to account for the added insight gained since the publication of [6].

3.1   Open loop validation

We compute an *unbiased* estimate $\tilde{G}(\hat{\theta}, q)$ of $\partial G(q)$. Thus, consider a model set $\mathcal{M}_{OL} = \{\tilde{G}(\theta, q) \mid \theta \in D_\theta \subset \mathbf{R}^k\}$, for some subset D_θ, and an independently parametrized noise model. The assumption on unbiasedness implies that $\tilde{G}(\theta_0, q) = \partial G(q)$ for some $\theta_0 \in D_\theta$. Using experimental data $[\epsilon \ u]$ collected in open loop (see (3)-(4)), one can then compute an unbiased estimate $\tilde{G}(\hat{\theta}, q)$ of $\partial G(q)$, as well as an estimate of the covariance matrix P_θ of $\hat{\theta}$. The true parameter θ_0 then lies with probability $\alpha(k, \chi^2_{ol})$ in the ellipsoidal uncertainty region

$$U_{OL} = \{\theta \mid (\theta - \hat{\theta})^T P_\theta^{-1} (\theta - \hat{\theta}) < \chi^2_{ol}\} \tag{5}$$

where $\alpha(k, \chi^2_{ol}) = Pr(\chi^2(k) \leq \chi^2_{ol})$ with $\chi^2(k)$ the chi-square probability distribution with k parameters. This parametric uncertainty region U_{OL} defines a corresponding uncertainty region in the space of transfer functions which we denote $\mathcal{D}_{OL}$:

$$\mathcal{D}_{OL} = \{\hat{G}(q) + \tilde{G}(\theta, q) \mid \tilde{G}(\theta, q) \in \mathcal{M}_{OL} \text{ and } \theta \in U_{OL}\} \tag{6}$$

We then have the following property.

Lemma 1: $G_0 \in \mathcal{D}_{OL}$ *with probability* $\alpha(k, \chi^2_{ol})$.
The proof follows directly from the properties of estimated models when variance errors only are concerned.   ∎

The importance of Lemma 1 is that our validation procedure has delivered a validated model set $\mathcal{D}_{OL}$, in which the true system is guaranteed to lie, at some probability level. We now introduce the following definition for the validation of the model $\hat{G}$.

Definition :   The model $\hat{G}$ is called *validated* if $\hat{G} \in \mathcal{D}_{OL}$ or, equivalently, if there exists $\theta^* \in U_{OL}$ such that $\tilde{G}(\theta^*, q) = 0$.

Comments

1. The estimated model $\tilde{G}$ is a correction to the prior model $\hat{G}$ that is under test. Thus, one could, in the application for which the model is to be used, replace $\hat{G}$ by the better model $\hat{G} + \tilde{G}$, or by a new low order model $\hat{G}$ in the validated set $\mathcal{D}_{OL}$. In the sequel, where we focus on the use of the model for control design, we assume that the control design is based on $\hat{G}$ (possibly a new one), but not on $\hat{G} + \tilde{G}$.

2. In fact, we shall see later that for control design it is not so much the validation of the model $\hat{G}$ that matters but the fact that the validation procedure described above yields a validated region $\mathcal{D}_{OL}$, in which the true system G_0 is known to lie. Thus, even if the model $\hat{G}$ is not validated, the controller design and controller validation procedure described in the sequel of this chapter still apply.

3. The validation procedure just described can be applied to any model $\hat{G}$, whether it is a full order or reduced order model of the true G_0.

3.2   Role of the experimental conditions

The validated model set $\mathcal{D}_{OL}$ depends very much on the experimental conditions under which the validation has been performed. This is perhaps not so apparent in the exact definition (6) of $\mathcal{D}_{OL}$ via the parameter covariance matrix P_θ. However, let us recall that a reasonable approximation for the covariance of the transfer function estimate $\tilde{G}(\hat{\theta}, q)$ is given by:

$$cov(\tilde{G}(\hat{\theta}, e^{j\omega})) \approx \frac{n}{N} \frac{\phi_v(\omega)}{\phi_u(\omega)} \tag{7}$$

This shows clearly the role of the signal spectra $\phi_u(\omega)$ and $\phi_v(\omega)$ in shaping the validated set $\mathcal{D}_{OL}$. Thus, two different validation data sets $[\epsilon^{(1)} \ u^{(1)}]$ and $[\epsilon^{(2)} \ u^{(2)}]$ will yield two different validated regions $\mathcal{D}_{OL}^{(1)}$ and $\mathcal{D}_{OL}^{(2)}$. The model $\hat{G}$ may well be validated by one of these two experiments and not by the other.

The role of the experimental conditions on the shape of the validated set, and the importance of tuning the validation experiment to the objective to which the model (or the model set) is to be used, are the central themes of this chapter. We shall see, in particular, how the validation experiment can be tuned when the objective is a robust model-based control design.

3.3 Closed loop validation

On the basis of these observations, we now show that a validated set of models $\mathcal{D} = \{\hat{G}(q) + \tilde{G}(\theta, q) \mid \theta \in U\}$ for some parameter set U can alternatively be computed from a closed loop validation experiment. Consider that the feedback control law $u(t) = C(q)[r(t) - y(t)]$ is applied to the true system $G_0(q)$, with some stabilizing controller $C(q)$. The closed loop system is :

$$y(t) = \frac{G_0 C}{1 + G_0 C} r(t) + \frac{1}{1 + G_0 C} v(t) \overset{\triangle}{=} T_0 r(t) + n(t) \tag{8}$$

The closed loop model is $\hat{T} = \frac{\hat{G}C}{1+\hat{G}C}$. We can then simulate $\hat{y}(t) = \hat{T}r(t)$ and define the closed loop model error

$$\epsilon(t) = (T_0 - \hat{T})r(t) + n(t) \overset{\triangle}{=} \partial T(q)r(t) + n(t) \tag{9}$$

Consider now a model set $\mathcal{M}_{CL} = \{\tilde{T}(\xi, q) \mid \xi \in D_\xi \subset \mathbf{R}^f\}$, for some subset D_ξ defining stable models, such that $\tilde{T}(\xi_0, q) = \partial T(q)$ for some $\xi_0 \in D_\xi$. Using experimental data $[\epsilon \; r]$ collected on the closed loop system, we can then compute an unbiased estimate $\tilde{T}(\hat{\xi}, q)$ of $\partial T(q)$, together with an estimate of the covariance matrix P_ξ of the parameter vector $\hat{\xi}$. The true parameter ξ_0 then lies with probability $\alpha(f, \chi^2_{cl})$ in the ellipsoidal uncertainty region

$$U_{CL} = \{\xi \mid (\xi - \hat{\xi})^T P_\xi^{-1}(\xi - \hat{\xi}) < \chi^2_{cl}\} \tag{10}$$

where $\alpha(f, \chi^2_{cl}) = Pr(\chi^2(f) \leq \chi^2_{cl})$ with $\chi^2(f)$ the chi-square probability distribution with f parameters. This parametric uncertainty region U_{CL} defines a corresponding uncertainty region in the space of closed loop transfer functions $T(\xi, q)$ which we denote $\mathcal{S}_{CL}$:

$$\mathcal{S}_{CL} = \{\hat{T}(q) + \tilde{T}(\xi, q) \mid \tilde{T}(\xi, q) \in \mathcal{M}_{CL} \text{ and } \xi \in U_{CL}\} \tag{11}$$

$\mathcal{S}_{CL}$ is the set of closed loop transfer functions that are validated by our closed loop experiment. From this set (in fact from U_{CL}) we can now define the set $\mathcal{D}_{CL}$ of transfer functions $G(\theta, q)$ that are validated by this closed loop experiment:

$$\mathcal{D}_{CL} = \{\hat{G} + \tilde{G}(\xi, q) \mid \tilde{G}(\xi, q) = \frac{1}{C(q)} \times \frac{\tilde{T}(\xi, q)(1 + \hat{G}C)}{1 - \hat{T} - \tilde{T}(\xi, q)} \text{ and } \xi \in U_{CL}\} \tag{12}$$

The notation $\tilde{G}(\xi, q)$ used in (12) denotes the rational transfer function model whose coefficients are uniquely determined from ξ by the inverse mapping

$$\tilde{G}(\xi, q) = \frac{1}{C(q)} \times \frac{\tilde{T}(\xi, q)(1 + \hat{G}C)}{1 - \hat{T} - \tilde{T}(\xi, q)}. \tag{13}$$

We then have the following property.

Lemma 2: $T_0 \in \mathcal{S}_{CL}$ and $G_0 \in \mathcal{D}_{CL}$ with probability $\alpha(f, \chi^2_{cl})$.   ■

Comments

1. Following our earlier definition of a validated model, we observe that the closed loop model $\hat{T}$ is validated if $\hat{T} \in S_{CL}$ or, equivalently, if there exists a $\xi^* \in U_{CL}$ such that $\tilde{T}(\xi^*, q) = 0$. Similarly, the open loop model $\hat{G}$ is validated by this closed loop experiment if $\hat{G} \in D_{CL}$, which is equivalent with the existence of $\xi^* \in U_{CL}$ such that $\tilde{G}(\xi^*, q) = 0$, with $\tilde{G}(\xi^*, q)$ defined by the mapping (13).

2. However, the most useful aspect of this closed loop validation procedure, from a control objective point of view, is not so much the validation of the initial model $\hat{G}$ as it is the validation of the uncertainty set D_{CL}. Sets validated by closed loop experiments typically have properties that allow for a larger set of stabilizing controllers than sets validated in open loop.

4   Controller validation and model validation for control

We now consider the situation where a controller is designed on the basis of the nominal model $\hat{G}$. For the theory that we develop, this model need not necessarily be inside the validated set D, but the typical situation is where $\hat{G} \in D$. Indeed, if the model $\hat{G}$ has failed a range of validation attempts, any sensible designer will want to replace $\hat{G}$ by a model that is contained in the validated set. We then introduce the concept of controller validation.

Definition : Let the validation procedure of a model $\hat{G}(q)$ result in a validated set D of transfer function models containing G_0, and let $C(q)$ be a controller designed from $\hat{G}(q)$. Then $C(q)$ is called a *validated controller* for the set D if it stabilizes all models in D.

Having defined a validated controller, we turn to the question of *model validation for control*. Consider first that two different validation experiments, performed on the same model $\hat{G}$, have led to two different validated sets $D^{(1)}$ and $D^{(2)}$. The same controller $C(q)$ may be validated for both sets, or for one of them, or for neither. More generally, denote by $C^{(1)}$ the set of controllers that are validated by the first experiment, and by $C^{(2)}$ the set of controllers that are validated by the second experiment. By this we mean that, for each $C \in C^{(1)}$, say, and for each $G \in D^{(1)}$, the closed loop made up of (G, C) is stable. Then we shall consider that the validated set $D^{(1)}$ is a better uncertainty set than $D^{(2)}$ for control design if the set of stabilizing controllers $C^{(1)}$ is "larger than" the set $C^{(2)}$ in some sense. Given that the validation results strongly depend on the experimental conditions, this will then lead us to the concept of *validation design for control*. To make these ideas precise, we introduce a metric on the size of the validated set, and we appeal to some basic tools and results of robust control theory.

5 The Vinnicombe gap metric and its stability result

Various measures exist to characterize the distance between two plants. We adopt here the Vinnicombe gap metric ([12,13]) denoted δ_ν. The Vinnicombe gap (or distance) between a scalar plant G and a model $\hat{G}$ is defined as

$$\delta_\nu(\hat{G}, G) = \begin{cases} \max_\omega \kappa\left(\hat{G}(e^{j\omega}), G(e^{j\omega})\right) & if \ (16) \ is \ satisfied \\ 1 & otherwise \end{cases} \quad (14)$$

where

$$\kappa\left(\hat{G}(e^{j\omega}), G(e^{j\omega})\right) \triangleq \frac{|\hat{G}(e^{j\omega}) - G(e^{j\omega})|}{\sqrt{1 + |\hat{G}(e^{j\omega})|^2}\sqrt{1 + |G(e^{j\omega})|^2}} \quad (15)$$

The condition to be fulfilled in order to have $\delta_\nu(\hat{G}, G) < 1$ is :

$$(1 + \hat{G}^* G)(e^{j\omega}) \neq 0 \quad \forall \omega \quad and \quad wno(1 + \hat{G}^* G) + \eta(G) - \tilde{\eta}(\hat{G}) = 0, \quad (16)$$

where $G^*(e^{j\omega}) = G(e^{-j\omega})$, $\eta(G)$ (resp. $\tilde{\eta}(G)$) denotes the number of poles of G in the complement of the closed (resp. open) unit disc, while $wno(G)$ denotes the winding number about the origin of $G(z)$ as z follows the unit circle indented into the exterior of the unit disc around any unit circle pole and zero of $G(z)$.

If the conditions (16) are satisfied, then the distance between two plants has a simple frequency domain interpretation (in the SISO case). Indeed, the quantity $\kappa(\hat{G}(e^{j\omega}), G(e^{j\omega}))$ is the chordal distance between the projections of $\hat{G}(e^{j\omega})$ and $G(e^{j\omega})$ onto the Riemann sphere of unit diameter [12]. The distance $\delta_\nu(\hat{G}, G)$ between $\hat{G}$ and G is therefore, according to (14), the supremum of these chordal distances over all frequencies.

The main interest of the Vinnicombe metric is its use as a tool for the robust stability analysis of feedback systems. Thus, consider a closed loop system made up of the negative feedback connection of a plant G and a controller C. For such feedback system one can define a generalized stability margin [13].

Definition: generalized stability margin.

$$b_{GC} = \begin{cases} \min_\omega \kappa\left(G(e^{j\omega}), -\frac{1}{C(e^{j\omega})}\right) & if \ [C \ G] \ is \ stable \\ 0 & otherwise \end{cases} \quad (17)$$

where $\kappa(G_1, G_2)$ was defined in (15). Note that $0 \leq b_{GC} \leq 1$.

The following is an important robust stability result based on the Vinnicombe metric between plants.

Proposition 1 [12]. Consider a model $\hat{G}$ and a controller C that stabilizes $\hat{G}$ with a stability margin $b_{\hat{G}C}$. Then C stabilizes all G such that

$$\delta_\nu(\hat{G}, G) < b_{\hat{G}C}. \tag{18}$$

∎

The condition (18) of Proposition 1 is rather conservative, since $\delta_\nu(\hat{G}, G) = \max_\omega \kappa(\hat{G}(e^{j\omega}), G(e^{j\omega}))$ while $b_{\hat{G}C} = \min_\omega \kappa(\hat{G}(e^{j\omega}), -\frac{1}{C(e^{j\omega})})$. Thus, it is a min-max type condition. A pointwise (i.e. frequency by frequency), and therefore less conservative condition is as follows.

Proposition 2 [12]. Consider a model $\hat{G}$ and a controller C that stabilizes $\hat{G}$. Then C stabilizes all G such that

$$\kappa\left(\hat{G}(e^{j\omega}), G(e^{j\omega})\right) < \kappa\left(\hat{G}(e^{j\omega}), -\frac{1}{C(e^{j\omega})}\right) \; \forall \; \omega \; and \; \delta_\nu(\hat{G}, G) < 1 \tag{19}$$

∎

6   The worst case Vinnicombe distance for validated model sets

In the validation context that is of interest to us here, the true system G_0 is unknown, but we have shown that it lies, with probability 0.95 say, in some validated set $\mathcal{D}$. In order to apply the robust stability results of Vinnicombe to our validation results, we introduce the concept of *worst case Vinnicombe distance* between a model $\hat{G}$ and a validated model set $\mathcal{D}$: it corresponds to the largest Vinnicombe distance between the model $\hat{G}$ and any plant inside the set $\mathcal{D}$.

Definition of the worst case Vinnicombe distance: The worst case Vinnicombe distance $\delta_{WC}(\hat{G}, \mathcal{D})$ between a model $\hat{G}$ and a model set $\mathcal{D}$ is defined as

$$\delta_{WC}(\hat{G}, \mathcal{D}) = \max_{G_\mathcal{D} \in \mathcal{D}} \delta_\nu(\hat{G}, G_\mathcal{D}) \tag{20}$$

Another important quantity is now defined: the **worst case chordal distance**. Its computation is the result of a convex optimization problem involving Linear Matrix Inequality (LMI) constraints [1].

Definition of the worst case chordal distance at frequency ω.
At a particular frequency ω, we define $\kappa_{WC}(\hat{G}(e^{j\omega}), \mathcal{D})$ as the maximum chordal distance between the projections on the Riemann sphere of $\hat{G}(e^{j\omega})$ and of the frequency responses of all plants in $\mathcal{D}$ at the same frequency:

$$\kappa_{WC}\left(\hat{G}(e^{j\omega}), \mathcal{D}\right) = \max_{G_\mathcal{D} \in \mathcal{D}} \kappa\left(\hat{G}(e^{j\omega}), G_\mathcal{D}(e^{j\omega})\right) \tag{21}$$

Having extended the distances between plants to worst case distances between a model and a model set, we can now also extend the robust stability results of Vinnicombe to validated model sets.

Theorem 1. Let $\hat{G}$ be a model, C a stabilizing controller for $\hat{G}$ yielding a generalized stability margin $b_{\hat{G}C}$, and $\mathcal{D}$ a validated set of transfer functions containing the true plant G_0. Then C stabilizes all plants in the set $\mathcal{D}$, and hence also G_0, if the following condition holds :

$$\delta_{WC}(\hat{G}, \mathcal{D}) < b_{\hat{G}C}. \tag{22}$$

Proof :   It follows immediately from the definitions that for any $G \in \mathcal{D}$, and hence for G_0,

$$\delta_\nu(\hat{G}, G) \leq \delta_{WC}(\hat{G}, \mathcal{D}) < b_{\hat{G}C}$$

and the stability then follows from Proposition 1.   ∎

Using the pointwise version of the robust stability result of Vinnicombe, we can now state our main stability result for validated model sets.

Theorem 2 (main stability theorem). Let $\hat{G}$ be a model, C a stabilizing controller for $\hat{G}$, and $\mathcal{D}$ a validated set of parametrized transfer functions containing the true plant G_0. Then C stabilizes all plants in the set $\mathcal{D}$, and hence also G_0, if the following condition holds :

$$\kappa_{WC}\left(\hat{G}(e^{j\omega}), \mathcal{D}\right) < \kappa\left(\hat{G}(e^{j\omega}), -\frac{1}{C(e^{j\omega})}\right) \quad \forall \, \omega \in [0, \pi] \tag{23}$$

Proof :   It follows from the definition of worst case chordal distance that, at any frequency ω and for any model $G \in \mathcal{D}$, we have

$$\kappa\left(\hat{G}(e^{j\omega}), G(e^{j\omega})\right) \leq \kappa_{WC}\left(\hat{G}(e^{j\omega}), \mathcal{D}\right) < \kappa\left(\hat{G}(e^{j\omega}), -\frac{1}{C(e^{j\omega})}\right).$$

It follows from Proposition 2 that any $G \in \mathcal{D}$ is stabilized by C.   ∎

We shall illustrate the application of these robust stability results for validated model sets in Section 8.

Computational aspects

Our two stability theorems are very powerful tools to check the stability of a designed controller C on the system G_0 before it is actually applied to that system. The only requirement is that G_0 be inside the validated region $\mathcal{D}$, which region is itself derived from the covariance matrix of the estimated parameters of the model error model. The results rely heavily on our ability to compute the worst case chordal distance at frequency ω, $\kappa_{WC}(\hat{G}(e^{j\omega}), \mathcal{D})$, between a model $\hat{G}$ and a set $\mathcal{D}$, defined in (21), and/or the worst case Vinnicombe distance $\delta_{WC}(\hat{G}, \mathcal{D})$ between these two objects, defined in (20). This is by no means a trivial matter. The solution to these problems has been obtained using LMI techniques: see [1].

7 Design issues: model validation for control design

We have developed a complete setup from model validation to controller validation, and the computational tools are available to check whether a controller designed from a model stabilizes the true plant, at least at some prespecified probability level. We have also explained the role of the experimental conditions on the shape of the validated sets, and we have developed tools to compute a measure of the size of these validated sets that is directly related to the capability of a controller to stabilize all models in such validated set. These tools can now be used for *validation design*. The following design guidelines can be proposed for the validation of a model that is to be used for control design.

- With some model $\hat{G}$ as the starting point, the validation for control procedure consists of the following steps:
  - Using $\hat{G}$ as the model, perform a validation experiment (see Section 3). This yields a validated set $\mathcal{D}$ containing the true G_0 with probability 0.95%, say. The model $\hat{G}$ may or may not lie in $\mathcal{D}$. Compute the worst case Vinnicombe distance $\delta_{WC}(\hat{G}, \mathcal{D})$ and, possibly also, the worst case chordal distance $\kappa_{WC}(\hat{G}(e^{j\omega}), \mathcal{D})$ at each frequency.
  - Design a controller C and compute its nominal stability margin $b_{\hat{G},C}$ or the chordal distance $\kappa(\hat{G}(e^{j\omega}), -\frac{1}{C(e^{j\omega})})$ at each frequency.
  - Check whether $\delta_{WC}(\hat{G}, \mathcal{D}) < b_{\hat{G},C}$ or, better, whether at each frequency $\kappa_{WC}(\hat{G}(e^{j\omega}), \mathcal{D}) < \kappa(\hat{G}(e^{j\omega}), -\frac{1}{C(e^{j\omega})})$. If so, then C stablizes the true system G_0.
- Given a choice between different experimental conditions for the validation procedure, one should give preference to a validation experiment that yields an uncertainty set $\mathcal{D}$ with the smallest possible worst case Vinnicombe gap.
- The projections of Nyquist plots on the Riemann sphere have maximal resolution around the equator, i.e. where the transfer functions have an amplitude close to one. This has important consequences for "validation for control" design: see [2] for more details on closed loop validation.
- Given a validated set $\mathcal{D}$ and a corresponding worst case gap $\delta_{WC}(\hat{G}, \mathcal{D})$, one can compute a sequence of controllers C_i to drive up the nominal performance of the $(\hat{G}, C_i)$ loop while keeping $b_{\hat{G},C_i} > \delta_{WC}(\hat{G}, \mathcal{D})$. This guarantees stability of the actual closed loop system.

8 A simulation example

Consider the following true system G_0 and model $\hat{G}$, respectively,

$$y = G_0 u + H_0 e = \frac{z^{-1} + 0.25z^{-2}}{1 - 1.4z^{-1} + 0.45z^{-2}} u + \frac{1}{1 - 1.4z^{-1} + 0.45z^{-2}} e$$
$$\hat{y} = \hat{G}u = \frac{1.0141z^{-1} + 0.2397z^{-2}}{1 - 1.4237z^{-1} + 0.4835z^{-2}} u$$

The actual Vinnicombe distance between $\hat{G}$ and G_0 is $\delta_\nu(\hat{G}, G_0) = 0.0163$. For this model $\hat{G}$, an open-loop and a closed-loop validation were achieved leading to two uncertainty regions $\mathcal{D}_{OL}$ and $\mathcal{D}_{CL}$ correponding to a probability level of 0.95. The controller chosen for closed-loop validation is a proportional controller $C(q) = 1$. The model was validated with 1000 data collected in open-loop and closed-loop, respectively, having the following statistics:

$$Open - loop: \quad \sigma_u^2 = 0.2 \quad and \quad \sigma_e^2 = 1 \Longrightarrow \sigma_y^2 = 23.4$$
$$Closed - loop: \quad \sigma_r^2 = 10 \quad and \quad \sigma_e^2 = 1 \Longrightarrow \sigma_y^2 = 26.9$$

Figure 1 presents the Nyquist plots of G_0, $\hat{G}$ and $\hat{G} + \tilde{G}$, as well as the smallest overbounding ellipsoids of the uncertainty regions $\mathcal{D}_{OL}$ and $\mathcal{D}_{CL}$ at each frequency. Observe that G_0 and $\hat{G}$ lie inside both $\mathcal{D}_{OL}$ and $\mathcal{D}_{CL}$ for all frequencies. Thus, $\hat{G}$ is validated by both experiments here.

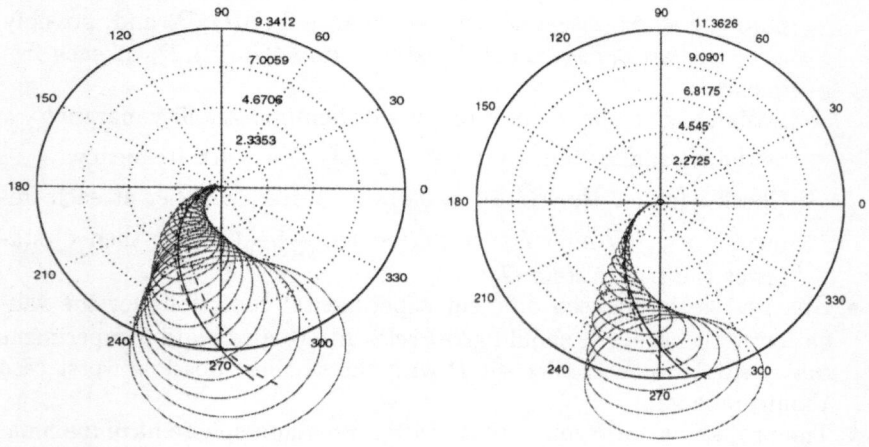

Fig. 1. Nyquist plot of G_0 (solid), $\hat{G}$ (dash), and $\hat{G} + \tilde{G}$ (dashdot), with ellipsoidal estimates of $\mathcal{D}_{OL}$ and $\mathcal{D}_{CL}$. Left: open loop validation. Right: closed loop validation.

The worst case Vinnicombe distances are:

$$\delta_{WC}(\hat{G}, \mathcal{D}_{OL}) = 0.2604 > \delta_{WC}(\hat{G}, \mathcal{D}_{CL}) = 0.0572 > \delta_\nu(\hat{G}, G_0) = 0.0163.$$

Note that the worst case Vinnicombe distance is much smaller with the closed-loop validated set than with the open-loop set. Thus, the validated set $\mathcal{D}_{CL}$ should allow for less conservative control designs.

We consider a proportional controller $C(q) = 1.5$ which stabilizes the nominal model $\hat{G}$, yielding a nominal stability margin $b_{\hat{G}C} = 0.0461$. This controller also stabilizes G_0, but in practice G_0 is unknown and the stabilization of G_0 by the controller C can only be ascertained by the use of one of the stability theorems of Section 6. We first check whether the Min-Max type condition of Theorem 1 is verified. We have:

$$
\hat{b}_{\hat{G}C} = 0.0461 < \overbrace{\delta_{WC}(\hat{G}, \mathcal{D}_{CL})}^{=0.0572} < \overbrace{\delta_{WC}(\hat{G}, \mathcal{D}_{OL})}^{=0.2604}
$$

Thus, the robust stability condition of Theorem 1 is violated with both of the validated regions. We now check the less conservative condition of Theorem 2. Figure 2 compares the worst case chordal distances (for $\mathcal{D}_{OL}$ and $\mathcal{D}_{CL}$) and the pointwise stability margin $\kappa(\hat{G}(e^{j\omega}), -\frac{1}{C(e^{j\omega})})$.

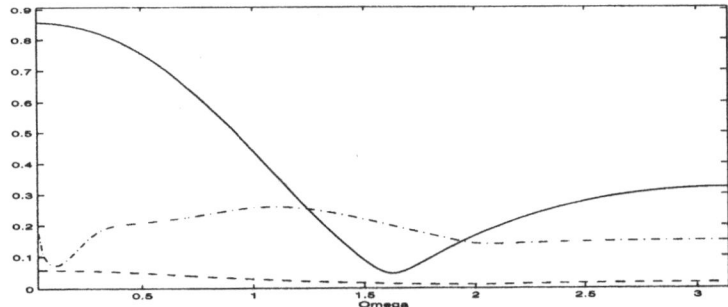

Fig. 2. Frequency by frequency comparison of $\kappa_{WC}(\hat{G}(e^{j\omega}), \mathcal{D}_{OL})$ (dashdot), $\kappa_{WC}(\hat{G}(e^{j\omega}), \mathcal{D}_{CL})$ (dash) and $\kappa(\hat{G}(e^{j\omega}), -\frac{1}{C(e^{j\omega})})$ (solid)

It shows that, even with this less conservative condition, the stability condition of Theorem 2 is violated when the set $\mathcal{D}_{OL}$ is used. However, the stabilization of the true G_0 is guaranteed by the stability condition (23) when the set $\mathcal{D}_{CL}$ is used.

9   Conclusions

We have displayed the role of experimental conditions in the validation of a model. We have then developed tools that allow one to connect some measures of the "size" of a validated model set to the stability margin of a controller designed from the nominal model. This has then led us to propose validation design guidelines, when the validation is performed for the purpose of designing a robust controller.

References

1. X. Bombois, M. Gevers, and G. Scorletti. Controller validation based on an identified model. *submitted to IEEE Transactions on Automatic Control*, 1999.
2. M. Gevers, B. Codrons, and F. De Bruyne. Model validation in closed-loop. In *Proc. American Control Conference 99, to appear*, San Diego, USA, 1999.
3. L. Giarré, M. Milanese, and M. Taragna. H_∞ identification and model quality evaluation. *IEEE Trans. Automatic Control*, 42(2):188–199, 1997.
4. G.C. Goodwin, M. Gevers, and B. Ninness. Quantifying the error in estimated transfer functions with application to model order selection. *IEEE Trans. Automatic Control*, 37:913–928, 1992.
5. R.G. Hakvoort. *System Identification for Robust Process Control - PhD Thesis*. Delft University of Technology, Delft, The Netherlands, 1994.
6. L. Ljung. Identification for control - what is there to learn ? *Workshop on Learning, Control and Hybrid Systems, Bangalore*, 1998.
7. L. Ljung and L. Guo. Classical model validation for control design purposes. *Mathematical Modelling of Systems*, 3:27–42, 1997.
8. L. Ljung and L. Guo. The role of model validation for assessing the size of the unmodelled dynamics. *IEEE Trans. on Automatic Control*, AC-42(9):1230–1239, September 1997.
9. K. Poolla, P.P. Khargonekar, A. Tikku, J. Krause, and K. Nagpal. A time-domain approach to model validation. *IEEE Trans. Automatic Control*, 39:951–959, May 1994.
10. M.G. Safonov and T.C. Tsao. The unfalsified control concept and learning. *IEEE Trans. Automatic Control*, 42(6):843–847, June 1997.
11. R.S. Smith and J.C Doyle. Model invalidation: A connection between robust control and identification. *IEEE Trans. Automatic Control*, 37:942–952, July 1992.
12. G. Vinnicombe. Frequency domain uncertainty and the graph topology. *IEEE Trans Automatic Control*, AC-38:1371–1383, 1993.
13. G. Vinnicombe. *Uncertainty and Feedback (H_∞ loop-shaping and the ν-gap metric)*. Book to be published, 1999.

Modeling and Validation of Nonlinear Feedback Systems

Roy Smith[1] and Geir Dullerud[2]

[1] Electrical & Computer Engineering Dept., University of California,
Santa Barbara, CA 93106, USA.
[2] Mechanical & Industrial Eng. Dept., University of Illinois,
Urbana, IL 61801, USA.

Abstract. Model validation provides a useful means of assessing the ability of a model to account for a specific experimental observation, and has application to modeling, identification and fault detection. Prior theoretical and application work in the area of model validation for robust control models focussed on linear fractional models. In this paper we discuss the extension of these methods to certain classes of nonlinear models. The Moore-Greitzer model of rotating stall is used as a simple example to illustrate the underlying ideas.

1 Introduction

The models used in robust control include unknown, norm bounded perturbations; the inclusion of which have the effect of describing a model set. Analysis and design is performed with respect to all models in the set, thereby providing a margin of safety with respect to the designer's uncertainty about the true variety of system behaviors. Smaller model sets allow potentially higher performing designs but also imply greater knowledge about the system behavior. Obtaining such models is usually an iterative process requiring a significant amount of engineering judgment. We are interested in the extent to which robust control models can describe observed experimental behavior; and in the current work will focus on simple nonlinear extensions of this modeling framework.

Classical system identification methods deliver a model in which our uncertainty about the system behavior is often characterized in terms of an additive probabilistic noise signal. Such models are well suited to open-loop problems like filtering, estimation, and prediction. Perturbation models are better suited to closed-loop problems as they are able to account for unmodeled and potentially destabilizing dynamics.

Naturally our model framework must be compatible with our end use. We consider closed-loop control system design as the objective and will focus on $\mathbf{H}_\infty$ robust control models: those including specified $\mathbf{H}_\infty$ norm bounded perturbations and unknown signals [1,2]. Figure 1 gives an example of such a model, using a multiplicative perturbation Δ, and an output disturbance d. Note that the observed input-output behavior contains both a noise and perturbation response.

$$
\begin{aligned}
y &= (I + W_\Delta \Delta) P_{nom}\, u \ + \ W_d\, d \\
&= \underbrace{P_{nom}\, u}_{\substack{\text{nominal} \\ \text{response}}} \ + \ \underbrace{W_\Delta \Delta P_{nom}\, u}_{\substack{\text{perturbation} \\ \text{response}}} \ + \ \underbrace{W_d\, d}_{\substack{\text{disturbance} \\ \text{response}}}
\end{aligned}
$$

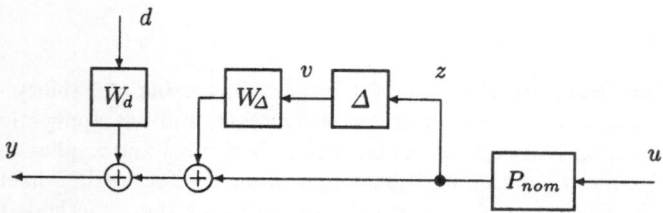

Fig. 1. Robust control frameworks: a multiplicative perturbation example

It is not possible, on the basis of a single experiment, to accurately distinguish between the perturbation and noise responses. If the designer is prepared to make assumptions about the system—for example, that the disturbance, $d(t)$, is Gaussian with known variance—then it is possible to draw stronger conclusions about the various responses. However, these conclusions are a function of, and therefore sensitive to, the assumptions. Our work focuses more on testing such assumptions against the physical observations.

The distinction between the perturbation response and the noise response is critical for control system design. Consider the case where P_{nom} is an LTI SISO system, in a unity gain negative feedback configuration with a controller C. The closed-loop output is given by

$$
y = \frac{(1 + W_\Delta \Delta) P_{nom} C}{1 + (1 + W_\Delta \Delta) P_{nom} C}\, r \ + \ \frac{W_d}{1 + (1 + W_\Delta \Delta) P_{nom} C}\, d,
$$

where r is the reference input. The weighted perturbation, $W_\Delta \Delta$, affects the closed-loop stability of the system. On the other-hand the disturbance weight W_d affects only the closed-loop disturbance response. It is clearly important to distinguish between the uncertain parts of the response which are potentially destabilizing and those that are not. Uncertainty about potentially destabilizing dynamics should be modeled by the effect of $W_\Delta \Delta$.

It is equally important to consider the ultimate application of the model. If the objective is to use the model for control design purposes then this should be reflected in the criteria uses to evaluate the model. An illustrative example will make this clear. For simplicity, consider the problem of selecting between two models, $P_1(s)$ and $P_2(s)$. Figure 2 illustrates the frequency responses of each model, and the "true" system, $P(s)$.

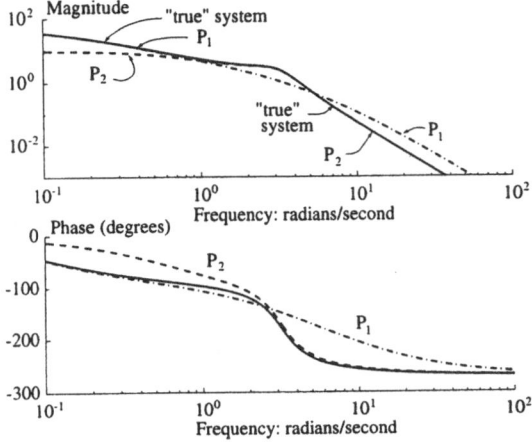

Fig. 2. Frequency response comparison between the "true" system (*solid*), and the models $P_1(s)$ (*dot-dashed*) and $P_2(s)$ (*dashed*)

The model $P_1(s)$ is closer match to $P(s)$ over most of the frequency range, and this is also borne out by the comparison of the step responses of $P(s)$ and the models $P_1(s)$ and $P_2(s)$ in Figure 3.

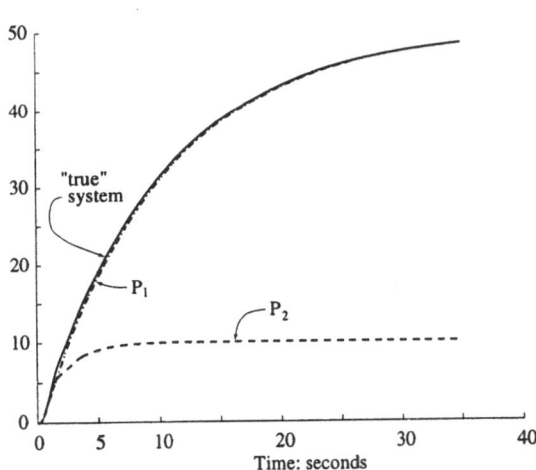

Fig. 3. Open-loop prediction comparison between the "true" system (*solid*), and the models $P_1(s)$ (*dot-dashed*) and $P_2(s)$ (*dashed*). Step responses are shown for each case

On this basis one could select $P_1(s)$ as a suitable model and proceed with a control design. The controller,

$$C(s) = \frac{0.01(10s + 1)}{(0.1s + 1)},$$

gives a reasonable design. The closed-loop frequency magnitude response of all three systems, using this choice for $C(s)$, is shown in Figure 4.

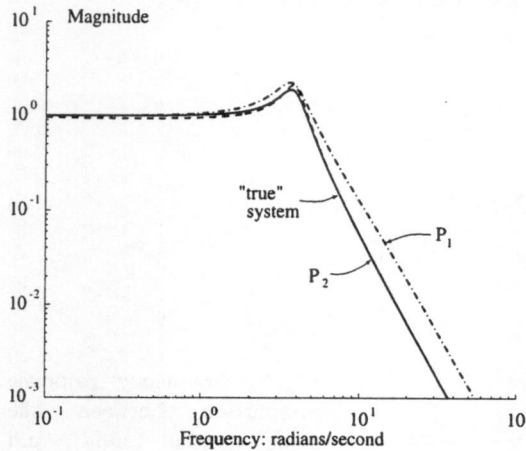

Fig. 4. Closed-loop prediction comparison between the "true" system (*solid*), and the models $P_1(s)$ (*dot-dashed*) and $P_2(s)$ (*dashed*). The magnitude of the frequency response is shown for each case

Unfortunately, both the "true" system $P(s)$ and the model $P_2(s)$ are destabilized by this choice of $C(s)$. The model $P_1(s)$ on the other hand is stable. Clearly the closed-loop predictive abilities of $P_2(s)$ are far better than $P_1(s)$. The key point is that $P_2(s)$ is a closer approximation to $P(s)$ in the frequency range that will eventually become the loop cross-over frequency. Selecting a control design model on the basis of open-loop experiments suffers from two potentially serious drawbacks: the critical parts of the response (in this case those close to the cross-over frequency) may be swamped by other benign aspects of the open-loop response; and the critical part of responses may not be known in advance of implementing an actual control design. In our simple example, the eventual cross-over frequency is not known until after a design has been implemented.

These issues can be addressed by careful, iterative, application of a series of closed-loop identification and design steps. We note that the model validation approaches that we will detail in subsequent sections are equally applicable to open- and closed-loop scenarios.

2   Model Validation: Linear Robust Control Models

Model validation is the assessment of the quality of a given model with respect to experimental data. The problem of interest is to determine whether or not the model—which in this context includes assumptions about the unknown perturbations and signals—is consistent with the experimental observation. One can never *validate* models, simply because of the impossibility of testing all experimental conditions and inputs. A model is said to be *invalidated* if a particular input-output datum is not consistent with the model.

We consider the model, including norm bounds on unknown perturbations and unknown signals, to be given. An input-output experiment has been conducted and we wish to determine whether or not the input-output record is consistent with the perturbation model of the plant. This amounts to determining the size of the smallest perturbation, and smallest unknown signal, which allows the model to reproduce the observed input-output datum. If the smallest perturbation and unknown signal required to account for the observed datum are larger than the assumed bounds in the specification of the model then the datum has *invalidated* the model.

Several problem formulations are possible. Original work in this area used a frequency domain setting [3,4]. This has the advantage of being able to handle multiple perturbations in a linear fractional framework. However, the connection with time domain experimental data must be considered as approximate, except in special cases. More recent work [5,6] has dealt with a purely discrete time domain setting. The most recent results deal with a sampled-data formulation [7,8], and have the closest connection with practical experimental problems.

We will focus on the discrete time domain setting as it is straightforward to present and still allows us to distinguish between time-invariant and time-varying perturbations. Consider the model framework to be the multiplicative perturbation structure illustrated in Figure 1. Note however that the framework and the methods also apply to more general linear fractional transformation (LFT) models and block structured perturbations.

It is first necessary to be a little more precise in the statement of the model. The nominal model, P_{nom}, is an LTI discrete-time system, typically specified by its pulse response coefficients, $p_0, p_1, \cdots, p_n$. The complete specification of the model includes the assumptions that Δ is a member of a class of perturbations, $\boldsymbol{\Delta}$ with $\|\Delta\|_\infty \leq 1$, and $d \in l_2$, with $\|d\|_2 \leq 1$. The class of perturbations, $\boldsymbol{\Delta}$, is typically LTI or linear time-varying (LTV) $\mathbf{H}_\infty$ bounded operators. The unity normalization is without loss of generality because of the inclusion of the weights W_Δ and W_d.

The experimental observation consists of N input/output time-domain measurements, $(u_0, u_1, \cdots, u_{N-1})$, and $(y_0, y_1, \cdots, y_{N-1})$. We can now formally state the model validation problem, for this particular framework.

Problem 1 (Model validation) *Given a perturbation model and an input-output datum (y_k, u_k), $k = 0, 1, \cdots, N - 1$; does there exist a discrete-time signal d_k and an operator Δ, satisfying $\|d\|_2 \leq 1$ and $\Delta \in \boldsymbol{\Delta}$, $\|\Delta\|_\infty \leq 1$, such that,*

$$y = (I + W_\Delta \Delta) P_{nom} u + W_d d.$$

The underlying key step in solving the model validation problem is to pose the condition, $\Delta \in \boldsymbol{\Delta}$, with $\|\Delta\|_\infty \leq 1$, as a computable condition upon the signals z_k and v_k. This form of result is known as an extension condition, and is now presented for the LTI and LTV cases.

Before doing this we must introduce some nonstandard notation. For a sequence of vectors $v = \{v_0, v_1, \cdots, v_{N-1} \in \mathbf{R}^m\}$, let $V \in \mathbf{R}^{mN \times N}$ denote the associated lower block Toeplitz matrix defined as

$$V = \begin{bmatrix} v_0 & 0 & 0 & \cdots & 0 \\ v_1 & v_0 & 0 & \cdots & 0 \\ \vdots & \vdots & \vdots & \cdots & \vdots \\ v_{N-1} & v_{N-2} & v_{N-3} & \cdots & v_0 \end{bmatrix}.$$

Let $\mathcal{S}^m$ denote the set of one sided sequences with elements in $\mathbf{R}^m$. Define the l-step truncation operator, $\pi_l : \mathcal{S}^m \longrightarrow \mathcal{S}^m$, by

$$\pi_l u_k = \begin{cases} u_k & k = 0, \cdots, l \\ 0 & k > l \end{cases}$$

The extension condition for the existence of linear time-invariant operators is as follows.

Theorem 2 ([5]). *Given sequences, $v_k \in \mathbf{R}^{\dim(v)}$ and $z_k \in \mathbf{R}^{\dim(z)}$, $k = 0, \cdots, N-1$, there exists a stable, linear, time-invariant, causal operator Δ, with $\|\Delta\|_\infty \leq \alpha$, such that, $\pi_{N-1} v_k = \Delta \, \pi_{N-1} z_k$ if and only if,*

$$V'V \leq \alpha^2 Z'Z.$$

The case of LTV perturbations can be considered by using the following extension condition.

Theorem 3. *Given sequences, $v_k \in \mathbf{R}^m$ and $z_k \in \mathbf{R}^p$, $k = 0, \cdots, N-1$, there exists a stable, linear, time-varying, causal operator Δ, with $\|\Delta\|_\infty \leq \alpha$, such that, $\pi_{N-1} v_k = \Delta \, \pi_{N-1} z_k$, if and only if,*

$$\|\pi_l v\|_2 \leq \alpha \|\pi_l z\|_2, \quad \text{for all } l = 1, \cdots, N.$$

The computational requirements imposed by the LTV condition are significantly less than those imposed by the LTI extension condition.

The requirement that the perturbation model output matches the observed datum is simply a linear constraint in the unknown signals v_k and d_k, which we combine with the appropriate extension condition in the following matrix optimization problem.

$$\beta = \min_{v_k, d_k} \alpha, \qquad \text{subject to,} \qquad y_k = P_{nom} u_k + W_\Delta v_k + W_d d_k,$$

$$z_k = P_{nom} u_k,$$

$$V'V \leq \alpha^2 Z'Z,$$

$$d_k' d_k \leq \alpha^2.$$

Clearly, β is the size of the smallest norm perturbation, Δ, and unknown signal, d, accounting for the observation. If $\beta > 1$, the datum invalidates the model. Useful engineering information can be obtained from the size of β and this may be helpful in modifying the model to more closely describe the observed system behaviors.

This approach can be applied to multiple perturbation structures by applying the block Toeplitz conditions to the input-output components of each Δ block perturbation. Note that the problem as stated is convex in d_k and v_k. Having $Z'Z$ known and constant is critical to convexity. The formulation can be applied to general LFT perturbation interconnections but will not always result in convex problems. See for example [9,10] for efficient algorithmic approaches to optimization problems of this kind.

3   Model Validation: Nonlinear Models

We would now like to outline a direction for extending the above model validation approaches to nonlinear systems, in a way takes as much advantage of the powerful linear theory and computational methods as possible. We will exploit the fact that the extension conditions are applied directly to signals (measured, calculated, or unknown) in the model and it makes no difference whether these signals are generated by a linear or a nonlinear system. In order to do this we consider a nonlinear model framework which is a simple extension of linear fractional transformation (LFT) models.

3.1   Nonlinear Feedback Models

Our nonlinear model framework is illustrated in Figure 5, where we again use the perturbation, Δ, to allow the model to describe a set of behaviors. To obtain the maximum benefit of the existing linear theory, as much of the dynamics as possible should be put into the linear part of the structure, $M(s)$.

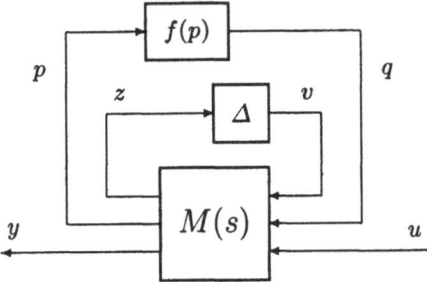

Fig. 5. Generic nonlinear LFT modeling framework. $M(s)$ is an LTI system; $f(p)$ is a static nonlinearity; and Δ is a block structured, norm bounded perturbation (LTI or LTV)

This can be used to model systems which exhibit complex nonlinear behavior. For example: hysteresis, limit cycles, sub-harmonics, frequency entrainment (Rayleigh equation), jump resonances (Duffing equation), and nonlinear damping (Froude equation). To further illustrate this, consider the Rayleigh equation,

$$\ddot{y} - 2\zeta(1 - \alpha\dot{y}^2)\dot{y} + y = u.$$

This can be expressed as a nonlinear LFT with,

$$M(s) = \left[\begin{array}{c|c} A & B \\ \hline C & D \end{array}\right] = \left[\begin{array}{cc|cc} 2\zeta & -1 & -2\zeta\alpha & 1 \\ 1 & 0 & 0 & 0 \\ \hline 1 & 0 & 0 & 0 \\ 0 & 1 & 0 & 0 \end{array}\right] \qquad \text{and} \qquad q = f(p) = p^3.$$

Figure 6 illustrates some of the rich nonlinear behavior possible. In this case a response to a sinusoid which contains higher frequency harmonics in the output. This system contains a stable limit cycle which can be seen in the phase portraits shown in Figure 7.

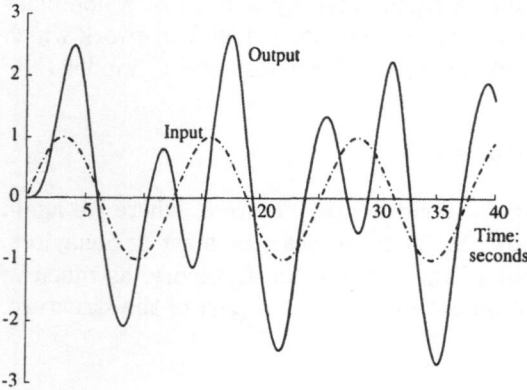

Fig. 6. Rayleigh system response (*solid*) to the input, $u(t) = \sin(0.5t)$ (*dashed*). The system parameter values are $\zeta = 0.4$ and $\alpha = 0.5$

Hysteresis can be observed in a simple example in this framework. Consider,

$$M(s) = \left[\begin{array}{cc} \dfrac{1}{\tau s + 1} & \dfrac{1}{\tau s + 1} \\ 1 & 0 \end{array}\right], \qquad \text{and} \qquad f(p) = \dfrac{2}{\pi}\arctan(10p).$$

This system has two stable equilibria, and an unstable equilibrium at the origin. Figure 8 illustrates the system response to a triangular input waveform and the hysteresis is evident. Notice that the initial and final values of the input are not zero and not equal, for a zero input value.

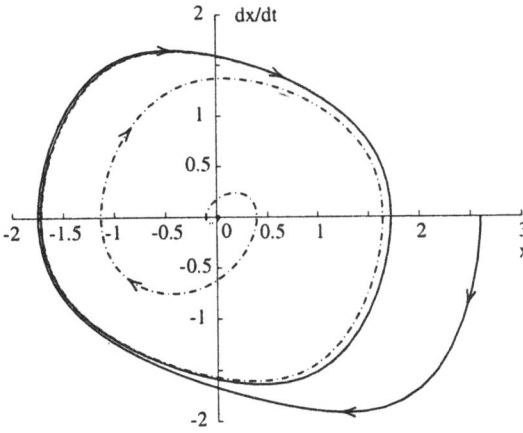

Fig. 7. Phase portraits of the Rayleigh system illustrating a stable limit cycle

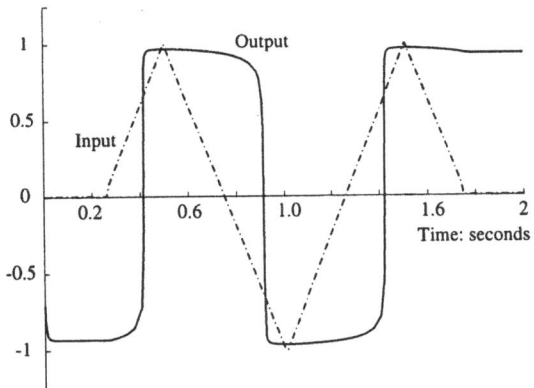

Fig. 8. Time domain response of the hysteresis system

3.2   A Model Validation Framework

The model framework, illustrated in Figure 5, is described by the equations,

$$\begin{bmatrix} z \\ p \\ y \end{bmatrix} = M \begin{bmatrix} v \\ q \\ u \end{bmatrix}, \qquad v = \Delta z, \quad \text{and} \quad q = f(p),$$

where $\Delta : L_2 \longrightarrow L_2$ is the unknown block structured (LTI/LTV) perturbation and $f(p)$ is a static nonlinear function. The nonlinearity is a known part of the model in this formulation, although it is possible to describe some degree of uncertainty in the nonlinearity by appropriately augmenting it with one of the perturbation blocks in Δ.

As in the linear case, the inputs and outputs, u and y, are assumed to be known. However the nonlinearity introduces several additional subtleties into both the model and the experimental configuration. In the purely linear model case a useful experimental protocol is to excite the system starting at rest and allow it to return to rest. This makes it easy to compare the results of multiple experiments.

In the nonlinear LFT model framework outlined here it is more difficult to implement such a protocol[1]. Note that equilibrium solutions of our framework satisfy

$$p = M_{21}(0)v + M_{22}(0)f(p) + M_{23}(0)u,$$

and in the nonlinear case there may be more than one stable equilibrium. In this case experiments may drive the system from one equilibrium to another. A more significant difficulty is that the equilibrium itself may be a function of the perturbation, Δ. Under such circumstances the running of the experiment and the interpretation of the results require some care.

We will give an example which obviates addressing some of these issues by selecting a model structure in which Δ does not affect the equilibria; however a general approach to model validation must unavoidably confront these. The next section introduces a nonlinear Moore-Greitzer model of compressor rotating stall. We then develop a nonlinear LFT model, approximating the full nonlinear model, as use model validation techniques to assess the quality of the approximation.

4   An Example: Nonlinear Moore-Greitzer Model

We use the Moore-Greitzer model of rotating stall in jet engine compressors, detailed in [11], as an example system. This is a nonlinear three state model which contains a transcritical bifurcation. The model describes the operation of an axial compressor under both stalled and non-stalled conditions. There is a hysteresis effect involved in using the throttle to move the operating point in and out of the rotating stall condition.

For this work we are interested in comparing the nonlinear LFT model and the original nonlinear model. The issue of how realistic a description the Moore-Greitzer model provides of rotating compressor stall is beyond the scope of this paper. For more detailed work on the control design problem itself see [12].

The model is expressed in terms of the dimensionless variables: Φ (averaged flow through the compressor); Ψ (averaged pressure rise across the compressor); R (square of the stall amplitude); and γ (throttle parameter). The variable R gives a measure of the severity of the stall, with $R = 0$ corresponding to the non-stalled condition. The model equations are

$$\dot{\Phi} = \frac{1}{l_c}\left(\Psi_c(\Phi) - \Psi - \frac{3RH}{4}\left(\frac{\Phi}{W} - 1\right)\right)$$

---

[1] In some cases it is not possible. Consider for example, the Rayleigh system outlined earlier which does not have a stable equilibrium, but has instead a stable limit cycle.

$$\dot{R} = \sigma R \left(1 - \left(\frac{\Phi}{W} - 1 \right)^2 - \frac{R}{4} \right)$$

$$\dot{\Psi} = \frac{1}{4l_c B^2} (\Phi - \Phi_T),$$

where Φ_T is the flow through the throttle. The compressor characteristic, $\Psi_c(\Phi)$, is modeled by

$$\Psi_c(\Phi) = \psi_{c0} + H \left(1 + \frac{3}{2} \left(\frac{\Phi}{W} - 1 \right) - \frac{1}{2} \left(\frac{\Phi}{W} - 1 \right)^3 \right).$$

The details of the constants in this equation are given in [11]. Under the usual assumption of a sufficiently short throttle duct length the throttle characteristic is modeled by,

$$\Psi = \Phi_T^2 / \gamma^2.$$

We show the equilibrium solutions graphically to illustrate the relationships present: the values of Ψ which satisfy the *two* conditions $\dot{\Phi} = 0$ and $\dot{R} = 0$, are plotted with respect to Φ in Figure 9. Note that the dependence on R is not explicitly shown, and for certain values of Φ this plot has two branches. Also shown are the steady state throttle characteristic curves for two different values of γ. The operating equilibrium points for the entire compressor system are thus given by the intersection points of the plot for Φ and the steady state throttle characteristic, which occur exactly when the third equilibrium equation $\dot{\Psi} = 0$ is satisfied.

For $\Phi > 2W$, there is a single compressor equilibrium curve corresponding to $R = 0$, and these equilibria are stable. Note that this curve corresponds to $\Psi = \Psi_c(\Phi)$. For $\Phi \leq 2W$, the $R = 0$ equilibrium curve, again corresponding to $\Psi = \Psi_c(\Phi)$, is now unstable. There is a second equilibrium curve, with $R \neq 0$ which has both stable and unstable parts.

For $\gamma > 0.4375$, the only intersection is on the stable compressor curve with $\Phi > 2W$ and $R = 0$. For γ between 0.3860 and 0.4375 there are three intercepts, two are stable and of these one has no stall ($R = 0$) and still satisfies $\Phi > 2W$. The second of the stable equilibrium curves corresponds to a stalled condition ($R > 0$) with a significant drop in pressure, Ψ. There is also an unstable equilibrium curve with an intermediate stall level. For $\gamma < 0.3860$ the only stable equilibrium curve corresponds to a stalled condition ($R > 0$). Moving γ from greater than 0.4375 to less than 0.3860 and back again produces a hysteresis effect.

4.1 The Nonlinear LFT Model

We have chosen to represent the Moore-Greitzer model with a two state approximation in the nonlinear LFT approach. The two states retained are Φ and Ψ and our model has the same stable (Φ, Ψ) equilibria as the original.

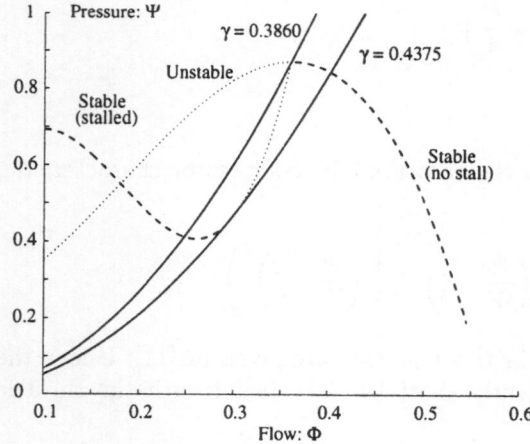

Fig. 9. Moore-Greitzer model equilibrium curves. The compressor constants used are: $l_c = 4.0$; $H = 0.32$; $W = 0.18$; $\psi_{c0} = 0.23$; $B = 0.20$; and $\sigma = 1.7778$

The input, γ, has been replaced by two inputs, $\gamma = u_0 + u$, where u_0 is the desired constant throttle bias, and would normally specify the desired operating equilibrium. The structure of our model is detailed in Figure 10.

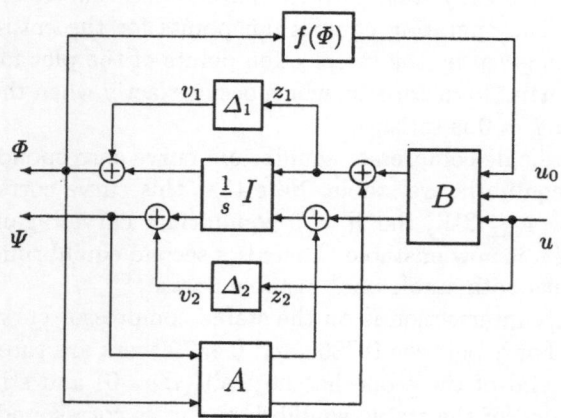

Fig. 10. Nonlinear LFT model of the Moore-Greitzer system

The nonlinear feedback function, $f(\Phi)$, is based on the compressor characteristic including the equilibrium effect of stall. Note that the equilibrium values of Φ and Ψ satisfy,

$$A \begin{bmatrix} \Phi \\ \Psi \end{bmatrix} + B \begin{bmatrix} f(\Phi) \\ u_0 \\ 0 \end{bmatrix} = 0,$$

and are independent of the perturbations $\Delta_{1,2}$. This deliberate modeling choice has been discussed in Section 3.2. The perturbations are assumed to satisfy as yet unknown bounds, $\|\Delta_1\|_\infty \leq \alpha_1$ and $\|\Delta_2\|_\infty \leq \alpha_2$. The objective of this work is to estimate suitable values for α_1 and α_2 using a model validation approach.

4.2   Model Validation Experiments

Four simulation "experiments" have been conducted at two equilibria. These are shown graphically for both the LFT model and the Moore-Greitzer model, in Figure 11. In all cases the throttle was set to $\gamma = u_0 = 0.41$. Experiments 1, 2, and 3 begin at the unstalled equilibrium and are excited by a pulse disturbances on the input u. Note that the trajectories illustrate different operating dynamics, and in the Expt. 3 case the disturbance is large enough to transition the system to the stalled equilibrium. Expt. 4 begins at the stalled equilibrium and transitions to the unstalled equilibrium. The agreement between the Moore-Greitzer model and the nominal nonlinear LFT model (with $\Delta_{1,2} = 0$) varies between experiments. We will find the smallest Δ for each experiment, which when included in the LFT model, allows the two model behaviors to match exactly.

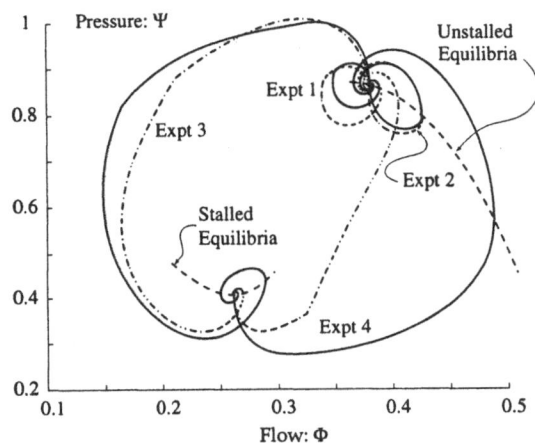

Fig. 11. State trajectories of simulation experiments. Moore-Greitzer model (*solid*); nonlinear LFT (*dot-dash*). Also shown are the stable equilibria (*dashed*)

The known signals from each Moore-Greitzer simulation are Φ, Ψ, u_0, and u, and in the model structure given in Figure 10 it is a simple matter to calculate the perturbation inputs (z_1 and z_2) and perturbation outputs (v_1 and v_2).

The linear model validation methods outlined in Section 2 are now used to find the size of the smallest perturbation, of a particular type (LTI or LTV), that allows the nonlinear LFT perturbation model to exactly match the full nonlinear simulation. The results of solving Problems 1 and 2 for the above experimental data are summarized in Table 1.

Table 1. Model validation results for LTI and LTV perturbations

Experiment	α_1 LTI	α_2 LTI	α_1 LTV	α_2 LTV
1	508	1.97	1.25	2.30
2	1.11	0.44	0.29	0.24
3	*	3.47	0.56	1.48
4	*	5.84	2.17	5.16

The results clearly indicate that the LTV perturbations give a much closer fit to the unmodeled behavior. In certain of the analyses the LTI lower bounds gave no indication of convergence for values of less than 1000 (denoted by * in the above table). The appropriate value to select for $\alpha_{1,2}$ in a robust design problem will depend on the anticipated region of operation. More extensive simulation over a wider range of operating conditions would be required to give a bound with any degree of confidence.

5   Future Directions

In this paper we have introduced a modeling framework for nonlinear systems based on the linear fractional transformation used in robust control theory. This framework involves augmenting the usual paradigm, which contains a linear system with uncertain dynamical perturbations, with a *memoryless* nonlinearity. This addition has a profound effect on the possible behaviors modeled, and results in a broad and rich range of nonlinear dynamical systems, examples of which have been provided in the text. Future research is aimed at: (1) further research into model validation in this context; and (2) robustness analysis of these LFT systems. Below we outline the focus of each of these investigations.

The first of these directions is purely a model validation and identification issue. If the static nonlinearity which has been added to the model presented here is permitted to reside in an arbitrary function class, then nearly all finite dimensional nonlinear dynamical systems can be characterized. This LFT configuration is therefore only useful if the class of nonlinearities is small enough to make computation possible, but large enough to admit a range of complex behaviors. In addition to this fundamental issue, computational algorithms must be developed to solve the model validation problem in this context. Here a special case of such a general algorithm has been presented in terms of model validation of a Moore-Greitzer model.

The second major issue to be pursued with regard to the model presented here is robustness analysis of both a quantitative and qualitative nature. With the nonlinear systems that this LFT model can produce a more detailed understanding of issues such as hysteresis, bifurcation, point stability and orbit stability is required. In particular how such phenomena can be predictably modeled and captured in the LFT framework. Quantitative questions must also be answered, for instance direct testing criteria for robustness against the above phenomena.

Acknowledgements

Roy Smith is supported by NSF under grant ECS–9634498 and Geir Dullerud is supported by NSERC. The authors also acknowledge very useful discussions on the details of the Moore-Greitzer model with Dan Fontaine, Petar Kokotović and Raff D'Andrea.

References

1. John Doyle, "Structured uncertainty in control system design", in *Proc. IEEE Control Decision Conf.*, 1985, pp. 260–265.
2. The MathWorks, Inc., Natick, MA, *μ-Analysis and Synthesis Toolbox (μ-Tools)*, 1991.
3. Roy S. Smith and John C. Doyle, "Model validation: A connection between robust control and identification", *IEEE Trans. Auto. Control*, vol. 37, no. 7, pp. 942–952, July 1992.
4. Roy S. Smith, "Model validation for robust control: an experimental process control application", *Automatica*, vol. 31, no. 11, pp. 1637–1647, Nov. 1995.
5. Kameshwar Poolla, Pramod Khargonekar, Ashok Tikku, James Krause, and Krishan Nagpal, "A time-domain approach to model validation", *IEEE Trans. Auto. Control*, vol. 39, no. 5, pp. 951–959, 1994.
6. T. Zhou and H. Kimura, "Time domain identification for robust control", *Syst. and Control Letters*, vol. 20, pp. 167–178, 1993.
7. Roy Smith and Geir Dullerud, "Validation of continuous-time control models by finite experimental data", *IEEE Trans. Auto. Control*, vol. 41, no. 8, pp. 1094–1105, Aug. 1996.
8. Sundeep Rangan and Kameshwar Poolla, "Time-domain validation for sampled-data uncertainty models", *IEEE Trans. Auto. Control*, vol. 41, no. 7, pp. 980–991, 1996.
9. Yu. E. Nesterov and A. S. Nemirovskii, *Interior-Point Polynomial Algorithms in Convex Programming*, SIAM, Philadelphia, 1994.
10. Geir Dullerud and Roy Smith, "Experimental application of time domain model validation: Algorithms and analysis", *Int. J. Robust & Nonlinear Control*, vol. 6, pp. 1065–1078, 1996.
11. F.K. Moore and E. M. Greitzer, "A theory of post-stall transients in axial compression systems—part i: Development of equations", *Journal of Turbomachinery*, vol. 108, pp. 68–76, 1986.
12. M. Krstic, D. Fontaine, P. Kokotovic, and J. Paduano, "Useful nonlinearities and global stabilization of bifurcations in a model of jet engine surge and stall", *IEEE Trans. Auto. Control*, vol. 43, no. 12, pp. 1739–1745, 1998.

Towards A Harmonic Blending of Deterministic and Stochastic Frameworks in Information Processing

Le Yi Wang[1] and George Yin[2]

[1] Department of Electrical and Computer Engineering, Wayne State University, Detroit, MI 48202
[2] Department of Mathematics, Wayne State University, Detroit, MI 48202

Abstract. This paper presents some preliminary efforts towards a new methodology which harmonically blends deterministic and stochastic frameworks in information processing, including identification, signal processing, communications, system design, etc. We begin with a discussion on distinctive features of the two frameworks and explanation of compelling reasons and motivating issues for introducing such a combined framework. Using persistent identification as an example, we demonstrate the application and utility of the methodology. Lower and upper bounds on identification errors are obtained for systems subject to both deterministic unmodelled dynamics and random external disturbances.

1 Introduction

An ancient Chinese folklore philosophy predicts that "Those who separate too long will unite; those who unite too long will separate." It seems that after a long history of parallel developments of stochastic and deterministic frameworks in identification, signal processing, communications, and control system design, the time is mature to investigate their mergers. But are there any compelling reasons for pursuing such a formidable goal? To answer this question, we shall have a quick tour on the foundations of these frameworks.

1.1 Deterministic and Stochastic Frameworks

The main difference between these two frameworks is reflected in their distinctive descriptions of uncertainty and expressions of system properties. The characterizes uncertainties in terms of their chances of occurrences and expresses system properties with a probability measure attached. In contrast, the characterizes uncertainty by the concept of deterministic uncertainty sets and expresses system properties as guaranteed even in the worst case of uncertainties.

While the probability theory is now developed rigorously on the basis of Kolmogorov axioms, its practical foundation remains the intuitive concept of "chances." Basically, on average the number of occurrences of an event in repeated experience approaches its "probabilistic expectation" (Laws of Large Numbers). This can be further extended to general ergodic processes when deeper properties of random variables or stochastic processes (such as Central Limit Theorem, Large Deviations, etc.) are involved. For information processing, such as system identification and communications, this property of "time averages approach space averages" can then be used to extract statistical information when a large sample of data becomes available. As a result, stochastic information processing often aims at establishing asymptotic properties. Examples include: (1) In system identification problems, disturbances are assumed to be i.i.d. white noises. Systems with unknown parameters are modelled via inputs and outputs, and consistency of the parameter estimates are pursued. (2) In discrete-event dynamic systems, events are represented as Markov processes. Discounted averages of certain penalty functions are used to evaluate system performance.

The deterministic framework of information processing is developed based on the idea of "guaranteed properties" or "robustness". Characterizing uncertainties as belonging to a fixed set without any statistical information, the deterministic worst-case doctrine demands that system properties, such as estimation errors in system identification (worst-case identification), stability and performance in control systems (robust control), to be valid for all possible uncertainties in the set. Examples include: (1) In system identification problems, disturbances are assumed to be uniformly bounded. Unknown system parameters are modelled as deterministic and belonging to prior uncertainty sets. Parameter estimates are evaluated by worst-case bounds. (2) In robust control, plant uncertainties are described as an uncertainty set, such as unstructured unmodelled dynamics, parametric or structured uncertainties. Controllers are to be designed such that stability and performance are maintained for all possible plants in the set. (3) In discrete-event systems, uncontrollable events are represented by deterministic actions. The set of acceptable states is specified and supervisors are to be designed such that discrete-event states will stay in the set under all possible uncontrollable events.

1.2 Motivating Issues

Both frameworks have demonstrated substantial utility in modeling uncertainties and in studying system performance and control design. On the other hand, each has its own weakness as well.

Issues Concerning Stochastic Frameworks:

1. There are system components such as modelling errors, which are not repeated during identification experiments or system evolution. As a result, they are fundamentally deterministic.

2. Traditionally, stochastic analysis often concentrates on asymptotic behavior. As a result, some powerful design tools, such as robust control design that relies on an uncertainty set description of the plant to construct controllers, cannot be directly applied when the information from system identification is given in terms of asymptotic statistical properties.

3. Any random event with a nonzero probability, no matter how small, will have a chance to occur. Engineers may be uncomfortable with a statement that the system will escape from stability region occasionally.

Issues Concerning Worst-Case Frameworks

1. Deterministic approaches pursue system properties that are preserved under the worst-case scenarios of deterministic uncertainties. This often leads to extremely high complexities. This is evidenced by computational complexities of optimal radius and time complexity in worst-case identification, computational complexity of designing discrete-event supervisors, and large irreducible identification errors due to disturbances.

2. It is very difficult to find reasonable bounds for disturbances that demonstrate bursts occasionally. It is similarly difficult to give accurate hard bounds on system modeling errors due to many non-ideal conditions.

3. When relatively large samples are available, it is unnecessarily conservative to insist on worst-case scenarios. Anyway, how many people have seen noises that are all constants over a sample path of 1000 points, or that follow exactly control signals?

2   Basic Principles and Formulation

2.1   Basic Principles

The framework introduced here is developed on the basis of the following observations and principles:

1. *An uncertain component can be modelled as stochastic only when it occurs repeatedly in the experiment, and the intended applications allow averaging on its occurrences. Otherwise, it should be modelled as deterministic.* This implies that many practical systems contain both types of components. Hence, a combined framework is indeed needed.

   For example, parameter deviations of a circuit board can be modelled as stochastic if the intended application is its average performance. However, when the performance of an individual board is considered, the parameter variations should be represented as deterministic since its occurrence is not repeated in this situation. Based on this principle, for system identification, unmodelled dynamics should be modelled as deterministic and disturbances as stochastic.

2. *Unlike the common perception, the worst-case doctrine, such as worst-case identification and robust control, is not risk-free and performance-guaranteed. It is a threshold methodology which bears risk (most likely in a probability sense) on its assumptions. In contrast, stochastic frameworks include stochastic properties in their assumptions.* This understanding indicates that one should not overly obsessed with or confined by 100% guarantees which the worst-case doctrine appears to offer.

A real life example is offered by design of buildings that can sustain earthquakes. A deterministic approach will set a threshold, say, Degree 7 Richter Scale, and design buildings that can stand against any earthquakes below that threshold. This worst-case scenario does not alter the fact that earthquakes may exceed the threshold and buildings may still collapse with a probability α. Here, the risk is taken on the assumptions. In a stochastic framework, the design criterion will be that "Designing buildings that will stand during an earthquake with a chance larger than $1 - \alpha$." Although people may not be comfortable in the latter statement (who wants to take that chance?), it is the actual reality.

In system identification, the assumption that the disturbance d is uniformly bounded by $\Omega = \{d : \sup_t |d(t)| \le \delta\}$ always bears a risk to be wrong, say, with a probability 0.1%. Suppose that a robust control, designed based on worst-case system identification using this assumption, guarantees stability and performance. While the statement is that "if the assumption is valid, the system properties are 100% guaranteed," it dose not alter the fact that there is a 0.1% chance that these properties do not hold. Here, the risk comes from the assumptions. Recently, there are efforts in characterizing stochastic properties of disturbances in deterministic frameworks. By eliminating "bad" noises in a deterministically defined subset of Ω, complexities are greatly reduced. This method again shifts chances to assumptions that disturbances outside the subset may occur with a certain probability.

3. *Due to system dynamics, integral types of performance and asymptotic nature of stability, an occasional failure of system properties does not necessarily lead to disasters.* As a result, probability descriptions of system performance might be acceptable, as long as it is appropriately related to system dynamics and performance measures.

Due to the existence of stochastic components, system properties will have a certain chance to fail. Fortunately, stability is a long term property. A sudden burst of noise over a short period will not make the system unbounded from that moment on as long as it returns to stability quickly. Similarly, in most performance specifications such as frequency domain criteria and integral types of performance indices, an occasional burst of bad performance will have only limited effects on total performance.

This understanding, however, points out the importance in expressing the probability nature of stability and system performance in relation to system dynamics and performance measures. For instance, systems with slow dynamics are capable of tolerating larger and longer deviations. Consequently, they will allow larger chance of performance failures.

2.2   Basic Formulation

The basic formulation of this framework is, surprisingly, very simple. Suppose that $y(v, d)$ is a mapping from the domain $\Omega_v \times \Omega_d$ to the range Ω_y. Assume that v is a deterministic vector and d is a random vector with probability P_d.

For any given $v \in \Omega_v$ and a subset $M \subseteq \Omega_y$, we define the co-image of M in Ω_d as

$$D_v(M) = \{d \in \Omega_d : y(v, d) \in M\}$$

Then, $F(M|v)$ is defined by

$$F(M|v) = P_d(D_v(M))$$

and

$$F(M) = \sup_{v \in \Omega_v} F(M|v)$$

$F(M)$ will be called *the of M*. In other words, $F(M)$ is the worst-case probability of the set M. $F(M)$ has the following basic properties:

1. $F(M) \geq 0$. $F(\Omega_y) = 1$.
2. If $M_1, M_2 \subseteq \Omega_y$ and $M = M_1 \cup M_2 \subseteq \Omega_y$, then $F(M) \leq F(M_1) + F(M_2)$.

We shall point out that pure worst-case and stochastic frameworks can be imbedded as special cases. For instance, if $y(v, d)$ represents a system performance such as identification errors or a performance index, then the deterministic framework seeks the worst-case bound

$$\mu = \sup_{v \in \Omega_v} \sup_{d \in \Omega_d} y(v, d)$$

This can be expressed equivalently, except on a set of probability zero, to

$$\mu = \inf\{\beta : F(\{y > \beta\}) = 0\}$$

On the other hand, if v is also modelled as stochastic, then $F(M)$ is reduced to regular probability.

Example 1. (1) $y = v + d$, where v is a scalar and deterministic with $v_{\min} \leq v \leq v_{\max}$, d is random with probability P_d. Let $M_1 = \{y \geq \alpha\}$ and $M_2 = \{y < \alpha\}$. Then

$$F(M_1) = \sup_{v_{\min} \leq v \leq v_{\max}} P_d(d \geq \alpha - v) = P_d(d \geq \alpha - v_{\max})$$

$$F(M_2) = \sup_{v_{\min} \le v \le v_{\max}} P_d(d < \alpha - v) = P_d(d < \alpha - v_{\min})$$

Note that although M_1 and M_2 are disjoint sets and $M_1 \cup M_2 = \mathbf{R}$

$$F(\mathbf{R}) = 1 \le P_d(d \ge \alpha - v_{\max}) + P_d(d < \alpha - v_{\min})$$
$$= 1 + P_d(\alpha - v_{\max} \le d < \alpha - v_{\min})$$

implying that in general F is not additive.

(2) In system identification, estimation errors can sometimes be expressed as

$$\eta = L\Phi\tilde{\theta} + L\widehat{E}$$

where Φ is a given regression matrix, L an identification mapping, $\tilde{\theta}$ infinite dimensional deterministic unmodelled dynamics with $\|\tilde{\theta}\|_1 \le \varepsilon$, and $\widehat{E}$ an N-dimensional random disturbance vector. Now, $\|\eta\|_1$ is a mapping from $\mathbf{R}^\infty \times \mathbf{R}^N$ to $\mathbf{R}_+$ and

$$F(\{\|\eta\|_1 \ge \beta\}) = \sup_{\|\tilde{\theta}\|_1 \le \varepsilon} P(D_{\tilde{\theta}})$$

where $D_{\tilde{\theta}} = \{\widehat{E} \in \mathbf{R}^N : \eta = L\Phi\tilde{\theta} + L\widehat{E}, \|\eta\|_1 \ge \beta\}$.

(3) A plant is expressed as $P = P_0 + W\Delta$, $\|\Delta\|_{H^\infty} \le \varepsilon$ where $\|\cdot\|_{H^\infty}$ is the H^∞ norm. The disturbance d has a normal distribution. Suppose that C is a robustly stabilizing controller. Then the output can be expressed as

$$y = (1 + PC)^{-1}d = (1 + (P_0 + W\Delta)C)^{-1}d$$

Let $M = \{\|y\|_2 \ge \beta\}$. Then

$$F(M) = \sup_{\|\Delta\|_{H^\infty} \le \varepsilon} P(D_\Delta)$$

where $D_\Delta = \{d : \|(1 + (P_0 + W\Delta)C)^{-1}d\|_2 \le \beta\}$, where $\|\cdot\|_2$ denotes the l^2 norm.

3   A Case Study: of Systems with Deterministic Unmodelled Dynamics and Stochastic Disturbances

This section presents an example of the basic ideas presented in the previous section. In this example, a framework of system identification is introduced that captures the hybrid features of systems subject to both deterministic unmodelled dynamics and stochastic disturbances.

It has long been debated that the stochastic framework is somewhat not control oriented and the worst-case identification is overly conservative. Stochastic identification concentrates on asymptotic behavior of estimates without quantifying error bounds at fixed finite time. On the other hand, the deterministic worst-case identification errors cannot be reduced below noise/signal ratios [35]. Moreover, time complexity of reducing identification errors towards the noise/signal ratios expands exponentially [29], and hence is forbiddingly high for applications to adaptive control of time varying systems. A question arises: Can identification problems be presented in a framework that captures the hybrid features of practical identification settings?

The framework introduced here are characterized by the following features: (1) Prior information is deterministic on plants (deterministic uncertainty sets) and stochastic on disturbances (stochastic distributions). It follows that in measuring the performance of identification algorithms, identification errors are evaluated against worst-case unmodelled dynamics but statistical effects from disturbances. (2) For applications to adaptive control, identification must persist in time in the sense that input signals and identification algorithms must provide satisfactory models for control design for all possible observation windows. This leads to the problem of persistent identification in which identification errors are evaluated for the worst-case over all possible starting time of observation windows of a fixed length.

The main ideas of the framework can be summarized as follows. For a given observation window of length N and starting time t_0, identification errors are measured by the l^1 norm. The probability for the errors to be larger than a given tolerance level β is then expressed as a function of disturbance, unmodelled dynamics, input signal, identification algorithm, as well as N and t_0. By taking the worst case over all possible unmodelled dynamics and t_0, and by selecting optimal inputs and identification mappings, an intrinsic relationship is established among the observation length N, identification error tolerance and confidence levels. First, for a given β and confidence level, the corresponding N is a measure of time complexity. Second, for a given N and confidence level, the corresponding β describes the optimal posterior uncertainty, suitable for robust control design. Finally, for a given β and N, the related probability conforms to the one studied in the classical stochastic identification framework.

The framework includes the main ingredients of both stochastic approaches and worst-case deterministic identification. The worst-case identification corresponds to the case where the disturbances are i.i.d., uniformly distributed in $[-\delta, \delta]$, and zero confidence levels. When unmodelled dynamics are zero, the problem is reduced to purely stochastic (persistent) identification problems. However, the main benefit of the framework is to allow a coherent treatment of issues involved in such hybrid identification problems, including inherent irreducible identification errors, optimal and suboptimal probing inputs, desirable identification algorithms, and relationships to closed-loop identification in adaptive systems.

We derive upper and lower bounds on identification errors, and obtain the speed of persistent identification. It is revealed that the class of full-rank periodic inputs and the standard least-squares estimation possesses certain appealing properties for the problems under study.

Related Literature. The problems and approaches discussed here are related to many topics which have been extensively studied in the literature. The following review will be far from exhaustive.

The concept of persistent identification in deterministic identification problems was introduced in Wang [37], and Zames, Lin and Wang [43]. Complexity issues in identification have been pursued by many researchers. The concepts of ε-net and ε-dimension in the Kolmogorov sense [18] were first employed by Zames [42] in studies of model complexity and system identification. Time complexity in worst-case identification was studied by Tse, Dahleh and Tsitsiklis [35], Dahleh, Theodosopoulos and Tsitsiklis [9], and Poolla and Tikku [29]. Results of n-widths of many other classes of functions and operators were summarized in Pinkus [27]. A general and comprehensive framework of information-based complexity was developed in Traub, Wasilkowski and Wozniakowski [34].

The problem of worst-case identification is now a very active research area. Milanes is one of the first researchers in recognizing the importance of worst-case identification. Milanes and Belforte [23], and Milanese and Vicino [24] introduced the problem of set-membership identification and produced many interesting results on the subject. Algorithms for worst-case identification were developed in Gu and Khargonekar [14], Makila [22], and Chen, Nett and Fan [7].

A new methodology which characterizes stochastic properties of disturbances in a deterministically defined uncertainty set was recently introduced by Venkatesh and Dahleh [36]. The issues of estimation consistency in worst-case identification were treated by Kakvoort and Van den Hof [15] in which main ideas from probability frameworks were extended to set-membership descriptions for disturbances. The paper by Paganini [26] derived set descriptions of white noise which are important for restoring consistency in worst-case identification.

Like its deterministic counterpart, many significant results have been obtained for identification and adaptive control involving random disturbances in the past a few decades. For instance, model validation approaches provide means of obtaining identification error bounds in both deterministic and statistical frameworks. There is a large amount of literature available on stochastic control. Here, we cite only the books by Aström and Wittenmark [2], Caines [5], Chen and Guo [6], Kumar and Varaiya [19], Ljung and Söderström [21], and Solo and Kong [33], among others. Further references of recent work can be found in the references cited in the aforementioned books. For related work in analyzing recursive stochastic algorithms, we refer the reader to the most recent work of Kushner and Yin [20] and the references therein.

3.1   Problem Formulation

For $p = 1$ or ∞, l^p is the normed space that consists of sequences $\{u(t) \in \mathbb{R}, t = 0, 1, \ldots\}$, for which

$$\|u\|_p := \begin{cases} \sum_{t=0}^{\infty} |u(t)| < \infty, & p = 1; \\[2ex] \sup_{0 \leq t < \infty} |u(t)| < \infty, & p = \infty. \end{cases}$$

For a positive integer n and a sequence $h = \{h(0), h(1), \ldots\} \in l^1$, T_n is the n-truncation operator $T_n : l^1 \mapsto \mathbb{R}^n$, defined by

$$T_n(h) = h_n = \{h(0), \ldots, h(n-1)\}.$$

Denote $M_n = T_n(l^1) = \{T_n(h) : h \in l^1\}$, and $\widetilde{M}_n = (1 - T_n)(l^1)$. We will use E to denote the expectation, and $P(\cdot)$ the probability.

Plants to be identified are single-input single-output, discrete-time, stable, linear-time-invariant (LTI) systems $h \in l^1$ with input/output relationships

$$y(t) = (h * u)(t) + d(t)$$

where $*$ denotes convolution, the probing input $u \in l^\infty$ with $\|u\|_\infty \leq \kappa_u$, and the exogenous disturbance $\{d(t)\}$ is a sequence of random variables. We are concerned with open-loop identification problems. Hence, the probing input u is deterministic and can be selected arbitrarily, albeit $\|u\|_\infty \leq \kappa_u$, by the designer.

A priori information on h is given by a deterministic uncertainty set $\mathcal{A} \subset l^1$ which contains h. $h \in \mathcal{A}$ can be decomposed into $\theta_n = T_n(h)$; and $\widetilde{\theta}_n = (1 - T_n)(h)$. θ_n is the modelled part, $\widetilde{\theta}_n$ is the unmodelled dynamics, and n is a measure of model complexity.

Here, we assume that for a given model complexity n, the *a priori* uncertainty on the plant is

$$\mathcal{A} = \{h \in l^1 : \quad \|(1 - T_n)(h)\|_1 \leq \varepsilon_n\}.$$

After applying an input u to the system and taking $N \geq n$ output observations in the time interval $t_0, \ldots, t_0 + N - 1$, we obtain the observation equations

$$y(t) = \phi(t)\theta_n + \widetilde{\phi}(t)\widetilde{\theta}_n + d(t), \quad t = t_0, \ldots, t_0 + N - 1,$$

where $\phi(t) = [u(t), \ldots, u(t - n + 1)]$, $\widetilde{\phi}(t) = [u(t-n), u(t-n-1), \ldots]$. Or, in a vector form $Y_N(t_0) = \Phi_N(t_0)\theta_n + \widetilde{\Phi}_N(t_0)\widetilde{\theta}_n + D_N(t_0)$.

Assumptions:

(**A1**) For all t_0, $\Phi_N(t_0)$ is full column rank.

(A2) The disturbance $\{d(t)\}$ is a sequence of i.i.d. random variables, with a common distribution d that is symmetric with respect to the origin and $E|d|^2 < \infty$.

(A3) The estimators are linear and unbiased. That is, for some matrix $L(t_0) \in \mathbb{R}^{n \times N}$, depending on t_0, $\widehat{\theta}_n(t_0) = L(t_0)Y_N(t_0)$ such that $E\widehat{\theta}_n(t_0) = \theta_n$ for all $\theta_n \in \mathbb{R}^n$ and all t_0, when $\widetilde{\theta}_n = 0$. In addition, the moment generating function $G(z) = E\exp(zd)$ exists.

Define the set of estimators as

$$\mathbb{L} = \{L(t_0) \in \mathbb{R}^{n \times N} : \quad L(t_0)\Phi_N(t_0) = I_n\}.$$

It should be emphasized that $\mathbb{L}$ depends on $\Phi_N(t_0)$ and t_0, although this dependence is suppressed from the notation.

Now, for any $L(t_0) \in \mathbb{L}$, we have the estimation error

$$\eta = \widehat{\theta}_n(t_0) - \theta_n = L(t_0)\widetilde{\Phi}_N(t_0)\widetilde{\theta}_n + L(t_0)D_N(t_0)$$

The first term in η is deterministic and the second is stochastic. We will denote $\eta_d = L(t_0)\widetilde{\Phi}_N(t_0)\widetilde{\theta}_n$, $\eta_s = L(t_0)D_N(t_0)$. As a result, $\eta = \eta_d + \eta_s$. In what follows, we present a number of results. The detailed proofs can be found in [40].

3.2 Persistent Identification

For a given tolerable identification error $\beta > 0$, denote the extremal probability by

$$Q_{(N,L,u,t_0,\varepsilon_n)}(\|\eta\|_1 \geq \beta) = \sup_{\|\widetilde{\theta}_n\|_1 \leq \varepsilon_n} P_{(N,L,u,t_0,\widetilde{\theta}_n)}(\|\eta\|_1 \geq \beta). \tag{1}$$

which represents the worst-case probability for estimation errors to exceed β. After minimization of the probability by using identification mappings

$$Q_{(N,u,t_0,\varepsilon_n)}(\beta) = \inf_{L \in \mathbb{L}} \quad Q_{(N,L,u,t_0,\varepsilon_n)}(\|\eta\|_1 \geq \beta),$$

followed by the worst-case over all t_0

$$Q_{(N,u,\varepsilon_n)}(\beta) = \sup_{t_0} \quad Q_{(N,u,t_0,\varepsilon_n)}(\beta),$$

and selection of optimal inputs, we have the following definition.

Definition 1.

$$Q_{(N,\varepsilon_n)}(\beta) = \inf_{\|u\|_\infty \leq \kappa_u} \sup_{t_0} \inf_{L \in \mathbb{L}} \sup_{\|\widetilde{\theta}_n\|_1 \leq \varepsilon_n} P_{(N,L,u,t_0,\widetilde{\theta}_n)}(\|\eta\|_1 \geq \beta) \tag{2}$$

is called optimal persistent identification error, which is a function of the signal space $U = \{u \in l^\infty : \|u\|_\infty \leq \kappa_u\}$ and identification mapping set $\mathbb{L}$.

$Q_{(N,\varepsilon_n)}(\beta)$ is an intrinsic relationship among estimation errors, observation lengths, and the corresponding probabilities. Other than its defining viewpoint (2), $Q_{(N,\varepsilon_n)}(\beta)$ can also be interpreted from other viewpoints. Especially, for a selected confidence level $0 \leq \alpha \leq 1$, we are seeking the minimal observation length N defined by

$$N_\alpha(\varepsilon_n, \beta) = \inf\{N: \quad Q_{(N,\varepsilon_n)}(\beta) \leq \alpha\}. \tag{3}$$

The quantity $N_\alpha(\varepsilon_n, \beta)$ is a complexity measure of the identification problem, which indicates how fast one can reduce the size of uncertainty on θ_n to β with confidence α, when the size of unmodelled dynamics is ε_n.

Denote by $\mathbf{N}$ the following class of input signals:

$$\mathbf{N} := \{u \in l^\infty: \quad \|u\|_\infty \leq \kappa_u, \ u \text{ is } n\text{-periodic and full rank}\}$$

where "full rank" means that the $n \times n$ Toeplitz matrix $\Phi_n(0)$ is full rank. Also, we will use LS to denote the standard least-squares estimation. When the input is limited to $\mathbf{N}$ and the identification mapping is specified to the LS estimation, we introduce the following definition.

Definition 2.

$$Q^0_{(N,\varepsilon_n)}(\beta) = \inf_{u \in \mathbf{N}} \ \sup_{t_0} \ \sup_{\|\tilde{\theta}_n\|_1 \leq \varepsilon_n} P_{(N,\mathrm{LS},u,t_0,\tilde{\theta}_n)}(\|\eta\|_1 \geq \beta),$$

and

$$N^0_\alpha(\varepsilon_n, \beta) = \inf\{N: \quad Q^0_{(N,\varepsilon_n)}(\beta) \leq \alpha\}.$$

Obviously,

$$Q_{(N,\varepsilon_n)}(\beta) \leq Q^0_{(N,\varepsilon_n)}(\beta), \quad N_\alpha(\varepsilon_n, \beta) \leq N^0_\alpha(\varepsilon_n, \beta). \tag{4}$$

3.3 Upper Bounds on $Q^0_{(N,\varepsilon_n)}(\beta)$ and $N^0_\alpha(\varepsilon_n, \beta)$

Theorem 1. *Let* $\beta = \varepsilon_n + \varepsilon$, $\varepsilon > 0$. *An upper bound of* $N^0_\alpha(\varepsilon_n, \beta)$ *is given by* $N^0_\alpha(\varepsilon_n, \beta) \leq \kappa^0 n$, *where*

$$\kappa^0 = \inf\left\{\kappa: \quad \kappa \geq \frac{\log(\alpha/(2n))}{\log[g(\varepsilon/n)]}\right\}.$$

Example 2. (1) If $d \sim \mathcal{N}(0, \sigma^2)$, then

$$\kappa^0 = \inf\left\{\kappa: \quad \kappa \geq \frac{\log(\alpha/(2n))}{\log[g(\varepsilon/n)]}\right\} = \left\lceil \frac{2[-\log(\alpha/2n)]\sigma^2 n^2}{\varepsilon^2} \right\rceil,$$

and

$$N_\alpha(\varepsilon_n, \varepsilon_n + \varepsilon) \leq \left\lceil \frac{2[-\log(\alpha/2n)]\sigma^2 n^2}{\varepsilon^2} \right\rceil n.$$

(2) If $d \sim U[-\delta, \delta]$, the uniform distribution, then

$$g(\tau) = \inf_z E \exp(z(d - \tau))$$
$$= \inf_z \frac{1}{z\delta} \sinh(\delta z) \exp(-z\tau).$$

The upper bound can then be computed via Theorem 1.

3.4   Lower Bounds

The estimation error contains two terms. The first term results from deterministic unmodelled dynamics, and the second from stochastic disturbances. The error depends also on the input u and identification algorithms.

Deterministic Lower Bounds. In the special case of noise-free observations, i.e., $d(t) = 0$, the estimation error is reduced to $\eta = \eta_d = L(t_0)\widetilde{\Phi}_N(t_0)\widetilde{\theta}_n$. Although intuitively, longer observation data, better selected inputs and better identification mappings may reduce identification errors, there exist certain inherent irreducible errors regardless all such efforts.

Theorem 2. *The optimal deterministic estimation error is bounded below by*

$$\inf_{\|u\|_\infty \leq \kappa_u} \sup_{0 \leq t_0 < \infty} \inf_{L(t_0) \in \mathbf{L}} \sup_{\|\widetilde{\theta}_n\|_1 \leq \varepsilon_n} \|\eta_d\|_1 \geq \varepsilon_n. \tag{5}$$

This theorem shows that no matter how long the observation windows are, how the input signals are selected, and how the identification algorithms are designed, persistent identification errors on θ_n cannot be reduced below ε_n, which is the size of unmodelled dynamics. For the special case $N = n$, this result was obtained by Wang [37]. The general result proved here turns out to be much more difficult to establish. It should also be pointed out that this conclusion is unique in persistent identification problems. If the starting time t_0 is fixed, then it can be easily shown that the lower bound is 0.

Probabilistic Lower Bounds.

Theorem 3. *Assume that (A1)–(A3) are satisfied. Denote the common distribution of the i.i.d. disturbance by d. For $\beta = \varepsilon_n + \varepsilon$, $Q_{(N,\varepsilon_n)}(\beta)$ is bounded below by*

$$Q_{(N,\varepsilon_n)}(\varepsilon_n + \varepsilon) \geq \frac{1}{2} \left(P(d \geq \varepsilon N \kappa_u | d \geq 0) + P(d \geq (2\varepsilon_n + \varepsilon) N \kappa_u | d \geq 0) \right).$$

In the special case of $d \sim \mathcal{N}(0, \sigma^2)$ or $d \sim U[-\delta, \delta]$, tighter lower bounds than those of Theorem 3 can be explicitly obtained.

Theorem 4. *Suppose that the conditions of Theorem 3 are satisfied and* $d \sim \mathcal{N}(0, \sigma^2)$. *Then,*

$$
\begin{aligned}
Q_{(N,\varepsilon_n)}(\varepsilon_n + \varepsilon) &\geq \left(\frac{1}{\sqrt{2\pi}} \right) m \left(\frac{\sigma}{\varepsilon \kappa_u \sqrt{N}} \right) \exp\left(-\frac{\varepsilon^2 \kappa_u^2 N}{2\sigma^2} \right) \\
&+ \left(\frac{1}{\sqrt{2\pi}} \right) m \left(\frac{\sigma}{(2\varepsilon_n + \varepsilon)\kappa_u \sqrt{N}} \right) \exp\left(-\frac{(2\varepsilon_n + \varepsilon)^2 \kappa_u^2 N}{2\sigma^2} \right).
\end{aligned}
\tag{6}
$$

where $m(x) = (x - x^3)$.

Theorem 5. *Suppose* $d \sim U[-\delta, \delta]$.

1. *The following lower bounds hold:*

$$
Q_{(N,\varepsilon_n)}(\varepsilon_n + \varepsilon) \geq
$$

$$
\begin{cases}
0, & \text{if } \varepsilon\kappa_u \geq \frac{\delta}{N} \\
\frac{1}{2}\left[\frac{\delta - \varepsilon N \kappa_u}{\delta} \right], & \text{if } \varepsilon\kappa_u < \frac{\delta}{N} \leq (2\varepsilon_n + \varepsilon)\kappa_u, \\
\frac{1}{2}\left(\left[\frac{\delta - (2\varepsilon_n + \varepsilon)N\kappa_u}{\delta} \right] + \left[\frac{\delta - \varepsilon N\kappa_u}{\delta} \right] \right), & \text{if } \frac{\delta}{N} > (2\varepsilon_n + \varepsilon)\kappa_u,
\end{cases}
$$

2. *If* $\varepsilon < \delta/\kappa_u$, *then* $Q_{(N,\varepsilon_n)}(\varepsilon_n + \varepsilon) > 0$.

Acknowledgement. *The research of L. Y. Wang was supported in part by the National Science Foundation under grants ECS-9412471 and ECS-9634375; The research of G. Yin was supported in part by the National Science Foundation under grant DMS-9529738.*

References

1. H. Akaike, Maximum likelihood identification of Gaussian autoregressive moving average models, *Biometrika* **60** (1973), 407-419.
2. K. Aström and B. Wittenmark, *Adaptive Control*, Addison-Wesley, 1989.
3. P. Billingsley, *Convergence of Probability Measures*, J. Wiley, New York, 1968.
4. R.L. Burden and J.D. Faires, *Numerical Analysis*, 5th Ed., PWS Publ. Co., Boston, 1993.
5. P. E. Caines, *Linear Stochastic Systems*, Wiley, New York, 1988.
6. H.-F. Chen and L. Guo, *Identification and Stochastic Adaptive Control*, Birkhäuser, Boston, 1991.
7. J. Chen, C.N. Nett, and M.K.H. Fan, Optimal non-parametric system identification from arbitrary corrupt finite time series, *IEEE Trans. Automatic Control*, **AC-40** (1995), 769-776.
8. H. Cramér, *Mathematical Methods of Statistics*, Princeton Univ. Press, Princeton, 1946.
9. M.A. Dahleh, T. Theodosopoulos, and J.N. Tsitsiklis, The sample complexity of worst-case identification of FIR linear systems, *System Control Lett.* **20** (1993).

10. S. N. Ethier and T.G. Kurtz, *Markov Processes, Characterization and Convergence*, Wiley, New York, 1986.
11. W. Feller, *An Introduction to Probability Theory and Its Applications Volume I*, Wiley, 3rd Ed., New York, 1968.
12. W. Feller, *An Introduction to Probability Theory and Its Applications Volume II*, Wiley, New York, 1966.
13. J. Gärtner, On large deviations from the invariant measure, *Theory Probab. Appl.* **22** (1977), 24-39.
14. G. Gu and P. P. Khargonekar, Linear and nonlinear algorithms for identification in H_∞ with error bounds, *IEEE Trans. Automat. Control*, **AC-37** (1992), 953-963.
15. R.G. Kakvoort and P.M.J. Van den Hof, Consistent parameter bounding identification for linearly parameterized model sets, *Automatica*, Vol. 31, pp. 957-969, 1995.
16. R. Z. Hasminskii and I. A. Ibragimov, On density estimation in the view of Kolmogorov's ideas in approximation theory, *Ann. Statist.* **18** (1990), 999-1010.
17. I. A. Ibragimov and R. Z. Hasminskii, *Statistical Estimation, Asymptotic Theory*, Springer-Verlag, New York, 1981.
18. A. N. Kolmogorov, On some asymptotic characteristics of completely bounded spaces, *Dokl. Akad. Nauk SSSR*, **108** (1956), 385-389.
19. P. R. Kumar and P. Varaiya, *Stochastic Systems: Estimation, Identification and Adaptive Control*, Prentice-Hall, Englewood Cliffs, NJ, 1986.
20. H. J. Kushner and G. Yin, *Stochastic Approximation Algorithms and Applications*, Springer-Verlag, New York, 1997.
21. L. Ljung and T. Söderström, *Theory and Practice of Recursive Identification*, MIT Press, Cambridge, MA, 1983.
22. P.M. Mäkilä, "Robust identification and Galois sequences," *Int. J. Contr.*, Vol. 54, No. 5, pp. 1189-1200, 1991.
23. M. Milanes and G. Belforte, Estimation theory and uncertainty intervals evaluation in the presence of unknown but bounded errors: Linear families of models and estimators, *IEEE Trans. Automat. Control* **AC-27** (1982), 408-414.
24. M. Milanese and A. Vicino, Optimal estimation theory for dynamic systems with set membership uncertainty: an overview, *Automatica*, **27** (1991), 997-1009.
25. D. C. Montgomery and E. A. Peck, *Introduction to Linear Regression Analysis*, J. Wiley, New York, 1982.
26. F. Paganini, A set-based approach for white noise modeling, *IEEE Trans. Automat. Control* **AC-41** (1996), 1453-1465.
27. A. Pinkus, *n-widths in approximation theory*, Springer-Verlag, 1985.
28. R. J. Serfling, *Approximation Theorems of Mathematical Statistics*, J. Wiley & Son, New York, 1980.
29. K. Poolla and A. Tikku, On the time complexity of worst-case system identification, *IEEE Trans. Automat. Control*, **AC-39** (1994), 944-950.
30. J. Rissanen, Estimation of structure by minimum description length, Workshop Ration. Approx. Syst., Catholic University, Louvain, France.
31. R. Shibata, Asymptotically efficient selection of the order of the model for estimating parameters of a linear process, *Ann. Statist.* **8** (1980), 147-164.
32. R. Shibata, An optimal autoregressive spectral estimate, *Ann. Statist.* **9** (1981), 300-306.

33. V. Solo and X. Kong, *Adaptive Signal Processing Algorithms*, Prentice-Hall, Englewood Cliffs, NJ, 1995.
34. J. F. Traub, G. W. Wasilkowski, and H. Wozniakowski, *Information-Based Complexity*, Academic Press, New York, 1988.
35. D.C.N. Tse, M. A. Dahleh and J.N. Tsitsiklis, Optimal asymptotic identification under bounded disturbances, *IEEE Trans. Auto. Control*, **AC-38** (1993), 1176-1190.
36. S.R. Venkatesh and M.A. Dahleh, Identification in the presence of classes of unmodeled dynamics and noise, *IEEE Trans. Automat. Control* **AC-42** (1997), 1620-1635.
37. L. Y. Wang, Persistent identification of time varying systems, *IEEE Trans. Automat. Control*, **AC-42** (1997), 66-82.
38. L.Y. Wang and J. Chen, Persistent identification of unstable LTV systems, Proc. 1997 CDC Conference, San Diego, 1997.
39. L. Y. Wang and L. Lin, Persistent identification and adaptation: Stabilization of slowly varying systems in H^∞, *IEEE Trans. Automat. Control*, Vol. 43, No. 9, pp. 1211-1228, 1998.
40. L.Y. Wang and G. Yin, Persistent identification of systems with unmodelled dynamics and exogenous disturbances, preprint, 1998.
41. G. Yin, A stopping rule for least-squares identification, *IEEE Trans. Automat. Control*, **AC-34** (1989), 659-662.
42. G. Zames, On the metric complexity of causal linear systems: ε-entropy and ε-dimension for continuous time, *IEEE Trans. Automat. Control*, **AC-24** (1979), 222-230.
43. G. Zames, L. Lin and L.Y. Wang, Fast identification n-widths and uncertainty principles for LTI and slowly varying systems, *IEEE Trans. Automat. Control*, **AC-39** (1994), 1827-1838.

Suboptimal Conditional Estimators for Restricted Complexity Set Membership Identification

Andrea Garulli[1], Boleslaw Kacewicz[2], Antonio Vicino[1], and Giovanni Zappa[3]

[1] Dipartimento di Ingegneria dell'Informazione, Università di Siena
Via Roma 56, 53100 Siena, Italy E-mail: garulli@ing.unisi.it,
vicino@ing.unisi.it
[2] Department of Applied Mathematics, University of Mining and Metallurgy
Al. Mickiewicza 30, Paw. A3/A4, III p., pok. 301, 30-059 Cracow, Poland
E-mail: kacewicz@mat.agh.edu.pl
[3] Dipartimento di Sistemi e Informatica, Università di Firenze
Via S. Marta 3, 50139 Firenze, Italy
E-mail: zappa@dsi.unifi.it

Abstract. When the problem of restricted complexity identification is addressed in a set membership setting, the selection of the worst-case optimal model requires the solution of complex optimization problems. This paper studies different classes of suboptimal estimators and provides tight upper bounds on their identification error, in order to assess the reliability level of the identified models. Results are derived for fairly general classes of sets and norms, in the framework of Information Based Complexity theory.

1 Introduction

A considerable amount of the recent literature on control oriented identification ([7,17,12]) is focused on the *restricted complexity* (or *conditional*) identification problem ([11,6,18,5]). The basic assumption is the following: the true system lives in the world of "real systems", which includes infinite dimensional and possibly nonlinear systems, while the set of models, within which we look for an "optimal" estimate, lives in a finite, possibly low dimensional world, whose structure and topology is compatible with modern robust control synthesis techniques. The problem is to find the model within the model set, whose distance from the real system is minimal.

In this paper the worst-case setting of the *Information Based Complexity* (IBC) paradigm is adopted [16]. The distance from the real (unknown) system is intended as the distance from the *feasible system set*, i.e. the set of all the systems which are compatible with the available information, either a priori or provided by the data collected in the identification experiment. Two main issues naturally arise in this context. The first is to analyze the relations between the different sources of errors in the identification process: finiteness of the data set, disturbances affecting the measurements and approximations introduced by the low dimensional model parameterization (in statistical identification this interplay is known as the bias-variance tradeoff). The second issue concerns the estimation algorithms. In the IBC setting, it is known that *conditional central algorithms* are optimal [9]. However, even if for specific situations they can be computed in a effective way [4], in general they require the solution of difficult min-max optimization problems and are computationally intractable. For this reason *suboptimal estimators* have been adopted [3]; they exhibit a lower computational complexity at the price of larger estimation errors. Surprisingly enough, little attention has been devoted to compute relative bounds on these *estimation errors*. The investigation of these issues is fundamental in the choice of the model parameterization as well as in the selection of the appropriate estimation algorithm, thus favoring the applicability of worst-case identification techniques (see [14,13] and the references therein).

The aim of this paper is to survey a number of results in this direction, obtained recently by the authors. These results are derived in a fairly general context, through the solution of approximation problems involving compact sets, for different norms in the measurement space and in the model space.

The paper is organized as follows. In Section 2, the conditional estimation problem is formulated in the IBC setting, and several estimators are introduced. Subsequently, the identification of low dimensional models for discrete-time systems, based on noisy input/output data, is considered. Several bounds on the estimation errors of the conditional algorithms are provided for a general class of norms. A closer inspection to the properties of suboptimal conditional estimators is finally provided in section 4. In particular, tight bounds on the estimation errors are provided for the ℓ_2 norms.

2   The Worst-Case Setting of Information Based Complexity

2.1   Formulation

First, let us introduce the setting and terminology used throughout the paper. It corresponds to one of the settings under which computational problems are studied in the information-based complexity theory (see [15], [16]). Let X be a linear space over a field of real or complex numbers equipped with a norm $\|\cdot\|_X$, and K be a subset of X. The elements $h \in X$ are called the *problem elements*. The goal is to compute approximations to the problem elements h, for all $h \in K$. To this purpose, some information about h is needed. The *information operator* is a mapping $\mathcal{N} : K \to Y$, where Y is a linear space over real or complex field with a norm $||\cdot||_Y$. Exact information about h is given by the value of the information operator $\mathcal{N}(h)$. In many situations, for example in practical applications where gaining information is related to measuring some physical quantities, exact information is not available. We have instead a corrupted value $y \in Y$ such that

$$||y - \mathcal{N}(h)||_Y \le \epsilon, \tag{1}$$

where ϵ is a given positive number. An approximation to h based on information y is computed as $\psi(y)$, where ψ is a mapping from Y to X, called an (idealized) *algorithm*. Given y, we denote by $\mathcal{E}$ the *feasible problem element set*, i.e. the set of all problem elements which are consistent with the corrupted information y,

$$\mathcal{E} = \{ h \in K : ||\mathcal{N}(h) - y||_Y \le \epsilon \}. \tag{2}$$

The worst-case *error* of an algorithm ψ is now defined by its worst performance for all $h \in \mathcal{E}$,

$$e_y(\psi) = \sup_{h \in \mathcal{E}} ||h - \psi(y)||_X. \tag{3}$$

Our goal is to find algorithms which have minimal, or nearly minimal, errors.

2.2   Central and Interpolatory Algorithms

We recall geometrical notions of the *radius* and *diameter* of a set A in the space X. They are defined by

$$\text{radius}(A) = \inf_{h \in X} \sup_{a \in A} ||h - a||_X \tag{4}$$

and

$$\text{diameter}(A) = \sup_{a_1, a_2 \in A} ||a_1 - a_2||_X. \tag{5}$$

Clearly

$$\text{radius}(A) \le \text{diameter}(A) \le 2\text{radius}(A). \tag{6}$$

The *Chebyshev center* of the set A is an element $c(A)$ in X (if it exists) such that the infimum in (4) is attained,

$$\sup_{a \in A} ||c(A) - a||_X = \text{radius}(A). \tag{7}$$

By means of the radius and diameter of a set we define two quantities that allow to characterize the quality of information provided by $\mathcal{N}$. The *radius* and *diameter of information* are given respectively by

$$rad_y = rad_y(\mathcal{N}) = \text{radius}(\mathcal{E}) \tag{8}$$

and

$$d_y(\mathcal{N}) = \text{diameter}(\mathcal{E}). \tag{9}$$

A useful property of the radius of information, is that it provides a lower bound on the error of any algorithm,

$$e_y(\psi) \geq rad_y(\mathcal{N}) \ , \ \forall \psi. \tag{10}$$

Furthermore, we can always choose an algorithm whose error is equal (or arbitrarily close) to this lower bound. Assuming that the Chebyshev center of $\mathcal{E}$ exists for all y, we define the *central algorithm* by

$$\psi^C(y) = c(\mathcal{E}). \tag{11}$$

Then, by definition, we have that

$$e_y(\psi^C) = rad_y(\mathcal{N}). \tag{12}$$

This means that the central algorithm (if it exists) is optimal. If the Chebyshev center of $\mathcal{E}$ does not exist, then we can always select an element in X for which the infimum in the definition of the radius is almost attained. We obtain in this way an algorithm whose error is arbitrarily close to $rad_y(\mathcal{N})$. In any case, we have that

$$rad_y(\mathcal{N}) = \inf_\psi e_y(\psi). \tag{13}$$

An algorithm ψ is said to be almost optimal within a factor α (*α-almost optimal*) if

$$e_y(\psi) \leq \alpha e_y(\psi^C) \quad \forall y.$$

Let us now consider an algorithm defined by

$$\psi^I(y) = \hat{h}, \tag{14}$$

where $\hat{h}$ is any element from the set $\mathcal{E}$ (i.e., $\hat{h}$ interpolates y). This algorithm is called *interpolatory algorithm* and has the property that

$$rad_y(\mathcal{N}) \leq e_y(\psi^I) \leq d_y(\mathcal{N}) \leq 2rad_y(\mathcal{N}), \tag{15}$$

which means that it is almost optimal within a factor 2.

2.3   Conditional Setting

In the formulation above, approximations provided by an algorithm ψ may be located in the entire space X. As follows from the definition, the radius of information is the radius of a minimal ball that contains the set $\mathcal{E}$, whose center is a point from X. It is interesting to consider a situation when an algorithm is constrained to provide approximations in a prescribed nonempty subset $\mathcal{M} \subset X$. In this case, the radius of information is meant to be the radius of the minimal ball centered at a point from $\mathcal{M}$. Such a restriction arises naturally in system identification problems, such as those considered in the Section 3. This setting is called *conditional*. A *conditional algorithm* is a mapping

$$\psi : Y \to \mathcal{M}, \tag{16}$$

and the *conditional radius of information* $rad_y^C = rad_y^C(\mathcal{N})$ is defined by

$$rad_y^C(\mathcal{N}) = \inf_{z \in \mathcal{M}} \sup_{h \in \mathcal{E}} ||h - z||. \tag{17}$$

Clearly, for any $\mathcal{M}$ it holds that $rad_y^C(\mathcal{N}) \geq rad_y(\mathcal{N})$.

An obvious generalization of the central algorithm is the *conditional central algorithm* given by

$$\psi^{CC}(y) = c^C(\mathcal{E}), \tag{18}$$

where $c^C(\mathcal{E})$ is the *conditional Chebyshev center* of $\mathcal{E}$, defined as a point in $\mathcal{M}$ (if it exists) such that

$$\sup_{h \in \mathcal{E}} ||h - c^C(\mathcal{E})||_X = rad_y^C(\mathcal{N}). \tag{19}$$

By (19), we have

$$e_y(\psi^{CC}) = rad_y^C(\mathcal{N}). \tag{20}$$

The conditional radius provides a lower bound on the error of any conditional algorithm

$$rad_y^C(\mathcal{N}) = \inf_{\psi} e_y(\psi), \tag{21}$$

where the infimum is taken with respect to all conditional algorithms (16).

In the conditional setting, several *suboptimal algorithms* are also considered. In order to introduce them, the notion of projection over $\mathcal{M}$ is needed. For $h \in X$ let

$$C(h) = \{\hat{z} \in \mathcal{M} : ||h - \hat{z}||_X = \inf_{z \in \mathcal{M}} ||h - z||_X\} \tag{22}$$

be the set of best approximation elements for h in $\mathcal{M}$ (assumed nonempty). A projection $P : X \to \mathcal{M}$ is defined to be any mapping such that $P(h) \in C(h)$, for all $h \in X$. Recall that if $\mathcal{M}$ is a finite dimensional subspace of X, then $C(h)$ is nonempty. If additionally $C(h)$ is a singleton for each h then P is continuous (see e.g. [10]). In general, P is a nonlinear operator. If X is a Hilbert space and $\mathcal{M}$ a finite dimensional subspace of X, then P is linear. The *central-projection algorithm* ψ^{CP} is given by

$$\psi^{CP}(y) = P(c(\mathcal{E})). \tag{23}$$

The central projection algorithm is thus defined by the projection on $\mathcal{M}$ of the unconditional center of $\mathcal{E}$.

An *interpolatory-projection algorithm* ψ^{IP} is defined by

$$\psi^{IP}(y) = P(\hat{h}), \tag{24}$$

where $\hat{h}$ is any element from the set $\mathcal{E}$.

Finally the *restricted-projection algorithm* ψ^{RP} is defined as

$$\psi^{RP}(y) = z^{RP} \tag{25}$$

where

$$z^{RP} = \arg \inf_{z \in \mathcal{M}} \|\mathcal{N}(z) - y\|_Y. \tag{26}$$

Notice that $\mathcal{N}(z^{RP})$ is the projection in the Y space of the measurement y over the subset $\mathcal{N}(\mathcal{M})$.

3   Identification in a General Class of Norms

3.1   Formulation

The setting and results presented in this section are based on [8]. Identification of causal, single-input single-output, linear time invariant, discrete-time systems represented by one-sided impulse response $h = [h_0, h_1, \dots]$, is considered. It is assumed that h belongs to a linear space X equipped with a norm $\| \cdot \|_X$, which contains as a subspace a space $X_N = \{h : h = [h_0, \dots, h_{N-1}, 0, 0, \dots]\}$, where $N \geq 1$. Let T_N denote the mapping $T_N : X \to X_N$ given by $T_N(h) = [h_0, h_1, \dots, h_{N-1}, 0, 0, \dots]$. The aim is to identify the system h, on the basis of the available information.

A priori knowledge about h is expressed by $h \in K$, where K is a (nonempty) subset of X. Next, it is assumed that information is given by measurements of the first N elements of the output sequence $y = [y_0, y_1, \dots]$. The measurements are related to an input sequence $u = [u_0, u_1, \dots]$ by

$$y_l = \sum_{k=0}^{l} h_k u_{l-k} + e_l, \quad l = 0, 1, \dots, N-1, \tag{27}$$

where e_l is an error corrupting the output. It is assumed that $|u_k| \leq 1$ for $k \geq 0$ and $u_0 \neq 0$. In the sequel, for a sequence $a = [a_0, a_1, \dots]$ we shall denote by a^N the sequence $[a_0, a_1, \dots, a_{N-1}, 0, 0, \dots]$ or the vector $[a_0, a_1, \dots, a_{N-1}]^T$, the meaning being always clear from the context. In a matrix form, (27) can be rewritten as

$$y^N = U_N h^N + e^N, \tag{28}$$

where U_N is a lower triangular $N \times N$ nonsingular Toeplitz matrix given by input u

$$
U_N = \begin{bmatrix}
u_0 & 0 & 0 & \cdots & 0 \\
u_1 & u_0 & 0 & \cdots & 0 \\
u_2 & u_1 & u_0 & \cdots & 0 \\
\vdots & \vdots & & \vdots & \vdots \\
u_{N-1} & u_{N-2} & \cdots & \cdots & u_0
\end{bmatrix} . \tag{29}
$$

The space Y is a normed space of vectors $y^N = [y_0, y_1, \ldots, y_{N-1}]^T$, and the information operator $\mathcal{N} : X \to Y$ is given by $\mathcal{N}(h) = U_N h^N$. The actual noisy information is provided by the vector y^N in (28). The measurement error is assumed to be bounded by $\|e^N\|_Y \le \epsilon$, where ϵ is a given number accounting for the accuracy of output measurements. The set $\mathcal{E}$ of systems consistent with the information y^N is given by $\mathcal{E} = \{h \in K : \|y^N - U_N h^N\|_Y \le \epsilon\}$.

In the following, conditional identification problems will be addressed, in which possible approximations of h can be selected from a linear n-dimensional subspace $\mathcal{M} = M_n$ of X, representing the model set. The number n corresponds to the number of parameters in the approximate model. In many applications, the desired number of parameters is small, and n is much less than N.

3.2 Upper and Lower Bounds on the Radius

Recall that the minimal error (central) algorithm is defined by $\psi^{CC}(y^N) = \hat{h}^*$, where $\hat{h}^* \in M_n$ is taken to be an element for which

$$
\sup_{h \in \mathcal{E}} \|h - \hat{h}^*\|_X = \inf_{\hat{h} \in M_n} \sup_{h \in \mathcal{E}} \|h - \hat{h}\|_X.
$$

To stress the dependence of the radius of information on M_n, we shall denote it, when convenient, by $rad_y^C(\mathcal{N}) = rad_y(M_n)$. For a discussion of the radius in the conditional context with a Hilbert norm in X, see [5] and [9]. Note that the radius of information is a nonincreasing function of the complexity of the model, i.e., $rad_y(M_{n+1}) \le rad_y(M_n)$ for $M_n \subset M_{n+1}$.

Both the central algorithm and the radius of information are in general difficult to compute, even in the unconditional context. We shall therefore concentrate on deriving possibly tight bounds on the minimal identification error and on finding suboptimal estimation algorithms. First, an upper bound on the local error of the interpolatory-projection algorithm ψ^{IP} is derived; next, a lower bound on the error of an arbitrary algorithm ψ with values in the space M_n is computed.

In the conditional context, the *diameter of information* is defined by

$$
d_y = d_y(M_n) = \sup_{h, \tilde{h} \in \mathcal{E}} \|P_n(h) - P_n(\tilde{h})\|_X. \tag{30}
$$

The conditional diameter of information measures the size of the projection of the set $\mathcal{E}$ on M_n. From now on, the symbol d_y will be always used in the sense of (30). The quality of models belonging to M_n, representing systems consistent with the output y, can be measured by the following quantity

$$me_y = me_y(M_n) = \sup_{h \in \mathcal{E}} ||h - P_n(h)||_X. \tag{31}$$

We call it the *model error* with respect to a model set M_n. The model error decreases with increasing complexity of the model set, i.e., $me_y(M_{n+1}) \leq me_y(M_n)$ for $M_n \subset M_{n+1}$. Obviously, for the interpolatory algorithm (24) one has $e_y(\psi^{IP}) \leq \sup_{h \in \mathcal{E}} ||h - P_n(h)||_X + \sup_{h \in \mathcal{E}} ||P_n(h) - P_n(\hat{h})||_X$, so that for any norms in X and Y it holds that $e_y(\psi^{IP}) \leq me_y + d_y$.

Turning to a lower bound on the error, since P_n is a projection, we have for arbitrary algorithm ψ with values in M_n that $e_y(\psi) \geq \sup_{h \in \mathcal{E}} ||h - P_n(h)||_X = me_y$. On the other hand, since $||P_n(h) - \psi(y^N)||_X \leq ||P_n(h) - h||_X + ||h - \psi(y^N)||_X$, it holds that $\sup_{h \in \mathcal{E}} ||P_n(h) - \psi(y^N)||_X \leq me_y + e_y(\psi)$. Finally, the triangle inequality yields that $\sup_{h \in \mathcal{E}} ||P_n(h) - \psi(y^N)||_X \geq \frac{1}{2} d_y$. The following theorem summarizes the above results.

Theorem 1. *For arbitrary norms in X and Y, and any set K:*
(i) For the interpolatory-projection algorithm ψ^{IP},

$$e_y(\psi^{IP}) \leq me_y + d_y , \quad \forall y, \tag{32}$$

where the diameter d_y and the model error me_y are given by (30) and (31), respectively.
(ii) For any algorithm ψ,

$$e_y(\psi) \geq \max\{me_y, \frac{1}{2}d_y - me_y\} , \quad \forall y, \tag{33}$$

and, consequently,

$$e_y(\psi) \geq \max\{me_y, \frac{1}{4}d_y\} , \quad \forall y . \tag{34}$$

Both upper and lower bounds in (32) and (33) (UP and LOW, respectively) are expressed in terms of the diameter of information and the model error, and are almost optimal in the sense that their ratio is bounded by an absolute constant, independently of ϵ, M_n and U_N. In the non-trivial case, when $\max\{d_y, me_y\} > 0$, we have that $\frac{UP}{LOW} \leq 5$. Furthermore, it holds that $UP - LOW \leq d_y$. In the unconditional case with $X = X_N$ and $M_n = X_N$ we have $me_y = 0$, $UP = d_y$ and $LOW = \frac{1}{2}d_y$, which agrees with (6) and (15). If the measurements are noise free, $\epsilon = 0$, and $|| \cdot ||_X$ has the P-property, then $d_y = 0$ and $UP = LOW = me_y$, i.e., the radius of information reduces to the model error.

Theorem 1 also states that the error of interpolatory-projection algorithm differs from the minimal possible error at most by a factor of 5 (we do not address here possible further reduction of this constant). This algorithm is thus almost optimal.

3.3   Product Norms and Sets

In the identification setting introduced in subsection 3.1, the measurements y^N provide information only on the first N samples of the impulse response h. This fact has remarkable consequences on the structure of the identification error and on the choice of M_n, if suitable assumptions are made on the norm in X and the set K. In the following, the class of product norms in X and product sets K will be considered. To introduce these concepts, let R_N be a subspace of X given by

$$R_N = \{h \in X : \quad h = [0, \ldots, 0, h_N, h_{N+1}, \cdots]\}. \tag{35}$$

Each element $h \in X$ has a unique decomposition $h = h^N + r^N$, where $h^N \in X_N$ and $r^N \in R_N$. Let $|| \cdot ||_{X_N}$ and $|| \cdot ||_{R_N}$ be arbitrary norms in X_N and R_N, respectively, and $|| \cdot ||$ be a norm in $\mathbb{R}^2$. We assume about the latter that $||[1, 0]^T|| = ||[0, 1]^T|| = 1$ and that the function $f(a, b) = ||[a, b]^T||$ is nondecreasing with respect to each variable a and b for $a, b \geq 0$. It is easy to see that the functional defined by

$$||h||_X = f(||h^N||_{X_N}, ||r^N||_{R_N}) \tag{36}$$

is a norm in X, called the *product norm*. Some observations are now in order.
1) By definition, we have that $||h||_X = ||h||_{X_N}$ for $h \in X_N$, and $||h||_X = ||h||_{R_N}$ for $h \in R_N$.
2) The weighted l_p norms ($1 \leq p \leq \infty$) given by

$$||h||_{l_p} = (\sum_{k=0}^{\infty} \alpha_k^p |h_k|^p)^{1/p}, \tag{37}$$

where $\alpha_k > 0$, provide examples of product norms (with $||h^N||_{X_N} = (\sum_{k=0}^{N-1} \alpha_k^p |h_k|^p)^{1/p}$, $||r^N||_{R_N} = (\sum_{k=N}^{\infty} \alpha_k^p |h_k|^p)^{1/p}$ and $f(a, b) = (|a|^p + |b|^p)^{1/p}$).
    Product norms have the following useful property. The norm in X has the *projection property (P-property)* if for each $M_n \subset X_N$ there is a projection $P_n : X \to M_n$, such that $P_n(T_N h) = P_n(h)$. It will be assumed in the sequel that for norms possessing the P-property a projection P_n is always selected to satisfy the above condition.

Lemma 1 ([8]).   *Product norms have the (P) property.*

Let us now define a class of product sets. We call K a *product set* if there are nonempty sets $K_1 \subset X_N$ and $K_2 \subset R_N$ such that

$$K = \{h \in X : h^N \in K_1 \text{ and } r^N \in K_2 \}. \tag{38}$$

For example, the set $K = \{h : \max_{i \geq 0} |h_i| \leq 1\}$ is product, while $K = \{h : \sum_{i=0}^{\infty} |h_i| \leq 1\}$ is not. If $K_1 = X_N$ (i.e., no 'a priori' restrictions are imposed on the first N components of h) then the set K is called *residual*.

It can be shown that in several identification settings, attention can be restricted to subspaces M_n that are contained in X_N, under mild assumptions . The following theorem extends to product norms the result proven in [5] for the ℓ_2 identification.

Theorem 2 ([8]).   *Assume that the norm in X is a product norm and K is a product set, with K_2 centrally symmetric (i.e., $h \in K_2$ implies that $-h \in K_2$). Then, for any subspace $M_n \subset X$ and any algorithm $\psi : Y \to M_n$ we have for arbitrary norm in Y*

$$e_y(\psi) \geq e_y(T_N\psi) \ , \ \forall y.$$

Theorem 2 says that, as far as the minimal error is of interest, there is no advantage in considering algorithms ψ which give estimates outside X_N, since the error of truncated algorithm $T_N\psi$ with values in X_N is no greater than the error of ψ. For this reason, from now on we restrict ourselves to model sets $M_n \subset X_N$. For another result of a similar type as Theorem 2 see also [3].

3.4   The Model Error and the Diameter of Information

Bounds on the model error

Let us denote by re_y a measure of the "residual" part of systems in $\mathcal{E}$,

$$re_y = \sup_{h \in \mathcal{E}} \|h - T_N h\|_X, \tag{39}$$

and call it the *residual error*. The residual error describes the behavior of the components of h which are not involved in computation of the measurement vector y^N and it mainly depends on the 'a priori' information set K. Let

$$tme_y = tme_y(M_n) = \sup_{h \in \mathcal{E}} \|T_N h - P_n(h)\|_X \tag{40}$$

denote the *truncated model error*. For product sets K the residual error re_y is independent of y and takes a form

$$re_y = re = \sup_{h \in K} \|h - T_N h\|_X = \sup_{h \in K_2} \|h\|_X . \tag{41}$$

If additionally the norm in X has the P-property then the truncated model error can be expressed as

$$tme_y = \sup_{h^N \in K_1 : \|U_N h^N - y^N\|_Y \leq \epsilon} \|h^N - P_n(h^N)\|_X . \tag{42}$$

In a general case, the bounds on the model error in terms of the residual error and the truncated model error are shown in the following proposition. In the case of a product norm in X and a product set K an exact expression for the model error can be obtained.

Proposition 2 ([8]).
(i) For any norms in X, Y and any set K, it holds that

$$me_y \leq re_y + tme_y \quad , \quad \forall y , \tag{43}$$

where re_y and tme_y are given in (39) and (40), respectively.
(ii) For a product norm in X and a product set K,

$$me_y = f(tme_y, re) \quad , \quad \forall y , \tag{44}$$

where

$$tme_y = \sup_{h^N \in F_y} \|h^N - P_n(h^N)\|_{X_N} \quad and \quad re = \sup_{r^N \in K_2} \|r^N\|_{R_N}, \tag{45}$$

with $F_y = \{ h^N \in K_1 : \|U_N h^N - y^N\|_Y \leq \epsilon \}$.

Turning to the diameter of information, it is easy to check the following result

Proposition 3. *If the norm in X has the P-property and K is a product set, then*

$$d_y = \sup_{h^N, \hat{h}^N \in F_y} \|P_n(h^N) - P_n(\hat{h}^N)\|_X, \tag{46}$$

where F_y is given in Proposition 1.

Bounds on the radius of information
Combining Theorem 1 with Proposition 2 and 3, one gets the following result.

Theorem 3.
(i) For any norms in X, Y and any set K it holds that

$$e_y(\psi^{IP}) \leq re_y + tme_y + d_y \quad , \quad \forall y . \tag{47}$$

where re_y, tme_y, d_y are given respectively by (39), (40), (30).
(ii) For a product norm in X and a product set K,

$$e_y(\psi^{IP}) \leq f(tme_y, re) + d_y \quad , \quad \forall y \tag{48}$$

and, for any algorithm $\psi : Y \to M_n$,

$$e_y(\psi) \geq \max\{f(tme_y, re), \frac{1}{2}d_y - f(tme_y, re)\} \quad , \quad \forall y , \tag{49}$$

where re, tme_y and d_y are given in (45) and (46).

The above theorem provides sharp upper and lower bounds on the local radius of information for a wide class of norms in X, including most cases of interest. The bounds are expressed in terms of quantities that are naturally connected with possible sources of errors in the identification process: the diameter of information depends on measurement errors, the truncated model error reflects errors arising due to model simplification and the residual error measures the residual part of a system.

We end this section by noting that if the projection P_n is linear and K is a residual set, then the formulas for tme_y and d_y are particularly simple and can be rewritten as follows.

Proposition 4. *Let* $\| \cdot \|_X$ *have the P-property and K be residual. If the projection P_n is linear then*

$$tme_y = \sup_{z \in Y, \|z\|_Y \leq \epsilon} \|(I - P_n)(U_N^{-1} z + U_N^{-1} y^N)\|_X \tag{50}$$

and

$$d_y = 2\epsilon \sup_{z \in Y, \|z\|_Y \leq 1} \|P_n U_N^{-1} z\|_X . \tag{51}$$

Proof. It results from the linearity of P_n applied to (42) and (46) (a vector $v \in \mathbb{R}^N$ is identified here with the sequence $[v^T, 0, 0, \dots] \in X_N$). $\square$

4   Identification in the ℓ_2 norm

4.1   Problem Specification

In Section 3, bounds on the conditional radius of information have been derived for a general class of norms in the spaces X and Y. This section is devoted to studying suboptimal estimators under the following assumptions:
(i) the spaces $X = \mathbb{R}^k$ and $Y = \mathbb{R}^m$ $(m \geq k)$ are equipped with the ℓ_2 norm (in the sequel, we shall denote the ℓ_2 norm by $\| \cdot \|$);
(ii) $K = X$;
(iii) $\mathcal{M}$ is an n-dimensional linear manifold in X $(n < k)$,

$$\mathcal{M} = \{z \in \mathbb{R}^k : z = z^o + M\alpha , \ \alpha \in \mathbb{R}^n, \ n < k\} \tag{52}$$

where M is a full rank $k \times n$ matrix.

Noisy measurements $y \in Y$ are available and satisfy $y = Uh + e$, where e is an additive noise vector and $U : X \to Y$ is a linear mapping, which defines the information operator $\mathcal{N}(h) = Uh$. It is assumed that U is a one-to-one mapping (i.e., U as a matrix is of full rank). The noise vector satisfies $\|e\| \leq \epsilon$.

We shall consider the estimation errors of the four algorithms introduced in the first section, namely the conditional central algorithm ψ^{CC}, the central projection algorithm ψ^{CP}, the interpolatory-projection algorithm ψ^{IP}

and the restricted projection algorithm ψ^{RP} defined in (18), (23), (24), (25). Optimality and almost optimality of the estimation algorithms are studied in the ℓ_2 case (more details can be found in [2]). Under the assumptions (i) and (ii), the set $\mathcal{E}$ of systems consistent with the information y is given by the ellipsoid

$$\mathcal{E} = \{h \in \mathbb{R}^k : h^T U^T U h - 2y^T U h + y^T y \leq \epsilon^2\}. \tag{53}$$

and the estimates $z^{CP} = \psi^{CP}(y)$, $z^{IP} = \psi^{IP}(y)$, $z^{RP} = \psi^{RP}(y)$ specified in (23), (24) and (26) can be easily computed as

$$z^{CP} = z^o + M(M^T M)^{-1} M^T (U^T U)^{-1} U^T (y - U z^o) \tag{54}$$

$$z^{IP} = z^o + M(M^T M)^{-1} M^T (\hat{h} - z^o) \tag{55}$$

$$z^{RP} = z^o + M(M^T U^T U M)^{-1} M^T U^T (y - U z^o) . \tag{56}$$

On the other hand, quite a complex min-max optimization problem must be solved in order to compute the conditional central estimate $c^C(\mathcal{E})$ in (18). A complete characterization of the conditional center of $\mathcal{E}$ and an efficient procedure for computing it has been recently proposed in [4].

Through a suitable change of coordinates in the space X, without loss of generality the problem can be reformulated replacing the ellipsoid $\mathcal{E}$ in (53) with a standard form

$$\mathcal{E} = \{h \in \mathbb{R}^k : h^T Q h \leq 1\} \tag{57}$$

$$Q = \text{diag}\{q_i\}_{i=1}^k , \quad 0 < q_1 \leq q_2 \leq \dots \leq q_k . \tag{58}$$

It can also be assumed that

$$M^T M = I \quad \text{and} \quad M^T z^o = 0 , \tag{59}$$

where I is the identity matrix. Then $c(\mathcal{E}) = 0$, and the estimates (55)–(56) provided by the interpolatory projection, central projection and restricted projection algorithms are given by

$$z^{CP} = z^o \tag{60}$$

$$z^{IP} = z^o + M M^T \hat{h} \tag{61}$$

$$z^{RP} = [I - M(M^T Q M)^{-1} M^T Q] z^o. \tag{62}$$

4.2   Reliability Level of Projection Algorithms

In the following, the evaluation of the reliability level of the above estimators is addressed. The most interesting case is that of central projection algorithms. The following theorem assures that the error of this estimate is less than 16% greater than the optimal error provided by the conditional central algorithm. More precisely, we have

Theorem 4. *Let the assumptions (i) and (ii) hold. The central projection algorithm ψ^{CP} satisfies*

$$e_y(\psi^{CP}) \leq \sqrt{\frac{4}{3}} \, rad_y^C(\mathcal{N}).$$

Moreover, the upper bound is achieved.

Proof. First, let us show that one can assume without loss of generality that $n = \dim(\mathcal{M}) = k - 1$. In fact, suppose $n < k - 1$ and denote by $\mathcal{M}_e$ the $(k-1)$-dimensional linear manifold containing $z^{CP} = z^o$ and orthogonal to z^o. By definition, the estimate provided by ψ^{CP} and the error $e_y(\psi^{CP})$ do not change if we consider $\mathcal{M}_e$ instead of $\mathcal{M}$. Moreover, $\mathcal{M} \subset \mathcal{M}_e$ and hence the conditional radius for $\mathcal{M}_e$ is less than or equal to the conditional radius for $\mathcal{M}$. Therefore, the maximum ratio between $e_y(\psi^{CP})$ and the conditional radius of information is always achieved for a $(k-1)$-dimensional manifold. Define

$$x^o = \arg\max_{x \in \mathcal{E}} \|z^o - x\|. \tag{63}$$

Denote by z^+ and z^- the orthogonal projection of respectively x^o and $-x^o$ onto $\mathcal{M}$. Set $\alpha = \arccos(\|z^+ - z^-\|/\|x^o - z^-\|)$ and

$$z_m = z^- + \frac{1}{2\cos^2\alpha}(z^+ - z^-).$$

for $0 \leq \alpha < \frac{\pi}{2}$ (the case $\alpha = \frac{\pi}{2}$ corresponds to $z^+ = z^- = z^o$ and is trivial). It is easy to verify that z_m satisfies $\|z_m - x^o\| = \|z_m - z^-\|$. Then, define $h = \|x^o - z^+\|$, $l = \|x^o - z_m\|$ and $d = \|x^o - z^o\|$. Now, let us consider the set $\tilde{\mathcal{E}} = \{z^-, x^o\}$, containing only two points. By construction, the projection of its Chebyshev center onto $\mathcal{M}$ is z^o and hence (by definition of x^o) the error $e_y(\psi^{CP})$ is the same for both $\mathcal{E}$ and $\tilde{\mathcal{E}}$, and is given by d. Moreover, the conditional Chebyshev center of $\tilde{\mathcal{E}}$ is given by

$$\tilde{z}^{CC} = \begin{cases} z_m & \text{if } 0 \leq \alpha \leq \dfrac{\pi}{4} \\[2mm] z^+ & \text{if } \dfrac{\pi}{4} < \alpha \leq \dfrac{\pi}{2} \end{cases}$$

and the corresponding conditional radius for $\mathcal{M}$ is

$$\overline{rad}_y^C = \begin{cases} l & \text{if } 0 \leq \alpha \leq \dfrac{\pi}{4} \\[2mm] h & \text{if } \dfrac{\pi}{4} < \alpha \leq \dfrac{\pi}{2} \end{cases}.$$

Then, by noting that $-x^o \in \mathcal{E}$, one has

$$rad_y^C \geq \max\{\|z^{CC} - x^o\|, \|z^{CC} + x^o\|\}$$
$$\geq \max\{\|z^{CC} - x^o\|, \|z^{CC} - z^-\|\}$$
$$\geq \min_{z \in \mathcal{M}} \max\{\|z - x^o\|, \|z - z^-\|\} = \overline{rad}_y^C.$$

Hence,

$$\frac{e_y(\psi^{CP})}{rad_y^C} \leq \frac{e_y(\psi^{CP})}{\overline{rad_y^C}} = g(\alpha) = \begin{cases} \dfrac{d}{l} = \cos\alpha\sqrt{4 - 3\cos^2\alpha} & \text{if } 0 \leq \alpha \leq \frac{\pi}{4} \\[2ex] \dfrac{d}{h} = \dfrac{\sqrt{4\tan^2\alpha + 1}}{2\tan\alpha} & \text{if } \frac{\pi}{4} < \alpha \leq \frac{\pi}{2} \end{cases}$$

Finally, standard calculations lead to

$$\max_{0 \leq \alpha \leq \frac{\pi}{2}} g(\alpha) = \sqrt{\frac{4}{3}}$$

and the maximum is achieved for $\alpha = \arccos\sqrt{\frac{2}{3}}$. This means that the bound is tight. □

Remark 1. It is perhaps surprising that the above result is not related to the particular shape or the symmetry of the ellipsoid $\mathcal{E}$. In fact, it can be shown that the same tight upper bound holds if the ellipsoid $\mathcal{E}$ is replaced by any compact set (see [1]). This allows to establish that the central projection algorithm is $\sqrt{\frac{4}{3}}$-almost optimal for any norm in Y, provided that the ℓ_2 norm is adopted in X.

The following theorem states that the error of an interpolatory projection algorithm is at most twice the conditional radius of information.

Theorem 5 ([2]). *Let the assumptions (i) and (ii) hold. Then, the interpolatory projection algorithm ψ^{IP} satisfies*

$$e_y(\psi^{IP}) \leq rad_y^C + rad_y \ (\leq 2rad_y^C)$$

and the upper bound is achieved. □

Notice that the theorem remains true for any balanced set $\mathcal{E}$.

Now, the restricted projection algorithm introduced in (25) is considered. The following straightforward proposition gives a simple geometric interpretation of the estimate z^{RP}.

Proposition 5. *Let (i) and (ii) hold. If $\mathcal{E} \cap \mathcal{M} \neq \emptyset$, then z^{RP} in (26) is the symmetry center of the set $\mathcal{E} \cap \mathcal{M}$.*

The previous result allows one to derive an upper bound on the estimation error of the restricted projection algorithm, in the case when there exist admissible conditional estimates belonging to the feasible set $\mathcal{E}$. However, it is not possible to determine such a bound in the general case.

Theorem 6 ([2]). *Let (i) and (ii) hold. Then,*

(i) *If $\mathcal{E} \cap \mathcal{M} \neq \emptyset$, then the algorithm ψ^{RP} satisfies $e_y(\psi^{RP}) \leq 2rad_y$ (and hence it is 2-almost optimal).*

(ii) For any $D > 0$, there exists a restricted complexity estimation problem such that

$$e_y(\psi^{RP}) > D \; rad_y^C \; .$$

Since in Theorem 4 it has been proven that the central projection algorithm is $\sqrt{\frac{4}{3}}$-almost optimal, one could conjecture that the estimate provided by the algorithm ψ^{CP} is always "better" than that of the algorithm ψ^{RP}. This is not true, as it is shown in the following example.

Example 1. Consider the ellipsoid $\mathcal{E}$ defined by (57) with $Q = \mathrm{diag}\{0.05, 0.25, 2.5\}$, and the linear set $\mathcal{M}$ in (52), where $z^o = [-0.47, -0.1, 0.94]^T$, $M = \frac{v}{\|v\|}$ with $v = [-0.17, 1.269, 0.05]^T$. By computing the conditional central estimate z^{CC}, following the procedure described in [4], and the estimates z^{CP} and z^{RP} according to (60)–(62), one obtains the following estimation errors

$$e_y(\psi^{CC}) = 25.0691,$$
$$e_y(\psi^{CP}) = 25.3368,$$
$$e_y(\psi^{RP}) = 25.1121.$$

which disproves the conjecture that $e_y(\psi^{CP}) \leq e_y(\psi^{RP})$.

5   Conclusions

Information Based Complexity provides a useful framework for the formulation of restricted complexity identification problems. In the set membership setting, optimal estimation algorithms require the solution of complex min-max optimization problems. On the other hand, suboptimal estimators provide a viable alternative, as long as it is possible to compute tight upper bounds on their worst-case identification error.

In this paper a number of results along these lines has been provided, showing that different classes of suboptimal conditional estimators guarantee errors which are not much larger than the radius of information, which represents the minimum achievable error. Future research will concern a deeper study of the tradeoff between computational complexity and quality of the identified model, in order to provide further information for the choice of the most appropriate conditional estimator in specific identification settings.

References

1. Garulli A., B. Kacewicz, A. Vicino, and G. Zappa (1999a), "Reliability of Projection Algorithms in Conditional Estimation", *J. of Optimization Theory and Application*, vol. 101, no. 1.

2. Garulli A., B. Kacewicz, A. Vicino, and G. Zappa (1999b), "Error bounds for conditional algorithms in restricted complexity set membership identification", to appear in *IEEE TAC*.

3. Garulli A., A. Vicino and G. Zappa (1997a), "Optimal and suboptimal H_2 and H_∞ estimators for set membership identification", *Proc. of 36th IEEE CDC*, San Diego, CA.

4. Garulli A., A. Vicino, and G. Zappa (1997b), "Conditional central algorithms for worst-case estimation and filtering", *Proc. of 36th IEEE CDC*, San Diego, CA.

5. Giarré L., B. Kacewicz and M. Milanese (1997), "Model quality evaluation in Set Membership Identification", *Automatica* vol. **33**, No. 6, pp. 1133–1140.

6. Goodwin G., M. Gevers and B. Ninness (1992), "Quantifying the error in estimated transfer functions with application to model order selection", *IEEE TAC*, **37**, 913–928.

7. Helmicki A. J., C. A. Jacobson and C. N. Nett (1991), "Control oriented system identification: a worst-case/deterministic approach in H_∞", *IEEE TAC* **36**, No. 10. pp. 1163–1176.

8. Kacewicz B. (1999), "Worst-case conditional system identification in a general class of norms", to appear in *Automatica*.

9. Kacewicz B., M. Milanese and A. Vicino (1988), "Conditionally optimal algorithms and estimation of reduced order models", *J. of Complexity*, **4**, pp. 73–85.

10. Kowalski M.A., K.A. Sikorski and F. Stenger (1995), *Selected Topics in Approximation and Computation*, Oxford University Press.

11. Mäkilä P. (1991), "On identification of stable systems and optimal approximation", *Automatica*, **27**, No. 4, pp. 663–676.

12. Mäkilä P., J. Partington and T.K. Gustafsson (1995), "Worst-case control-relevant identification", *Automatica*, **31**, pp. 1799-1819.

13. Milanese M., J. Norton, H. Piet-Lahanier and E. Walter Eds. (1996), *Bounding Approaches to System Identification*, Plenum Press.

14. Milanese M. and A. Vicino. (1991), "Optimal estimation theory for dynamic systems with set membership uncertainty: an overview", *Automatica*, **27**, pp. 997–1009.

15. Plaskota L. (1996), *Noisy Information and Computational Complexity*, Cambridge University Press.

16. Traub J., G. Wasilkowski and H. Woźniakowski (1988), *Information Based Complexity*, Academic Press.

17. Tse D., M. Dahleh and J. Tsitsiklis (1993), "Optimal asymptotic identification under bounded disturbances", *IEEE TAC*, **38**, No. 8, pp. 1176–1189.

18. Zames G., L. Lin and L. Wang (1994), "Fast identification n-widths and uncertainty principles for LTI and slowly varying systems", *IEEE TAC*, **39**, pp. 1827-1838.

Worst-Case Simulation of Uncertain Systems

Laurent El Ghaoui[1] and Giuseppe Calafiore[2]

[1] Ecole Nationale Supérieure de Techniques Avancées – Paris
[2] Dipartimento di Automatica e Informatica, Politecnico di Torino – Italy.

Abstract. In this paper we consider the problem of worst-case simulation for a discrete-time system with structured uncertainty. The approach is based on the recursive computation of ellipsoids of confidence for the system state, based on semidefinite programming.

1 Introduction

This paper is concerned with the problem of estimating the state of an uncertain discrete-time system of the form

$$x_{k+1} = \begin{bmatrix} \mathbf{A}_k \ \mathbf{b}_k \end{bmatrix} \begin{bmatrix} x_k \\ 1 \end{bmatrix}, \quad k = 0, 1, 2, \ldots$$

where the initial state $x_0 \in \mathbb{R}^n$ is known to belong to a given ellipsoid, and the system matrix $[\mathbf{A}_k \ \mathbf{b}_k]$ is only known to belong to an uncertainty set $\mathcal{U}$ that will be specified shortly.

The basic problem we consider is to compute an *ellipsoid of confidence* for the state x_{k+1}, based on a deterministic uncertainty model for the system matrices, and the previous confidence ellipsoid for the state x_k. This setting corresponds to the classical problem of state prediction, but in the deterministic (or *set-membership*) framework. The problems of measurement-based prediction (smoothing and filtering) may also be treated by the method presented in this paper and are object of current research [6,7].

The idea of propagating ellipsoids of confidence for systems with deterministic uncertainty was first proposed by Schweppe [13] and Bertzekas [1], and later developed by several authors, including Kurzhanski [9] and Chernousko [4]. These authors consider the case with additive uncertainty, meaning that the system matrix [A b] is assumed to be exactly known. Of course, the assumption that the dynamic and measurement matrices are exactly known is very strong. The benefit of this simplification is that it yields recursive equations for the predictive filter that are simple to implement and have a structure similar to that of the Kalman filter equations. Recently, Savkin and Petersen [12] have considered a problem with a special kind of structured uncertainty, with the assumption that the uncertainty is bounded in an energy sense. These assumptions lead to recursive Riccati equations for the confidence ellipsoid, similar in spirit to the above-mentioned approach.

The main result of this paper is that ellipsoids of confidence, of size minimal in a certain geometrical sense, can be recursively computed in polynomial time via *semidefinite programming* (SDP). SDPs are convex optimization problems that generalize linear programming, and which can be solved with great theoretical and practical efficiency, using interior-point methods [10,15].

The considered uncertainty model encompasses a very wide variety of perturbation structures, for example it can be used for uncertain systems described by matrices depending rationally on unknown-but-bounded parameters. It can also be used with more classical uncertainty models, e.g. for systems with independent additive perturbations on the state and measurement equations (these are deterministic equivalents of systems with independent process and measurement noise, as used in Kalman filtering).

Our approach basically extends the existing results to the case with structured uncertainty on *every* system matrix. To understand why the problem is much more difficult when A_k is allowed to be uncertain, note that if A_k is exactly known, and if both x_k and b_k belong to a convex set, then x_{k+1} also belongs to a convex set; this is not true if A_k is uncertain. We pay a price for being able to handle more general perturbation models, of course: we do not end up with recursive equations, but recursive optimization problems. However, the price is not to too high, since computations can still be done in polynomial-time.

The methods developed here belong to the class of methods now known as *robust programming* in the field of optimization, and developed by Oustry, El Ghaoui and Lebret [11] and Ben Tal and Nemirovskii [2]. Robust optimization is concerned with decision (optimization) problems with unknown-but-bounded data, and tries to compute, via semidefinite programming, robust solutions, that is, solutions that are guaranteed to satisfy the uncertain constraints of the optimization problem, despite the perturbations.

2   Preliminaries

2.1   Notation

For a square matrix X, $X \succ 0$ (resp. $X \succeq 0$) means X is symmetric, and positive-definite (resp. semidefinite). For a square matrix U, $U^\dagger$ denotes the (Moore-Penrose) pseudo-inverse of U. For $P \in \mathbb{R}^{n,n}$, with $P \succ 0$, and $x \in \mathbb{R}^n$, the notation $E(P,x)$ denotes the ellipsoid

$$E(P,x) = \left\{ \xi \mid (\xi - x)^T P^{-1}(\xi - x) \leq 1 \right\},$$

where x is the center, and P determines the "shape" of the ellipsoid.

2.2 Measures of size of an ellipsoid

The size of an ellipsoid is a function of the shape matrix P; we will denote it by $\phi(P)$ in the sequel. There are many alternative measures of size for an ellipsoid: volume, largest semi-axis length, etc. Our method will work on any such size function ϕ, provided it is a (quasi-) convex function of the "shape" matrix P, or of its inverse, over the set of positive-definite matrices. Examples of such functions, and their geometrical interpretation, are given in Table 1. In the sequel, we concentrate on the measure of size given by the sum of squares of semi-axis lengths; the extension to the other measures given in Table 1 is left to the reader.

measure function ϕ	measure
$\log \det P$	volume (convex in P^{-1})
$\mathrm{Tr}\,P$	sum of squared semi-axis lengths (convex in both P and P^{-1})
$\lambda_{\max}(P)$	largest semi-axis length (convex in P, quasi-convex in P^{-1})

Table 1. Examples of functions related to the size of ellipsoid $E(P, x)$, that are (quasi-) convex functions of P or of P^{-1} over the cone of positive-definite matrices.

2.3 Semidefinite programming

We will seek to formulate our estimation problems in terms of semidefinite programming problems, which are convex optimization problems involving linear matrix inequalities (LMIs). An LMI is a constraint on a vector $x \in \mathbb{R}^m$ of the form

$$\mathcal{F}(x) = \mathcal{F}_0 + \sum_{i=1}^{m} x_i \mathcal{F}_i \succeq 0 \tag{1}$$

where the symmetric matrices $\mathcal{F}_i = \mathcal{F}_i^T \in \mathbb{R}^{N,N}$, $i = 0, \ldots, m$ are given. The minimization problem

$$\text{minimize } c^T x \text{ subject to } \mathcal{F}(x) \succeq 0 \tag{2}$$

where $c \in \mathbb{R}^m$, is called a semidefinite program (SDP). SDPs are convex optimization problems and can be solved in polynomial-time with *e.g.* primal-dual interior-point methods [10,15].

3 Models of Uncertain Systems

3.1 LFR models

In this paper, we will consider uncertain systems modelled as

$$x_{k+1} = \begin{bmatrix} \mathbf{A}(\Delta_k) | \mathbf{b}(\Delta_k) \end{bmatrix} \begin{bmatrix} x_k \\ 1 \end{bmatrix}, \quad k = 0, 1, 2, \ldots \tag{3}$$

where Δ_k is a possibly time-varying uncertainty matrix. We assume that the matrix-valued functions $\mathbf{A}(\Delta)$, $\mathbf{b}(\Delta)$, etc, are given by a *linear-fractional representation* (LFR):

$$[\mathbf{A}(\Delta)|\mathbf{b}(\Delta)] = [A|b] + L\Delta\,(I - H\Delta)^{-1}\,[R_A|R_b], \qquad (4)$$

where $A, b, L, R = [R_A\ R_b]$, and H are constant matrices, while $\Delta \in \Delta_1$, where

$$\Delta_1 = \{\Delta \in \Delta \mid \|\Delta\| \le 1\},$$

and Δ is a linear subspace. We denote by $\mathcal{U}$ the set

$$\mathcal{U} = \{[\mathbf{A}(\Delta)|\mathbf{b}(\Delta)] \mid \Delta \in \Delta_1\}.$$

The subspace Δ, referred to as the *structure set* in the sequel, defines the structure of the perturbation, which is otherwise only bounded in norm. Together, the matrices A, b, C, d, L, R, H, and the subspace Δ, constitute a *linear-fractional representation* (LFR) of our uncertain system.

The above LFR models are not necessarily well-posed over Δ_1, meaning that if might happen that $\det(I - H\Delta) = 0$ for some $\Delta \in \Delta_1$; we return to this issue in §3.1.

In the sequel, we denote by $\mathcal{B}(\Delta)$ a linear subspace constructed from the subspace Δ, referred to as the *scaling subspace*, as follows:

$$\mathcal{B}(\Delta) = \{(S, T, G) \mid S\Delta = \Delta T,\ G\Delta = -\Delta^T G^T \text{ for every } \Delta \in \Delta\}. \quad (5)$$

We will give examples of LFR models, and explicit representations of associated sets Δ and $\mathcal{B}(\Delta)$ shortly. This uncertainty framework includes the case when parameters perturb each coefficient of the data matrices in a (polynomial or) rational manner. This is thanks to the representation lemma given below.

Lemma 2. *For any rational matrix function* $\mathbf{M} : \mathbb{R}^k \to \mathbb{R}^{n,c}$, *with no singularities at the origin, there exist nonnegative integers* $r_1, \ldots, r_L$, *and matrices* $M \in \mathbb{R}^{n,c}$, $L \in \mathbb{R}^{n,N}$, $R \in \mathbb{R}^{N,c}$, $H \in \mathbb{R}^{N,N}$, *with* $N = r_1 + \ldots + r_L$, *such that* $\mathbf{M}$ *has the following Linear-Fractional Representation (LFR): For all* δ *where* $\mathbf{M}$ *is defined,*

$$\mathbf{M}(\delta) = M + L\Delta\,(I - H\Delta)^{-1}\,R, \text{ where } \Delta = \mathbf{diag}\,(\delta_1 I_{r_1}, \ldots, \delta_L I_{r_L}).$$

A Linear-Fractional Representation (LFR) is thus a matrix-based way to describe a multivariable rational matrix-valued function. It is a generalization, to the multivariable case, of the well-known state-space representation of transfer functions. A constructive proof of the above result is given in [5]. The proof is based on a simple idea: first devise LFRs for simple (*e.g.*, linear) functions, then use combination rules (such as multiplication, addition, etc), to devise LFRs for arbitrary rational functions.

Well-posedness. The LFRs introduced earlier are not necessarily well-posed over $\boldsymbol{\Delta}_1$, meaning that if might happen that $\det(I - H\Delta) = 0$ for some $\Delta \in \boldsymbol{\Delta}_1$. Checking well-posedness is a NP-hard problem, known in robust control theory as the μ analysis problem, that is addressed *e.g.*, in [8]. In [8], the authors have proved that if there exist a triple $(S, T, G) \in \mathcal{B}(\boldsymbol{\Delta})$ such that $S \succ 0$, $T \succ 0$, and

$$H^T T H + H^T G + G^T H \prec S, \tag{6}$$

then the LFR is well-posed over $\boldsymbol{\Delta}_1$.

In this paper we will make the assumption that the LFR is well-posed. It turns out that this is not a strong assumption in our context, since the conditions we will obtain always imply the above condition, which in turn guarantees well-posedness.

Robustness lemma. We will need the following results.

Lemma 3 (unstructured perturbations). *Let* $\mathcal{F} = \mathcal{F}^T$, $\mathcal{L}$, $\mathcal{R}$ *and* $\mathcal{H}$ *be given matrices of appropriate size. We have*

$$\mathcal{F} + \mathcal{L}\Delta(I - \mathcal{H}\Delta)^{-1}\mathcal{R} + (\mathcal{L}\Delta(I - \mathcal{H}\Delta)^{-1}\mathcal{R})^T \succeq 0 \text{ for every } \Delta, \; \|\Delta\| \leq 1$$

if and only if there exists a scalar τ *such that*

$$\begin{bmatrix} \mathcal{F} & \mathcal{L} \\ \mathcal{L}^T & 0 \end{bmatrix} \succeq \tau \begin{bmatrix} \mathcal{R} & \mathcal{H} \\ 0 & I \end{bmatrix}^T \begin{bmatrix} I & 0 \\ 0 & -I \end{bmatrix} \begin{bmatrix} \mathcal{R} & \mathcal{H} \\ 0 & I \end{bmatrix}, \; \tau \geq 0. \tag{7}$$

Lemma 4 (structured perturbations). *Let* $\mathcal{F} = \mathcal{F}^T$, $\mathcal{L}$, $\mathcal{R}$ *and* $\mathcal{H}$ *be given matrices of appropriate size. We have*

$$\mathcal{F} + \mathcal{L}\Delta(I - \mathcal{H}\Delta)^{-1}\mathcal{R} + (\mathcal{L}\Delta(I - \mathcal{H}\Delta)^{-1}\mathcal{R})^T \succeq 0 \text{ for every } \Delta \in \boldsymbol{\Delta}_1$$

if there exist block-diagonal matrices

$$S = \mathbf{diag}\,(S_1, \ldots, S_l), \; S_i = S_i^T \in \mathbb{R}^{r_i, r_i},$$
$$G = \mathbf{diag}\,(G_1, \ldots, G_l), \; G_i = -G_i^T \in \mathbb{R}^{r_i, r_i},$$

such that

$$\begin{bmatrix} \mathcal{F} & \mathcal{L} \\ \mathcal{L}^T & 0 \end{bmatrix} \succeq \begin{bmatrix} \mathcal{R} & \mathcal{H} \\ 0 & I \end{bmatrix}^T \begin{bmatrix} S & G \\ G^T & -S \end{bmatrix} \begin{bmatrix} \mathcal{R} & \mathcal{H} \\ 0 & I \end{bmatrix}, \; S \succeq 0. \tag{8}$$

4   LMI Conditions for Ellipsoid Update

In this section, we give the main results on the simulation problem in the general context of systems with structured uncertainty. For ease of notation, we will drop the time index k on quantities at time k, and the quantities at time $k + 1$ will be denoted with a subscript $+$. With this convention, x stands for x_k, and x_+ stands for x_{k+1}.

Our aim is to determine a confidence ellipsoid $\mathcal{E}_+ = E(P_+, \hat{x}_+)$ for the state at the next time instant x_+, given that x belongs to the ellipsoid $\mathcal{E} = E(P, \hat{x})$ at the present time. To avoid inverting the matrix P, we introduce the following equivalent representation for $\mathcal{E}$:

$$\mathcal{E} = \{\hat{x} + Ez \mid \|z\|_2 \leq 1\},$$

where $P = EE^T$ and E^T is the Cholesky factor of P.

4.1   The case with no uncertainty

Consider first the case when $\mathbf{A} = A$, $\mathbf{b} = b$ are perfectly known. We want

$$P_+ \succeq (x_+ - \hat{x}_+)(x_+ - \hat{x}_+)^T$$

to hold whenever

$$x_+ = Ax + b, \quad x = \hat{x} + Ez, \quad \|z\| \leq 1.$$

The two conditions above may be rewritten in following form:

$$\begin{bmatrix} P_+ & A\hat{x} + b - \hat{x}_+ + AEz \\ (A\hat{x} + b - \hat{x}_+ + AEz)^T & 1 \end{bmatrix} \succeq 0 \text{ whenever } \|z\|_2 \leq 1.$$

Using the robustness lemma (lemma 3), we obtain an equivalent condition: there exists τ such that

$$\begin{bmatrix} P_+ & A\hat{x} + b - \hat{x}_+ & AE \\ (A\hat{x} + b - \hat{x}_+)^T & 1 - \tau & 0 \\ E^T A^T & 0 & \tau I \end{bmatrix} \succeq 0.$$

4.2   Robust version

We now seek $P_+, \hat{x}_+, \tau$ such that

$$\begin{bmatrix} P_+ & A\hat{x} + b - \hat{x}_+ & AE \\ (A\hat{x} + b - \hat{x}_+)^T & 1 - \tau & 0 \\ E^T A^T & 0 & \tau I \end{bmatrix} \succeq 0 \text{ for every } [\mathbf{A}\ \mathbf{b}] \in \mathcal{U}.$$

To obtain the robust counterpart, we just apply the robustness lemma to the following LFR

$$\begin{bmatrix} P_+ & A\hat{x} + b - \hat{x}_+ & AE \\ (A\hat{x} + b - \hat{x}_+)^T & 1 - \tau & 0 \\ E^T A^T & 0 & \tau I \end{bmatrix} =$$
$$\mathcal{F} + \mathcal{L}\Delta(I - \mathcal{H}\Delta)^{-1}\mathcal{R} + (\mathcal{L}\Delta(I - \mathcal{H}\Delta)^{-1}\mathcal{R})^T,$$

where

$$\mathcal{F} = \begin{bmatrix} P_+ & A\hat{x} + b - \hat{x}_+ & AE \\ (A\hat{x} + b - \hat{x}_+)^T & 1 - \tau & 0 \\ E^T A^T & 0 & \tau I \end{bmatrix},$$

$$\mathcal{L} = \begin{bmatrix} L \\ 0 \\ 0 \end{bmatrix}, \quad \mathcal{R} = \begin{bmatrix} 0 \\ R_A \hat{x} + R_b \\ R_A E \end{bmatrix}^T, \quad \mathcal{H} = H.$$

This result is summarized in the following theorem.

Theorem 1. *The ellipsoid*

$$\mathcal{E}_+ = \{\hat{x}_+ + E_+ z \mid \|z\|_2 \le 1\}$$

is an ellipsoid of confidence for the new state if $\hat{x}_+$ and $P_+ := E_+ E_+^T$ satisfy the LMI

$$\begin{bmatrix} P_+ & A\hat{x} + b - \hat{x}_+ & AE & L \\ (A\hat{x} + b - \hat{x}_+)^T & 1 - \tau & 0 & 0 \\ E^T A^T & 0 & \tau I & 0 \\ L^T & 0 & 0 & 0 \end{bmatrix} \succeq$$

$$\begin{bmatrix} 0 & R_A\hat{x} + R_b & R_A E & H \\ 0 & 0 & 0 & I \end{bmatrix}^T \begin{bmatrix} S & G \\ G^T & -S \end{bmatrix} \begin{bmatrix} 0 & R_A\hat{x} + R_b & R_A E & H \\ 0 & 0 & 0 & I \end{bmatrix},$$

$$S \succeq 0,$$

for some block-diagonal matrices

$$S = \operatorname{diag}(S_1, \ldots, S_l), \quad S_i = S_i^T \in \mathbb{R}^{r_i, r_i},$$
$$G = \operatorname{diag}(G_1, \ldots, G_l), \quad G_i = -G_i^T \in \mathbb{R}^{r_i, r_i}.$$

□

Using the elimination lemma [3], we may eliminate the variable $\hat{x}$ and decouple the problem of determining the optimal shape matrix from the one of obtaining the optimal center as follows.

Theorem 2. *The ellipsoid*

$$\mathcal{E}_+ = \{\hat{x}_+ + E_+ z \mid \|z\|_2 \le 1\}$$

is an ellipsoid of confidence for the new state if τ and $P_+ := E_+ E_+^T$ satisfy the LMI

$$\begin{bmatrix} P_+ & AE & L \\ E^T A^T & \tau I & 0 \\ L^T & 0 & 0 \end{bmatrix} \succeq \begin{bmatrix} 0 & R_A E & H \\ 0 & 0 & I \end{bmatrix}^T \begin{bmatrix} S & G \\ G^T & -S \end{bmatrix} \begin{bmatrix} 0 & R_A E & H \\ 0 & 0 & I \end{bmatrix}, \quad S \succeq 0, \quad (9)$$

$$\begin{bmatrix} 1 - \tau & 0 & 0 \\ 0 & \tau I & 0 \\ 0 & 0 & 0 \end{bmatrix} \succeq \begin{bmatrix} \hat{x}^T R_A^T + R_b^T & 0 \\ E^T R_A^T & 0 \\ H^T & I \end{bmatrix} \begin{bmatrix} S & G \\ G^T & -S \end{bmatrix} \begin{bmatrix} \hat{x}^T R_A^T + R_b^T & 0 \\ E^T R_A^T & 0 \\ H^T & I \end{bmatrix}^T \quad (10)$$

for some block-diagonal matrices

$$S = \mathbf{diag}\,(S_1, \ldots, S_l), \quad S_i = S_i^T \in \mathbb{R}^{r_i, r_i},$$
$$G = \mathbf{diag}\,(G_1, \ldots, G_l), \quad G_i = -G_i^T \in \mathbb{R}^{r_i, r_i}. \tag{11}$$

The minimum-trace ellipsoid is obtained by solving the semidefinite programming problem

$$\text{minimize } \mathbf{Tr}P_+ \text{ subject to (9), (10), (11)}. \tag{12}$$

At the optimum, we have

$$P_+ = \left[\,AE\ L\,\right] X^\dagger \left[\,AE\ L\,\right]^T,$$

and a corresponding center is given by

$$\hat{x}_+ = A\hat{x} + b + \left[\,AE\ L\,\right] X^\dagger \begin{bmatrix} E^T R_A^T S(R_A \hat{x} + R_b) \\ (SH + G)^T (R_A \hat{x} + R_b) \end{bmatrix},$$

where

$$X = \begin{bmatrix} \tau I & 0 \\ 0 & 0 \end{bmatrix} - \begin{bmatrix} R_A E & H \\ 0 & I \end{bmatrix}^T \begin{bmatrix} S & G \\ G^T & -S \end{bmatrix} \begin{bmatrix} R_A E & H \\ 0 & I \end{bmatrix}.$$

$\square$

5 Numerical Implementation

In this section, we discuss the numerical implementation of the solution to problem (12) using interior-point methods for semidefinite programming (SDP). Numerous algorithms are available today for SDP; in our experiments we have used the SDPpack package.

In order to work, these methods require that the problem be strictly feasible, and that its dual (in the SDP sense, see [15]) be also strictly feasible. Primal strict feasibility means that the feasible set is not "flat" (contained in a hyperplane in the space of decision variables). Dual strict feasibility means, roughly speaking, that the objective of the primal problem is bounded below on the (primal) feasible set.

For simplicity, we will reduce our discussion to the case when the matrix H is zero; this means that the perturbations enter *affinely* in the state-space matrices. We will also assume that the affine term R_b is zero. Finally, we constrain the matrix variables G in problem (12) to be zero. The extension of our results to the general case is straightforward.

The reduced problem we examine now takes the form

$$\text{minimize } \mathbf{Tr}P_+ \text{ subject to}$$

$$\begin{bmatrix} P_+ & AE & L \\ E^T A^T & \tau I - E^T R_A^T S R_A E & 0 \\ L^T & 0 & S \end{bmatrix} \succeq 0,$$

$$\begin{bmatrix} 1 - \tau - \hat{x}^T R_A^T S R_A \hat{x} & -\hat{x}^T R_A^T S R_A E \\ -E^T R_A^T S R_A \hat{x} & \tau I - E^T R_A^T S R_A E \end{bmatrix} \succeq 0, \tag{13}$$

$$S = \mathbf{diag}\,(S_1, \ldots, S_l), \quad S_i = S_i^T \in \mathbb{R}^{r_i, r_i}.$$

In the above, we have used the fact that the constraint $S \succeq 0$ is implied by the above LMIs.

5.1 Strict feasibility of primal problem

We have the following result.

Theorem 3. *If the system is well-posed, and if the matrix $\begin{bmatrix} R_A \hat{x} & R_A E \end{bmatrix}$ is not zero, then the primal problem (13) is strictly feasible; a strictly feasible primal point is given by*

$$\tau = 0.5$$
$$S = \frac{1}{4 \left\| \begin{bmatrix} R_A \hat{x} & R_A E \end{bmatrix} \right\|^2} \cdot I,$$
$$P_+ = I + \begin{bmatrix} AE & L \end{bmatrix} X^\dagger \begin{bmatrix} AE & L \end{bmatrix}^T,$$

where

$$X = \begin{bmatrix} \tau I & 0 \\ 0 & 0 \end{bmatrix} - \begin{bmatrix} R_A E & H \\ 0 & I \end{bmatrix}^T \begin{bmatrix} S & 0 \\ 0 & -S \end{bmatrix} \begin{bmatrix} R_A E & H \\ 0 & I \end{bmatrix}.$$

$\square$

Remark 1. If the matrix $\begin{bmatrix} R_A \hat{x} & R_A E \end{bmatrix}$ is zero, then the new iterate x_+ is independent of perturbation, and the new ellipsoid of confidence reduces to the singleton $\{A\hat{x} + b\}$.

Remark 2. If each $n \times r_i$ block L_i of L is full rank, and if E is also full rank, then the optimal P_+ (and corresponding E_+) is full rank.

5.2 Strict feasibility of dual problem

The problem dual to the SDP (13) is

maximize $-2 \left(\mathbf{Tr} A E Z_{12}^T + \mathbf{Tr} L Z_{13}^T \right)$ subject to

$$Z = \begin{bmatrix} I & Z_{12} & Z_{13} \\ Z_{12}^T & Z_{22} & Z_{23} \\ Z_{13}^T & Z_{23}^T & Z_{33} \end{bmatrix} \succeq 0, \quad X = \begin{bmatrix} x_{11} & x_{12}^T \\ x_{12} & X_{22} \end{bmatrix} \succeq 0, \quad x_{11} = \mathbf{Tr}(X_{22} + Z_{22}),$$

$$Z_{33}^{(i)} = R_i \left(E(X_{22} + Z_{22}) E^T + x_{11} \hat{x} \hat{x}^T + \hat{x} x_{12}^T E^T + E x_{12} \hat{x}^T \right) R_i^T, \quad i = 1, \ldots, l.$$

In the above, the notation $Z^{(i)}$ refers to the i-th $r_i \times r_i$ block in matrix X, and R_i is the i-th $r_i \times n$ block in R_A.

Theorem 4. *If, for each i, $i = 1, \ldots, l$, the matrix $R_i E$ is full rank, then the dual, problem is strictly feasible. A strictly feasible dual point is obtained* *by setting the dual variables* X, Z *to be zero, except for the block-diagonal terms:*

$$X_{22} = I$$
$$Z_{22} = I$$
$$x_{11} = 2n$$
$$Z_{33}^{(i)} = R_i(EE^T + n\hat{x}\hat{x}^T)R_i^T, \quad i = 1, \dots, l.$$

□

The next theorem summarizes the sufficient conditions we have obtained to guarantee that our algorithm behaves numerically well.

Theorem 5. *If the initial ellipsoid is not "flat" (that is, the initial E is full rank), and if for each i, the blocks L_i, R_i are also full rank, then at each step of the simulation algorithm, the SDP to solve is both primal and dual strictly feasible.*

□

6   Examples

Consider a second-order, continuous-time uncertain system

$$\ddot{y} + \mathbf{a}_1(t)\dot{y} + \mathbf{a}_2(t)y = \mathbf{a}_2(t),$$

where the uncertain, time-varying parameters $\mathbf{a}_i$, $i = 1, 2$ are subject to bounded variation of given relative amplitude ρ, precisely:

$$\mathbf{a}_i(t) = a_i^{\text{nom}}(1 + \rho\delta_i(t)), \quad i = 1, 2, \ t \geq 0,$$

where $-1 \leq \delta_i(t) \leq 1$ for every t, and a_i^{nom}, $i = 1, 2$, is the nominal value of the parameters.

By discretizing this system using a forward-Euler scheme with discretization period h, we obtain a system of the form (3), with the following LFR

$$[\mathbf{A}|\mathbf{b}] = \begin{bmatrix} 1 & h & 0 \\ -h\mathbf{a}_2 & -h\mathbf{a}_1 & h\mathbf{a}_2 \end{bmatrix} = [A|b] + L\Delta R,$$

where

$$[A|b] = \begin{bmatrix} 1 & h & 0 \\ -ha_2^{\text{nom}} & -ha_1^{\text{nom}} & ha_2^{\text{nom}} \end{bmatrix},$$

$$L = -h\rho \begin{bmatrix} 0 & 0 \\ a_1^{\text{nom}} & a_2^{\text{nom}} \end{bmatrix}, R = \begin{bmatrix} 0 & 1 & 0 \\ 1 & 0 & -1 \end{bmatrix}, \Delta = \text{diag}(\delta_1, \delta_2).$$

In Figure 1, the response in output y for $h = 0.1$, $a_1^{\text{nom}} = 3$, $a_2^{\text{nom}} = 9$, $x_0 = [-1 \ 0]^T$, $E_0 = 0.1x_0x_0^T$, is compared with 100 Monte-Carlo simulations, for $\rho = 0.2$ and for a time horizon of $T = 200$ steps. From the plot of

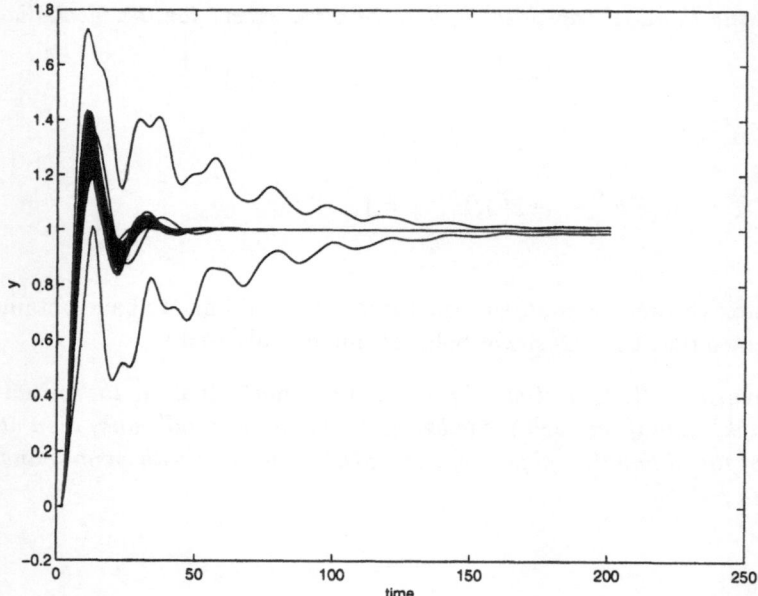

Fig. 1. Worst-case and random simulation for a second-order uncertain system ($\rho = 0.2$).

Figure 1, we notice that the Monte-Carlo and worst-case analyses agree on the qualitative behavior of the uncertain system (stability, in this case). The worst-case analysis seems to be somewhat conservative, but the reader should be aware that the actual worst-case behavior cannot be accurately predicted, in general, by taking random samples.

As a further example, we show in Figure 2 the worst-case simulation for a randomly generated fifth-order discrete time system with three uncertain parameters and $\rho = 0.1$.

7   Conclusions

In this paper we presented a recursive scheme for computing a minimal size ellipsoid (ellipsoid of confidence) that is guaranteed to contain the state at time $k + 1$ of a linear discrete-time system affected by deterministic uncertainty in all the system matrices, given a previous ellipsoid of confidence at time k. The ellipsoid of confidence can be recursively computed in polynomial time via semidefinite programming. We remark that the presented results are valid on a finite time horizon, while steady-state and convergence issues are currently under investigation.

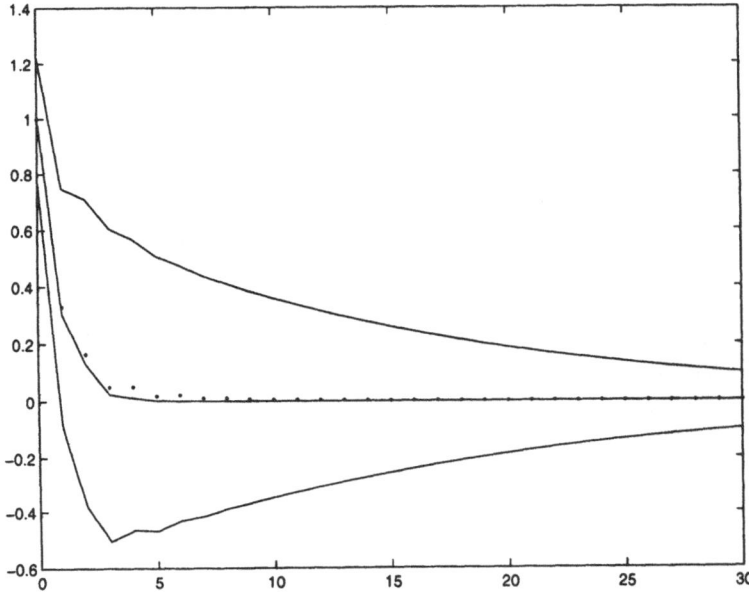

Fig. 2. Worst-case simulation for a random fifth-order uncertain system ($\rho = 0.1$). Inner solid line represents the nominal output; dots represent the projected centers of the confidence ellipsoids.

References

1. D.P. Bertzekas and I.B. Rhodes, "Recursive state estimation for a set-membership description of uncertainty," *IEEE Trans. Autom. Control*, 16: 117–124, 1971.
2. A. Ben-Tal and A. Nemirowski, "Robust truss topology design via semidefinite programming," *SIAM J. on Optimization*, (7)4: 991–1016, Nov. 1997.
3. S. Boyd, L. El Ghaoui, E. Feron, and V. Balakrishnan. *Linear Matrix Inequalities in System and Control Theory*. Studies in Applied Mathematics. Philadelphia: SIAM, June 1994.
4. F.L. Chernousko, *State Estimation of Dynamic Systems*. Boca Raton, Florida: CRC Press, 1994.
5. S. Dussy and L. El Ghaoui, "Measurement-scheduled control for the RTAC problem." To appear in *Int. J. Robust & Nonlinear Control*.
6. L. El Ghaoui, G. Calafiore, "Deterministic state prediction under structured uncertainty," in *Proceedings of the American Control Conference*, San Diego, California, June 1999.
7. L. El Ghaoui, G. Calafiore, "Worst-case estimation for discrete-time systems with structured uncertainty," In preparation, 1999.
8. M. K. H. Fan, A. L. Tits, and J. C. Doyle, "Robustness in the presence of mixed parametric uncertainty and unmodeled dynamics," *IEEE Trans. Aut. Control*, 36(1):25–38, Jan 1991.
9. A. Kurzhanski, I. Vályi, *Ellipsoidal Calculus for Estimation and Control*. Boston: Birkhäuser, 1997.

10. Yu. Nesterov and A. Nemirovsky, *Interior point polynomial methods in convex programming: Theory and applications*. Philadelphia: SIAM, 1994.
11. L. El Ghaoui, F. Oustry, and H. Lebret, "Robust solutions to uncertain semidefinite programs," *SIAM J. on Optimization*, (9)1: 33–52, 1998.
12. A.V. Savkin and I.R. Petersen, "Robust state estimation and model validation for discrete-time uncertain systems with a deterministic description of noise and uncertainty," *Automatica,* 34(2): 271–274, 1998.
13. F.C. Schweppe, "Recursive state estimation: Unknown but bounded errors and system inputs," *IEEE Trans. Autom. Control,* 13: 22–28, 1968.
14. L. El Ghaoui and J.-L. Commeau. lmitool, *version* 2.0. http://www.ensta.fr/~gropco, January 1999.
15. L. Vandenberghe and S. Boyd, "Semidefinite programming," *SIAM Review,* 38(1):49–95, March 1996.

On Performance Limitations in Estimation

Zhiyuan Ren[1], Li Qiu[2], and Jie Chen[3]

[1] Department of Electrical & Computer Engineering, Carnegie Mellon University, 5000 Forbes Avenue, Pittsburgh, PA 15213-3890 USA
[2] Department of Electrical & Electronic Engineering, Hong Kong University of Science & Technology, Clear Water Bay, Kowloon, Hong Kong
[3] Department of Electrical Engineering, College of Engineering, University of California, Riverside, CA 92521-0425 USA

Abstract. In this paper, we address the performance limitation issues in estimation problems. Our purpose is to explicitly relate the best achievable estimation errors with simple plant characteristics. In particular, we study performance limitations in achieving two objectives: estimating a signal from its version corrupted by a white noise and estimating a Brownian motion from its version distorted by an LTI system.

1 Introduction

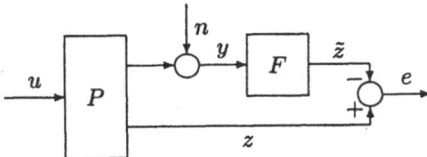

Fig. 1. A general estimation problem

A standard estimation problem can often be schematically shown by Fig. 1. Here $P = \begin{bmatrix} G \\ H \end{bmatrix}$ is an LTI plant, u is the input to the plant, n is the measurement noise, z is the signal to be estimated, y is the measured signal, $\tilde{z}$ is the estimate of z. Often u and n are modelled as stochastic processes with known means and covariances. We can assume, without loss of generality, that the means of the stochastic processes are zero. The objective is to design LTI filter F so that the steady state error variance

$$V = \lim_{t \to \infty} E[e(t)'e(t)]$$

is small. Clearly, for V to be finite for nontrivial u and n, it is necessary that $F \in \mathcal{RH}_\infty$ and $H - FG \in \mathcal{RH}_\infty$. This condition is also necessary and sufficient for the error to be bounded for arbitrary initial conditions of P and F, i.e., for the filter to be a bounded error estimator (BEE). There is an extensive theory on the optimal design of the filter F to minimize V, see for example [1,2,6]. The optimal error variance is then given by

$$V^* = \inf_{F,H-FG\in\mathcal{RH}_\infty} V.$$

Our interest in this paper is not on how to find the optimal filter F, which is addressed by the standard optimal filtering theory. Rather, we are interested in relating V^* with some simple characteristics of the plant P in some important special cases. Since V^* gives a fundamental limitation in achieving certain performance objectives in filtering problems, the simple relationship between V^* and the plant characteristics, in addition to providing deep understanding and insightful knowledge on estimation problems, can be used to access the quality of different designs and to ascertain impossible design objectives before a design is carried out.

The variance V gives an overall measure on the size of the steady state estimation error. Sometimes, we may wish to focus on some detailed features of the error. For example we may wish to investigate the variance of the projection of the estimation error on certain direction. This variance then gives a measure of the error in a particular direction. Assume that $z(t), \tilde{z}(t), e(t) \in \mathbb{R}^m$. Let $\xi \in \mathbb{R}^m$ be a vector of unit length representing a direction in $\mathbb{R}^m$. Then the projection of $e(t)$ to the direction represented by ξ is given by $\xi'e(t)$ and its steady state variance is given by

$$V_\xi = \lim_{t\to\infty} \boldsymbol{E}[(\xi'e(t))^2].$$

The best achievable error in ξ direction is then given by

$$V_\xi^* = \inf_{F,H-FG\in\mathcal{RH}_\infty} V_\xi.$$

The optimal or near optimal filter in minimizing V_ξ in general depends on ξ. This very fact may limit the usefulness of V_ξ^*, since we are usually more interested in the directional error information under an optimal or near optimal filter designed for all directions, i.e., designed to minimize V. Let $\{F_k\}$ be a sequence of filters satisfying $F_k, H - F_kG \in \mathcal{RH}_\infty$ such that the corresponding sequence of errors $\{e_k\}$ satisfies

$$V = \lim_{k\to\infty} \lim_{t\to\infty} \boldsymbol{E}[e_k(t)e_k(t)'].$$

Then we are more interested in

$$V^*(\xi) = \lim_{k\to\infty} \boldsymbol{E}[(\xi'e_k(t))^2].$$

In this paper, we will also give the relationship between V_ξ^*, $V^*(\xi)$ and simple characteristics of the plant P for the same cases when that for V^* is considered.

The performance limitations in estimation have been studied recently in [4,5,9,10] in various settings. In [4,5,9], sensitivity and complimentary sensitivity functions of an estimation problem are defined and it is shown that they have to satisfy certain integral constraints independent of filter design. In [10], a time domain technique is used to study the performance limitations in some special cases when one of n and u is diminishingly small and the other one is either a white noise or a Brownian motion.

This paper addresses similar problems as in [10], but studies them from a pure input output point of view using frequency domain techniques. We also study them in more detail by providing directional information on the best errors. The results obtained are dual to those in [3,7] where the performance limitations of tracking and regulation problems are considered. The new investigation provides more insights into the performance limitations of estimation problems.

This paper is organized as follows: Section 2 provides background materials on transfer matrix factorizations which exhibit directional properties of each nonminimum phase zero and antistable pole. Section 3 relates the performance limitation in estimating a signal from its corrupted version by a white noise to the antistable modes, as well as their directional properties, of the signal. Section 4 relates the performance limitation in estimating a Brownian motion from its version distorted by an LTI system to the nonminimum phase zeros of the system, as well as their directional properties. Section 5 gives concluding remarks.

2   Preliminaries

Let G be a continuous time FDLTI system. We will use the same notation G to denote its transfer matrix. Assume that G is left invertible. The poles and zeros of G, including multiplicity, are defined according to its Smith-McMillan form. A zero of G is said to be nonminimum phase if it has positive real part. G is said to be minimum phase if it has no nonminimum phase zero; otherwise, it is said to be nonminimum phase. A pole of G is said to be antistable if it has a positive real part. G is said to be semistable if it has no antistable pole; otherwise strictly unstable.

Suppose that G is stable and z is a nonminimum phase zero of G. Then, there exists a vector u of unit length such that

$$G(z)u = 0.$$

We call u a (right or input) zero vector corresponding to the zero z. Let the nonminimum phase zeros of G be ordered as $z_1, z_2, \ldots, z_\nu$. Let also η_1 be a zero vector corresponding to z_1. Define

$$G_1(s) = I - \frac{2 \operatorname{Re} z_1}{s + z_1^*} \eta_1 \eta_1^*.$$

Note that G_1 is so constructed that it is inner, has only one zero at z_1 with η_1 as a zero vector. Now GG_1^{-1} has zeros $z_2, z_3, \ldots, z_\nu$. Find a zero vector η_2 corresponding to the zero z_2 of GG_1^{-1}, and define

$$G_2(s) = I - \frac{2 \operatorname{Re} z_2}{s + z_2^*} \eta_2 \eta_2^*.$$

It follows that $GG_1^{-1}G_2^{-1}$ has zeros $z_3, z_4, \ldots, z_\nu$. Continue this process until $\eta_1, \ldots, \eta_\nu$ and $G_1, \ldots, G_\nu$ are obtained. Then we have one vector corresponding to each nonminimum phase zero, and the procedure yields a factorization of G in the form of

$$G = G_0 G_\nu \cdots G_1, \tag{1}$$

where G_0 has no nonminimum phase zeros and

$$G_i(s) = I - \frac{2 \operatorname{Re} z_i}{s + z_i^*} \eta_i \eta_i^*. \tag{2}$$

Since G_i is inner, has the only zero at z_i, and has η_i as a zero vector corresponding to z_i, it will be called a matrix Blaschke factor. Accordingly, the product

$$G_z = G_\nu \cdots G_1$$

will be called a matrix Blaschke product. The vectors $\eta_1, \ldots, \eta_\nu$ will be called zero Blaschke vectors of G corresponding to the nonminimum phase zeros $z_1, z_2, \ldots, z_\nu$. Keep in mind that these vectors depend on the order of the nonminimum phase zeros. One might be concerned with the possible complex coefficients appearing in G_i when some of the nonminimum phase zeros are complex. However, if we order a pair of complex conjugate nonminimum phase zeros adjacently, then the corresponding pair of Blaschke factors will have complex conjugate coefficient and their product is then real rational and this also leads to real rational G_0.

The choice of G_i as in (2) seems *ad hoc* notwithstanding that G_i has to be unitary, have the only zero at z_i and have η_i as a zero vector corresponding to z_i. Another choice, among infinite many possible ones, is

$$G_i(s) = I - \frac{2 \operatorname{Re} z_i}{z_i} \frac{s}{s + z_i^*} \eta_i \eta_i^*, \tag{3}$$

and if this choice is adopted, the same procedure can be used to find a factorization of the form (1). Of course, in this case the Blaschke vectors are not the same. We see that for the first choice $G_i(\infty) = I$, whereas for the second choice $G_i(0) = I$. We will use both choices in the following. For this purpose, we will call the factorization resulting from the first choice of Type I and that from the second choice of type II.

For an unstable G, there exist stable real rational matrix functions

$$\begin{bmatrix} \tilde{X} & -\tilde{Y} \\ -\tilde{N} & \tilde{M} \end{bmatrix}, \begin{bmatrix} M & Y \\ N & X \end{bmatrix}$$

such that

$$G = NM^{-1} = \tilde{M}^{-1}\tilde{N}$$

and

$$\begin{bmatrix} \tilde{X} & -\tilde{Y} \\ -\tilde{N} & \tilde{M} \end{bmatrix}\begin{bmatrix} M & Y \\ N & X \end{bmatrix} = I.$$

This is called a doubly coprime factorization of G. Note that the nonminimum phase zeros of G are the nonminimum phase zeros of $\tilde{N}$ and the antistable poles of G are the nonminimum phase zeros of $\tilde{M}$. If we order the antistable poles of G as $p_1, p_2, \ldots, p_\mu$ and the nonminimum phase zeros of G as $z_1, z_2, \ldots, z_\nu$, then M and N can be factorized as

$$\tilde{M} = \tilde{M}_0\tilde{M}_\mu \cdots \tilde{M}_1$$
$$\tilde{N} = \tilde{N}_0\tilde{N}_\nu \cdots \tilde{N}_1$$

with

$$\tilde{M}_i(s) = I - \frac{2\operatorname{Re}p_i}{s + p_i^*}\zeta_i\zeta_i^*$$
$$\tilde{N}_i(s) = I - \frac{2\operatorname{Re}z_i}{z_i}\frac{s}{s + z_i^*}\eta_i\eta_i^*$$

where $\zeta_1, \zeta_2, \ldots, \zeta_\nu$ are zero Blaschke vectors of M and $\eta_1, \eta_2, \ldots, \eta_\nu$ are those of N. Here also $\tilde{N}_0$ and $\tilde{M}_0$ have no nonminimum phase zeros. Notice that we used type I factorization for $\tilde{M}$ and type II factorization for $\tilde{N}$. The reason for this choice is solely for the convenience of our analysis in the sequel.

Consequently, for any real rational matrix G with nonminimum phase zeros $z_1, z_2, \ldots, z_\nu$ and antistable poles $p_1, p_2, \ldots, p_\mu$, it can always be factorized to

$$G = G_p^{-1}G_0G_z, \tag{4}$$

as shown in Fig. 2, where

$$G_p(s) = \prod_{i=1}^{\mu} \left[I - \frac{2\operatorname{Re}p_i}{s + p_i^*} \zeta_i \zeta_i^* \right]$$

$$G_z(s) = \prod_{i=1}^{\nu} \left[I - \frac{2\operatorname{Re}z_i}{z_i} \frac{s}{s + z_i^*} \eta_i \eta_i^* \right]$$

and G_0 is a real rational matrix with neither nonminimum phase zero nor antistable pole. Although coprime factorizations of G are not unique, this nonuniqueness does not affect factorization (4). Here $\eta_1, \eta_2, \ldots, \eta_\nu$ are called zero Blaschke vectors and $\zeta_1, \zeta_2, \ldots, \zeta_\nu$ pole Blaschke vectors of G.

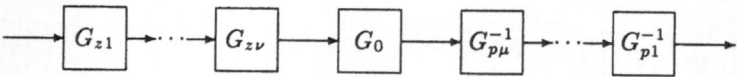

Fig. 2. Cascade factorization

3   Estimation under White Measurement Noise

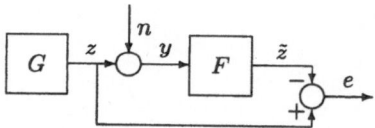

Fig. 3. Estimation under white measurement noise

Consider the estimation problem shown in Fig. 3. Here G is a given FDLTI plant, and n is a standard white noise. The purpose is to design a stable LTI filter F such that it generates an estimate $\tilde{z}$ of the true output z using the corrupted output y. This problem is clearly a special case of the general estimation problem stated in Sect. 1 with $P = \begin{bmatrix} G \\ G \end{bmatrix}$ and $u = 0$. The error of estimation is given by Fn. Since n is a standard white noise, the steady state variance of the error is given by

$$V = \|F\|_2^2$$

where $\|\cdot\|_2$ is the $\mathcal{H}_2$ norm. If we want V to be finite, we need to have $F(\infty) = 0$, in addition to $F, G - FG \in \mathcal{RH}_\infty$. Therefore

$$V^* = \inf_{F, G - FG \in \mathcal{RH}_\infty, F(\infty)=0} \|F\|_2^2.$$

Let $G = \tilde{M}^{-1}\tilde{N}$ be a left coprime factorization of G. Then $F \in \mathcal{RH}_\infty$ and $G - FG = (I - F)G = (I - F)\tilde{M}^{-1}\tilde{N} \in \mathcal{RH}_\infty$ if and only if $I - F = Q\tilde{M}$ for some $Q \in \mathcal{RH}_\infty$. Therefore

$$V^* = \inf_{Q \in \mathcal{RH}_\infty, Q(\infty)\tilde{M}(\infty)=I} \|I - Q\tilde{M}\|_2^2.$$

Now assume that G has antistable poles $p_1, p_2, \ldots, p_\mu$ with $\zeta_1, \zeta_2, \ldots, \zeta_\mu$ be the corresponding pole Blaschke vectors of type I. Then $\tilde{M}$ has factorization

$$\tilde{M} = \tilde{M}_0 \tilde{M}_\mu \cdots \tilde{M}_1$$

where

$$\tilde{M}_i(s) = I - \frac{2 \operatorname{Re} p_i}{s + p_i^*} \zeta_i \zeta_i^*.$$

Since $\tilde{M}_i(\infty) = I$, $i = 1, 2, \ldots, \mu$, it follows that $Q(\infty)\tilde{M}(\infty) = I$ is equivalent to $Q(\infty)\tilde{M}_0(\infty) = I$. Hence, by using the facts that $\tilde{M}_i$, $i = 1, 2, \ldots, \mu$, are unitary operators in $\mathcal{L}_2$ and that $\tilde{M}_1^{-1} \cdots \tilde{M}_\mu^{-1} - I \in \mathcal{H}_2^\perp$ and $I - Q\tilde{M}_0 \in \mathcal{H}_2$, we obtain

$$
\begin{aligned}
V^* &= \inf_{Q \in \mathcal{RH}_\infty, Q(\infty)\tilde{M}_0(\infty)=I} \|I - Q\tilde{M}_0\tilde{M}_\mu \cdots \tilde{M}_1\|_2^2 \\
&= \inf_{Q \in \mathcal{RH}_\infty, Q(\infty)\tilde{M}_0(\infty)=I} \|\tilde{M}_1^{-1} \cdots \tilde{M}_\mu^{-1} - I + I - Q\tilde{M}_0\|_2^2 \\
&= \|\tilde{M}_1^{-1} \cdots \tilde{M}_\mu^{-1} - I\|_2^2 + \inf_{Q \in \mathcal{RH}_\infty, Q(\infty)\tilde{M}_0(\infty)=I} \|I - Q\tilde{M}_0\|_2^2.
\end{aligned}
$$

Since $\tilde{M}_0$ is co-inner with invertible $\tilde{M}_0(\infty)$, there exists a sequence $\{Q_k\} \in \mathcal{RH}_\infty$ with $Q_k(\infty)\tilde{M}_0(\infty) = I$ such that $\lim_{k \to \infty} \|I - Q\tilde{M}_0\| = 0$. This shows

$$
\begin{aligned}
V^* &= \|\tilde{M}_1^{-1} \cdots \tilde{M}_\mu^{-1} - I\|_2^2 \\
&= \|\tilde{M}_2^{-1} \cdots \tilde{M}_\mu^{-1} - I + I - \tilde{M}_1\|_2^2 \\
&= \|\tilde{M}_2^{-1} \cdots \tilde{M}_\mu^{-1} - I\|_2^2 + \|I - \tilde{M}_1\|_2^2 \\
&= \sum_{i=1}^{\mu} \|I - \tilde{M}_i\|_2^2 \\
&= 2 \sum_{i=1}^{\mu} p_i.
\end{aligned}
$$

Here the first equality follows from that $\tilde{M}_1$ is a unitary operator in $\mathcal{L}_2$, the second from that $\tilde{M}_2^{-1} \cdots \tilde{M}_\mu^{-1} - I \in \mathcal{H}_2^\perp$ and $I - \tilde{M}_1 \in \mathcal{H}_2$, the third from repeating the underlying procedure in the first and second equalities, and the last from straightforward computation. The above derivation shows that an arbitrarily near optimal Q can be chosen from the sequence $\{Q_k\}$. Therefore

$$V^*(\xi) = \lim_{k \to \infty} \|\xi'(I - Q_k \tilde{M}_0 \tilde{M}_\mu \cdots \tilde{M}_1)\|_2^2.$$

The same reasoning as in the above derivation gives

$$V^*(\xi) = \sum_{i=1}^{\mu} \|\xi(I - \tilde{M}_i)\|_2^2 = 2 \sum_{i=1}^{\mu} p_i \cos^2 \angle(\xi, \zeta_i).$$

The last equality follows from straightforward computation.

The directional steady state error variance with an arbitrary F is

$$V_\xi = \|\xi'F\|_2^2$$

and the optimal directional steady state error variance is

$$
\begin{aligned}
V_\xi^* &= \inf_{F,G-FG \in \mathcal{RH}_\infty} \|\xi'F\|_2^2 \\
&= \inf_{Q \in \mathcal{RH}_\infty, Q(\infty)\tilde{M}_0(\infty)=I} \|\xi'(I - Q\tilde{M}_0\tilde{M}_\mu \cdots \tilde{M}_1)\|_2^2.
\end{aligned}
$$

By following an almost identical derivation as the non-directional case, we can show that the same sequence $\{Q_k\}$ giving near optimal solutions there also gives near optimal solutions here for every $\xi \in \mathbb{R}^m$. Hence,

$$V_\xi^* = V^*(\xi) = 2 \sum_{i=1}^{\mu} p_i \cos^2 \angle(\xi, \zeta_i).$$

We have thus established the following theorem.

Theorem 1. *Let G's antistable poles be $p_1, p_2, \ldots, p_\mu$ with $\zeta_1, \zeta_2, \cdots, \zeta_\mu$ being the corresponding pole Blaschke vectors of type I. Then*

$$V^* = 2 \sum_{i=1}^{\mu} p_i$$

and

$$V_\xi^* = V^*(\xi) = 2 \sum_{i=1}^{\mu} p_i \cos^2 \angle(\xi, \zeta_i).$$

This theorem says that to estimate a signal from its version corrupted by a standard while noise, the best achievable steady state error variance depends, in a simple way, only on the antistable modes of the signal to be estimated. The best achievable directional steady state error variance depends, in addition, on the directional characteristics of the antistable modes.

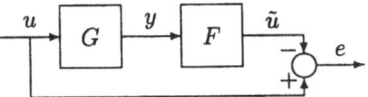

Fig. 4. Estimation of a Brownian motion process

4   Estimation of Brownian Motion

Consider the estimation problem shown in Fig. 4. Here G is a given FDLTI plant, u is the input to the plant which is assumed to be a Brownian motion process, i.e., the integral of a standard white noise, which can be used to model a slowly varying "constant". Assume that $G(0)$ is left invertible. The objective is to design an LTI filter F such that it measures the output of G and generates an estimate $\tilde{u}$ of u. This problem is clearly a special case of the general estimation problem stated in Sect. 1 with $P = \begin{bmatrix} G \\ I \end{bmatrix}$ and $n = 0$. The error of estimation is given by $(I - FG)u$. Since u is a Brownian process, the variance of the error is given by

$$V = \|(I - FG)U\|_2$$

where $U(s) = \frac{1}{s}I$ is the transfer matrix of m channels of integrators. If we want V to be finite, we need to have $I - F(0)G(0) = 0$, in addition to $F, I - FG \in \mathcal{RH}_\infty$. This requires $G(0)$ to be left invertible, which will be assumed. Equivalently, we need to have $F, FG \in \mathcal{H}_\infty$ and $F(0)G(0) = I$. Therefore,

$$V^* = \inf_{F, FG \in \mathcal{RH}_\infty, F(0)G(0)=I} \|(I - FG)U\|_2^2.$$

Let $G = \tilde{M}^{-1}\tilde{N}$ be a left coprime factorization of G. Then it is easy to see that $F, FG \in \mathcal{H}_\infty$ is equivalent to $F = Q\tilde{M}$ for some $Q \in \mathcal{H}_\infty$. Hence

$$V^* = \inf_{Q \in \mathcal{RH}_\infty, Q(0)\tilde{N}(0)=I} \|(I - Q\tilde{N})U\|_2^2.$$

Now let G have nonminimum phase zeros $z_1, z_2, \ldots, z_\nu$ with $\eta_1, \eta_2, \ldots, \eta_\nu$ being the corresponding input Blaschke vectors of type II. Then $\tilde{N}$ has factorizations

$$\tilde{N} = \tilde{N}_0\tilde{N}_\nu, \ldots, \tilde{N}_1$$

where

$$\tilde{N}_i = I - \frac{2\operatorname{Re} z_i}{z_i} \frac{s}{s + z_i^*} \eta_i\eta_i^*.$$

Since $\tilde{N}_i(\infty) = I$, $i = 1, 2, \ldots, \nu$, it follows that $Q(0)\tilde{N}(0) = I$ is equivalent to $Q(0)\tilde{N}_0(0) = I$. Hence, by using the facts that $\tilde{N}_i$, $i = 1, 2, \ldots, \nu$, are unitary operators in $\mathcal{L}_2$ and that $\tilde{N}_1^{-1} \cdots \tilde{N}_\nu^{-1} - I \in \mathcal{H}_2^\perp$ and $I - Q\tilde{N}_0 \in \mathcal{H}_2$, we obtain

$$
\begin{aligned}
V^* &= \inf_{Q \in \mathcal{RH}_\infty, Q(0)\tilde{N}_0(0)=I} \|(I - Q\tilde{N}_0\tilde{N}_\nu, \ldots, \tilde{N}_1)U\|_2^2 \\
&= \inf_{Q \in \mathcal{RH}_\infty, Q(0)\tilde{N}_0(0)=I} \|(\tilde{N}_1^{-1}\tilde{N}_2^{-1} \cdots \tilde{N}_\nu - I)U + (I - Q\tilde{N}_0)U\|_2^2 \\
&= \|(\tilde{N}_1^{-1}\tilde{N}_2^{-1} \cdots \tilde{N}_\nu - I)U\|_2^2 + \inf_{Q \in \mathcal{RH}_\infty, Q(0)\tilde{N}_0(0)=I} \|(I - Q\tilde{N}_0)U\|_2^2 .
\end{aligned}
$$

Since $\tilde{N}_0$ is co-inner with invertible $\tilde{N}(0)$, there exists a sequence $\{Q_k\} \in \mathcal{RH}_\infty$ with $Q_k(0)\tilde{N}_0(0) = I$ such that $\lim_{k\to\infty} \|(I - Q\tilde{N}_0)U\| = 0$. This shows

$$
\begin{aligned}
V^* &= \|(\tilde{N}_1^{-1} \cdots \tilde{N}_\nu^{-1} - I)U\|_2^2 \\
&= \|(\tilde{N}_2^{-1} \cdots \tilde{N}_\nu^{-1} - I + I - \tilde{N}_1)U\|_2^2 \\
&= \|(\tilde{N}_2^{-1} \cdots \tilde{N}_\nu^{-1} - I)U\|_2^2 + \|(I - \tilde{M}_1)U\|_2^2 \\
&= \sum_{i=1}^{\nu} \|(I - \tilde{N}_i)U\|_2^2 \\
&= 2\sum_{i=1}^{\nu} \frac{1}{z_i} .
\end{aligned}
$$

Here the first equality follows from that $\tilde{N}_1$ is a unitary operator in $\mathcal{L}_2$, the second from that $(\tilde{N}_2^{-1} \cdots \tilde{N}_\nu^{-1} - I)U \in \mathcal{H}_2^\perp$ and $(I - \tilde{N}_1)U \in \mathcal{H}_2$, the third from repeating the underlying procedure in the first and second equalities, and the last from straightforward computation.

The above derivation shows that an arbitrarily near optimal Q can be chosen from the sequence $\{Q_k\}$. Therefore

$$
V^*(\xi) = \lim_{k\to\infty} \|\xi'(I - Q_k\tilde{N}_0\tilde{N}_\nu \cdots \tilde{N}_1)U\|_2^2 .
$$

The same reasoning as in the above derivation gives

$$
V^*(\xi) = \sum_{i=1}^{\mu} \|\xi(I - \tilde{N}_i)U\|_2^2 = 2\sum_{i=1}^{\mu} \frac{1}{z_i} \cos^2 \angle(\xi, \eta_i) .
$$

The last equality follows from straightforward computation.

The directional steady state error variance with an arbitrary F is

$$
V_\xi = \|\xi'(I - FG)U\|_2^2
$$

and the optimal directional steady state error variance is

$$
\begin{aligned}
V_\xi^* &= \inf_{F, G - FG \in \mathcal{RH}_\infty} \|\xi'(I - FG)U\|_2^2 \\
&= \inf_{Q \in \mathcal{RH}_\infty, Q(0)\tilde{N}_0(0)=I} \|\xi'(I - Q\tilde{N}_0\tilde{N}_\nu \cdots \tilde{N}_1)U\|_2^2 .
\end{aligned}
$$

By following an almost identical derivation as the non-directional case, we can show that the same sequence $\{Q_k\}$ giving near optimal solutions there also gives near optimal solutions here for every $\xi \in \mathbb{R}^m$. Hence,

$$V_\xi^* = V^*(\xi) = 2 \sum_{i=1}^{\nu} \frac{1}{z_i} \cos^2 \angle(\xi, \eta_i).$$

We have thus established the following theorem.

Theorem 2. *Let G's nonminimum phase zeros be $z_1, z_2, \cdots, z_\nu$ with $\eta_1, \eta_2,$* $\cdots, \eta_\nu$ *being the corresponding Blaschke vectors of type II, then*

$$V^* = 2 \sum_{i=1}^{\nu} \frac{1}{z_i}$$

and

$$V_\xi^* = V^*(\xi) = 2 \sum_{i=1}^{\nu} \frac{1}{z_i} \cos^2 \angle(\xi, \eta_i).$$

This theorem says that to estimate a Brownian motion from its version distorted by an LTI system, the best achievable steady state error variance depends, in a simple way, only on the nonminimum phase zeros of the LTI system. The best achievable directional steady state variance depends, in addition, on the directional characteristics of the nonminimum phase zeros.

5   Concluding Remarks

This paper relates the performance limitations in two typical estimation problems to simple characteristics of the plants involved. By estimation problems we mean actually filtering problems here. The general estimation problems can include prediction and smoothing problems. We are now trying to extend the results in this paper to smoothing and prediction problems.

In the problem considered in Sect. 3, the noise is modelled by a white noise. In the problem considered in Sect. 4, the signal to be estimated is modelled as a Brownian motion. We are trying to extending our results to possibly other types of noises and signals.

References

1. B. D. O. Anderson and J. B. Moore, *Optimal Filtering*, Prentice-Hall, 1979.
2. K. J. Åström, *Introduction to Stochastic Control Theory*, Academic Press, 1970.
3. J. Chen, L. Qiu, and O. Toker, "Limitation on maximal tracking accuracy," *Proc. 35th IEEE Conf. on Decision and Control*, pp. 726-731, 1996, also to appear in *IEEE Trans. on Automat. Contr.*.

4. G.C. Goodwin, D.Q. Mayne, and J. Shim, "Trade-offs in linear filter design", *Automatica*, vol. 31, pp. 1367–1376, 1995.

5. G.C. Goodwin, M.M. Seron, "Fundamental design tradeoffs in filtering, prediction, and smoothing," *IEEE Trans. Automat. Contr.*, vol. 42, pp. 1240-1251, 1997.

6. H. Kwakernaak and R. Sivan, *Linear Optimal Control Systems*, Wiley- Interscience, New York, 1972.

7. L. Qiu and J. Chen, "Time domain characterizations of performance limitations of feedback control", *Learning, Control, and Hybrid Systems*, Y. Yamamoto and S. Hara, editors, Springer-Verlag, pp. 397-415, 1998.

8. L. Qiu and E. J. Davison, "Performance limitations of non-minimum phase systems in the servomechanism problem", *Automatica*, vol. 29, pp. 337-349, 1993.

9. M.M. Seron, J.H. Braslavsky, and G.C. Goodwin, *Fundamental Limitations in Filtering and Control*, Springer, 1997.

10. M.M. Seron, J.H. Braslavsky, D.G. Mayne, and P.V. Kokotovic, "Limiting performance of optimal linear filters," *Automatica*, 1999.

Design Criteria for Uncertain Models with Structured and Unstructured Uncertainties*

Ali H. Sayed and Vitor H. Nascimento

Adaptive and Nonlinear Systems Laboratory, Electrical Engineering
Department, University of California, Los Angeles, CA 90024

Abstract. This paper introduces and solves a weighted game-type cost criterion for estimation and control purposes that allows for a general class of uncertainties in the model or data. Both structured and unstructured uncertainties are allowed, including some special cases that have been used in the literature. The optimal solution is shown to satisfy an orthogonality condition similar to least-squares designs, except that the weighting matrices need to be modified in a certain optimal manner. One particular application in the context of state regulation for uncertain state-space models is considered. It is shown that in this case, the solution leads to a control law with design equations that are similar in nature to LQR designs. The gain matrix, however, as well as the Riccati variable, turn out to be state-dependent in a certain way. Further applications of these game-type formulations to image processing, estimation, and communications are discussed in [1–3].

1 Introduction

This paper develops a technique for estimation and control purposes that is suitable for models with bounded data uncertainties. The technique will be referred to as a BDU design method for brevity, and it expands on earlier works in the companion articles [1–4]. It is based on a constrained game-type formulation that allows the designer to explicitly incorporate into the problem statement a-priori information about bounds on the sizes of the uncertainties in the model. A key feature of the BDU formulation is that geometric insights (such as orthogonality conditions and projections), which are widely appreciated for classical quadratic-cost designs, can be pursued in this new framework. This geometric viewpoint was discussed at some length in the article [1] for a special case of the new cost function that we introduce in this paper.

* This material was based on work supported in part by the National Science Foundation under Award No. CCR-9732376. The work of V. H. Nascimento was also supported by a fellowship from CNPq - Brazil, while on leave from Escola Politécnica da Universidade de São Paulo.

The optimization problem (1) that we pose and solve here is of independent interest in its own right and it can be applied in several contexts. Examples to this effect can be found in [1] where similar costs were applied to problems in image restoration, image separation, array signal processing, and estimation. Later in this paper we shall discuss one additional application in the context of state regulation for state-space models with parametric uncertainties.

2   Formulation of the BDU problem

We start by formulating a general optimization problem with uncertainties in the data. Thus consider the cost function

$$J(x, y) = x^T Q x + R(x, y) \, ,$$

where $x^T Q x$ is a regularization term, while the residual cost $R(x, y)$ is defined by

$$R(x, y) \triangleq \left(A x - b + H y \right)^T W \left(A x - b + H y \right) .$$

Here, $Q > 0$ and $W \geq 0$ are given Hermitian weighting matrices, x is an n−dimensional column vector, A is an $N \times n$ known or nominal matrix, b is an $N \times 1$ known or nominal vector, H is an $N \times m$ known matrix, and y denotes an $m \times 1$ unknown perturbation vector. We now consider the problem of solving:

$$\hat{x} = \arg \min_x \max_{\|y\| \leq \phi(x)} J(x, y) \, , \tag{1}$$

where the notation $\| \cdot \|$ stands for the Euclidean norm of its vector argument or the maximum singular value of its matrix argument. The non-negative function $\phi(x)$ is a known bound on the perturbation y and it is only a function of x (it can be linear or nonlinear).

Problem (1) can be regarded as a constrained two-player game problem, with the designer trying to pick an $\hat{x}$ that minimizes the cost while the opponent $\{y\}$ tries to maximize the cost. The game problem is constrained since it imposes a limit (through $\phi(x)$) on how large (or how damaging) the opponent can be. Observe further that the strength of the opponent can vary with the choice of x.

2.1   Special Cases

The formulation (1) allows for both structured and unstructured uncertainties in the data. Before proceeding to its solution, let us exhibit two special cases. Consider first the problem

$$\min_x \max_{\substack{\|\delta A\| \leq \eta \\ \|\delta b\| \leq \eta_b}} \left[x^T Q x + \left((A + \delta A) x - (b + \delta b) \right)^T W \left((A + \delta A) x - (b + \delta b) \right) \right]$$

where $\{\delta A\}$ denotes an $N \times n$ perturbation matrix to the nominal matrix A, and δb denotes an $N \times 1$ perturbation vector to the nominal vector b. We showed in the companion article [3] that the above problem is equivalent to one of the following form:

$$\min_x \max_{\|y\| \le \eta \|x\| + \eta_b} \left[x^T Q x + \left(Ax - b + y \right)^T W \left(Ax - b + y \right) \right] ,$$

which is a special case of (1), with $H = I$ and $\phi(x) = \eta\|x\| + \eta_b$. In this example, the uncertainties $\{\delta A, \delta b\}$ are not related in any way and we shall say that they are unstructured. The special case $Q = 0$ and $W = I$ was treated in [1,4,5]. In particular, a geometric framework was developed in [1] for such problems that is similar in nature to the geometry of least-squares problems. We shall comment briefly on this aspect further ahead. On the other hand, reference [5] solves the case $Q = 0$ and $W = I$ by using LMI techniques, which for this particular problem turn out to be more costly than the direct solution methods proposed in [1,4]. When W is non-unity, the problem becomes more rich, and also more involved, even when $Q = 0$.

Consider now the alternative problem

$$\min_x \max_{\substack{\delta A \\ \delta b}} \left[x^T Q x + \left((A + \delta A)x - (b + \delta b) \right)^T W \left((A + \delta A)x - (b + \delta b) \right) \right]$$

where the perturbations $\{\delta A, \delta b\}$ are now assumed to be generated by a model of the form

$$\begin{bmatrix} \delta A & \delta b \end{bmatrix} = HS \begin{bmatrix} E_a & E_b \end{bmatrix} , \tag{2}$$

where S is a contraction, $\|S\| \le 1$, and $\{H, E_a, E_b\}$ are known. Then it can be easily seen that this problem is equivalent to the following

$$\min_x \max_{\|y\| \le \|E_a x - E_b\|} \left[x^T Q x + \left(Ax - b + Hy \right)^T W \left(Ax - b + Hy \right) \right] ,$$

which is again a special case of (1) with $\phi(x) = \|E_a x - E_b\|$. Here, the perturbations $\{\delta A, \delta b\}$ are related (for example, they both lie in the range space of H). We shall say that they are structured. Such structured perturbations have been used in robust control design (see, e.g., [6]).

The formulation (1) that we consider in this paper is more general in that it allows for other classes of perturbations through the choice of the function $\phi(x)$.

3   Solution of the BDU Problem

We now proceed to the solution of (1). It turns out that the derivation given in [3] for a special case of (1) extends to this more general scenario with the appropriate modifications.

First we note that for any given y, the residual cost $R(x,y)$ is convex in x. Therefore, the maximum

$$C(x) \triangleq \max_{\|y\| \le \phi(x)} R(x,y) , \qquad (3)$$

is a convex function in x. Now since $x^T Q x$ is strictly convex in x when $Q > 0$, we conclude that $x^T Q x + C(x)$ is strictly convex in x, which shows that problem (1) has a unique global minimum $\hat{x}$.[1] To determine $\hat{x}$ we proceed in steps.

3.1 The Maximization Problem

We now solve (3) for any fixed x. Note first that the cost $R(x,y)$ is convex in y, so that the maximum over y is achieved at the boundary, $\|y\| = \phi(x)$. We can therefore replace the inequality constraint in (3) by an equality. Introducing a Lagrange multiplier λ, the solution to (3) can then be found from the unconstrained problem:

$$\max_{y,\lambda} \left[(Ax - b + Hy)^T W (Ax - b + Hy) - \lambda(\|y\|^2 - \phi^2(x)) \right]. \qquad (4)$$

Note that since the original problem has an inequality constraint, the Lagrange multiplier must be nonnegative: $\lambda \ge 0$ [7]. Differentiating (4) with respect to y and λ, and denoting the optimal solutions by $\{y^o, \lambda^o\}$, we obtain the equations

$$(\lambda^o I - H^T W H) y^o = H^T W (Ax - b) , \quad \|y^o\| = \phi(x) . \qquad (5)$$

It turns out that the solution λ^o should satisfy $\lambda^o \ge \|H^T W H\|$. This is because the Hessian of the cost in (4) w.r.t y must be nonpositive-definite [7].[2] We should further stress that the solutions $\{y^o, \lambda^o\}$ are functions of x and we shall therefore sometimes write $\{y^o(x), \lambda^o(x)\}$.

At this stage, we do not need to solve the equations (5) for $\{y^o, \lambda^o\}$. It is enough to know that the optimal $\{y^o, \lambda^o\}$ satisfy (5).[3] Using this fact, we can verify that the maximum cost in (4) is equal to

$$C(x) = (Ax - b)^T \left[W + W H (\lambda^o(x)I - H^T W H)^\dagger H^T W \right] (Ax - b)$$
$$+ \lambda^o(x)\phi^2(x) , \qquad (6)$$

where $X^\dagger$ denotes the pseudo-inverse of X.

---

[1] It can be easily seen that in the special case $\phi(0) = 0$ and $Wb = 0$, the unique solution of (1) is $\hat{x} = 0$. In the sequel we shall therefore assume that $\phi(0)$ and Wb are not zero simultaneously.

[2] We refer to the case $\lambda^o = \|H^T W H\|$ as the singular case, while $\lambda^o > \|H^T W H\|$ is called the regular case. Both cases are handled simultaneously in our framework through the use of the pseudo-inverse notation.

[3] In fact, we can show that the solution λ^o is always unique while there might be several y^o.

3.2   The Minimization Problem

The original problem (1) is therefore equivalent to:

$$\min_{x} \left[x^T Q x + C(x) \right] . \tag{7}$$

However, rather than minimizing the above cost over n variables, which are the entries of the vector x, we shall instead show how to reduce the problem to one of minimizing a certain cost function over a single scalar variable (see (14) further ahead). For this purpose, we introduce the following function of two independent variables x and λ,

$$C(x, \lambda) = (Ax - b)^T \left[W + WH(\lambda I - H^T W H)^\dagger H^T W \right] (Ax - b) + \lambda \phi^2(x) .$$

Then it can be verified, by direct differentiation with respect to λ and by using the expression for $\lambda^o(x)$ from (5), that

$$\lambda^o(x) = \arg \min_{\lambda \geq \|H^T W H\|} C(x, \lambda) .$$

This means that problem (1) is equivalent to

$$\min_{\lambda \geq \|H^T W H\|} \min_{x} \left[x^T Q x + C(x, \lambda) \right] . \tag{8}$$

The cost function in the above expression, viz., $J(x, \lambda) = x^T Q x + C(x, \lambda)$, is now a function of two independent variables $\{x, \lambda\}$. This should be contrasted with the cost function in (7). Now define, for compactness of notation, the quantities $M(\lambda) = Q + A^T W(\lambda) A$ and $d(\lambda) = A^T W(\lambda) b$, where

$$W(\lambda) = W + WH(\lambda I - H^T W H)^\dagger H^T W .$$

To solve problem (8), we first search for the minimum over x for every fixed value of λ, which can be done by setting the derivative of $J(x, \lambda)$ w.r.t. x equal to zero. This shows that any minimum x must satisfy the equality

$$M(\lambda) x + \frac{1}{2} \lambda \nabla \phi^2(x) = d(\lambda) , \tag{9}$$

where $\nabla \phi^2(x)$ is the gradient of $\phi^2(x)$ w.r.t. x.

Special Cases

Let us reconsider the special cases $\phi(x) = \|E_a x - E_b\|$ and $\phi(x) = \eta \|x\| + \eta_b$. For the first choice we obtain $\nabla \phi^2(x) = 2 E_a^T (E_a x - E_b)$ so that the solution of Eq. (9), which is dependent on λ, becomes

$$x^o(\lambda) = \left[M(\lambda) + \lambda E_a^T E_a \right]^{-1} (d(\lambda) + \lambda E_a^T E_b) . \tag{10}$$

The second choice, $\phi(x) = \eta\|x\| + \eta_b$, was studied in the companion article [3]. In this case, solving for x^o is not so immediate since Eq. (9) now becomes, for any nonzero x,

$$x = \left[M(\lambda) + \lambda\eta \left(\eta + \frac{\eta_b}{\|x\|} \right) \right]^{-1} d(\lambda) . \tag{11}$$

Note that x appears on both sides of the equality (except when $\eta_b = 0$, in which case the expression for x is complete in terms of $\{M, \lambda, \eta, d\}$). To solve for x in the general case we define $\alpha = \|x\|^2$ and square the above equation to obtain the scalar equation in α:

$$\alpha^2 - d^T(\lambda) \left[M(\lambda) + \lambda\eta \left(\eta + \frac{\eta_b}{\alpha} \right) \right]^{-2} d(\lambda) = 0 . \tag{12}$$

It can be shown that a unique solution $\alpha^o(\lambda) > 0$ exists for this equation if, and only if, $\lambda\eta\eta_b < \|d(\lambda)\|^2$. Otherwise, $\alpha^o(\lambda) = 0$. In the former case, the expression for x^o, which is a function of λ, becomes

$$x^o(\lambda) = \left[M(\lambda) + \lambda\eta \left(\eta + \frac{\eta_b}{\alpha^o(\lambda)} \right) \right]^{-1} d(\lambda) . \tag{13}$$

In the latter case we clearly have $x^o(\lambda) = 0$.

The General Case

Let us assume that (9) has a unique solution $x^o(\lambda)$, as was the case with the above two special cases. This will also be always the case whenever $\phi(x)$ is a differentiable and strictly convex function (since then $J(x, \lambda)$ will be differentiable and strictly convex in x). We thus have a procedure that allows us to determine the minimizing x^o for every λ. This in turn allows us to re-express the resulting cost $J(x^o(\lambda), \lambda)$ as a function of λ alone, say $G(\lambda) = J(x^o(\lambda), \lambda)$. In this way, we conclude that the solution $\hat{x}$ of the original optimization problem (1) can be solved by determining the $\hat{\lambda}$ that solves

$$\min_{\lambda \geq \|H^T W H\|} G(\lambda) , \tag{14}$$

and by taking the corresponding $x^o(\hat{\lambda})$ as $\hat{x}$. That is, $\hat{x}$ solves (9) when $\lambda = \hat{\lambda}$. We summarize the solution in the following statement.

Theorem 1 (Solution). *The unique global minimum of (1) can be determined as follows. Introduce the cost function*

$$G(\lambda) = x^{oT}(\lambda) Q x^o(\lambda) + C[x^o(\lambda), \lambda] , \tag{15}$$

where $x^o(\lambda)$ is the unique solution of (9). Let $\hat{\lambda}$ denote the minimum of $G(\lambda)$ over the interval $\lambda \geq \|H^T W H\|$. Then the optimum solution of (1) is $\hat{x} = x^o(\hat{\lambda})$.

◇

We thus see that the solution of (1) requires that we determine an optimal scalar parameter $\hat{\lambda}$, which corresponds to the minimizing argument of a certain nonlinear function $G(\lambda)$ (or, equivalently, to the root of its derivative function). This step can be carried out very efficiently by any root finding routine, especially since the function $G(\lambda)$ is well defined and, moreover, $\hat{\lambda}$ is unique. We obtain as corollaries the following two special cases.

Corollary 1 (Structured Uncertainties). *When $\phi(x) = \|E_a x - E_b\|$, $x^o(\lambda)$ is given by (10) and the global minimum of (1) becomes*

$$\hat{x} = \left[\hat{Q} + A^T \hat{W} A\right]^{-1} A^T \hat{W} b \triangleq Kb \tag{16}$$

where $\hat{Q} = Q + \hat{\lambda} E_a^T E_a$ and $\hat{W} = W + W H(\hat{\lambda} I - H^T W H)^\dagger H^T W$.

◇

Corollary 2 (Unstructured Uncertainties). *When $\phi(x) = \eta\|x\| + \eta_b$, $x^o(\lambda)$ is given by (13) if $\lambda\eta\eta_b < \|d(\lambda)\|^2$ (otherwise it is zero). Moreover, $\alpha^o(\lambda)$ in (13) is the unique positive root of (12). Let $\hat{\lambda}$ denote the minimum of $G(\lambda)$ over the interval $\lambda \geq \|W\|$. Then*

$$\hat{x} = \left[\hat{Q} + A^T \hat{W} A\right]^{-1} A^T \hat{W} b \triangleq Kb, \tag{17}$$

if $\hat{\lambda}\eta\eta_b < \|d(\hat{\lambda})\|^2$ (otherwise $\hat{x} = 0$), where

$$\hat{Q} = Q + \hat{\lambda}\eta\left(\eta + \frac{\eta_b}{\alpha^o(\hat{\lambda})}\right) I, \quad \hat{W} = W + W(\hat{\lambda} I - W)^\dagger W.$$

◇

3.3 The Orthogonality Condition

Observe that the optimal solution $\hat{x}$ in the above cases satisfies an orthogonality condition of the form $\hat{Q}\hat{x} + A^T \hat{W}(A\hat{x} - b) = 0$, for some $\{\hat{Q}, \hat{W}\}$. Compared with the solution to the standard regularized least-squares problem,

$$\min_x \left[x^T Q x + (Ax - b)^T W(Ax - b)\right],$$

whose unique solution satisfies $Q\hat{x} + A^T W(A\hat{x} - b) = 0$, we see that the solution to the BDU problem satisfies a similar orthogonality condition, with the given weighting matrices $\{Q, W\}$ replaced by new matrices $\{\hat{Q}, \hat{W}\}$! To determine the necessary corrections to $\{Q, W\}$, one determines the optimal scalar $\hat{\lambda}$ from the minimization (14). The convenience of such a geometric viewpoint is discussed in [1] for the special case $Q = 0$ and $W = I$.

4 Application to state regulation

As mentioned earlier, the BDU cost functions can be useful in different contexts, including image restoration, image separation, array signal processing, and estimation (see [1] for some examples). Here we discuss another application for the weighted BDU problem in the context of state regulation for state-space models with parametric uncertainties.

Thus consider the linear state-space model $x_{i+1} = F_i x_i + G_i u_i$, where x_0 denotes the value of the initial state, and the $\{u_i\}$ denote the control (input) sequence. The classical linear quadratic regulator (LQR) problem seeks a control sequence $\{u_i\}$ that regulates the state vector towards zero while keeping the control cost low. This is achieved as follows. Introduce, for compactness of notation, the local cost

$$V_i(x_{i+1}, u_i) \triangleq \left(x_{i+1}^T R_{i+1} x_{i+1} + u_i^T Q_i u_i \right) , \quad R_{N+1} = P_{N+1} .$$

Then the optimal control is determined by solving

$$\min_{\{u_0, u_1, \ldots, u_N\}} \left(x_{N+1}^T P_{N+1} x_{N+1} + \sum_{j=0}^{N} \left[u_j^T Q_j u_j + x_j^T R_j x_j \right] \right) ,$$

with $Q_j > 0$, $R_j \geq 0$, and $P_{N+1} \geq 0$. We shall write the above problem more compactly as (note that x_0 does not really affect the solution):

$$x_0^T R_0 x_0 + \min_{\{u_0, u_1, \ldots, u_N\}} (V_0 + V_1 + \ldots + V_N) , \tag{18}$$

It is well known that the LQR problem can be solved recursively by re-expressing the LQR cost as nested minimizations of the form:

$$x_0^T R_0 x_0 + \min_{u_0} \left\{ V_0 + \min_{u_1} \left\{ V_1 + \ldots + \min_{u_N} \{V_N\} \right\} \right\}, \tag{19}$$

where only the last term, through the state-equation for x_{N+1}, is dependent on u_N. Hence we can determine $\hat{u}_N$ by solving

$$\min_{u_N} V_N , \quad \text{given } x_N , \tag{20}$$

and then progress backwards in time to determine the other control values. By carrying out this argument one finds the well-known state-feedback solution:

$$\begin{cases} \hat{u}_i = -K_i x_i \, , \\ K_i = (Q_i + G_i^T P_{i+1} G_i)^{-1} G_i^T P_{i+1} F_i \, , \\ P_i = R_i + K_i^T Q_i K_i + (F_i - G_i K_i)^T P_{i+1} (F_i - G_i K_i) \, . \end{cases} \qquad (21)$$

It is well known that the above LQR controller is sensitive to modeling errors. Robust design methods to ameliorate these sensitivity problems include the $\mathcal{H}_\infty$ design methodology (e.g., [8–11]) and the so-called guaranteed-cost designs (e.g., [12–14]). We suggest below a procedure that is based on the BDU problem solved above. At the end of this exposition, we shall compare our result with a guaranteed-cost design. [A comparison with an $\mathcal{H}_\infty$ design is given in [1] for a special first-order problem.]

4.1   State Regulation

Consider now the state-equation with parametric uncertainties:

$$x_{i+1} = (F_i + \delta F_i) x_i + (G_i + \delta G_i) u_i \, , \qquad (22)$$

with known x_0, and where the uncertainties $\{\delta F_i, \delta G_i\}$ are assumed to be generated via

$$\begin{bmatrix} \delta F_i & \delta G_i \end{bmatrix} = HS \begin{bmatrix} E_f & E_g \end{bmatrix} \, , \qquad (23)$$

for known H, E_f, E_g, and for any contraction $\|S\| \leq 1$. The solution of the case with unstructured uncertainties $\{\delta F_i, \delta G_i\}$, say $\|\delta F_i\| \leq \eta_{f,i}$ and $\|\delta G_i\| \leq \eta_{g,i}$, is very similar and is treated in [3]. We focus here on the above structured case (23) for the sake of demonstration. Still, we should mention that by choosing different $\phi(x)$, the approach described in the earlier sections can handle other classes of uncertainties as well.

Consider the problem of determining a control sequence $\{\hat{u}_j, 0 \leq j \leq N\}$ that solves the nested min-max optimizations:

$$x_0^T R_0 x_0 + \min_{u_0} \max_{\substack{\delta F_0 \\ \delta G_0}} \left\{ V_0 + \min_{u_1} \max_{\substack{\delta F_1 \\ \delta G_1}} \left\{ V_1 + \ldots + \min_{u_N} \max_{\substack{\delta F_N \\ \delta G_N}} \left\{ V_N \right\} \right\} \right\} \qquad (24)$$

where we are writing, for compactness of notation, $\{\delta F_i, \delta G_i\}$ under the max symbols instead of the complete notation.

In order to illustrate the structure of the solution of (24), let us consider the simple case $N = 1$, viz.,

$$x_0^T R_0 x_0 + \min_{u_0} \max_{\substack{\delta F_0 \\ \delta G_0}} \left\{ V_0 + \min_{u_1} \max_{\substack{\delta F_1 \\ \delta G_1}} \left\{ V_1 \right\} \right\} \, . \qquad (25)$$

To be even more explicit, recall that V_0 is a function of $\{x_1, u_0\}$ while V_1 is a function of $\{x_2, u_1\}$. Hence, V_0 is a function of $\{x_0, u_0, \delta F_0, \delta G_0\}$ and we shall denote this explicitly as $V_0(x_0, u_0, \delta F_0, \delta G_0)$. Likewise, we shall write $V_1(x_0, u_0, \delta F_0, \delta G_0, u_1, \delta F_1, \delta G_1)$. If we now solve the inner-most min-max problem in (25), for any $\{u_0, \delta F_0, \delta G_0\}$, i.e.,

$$\min_{u_1} \max_{\substack{\delta F_1 \\ \delta G_1}} V_1(x_0, u_0, \delta F_0, \delta G_0, u_1, \delta F_1, \delta G_1) , \qquad (26)$$

we obtain a representation for the solution $\{\hat{u}_1, \widehat{\delta F}_1, \widehat{\delta G}_1\}$ in terms of the unknowns $\{x_0, u_0, \delta F_0, \delta G_0\}$. That is, we find

$$\hat{u}_1 = f_1(x_0, u_0, \delta F_0, \delta G_0) ,$$
$$\widehat{\delta F}_1 = g_1(x_0, u_0, \delta F_0, \delta G_0) ,$$
$$\widehat{\delta G}_1 = h_1(x_0, u_0, \delta F_0, \delta G_0) ,$$

for some functions $\{f_1(\cdot), g_1(\cdot), h_1(\cdot)\}$. The resulting cost in (26) will also be a function of $\{x_0, u_0, \delta F_0, \delta G_0\}$, say

$$V_1^*(x_0, u_0, \delta F_0, \delta G_0) = \min_{u_1} \max_{\substack{\delta F_1 \\ \delta G_1}} V_1(x_0, u_0, \delta F_0, \delta G_0, u_1, \delta F_1, \delta G_1) .$$

Returning to (25), we now solve the outer-most min-max problem over $\{u_0, \delta F_0, \delta G_0\}$,

$$x_0^T R_0 x_0 + \min_{u_0} \max_{\substack{\delta F_0 \\ \delta G_0}} \{V_0 + V_1^*\} ,$$

which would then lead to a representation for $\{\hat{u}_0, \widehat{\delta F}_0, \widehat{\delta G}_0\}$ in terms of x_0,

$$\hat{u}_0 = f_0(x_0) , \quad \widehat{\delta F}_0 = g_0(x_0) , \quad \widehat{\delta G}_0 = h_0(x_0) .$$

Therefore, the optimal control values that solve (25) will be

$$\hat{u}_0 = f_0(x_0) \quad \text{and} \quad \hat{u}_1 = f_1(x_0, \hat{u}_0, \widehat{\delta F}_0, \widehat{\delta G}_0) ,$$

where the arguments of $f_1(\cdot)$ are now defined in terms of $\{\hat{u}_0, \widehat{\delta F}_0, \widehat{\delta G}_0\}$.

If we thus reconsider the original problem (24), and focus first on the inner-most optimization, say

$$\min_{u_N} \max_{\delta F_N, \delta G_N \atop \text{subject to (23)}} \left[u_N^T Q_N u_N + x_{N+1}^T P_{N+1} x_{N+1} \right] ,$$

then the above argument shows that in order to determine an expression for $\hat{u}_N$ from the above, the state vector x_N has to be taken as $\hat{x}_N$, which is the value that would result had the earlier optimal control signals $\{\hat{u}_j, 0 \leq$

$j \leq N - 1\}$ been determined already and using the worst-case disturbances. Then expanding the term $x_{N+1}^T P_{N+1} x_{N+1}$ by using the state equation for x_{N+1},

$$x_{N+1} = (F_N + \delta F_N)\hat{x}_N + (G_N + \delta G_N)u_N ,$$

the above problem reduces to a problem of the same form as the structured BDU problem (2) that we considered before with the identifications:

$$A \leftarrow G_N , \quad W \leftarrow P_{N+1} , \quad Q \leftarrow Q_N , \quad H \leftarrow H , \quad b \leftarrow -F_N\hat{x}_N ,$$

$$x \leftarrow u_N, \quad E_a \leftarrow E_g, \quad E_b \leftarrow E_f\hat{x}_N , \quad \delta A \leftarrow \delta G_N, \quad \delta b \leftarrow -\delta F_N\hat{x}_N.$$

Using (16), and the above identifications, we conclude that the optimal control value $\hat{u}_N$ is given by (compare with the LQR recursions)

$$
\begin{cases}
\hat{u}_N = -K_N\hat{x}_N , \\
K_N = \left(\hat{Q}_N + G_N^T \hat{W}_{N+1} G_N\right)^{-1} G_N^T \hat{W}_{N+1} F_N , \\
\hat{Q}_N = Q_N + \hat{\lambda}_N E_g^T E_g , \\
\hat{W}_{N+1} = P_{N+1} + P_{N+1} H \left(\hat{\lambda}_N I - H^T P_{N+1} H\right)^{\dagger} H^T P_{N+1} ,
\end{cases}
$$

where $\hat{\lambda}_N$ is the optimal parameter that corresponds to the above data $\{A, b, W, Q, H, E_a, E_b\}$, and which can be found as explained in Thm. 1 (or Cor. 1).

Moreover, using (6)–(7) and the above identifications again, we find that

$$\hat{x}_N^T R_N\hat{x}_N + \left(\min_{u_N} \max_{\substack{\delta F_N \\ \delta G_N}} V_N\right) = \hat{x}_N^T P_N\hat{x}_N ,$$

where P_N is given by (compare with the Riccati recursion in (21)):

$$P_N = R_N + K_N^T Q_N K_N + (F_N - G_N K_N)^T \hat{W}_{N+1}(F_N - G_N K_N) + $$
$$+ \hat{\lambda}_N \left[K_N^T E_g^T E_g K_N - K_N^T E_g^T E_f - E_f^T E_g K_N + E_f^T E_f\right] . \qquad (27)$$

We now proceed to determine an approximation for the optimal control value at time $N - 1$ by solving

$$\min_{u_{N-1}} \max_{\delta F_{N-1}, \delta G_{N-1}} \left[u_{N-1}^T Q_{N-1}u_{N-1} + x_N^T P_N x_N\right] ,$$

where we assume that $\hat{x}_{N-1}$ is available. We take the solution as $\hat{u}_{N-1}$, and so on. Note that this step is an approximation because we are employing the P_N found above, which is a function of $\hat{x}_N$. For optimality, we would need to determine the functional form $P_N(x_N)$ — this form is defined by the same equations as above with x_N replacing $\hat{x}_N$. It turns out that for single-state models, the value of P_N is independent of the state and therefore the above $\hat{u}_{N-1}$ agrees with the optimal value — see further ahead.

Remarks

Several remarks are due now.

1. The control values $\{\hat{u}_i\}$ found above are in terms of the worst-case state $\hat{x}_i$, which we show how to evaluate in the next section.

2. Compared with the solution to the LQR problem we see that there are three main differences in the recursions. First, the gain matrix K_N is not defined directly in terms of the original quantities $\{Q_N, P_{N+1}\}$ but in terms of modified quantities $\{\hat{Q}_N, \hat{W}_{N+1}\}$. Secondly, the term P_{N+1} in the LQR Riccati recursion is replaced by $\hat{W}_{N+1}$ in (27), in addition to a new correction term that is equal to $\hat{\lambda}_N \phi^2(\hat{u}_N)$. Finally, the above solution in fact has the form of a two-point boundary value problem (TPBVP). This is because the expressions for $\{K_N, P_N\}$ are dependent on the worst-case state $\hat{x}_N$ (through $\hat{\lambda}_N$). We can denote this dependency more explicitly by writing, for any i,

$$\hat{u}_i = -K_i(\hat{x}_i)\hat{x}_i \ . \tag{28}$$

A reasonable state-feedback implementation would be to choose $\hat{u}_i = -K_i(\hat{x}_i)x_i$ (see, e.g., [3] for a simulation in this case). We should mention that for single-state models, the state-dependency disappears (as we show in a later section).

3. Similar recursions and remarks are valid for the solution of problems with other kinds of uncertainties, e.g., unstructured uncertainties [3].

4.2   An Iterative Solution to the TPBVP

We are currently studying the TPBVP more closely. An iterative solution that we found performs reasonably well is the following.

I. Initialization. Choose initial values for all variables P_0 to P_N (for example, by running the LQR Riccati recursion or by using a suboptimal guaranteed-cost design). Choose also initial values for all $\hat{\lambda}_i$, say $\hat{\lambda}_i > \|H^T P_{i+1} H\|_F$.

II. Forwards Iteration. Given values $\{x_0, P_{i+1}, \hat{\lambda}_i\}$, we evaluate the quantities $\{\hat{W}_{i+1}, \hat{Q}_i, K_i, \hat{y}_i, \hat{u}_i\}$ by using the recursions derived above, as well as propagate the state-vectors $\{\hat{x}_i\}$ by using $\hat{x}_{i+1} = F_i\hat{x}_i + G_i\hat{u}_i + \hat{y}_i$ where, from (5), $\hat{y}_i$ is found by solving the equation

$$\left(\hat{\lambda}_i I - H^T P_{i+1} H\right) \hat{y}_i = H^T P_{i+1}(F_i\hat{x}_i + G_i\hat{u}_i) \ .$$

If the matrix $\left(\hat{\lambda}_i I - H^T P_{i+1} H\right)$ is singular, then among all possible solutions we choose one that satisfies $\|\hat{y}_i\|^2 = \|E_f \hat{x}_i + E_g \hat{u}_i\|^2$.

III. Backwards Iteration. Given values $\{P_{N+1}, \hat{u}_i, \hat{x}_i\}$ we find new approximations for $\{P_i, \hat{\lambda}_i\}$ by using the recursions derived above for the state regulation problem.

IV. Recursion. Repeat steps II and III.

4.3   The One-Dimensional Case

Several simplifications occur for one-dimensional systems. In particular, there is no need to solve a two-point boundary value problem. This is because for such models, the state-dependency in the recursions disappears and we can therefore *explicitly* describe the optimal control law.

To show this, let us verify that the value of $\hat{\lambda}_N$ (and more generally $\hat{\lambda}_i$) becomes independent of $\hat{x}_N$ ($\hat{x}_i$). Indeed, recall from (15) that $\hat{\lambda}_N$ is the argument that minimizes

$$ G(\lambda) = \left[K_N^2 Q_N + (F_N - G_N K_N)^2 \left(W_{N+1} + E_g^2 K_N^2 + E_f^2 \right) \right] \hat{x}_N^2 , $$

over $\lambda \geq \|H^2 P_{N+1}\|$. Here

$$ W_{N+1} = P_{N+1} + P_{N+1}^2 H^2 (\lambda - H^2 P_{N+1})^\dagger . $$

Therefore, the minimum of $G(\lambda)$ is independent of $\hat{x}_N$. It then follows that $\hat{\lambda}_N$ and K_N are independent of $\hat{x}_N$ and we can iterate the recursion for P_N backwards in time.

4.4   A Simulation

We compare below in Fig. 1 the performance of this design with a guaranteed-cost design. The example presented here is of a 2-state system. The nominal model is stable with only one control variable. Moreover, $H = I$, $E_f = 0$, $E_g = \begin{bmatrix} 0 & 0 & 0.4 \end{bmatrix}^T$, $G = \begin{bmatrix} 1 & -0.5 \end{bmatrix}^T$, and $N = 20$. The lower horizontal line is the worst-case cost that is predicted by our BDU construction. The upper horizontal line is an upper bound on the optimal cost. It is never exceeded by the guaranteed-cost design. The situation at the right-most end of the graph corresponds to the worst-case scenario. Observe (at the right-end of the graph) the improvement in performance in the worst-case.

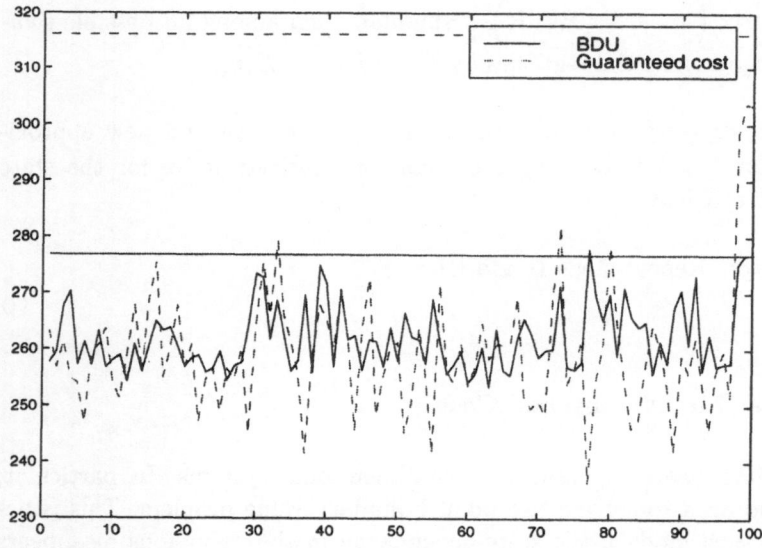

Fig. 1. 100 random runs with a stable 2-dimensional nominal model

5   Concluding remarks

Regarding the state-regulator application, earlier work in the literature on guaranteed-cost designs found either sub-optimal steady-state and finite-horizon controllers (e.g., [13]), or optimal steady-state controllers *over the class of linear control laws* [12]. Our solution has the following properties: i) It has a geometric interpretation in terms of an orthogonality condition with modified weighting matrices, ii) it does not restrict the control law to linear controllers, iii) it also allows for unstructured and other classes of uncertainties (see [3]), and iv) it handles both regular and degenerate situations. We are currently studying these connections more closely, as well as the TPBVP.

In this paper we illustrated one application of the BDU formulation in the context of state regulation. Other applications are possible [1].

Acknowledgment. The authors would like to thank Prof. Jeff Shamma of the Mechanical and Aerospace Engineering Department, UCLA, for his careful comments and feedback on the topic of this article.

References

1. A. H. Sayed, V. H. Nascimento, and S. Chandrasekaran. Estimation and control with bounded data uncertainties. *Linear Algebra and Its Applications*, vol. 284, pp. 259–306, Nov. 1998.

2. A. H. Sayed and S. Chandrasekaran. Estimation in the presence of multiple sources of uncertainties with applications. *Proc. Asilomar Conference*, vol. 2, pp. 1811-1815, Pacific Grove, CA, Nov. 1998.

3. V. H. Nascimento and A. H. Sayed. Optimal state regulation for uncertain state-space models. In *Proc. ACC*, San Diego, CA, June 1999.

4. S. Chandrasekaran, G. Golub, M. Gu, and A. H. Sayed. Parameter estimation in the presence of bounded data uncertainties. *SIAM J. Matrix Analysis and Applications*, 19(1):235–252, Jan. 1998.

5. L. E. Ghaoui and H. Hebret. Robust solutions to least-squares problems with uncertain data, *SIAM J. Matrix Anal. Appl.*, vol. 18, pp. 1035–1064, 1997.

6. M. Fu, C. E. de Souza, and L. Xie. $\mathcal{H}_\infty$ estimation for uncertain systems. *Int. J. Robust and Nonlinear Contr.*, vol. 2, pp. 87–105, 1992.

7. R. Fletcher. *Practical Methods of Optimization*. Wiley, 1987.

8. M. Green and D. J. N. Limebeer. *Linear Robust Control*. Prentice-Hall, Englewood Cliffs, NJ, 1995.

9. K. Zhou, J. C. Doyle, and K. Glover. *Robust and Optimal Control*. Prentice-Hall, NJ, 1996.

10. P. P. Khargonekar, I. R. Petersen, and K. Zhou. Robust stabilization of uncertain linear systems: Quadratic stabilizability and $\mathcal{H}_\infty$ control theory. *IEEE Transactions on Automatic Control*, vol. 35, no. 3, 1990.

11. B. Hassibi, A. H. Sayed, and T. Kailath. *Indefinite Quadratic Estimation and Control: A Unified Approach to $\mathcal{H}_2$ and $\mathcal{H}_\infty$ Theories*. SIAM, PA, 1999.

12. S. O. R. Moheimani, A. V. Savkin, and I. R. Petersen. Minimax optimal control of discrete-time uncertain systems with structured uncertainty. *Dynamics and Control*, vol. 7, no. 1, pp. 5–24, Jan. 1997.

13. L. Xie and Y. C. Soh. Guaranteed-cost control of uncertain discrete-time systems. *Control Theory and Advanced Technology*, vo. 10, no. 4, pp. 1235–1251, June 1995.

14. A. V. Savkin and I. R. Petersen. Optimal guaranteed-cost control of discrete-time nonlinear uncertain systems. *IMA Journal of Mathematical Control and Information*, vol. 14, no. 4, pp. 319–332, Dec. 1997.

15. D. G. Luenberger. *Optimization by Vector Space Methods*. Wiley, 1969.

16. T. Basar and G. J. Olsder. *Dynamic Noncooperative Game Theory*. Academic Press, 1982.

Robustness and Performance in Adaptive Filtering

Paolo Bolzern[1], Patrizio Colaneri[1], and Giuseppe De Nicolao[2]

[1] Politecnico di Milano, Dipartimento di Elettronica e Informazione
  Piazza Leonardo da Vinci 32, 20133 Milano, Italy
[2] Università di Pavia, Dipartimento di Informatica e Sistemistica
  Via Ferrata 1, 27100 Pavia, Italy

Abstract. Adaptive filtering of a scalar signal corrupted by noise is considered. In particular, the signal to be estimated is modeled as a linear regression depending on a drifting parameter. The mean-square and worst-case performances of the Normalized Least Mean Squares, Kalman, and central H_∞-filters are studied. The analysis shows that a compromise between performance and robustness should be pursued, for instance by applying the central H_∞-filter with the design parameter γ used as a tuning knob.

1 Introduction

Linear adaptive filtering is concerned with the on-line estimation of time-varying parameters in a linear regression model from noisy measurements. The most popular algorithms are Least Mean Squares (LMS), Recursive Least Squares (RLS) with forgetting, and the Kalman filter. The traditional assessment criterion is in terms of mean-square (H_2) performance, namely the filtering error variance assuming white noise disturbances (defined in a broad sense including the measurement noise as well as the parameter variations) [16], [10], [19].

However, a precise knowledge of the spectral characterization of the disturbances is seldom available. This motivates the interest for studying the worst-case performance of adaptive filters in the face of arbitrary disturbances. In particular, the H_∞-approach considers as a robustness index the maximum energy gain (attenuation level) from disturbances to the estimation error with respect to the set of admissible disturbances. In the general context of state-space estimation, the H_∞-theory is by now well established [17], [22], [14], [23]. By regarding adaptive filtering as a special state-space estimation problem, such an H_∞-machinery can be used to design robust H_∞-based algorithms [8], [9], [13], [18], [20], [21], [3]. For a fixed regressor sequence, necessary and sufficient conditions are available to verify whether or not a given filter achieves a certain attenuation level. However, a specific feature of real-time adaptive filtering is that the regressors are not known in advance, so that the problem arises of guaranteeing a certain attenuation level for all possible regressor sequences belonging to suitable classes (e.g. persistently exciting ones).

The purpose of the present paper is to analyze the mean-square performance and the robustness properties of a class of adaptive filtering algorithms in the face of both process- and measurement-disturbances. In particular, we will study the NLMS (Normalized Least Mean Squares), Kalman, and central H_∞-filters when the signal to be estimated is just the uncorrupted measured output. Most of the results are taken from [4], [5], [6].

After some preliminaries (Section 2), the NLMS filter is analyzed in Section 3. When the parameter vector is constant, it has been proven in [13] that NLMS is H_∞-optimal in the sense that it coincides with the (central) H_∞-filter that guarantees the minimum achievable attentuation level $\gamma = 1$. However, an objection to the meaningfulness of the associated minimization problem is that also the "simplistic estimator" (which constructs an estimate coincident with the last observed output ignoring all past information) achieves $\gamma = 1$. The answer given in [13] is that there are infinitely many estimators achieving $\gamma = 1$, and among them the simplistic estimator has the worst H_2-performance, whereas the (central) H_∞-filter is risk-sensitive optimal and also maximum-entropy. A second and more substantial objection is that in an adaptive context, a measure of robustness that does not account for the effect of parameter variations is insufficient. In this respect, we show that NLMS cannot guarantee a finite attenuation level for all possible regressor sequences.

As for the Kalman filter (Section 4), it is shown that an upper bound on the H_2-performance can be worked out in terms of an upper bound on the solution of the relevant Riccati difference equation. Assuming that the measurement noise has unit variance, it turns out that the H_2-performance is always less than 1. Moreover, in particular circumstances (corresponding to slow parameter drift) it can be arbitrarily close to zero. Concerning the H_∞-performance, Hassibi [12] has proven that the Kalman filter always guarantees an attenuation level not greater than 2. An attenuation level smaller than 2 can be obtained by suitably "detuning" the Kalman filter, i.e. by artificially increasing the variance of the process noise used in the design.

A natural way to attain robustness is to resort to the central H_∞-filter (Section 5). An upper bound for its mean-square performance can be found similarly to the Kalman filter. As for robustness, the central H_∞-filter can guarantee any desired attenuation level $\gamma > 1$. At the same time, however it is shown that, as γ tends to 1, the filter approaches the "simplistic estimator". This is a clear warning against the use of H_∞-performance as unique design criterion. We show that the filter based on a certain design parameter γ actually ensures an easy-to-compute attenuation level less than γ.

As a consequence of this result, it is possible to "recalibrate" the design of H_∞-filters in the sense that, for a desired attenuation level $\bar\gamma$, one can use a design parameter $\gamma > \bar\gamma$.

The analysis carried out in the paper points out that the H_∞-performance cannot be adopted as the unique design criterion. Therefore, as discussed in Section 6, it is necessary to look for a balance between H_2 and H_∞ performance, which is a nontrivial task given the time-varying and unpredictable nature of the regressors involved in the on-line estimation process. In particular, it seems that the central H_∞-filter, with the design parameter γ used as a tuning knob, can yield a satisfactory tradeoff. Alternatively, one can use a post-optimization procedure based on a scalar parameter optimization.

The paper ends with Section 7, where a simple numerical example is presented.

2   Preliminaries

Consider a scalar sequence $y(t)$ given by

$$y(t) = \varphi(t)'\vartheta(t) + v(t) \tag{1}$$
$$\vartheta(t+1) = \vartheta(t) + w(t) \quad , \quad t = 0, 1, \ldots \tag{2}$$

where $\vartheta(t) \in \Re^n$ is an unknown parameter vector, $\varphi(t) \in \Re^n$ is the regressor vector, and $v(t)$, $w(t)$ are unknown disturbances which represent the measurement noise and the parameter drift, respectively.

The adaptive filtering problem consists of estimating the uncorrupted output

$$z(t) = \varphi(t)'\vartheta(t) \tag{3}$$

using the measurements $y(i)$, and the regressors $\varphi(i), 0 \le i \le t$. In the sequel, the regressor sequence $\{\varphi(\cdot)\}$ is said to be *exciting* if

$$\lim_{T\to\infty} \sum_{i=0}^{T-1} \varphi(i)'\varphi(i) = \infty \tag{4}$$

ϵ-*exciting* if

$$\lim_{T\to\infty} \sum_{i=0}^{T-1} i^{-\epsilon}\varphi(i)'\varphi(i) = \infty \tag{5}$$

for some positive ϵ, *persistently exciting* if there exists $\alpha > 0$ such that

$$\lim_{T\to\infty} \frac{1}{T} \sum_{i=0}^{T-1} \varphi(i)\varphi(i)' > \alpha I \tag{6}$$

and *uniformly persistently exciting* (UPE) if there exist $\alpha > 0$, $\beta > 0$ and an integer $l > 0$ such that, for all integers $k \geq 0$

$$\alpha I < \sum_{i=k}^{k+l-1} \varphi(i)\varphi(i)' < \beta I \tag{7}$$

While (4), (6) and (7) are fairly standard excitation conditions, (5) is a slight reinforcement of (4) and amounts to requiring that the norm of the regressor vector $\varphi(i)$ does not tend to zero as fast as or faster than $1/i$. As for the relationships between these conditions, observe that $(7) \to (6) \to (4)$ and $(7) \to (5) \to (4)$.

A general recursive estimator with observer structure is given by

$$\hat{\vartheta}(t) = \hat{\vartheta}(t-1) + K(t)\left(y(t) - \varphi(t)'\hat{\vartheta}(t-1)\right) \quad , \quad \hat{\vartheta}(-1) = \hat{\vartheta}_0 \tag{8}$$

$$\hat{z}(t) = \varphi(t)'\hat{\vartheta}(t) \tag{9}$$

where $\hat{\vartheta}_0$ is an a-priori guess of the initial parameter vector and $K(t)$ is the gain of the adaptive filter to be properly designed.

Letting $x(t) = \vartheta(t) - \hat{\vartheta}(t-1)$, $e(t) = z(t) - \hat{z}(t)$ and $d(t) = [w(t)'\ v(t)']'$, from (1)-(3) and (8)-(9) it follows that

$$x(t+1) = A(t)x(t) + \begin{bmatrix} I & -K(t) \end{bmatrix} d(t) \tag{10}$$

$$e(t) = \varphi(t)'A(t)x(t) + \begin{bmatrix} 0 & -\varphi(t)'K(t) \end{bmatrix} d(t) \tag{11}$$

where $A(t) = I - K(t)\varphi(t)'$. System (10), (11) describes the relationship between the disturbance d and the estimation error e for a filter with gain $K(\cdot)$.

Two alternative design strategies are those based on the mean-square (H_2) and the worst-case (H_∞) criteria. The former aims at minimizing the H_2-norm of the (weighted) transference from the disturbances w, v, and $\vartheta(0)$ to the filtering error. More precisely, the problem is to find a filter gain $K(\cdot)$ such that the performance index J_2 is minimized, where

$$J_2 = \frac{1}{T} \sum_{t=0}^{T-1} \varphi(t)'[A(t)P_2(t)A(t)' + K(t)K(t)']\varphi(t) \tag{12}$$

and $P_2(\cdot)$ is the solution of

$$P_2(t+1) = A(t)P_2(t)A(t)' + \bar{q}I + K(t)K(t)', \quad P_2(0) = P_0 \tag{13}$$

In a stochastic framework, the index J_2 can be interpreted as the average variance of the filtering error when the disturbances $w(\cdot)$ and $v(\cdot)$ are independent white noises with variances $\bar{q}I$ and 1, respectively, and the initial state $\vartheta(0)$ is a random variable (independent of $w(\cdot)$ and $v(\cdot)$) with mean $\hat{\vartheta}_0$ and covariance P_0. In a deterministic setting, P_0 and $\bar{q}$ are to be regarded as weights used to balance the relative contributions of the disturbances. Hereafter, we assume $P_0 > 0$ and $\bar{q} > 0$.

As for the worst-case design strategy, it is based on the performance index

$$J_\infty(\bar{\gamma}) = \sum_{t=0}^{T-1} e(t)^2 - \bar{\gamma}^2 \left(\sum_{t=0}^{T-1} \frac{w(t)'w(t)}{\bar{q}} \right.$$

$$\left. + \sum_{t=0}^{T-1} v(t)^2 + \left(\vartheta(0) - \hat{\vartheta}_0 \right)' P_0^{-1} \left(\vartheta(0) - \hat{\vartheta}_0 \right) \right) \tag{14}$$

where $\bar{q} > 0$ and $P_0 > 0$ are suitable weights. More precisely, a filter guaranteeing $J_\infty(\bar{\gamma}) \leq 0$ for any $\vartheta(0)$ and any square summable sequences w and v is said to achieve a level of attenuation $\bar{\gamma}$. The fulfillment of such an inequality entails that the ratio of the l_2-norm of the filtering error to a weighted norm of the uncertain variables is not greater than $\bar{\gamma}$. Such a ratio is the *root-mean-square gain* from the disturbances to the filtering error.

It is an easy consequence of [22] to demonstrate that the generic filter (8)-(9) guarantees $J_\infty(\bar{\gamma}) \leq 0$ iff there exists a positive definite solution $\Pi(\cdot)$ of

$$\Pi(t+1) = A(t)\Pi(t)A(t)' + \bar{q}I$$

$$+ K(t)K(t)' + \frac{1}{\delta(t)} S(t)\varphi(t)\varphi(t)'S(t)' \tag{15}$$

$$\Pi(0) = P_0$$
$$S(t) = A(t)\Pi(t)A(t)' + K(t)K(t)'$$
$$\delta(t) = \bar{\gamma}^2 - \varphi(t)'S(t)\varphi(t)$$

such that the following *feasibility condition* holds:

$$\delta(t) > 0, \quad \forall t \in [0, T-1] \tag{16}$$

In this paper, the attention will be restricted to gains of the type

$$K(t) = \frac{P(t)\varphi(t)}{1 + \varphi(t)'P(t)\varphi(t)} \tag{17}$$

where $P(t)$ is a matrix function to be properly designed. In particular, we will focus on the following three cases:

Normalized least mean squares (NLMS) :

$$P(t) = \mu I \quad , \quad \mu > 0 \quad , \quad \forall t \geq 0 \tag{18}$$

Kalman filter :
$P(\cdot)$ is the solution of

$$P(t+1) = [P(t)^{-1} + \varphi(t)\varphi(t)']^{-1} + qI, \quad P(0) = P_0 \tag{19}$$

Central H_∞-filter :
$P(\cdot)$ is the solution of

$$P(t+1) = [P(t)^{-1} + \varphi(t)\varphi(t)'(1 - \frac{1}{\gamma^2})]^{-1} + qI, \quad P(0) = P_0 \qquad (20)$$

Note that in (19) and (20), γ and q must be regarded as design parameters which do not necessarily coincide with $\bar{\gamma}$ and $\bar{q}$ in (13) and (14). In the sequel, when $\gamma = \bar{\gamma}$ and $q = \bar{q}$, the central H_∞-filter and the Kalman filter will be said to be *tuned*. Finally observe that the Kalman filter can be seen as the limit of the central H_∞-filter when $\gamma \to \infty$.

3   Normalized LMS

From (17), (18), the gain of the normalized LMS filter is

$$K(t) = \frac{\mu\varphi(t)}{1 + \mu\varphi(t)'\varphi(t)} \qquad (21)$$

Under the assumptions that $\bar{q} = 0$ (no parameter drift), $P_0 = \mu I$, and the regressors $\varphi(\cdot)$ are exciting, Hassibi, Sayed and Kailath [13] demonstrated that such a filter is H_∞-optimal, namely it guarantees the minimum achievable attenuation level $\bar{\gamma} = 1$ over an infinite horizon ($T = \infty$). As a matter of fact, under these assumptions, the NLMS filter coincides with the tuned central H_∞-filter, so that it inherits the corresponding robustness properties.

A first objection to the importance of this H_∞-optimality property is already discussed in [13]. In fact, it is easy to see that the "simplistic estimator"

$$\hat{z}(t) = y(t) \qquad (22)$$

is capable of guaranteeing an attenuation level $\bar{\gamma} = 1$ as well. However, in [13] it is argued that, besides H_∞-robustness, the designer should be concerned with other desirable properties. In this respect the estimator (22) has the worst H_2-performance ($J_2 = 1$) among all the estimators yielding $J_\infty(1) \leq 0$, whereas the NLMS filter is just the central H_∞-filter which enjoys the properties of being risk-sensitive optimal and also maximum entropy.

A second criticism regards the assumption, implicit in the above analysis, that ϑ is a constant parameter vector. Indeed, if ϑ is constant it is questionable to use NLMS, which, though well suited to track time-varying parameters, does not yield consistent estimates in the constant parameter case. On the other hand, if ϑ is time-varying, the robustness analysis should explicitly account for the effects of the parameter drift. In principle, good robustness performance in the face of the measurement error signal $v(\cdot)$, may coexist with poor robustness against parameter variations. This motivates the analysis of the H_∞-performance of NLMS when $\bar{q} \neq 0$. To this purpose, note that, if $K(\cdot)$ is as in (21), then the feasibility condition (16) becomes

$$\varphi(t)'\Pi(t)\varphi(t) < \bar{\gamma}^2(1 + \mu\varphi(t)'\varphi(t))^2 - (\mu\varphi(t)'\varphi(t))^2 \tag{23}$$

for $t \in [0, T-1]$. Roughly speaking, this condition shows that the solution $\Pi(t)$ of (15) is not allowed to become "too large" along the regressor direction. As stated in the following theorem, persistent excitation may not suffice to ensure the fulfillment of (23).

Theorem 1. *[3],[4] For any given $\bar{\gamma} > 0$ and $\mu > 0$, there exist a sufficiently large T and a persistently exciting sequence $\{\varphi(\cdot)\}$ such that NLMS does not guarantee $J_\infty(\bar{\gamma}) \leq 0$.*

This result highlights the fact that, when the parameter drift is taken into account, NLMS does not ensure a finite level of attenuation $\bar{\gamma}$ for all possible regressor sequences (even assuming persistent excitation). This is a main difference with respect to the no-drift case studied in [13]. An open issue is whether a more stringent assumption on the regressors (such as uniform persistent excitation) can guarantee a finite level of attenuation for NLMS.

4   Kalman filter

First, we consider the mean–square performance of the Kalman filter, measured by the index J_2 defined in (12). It is well known that the tuned Kalman filter minimizes J_2. Obviously, the optimal value of J_2 depends on the particular regressor sequence. In the following, we derive upper bounds for J_2 under suitable assumptions on regressor excitation.

In order to derive the main results, we first investigate on some technical properties concerning the general Riccati recursion/indexRiccati equation

$$X(t+1) = \left(X(t)^{-1} + r^{-1}\varphi(t)\varphi(t)\right)^{-1} + qI \quad , \quad X(0) = X_0 \tag{24}$$

where q and r are positive scalars and $X_0 > 0$. The following lemma provides some upper bounds for the solution $X(\cdot)$ of (24), which improve on previous results worked out in [11] and [4].

Lemma 5. *[5] Consider the Riccati recursion (24) and assume that $\{\varphi(\cdot)\}$ is UPE with parameters α, β, l as defined in (7). Then, the solution $X(\cdot)$ satisfies the inequalities*

$$X(t) \leq X_0 + \frac{r + k\beta q(kl - 1) + k^2\alpha ql}{k\alpha}I \quad , \quad \forall t \geq 0 \tag{25}$$

$$X(t) \leq \frac{r + k\beta q(kl - 1) + k^2\alpha ql}{k\alpha}I \quad , \quad \forall t \geq kl \tag{26}$$

where $k \geq 1$ is an arbitrary integer.

By letting $r = 1$, the above bounds apply to the Riccati equation (19). Then, by exploiting the fact that for the tuned Kalman filter $P_2(t) = P(t)$, an upper bound for the mean-square performance J_2 can be obtained

Theorem 2. *[5] Assume that ηI is an upper bound for the solution $P(\cdot)$ of (19) on the interval $[0, T - 1]$, namely $P(t) \leq \eta I, \forall t \in [0, T - 1]$. Moreover, suppose that $\{\varphi(\cdot)\}$ is UPE. Then, the mean-square performance of the tuned Kalman filter satisfies the inequality*

$$J_2 \leq \left(1 + \frac{1}{\eta\beta}\right)^{-1}$$

It is interesting to assess the range of possible mean-square performance of the Kalman filter. Obviously, $J_2 \leq 1$ (even without any UPE assumption) from optimality arguments, recalling that the simplistic estimator (22) yields $J_2 = 1$. On the other hand, the next theorem shows that in particular circumstances (corresponding to slow parameter drift) the performance index J_2 can be arbitrarily close to zero, see also [10].

Theorem 3. *[5] Let J_2 be the mean-square performance of the tuned Kalman filter over $[0, \infty)$, and assume that $\{\varphi(\cdot)\}$ is UPE. Then, $J_2 \to 0$ as $q \to 0$.*

Turning now to the H_∞-performance, the robustness of the Kalman filter can be evaluated in terms of the guaranteed level of attenuation from the disturbances to the estimation error. In this respect, Hassibi [12] has recently shown that the Kalman filter always guarantees a level of attenuation $\bar{\gamma} \leq 2$.

Theorem 4. *[12] If the design parameter q is such that $q \geq \bar{q}$, then the Kalman filter guarantees $J_\infty(2) \leq 0$ over an arbitrarily long horizon T, irrespective of the regressor sequence.*

This result points out that the tuned Kalman filter, though designed according to a mean-square criterion, enjoys some (perhaps unexpected) guaranteed robustness properties. However, if the prescribed level of attenuation is less than 2, one may try to properly "detune" the filter through a suitable choice of the parameter q. As a matter of fact, the scalar q is a design knob which is commonly tuned to trade adaptation promptness against insensitivity to measurement noise. It is customary to increase q, and hence the filter gain, to improve "robustness" against parameter variations. As a matter of fact the following result shows that, under proper assumptions on the regressors, there always exists a suitable q ensuring a prescribed attenuation level $\bar{\gamma}$, even strictly less than 2.

Theorem 5. *[4] Assume that the sequence $\{\varphi(\cdot)\}$ is UPE and $\|\varphi(t)\|^2 > \epsilon > 0, \forall t$. Let $\bar{p}$ be the largest eigenvalue of P_0. Then, for any prescribed attenuation level $\bar{\gamma} > 1$, the Kalman filter with*

$$q \geq \bar{q} + max\{\frac{\beta\bar{p}^2}{\bar{\gamma}^2}, \frac{\beta[(\alpha + \beta)l + \frac{(\alpha\bar{p}+1)\bar{\gamma}^2}{\beta\bar{p}^2}]^2}{(\bar{\gamma}^2 - 1)\alpha^2\epsilon^2}\}$$

ensures $J_\infty(\bar\gamma) \leq 0$.

Summarizing, the design parameter q controls the robustness of the Kalman filter or, from a different point of view, its promptness to track time-varying parameters. The limit case $q = 0$ corresponds to the classical RLS (Recursive Least Squares) algorithm, which does not have tracking capabilities unless a proper forgetting factor is included. Conversely, as $q \to \infty$, the filter tends to the simplistic estimator (22), which overtrusts the observations, regardless of the noise level.

Note also that the increase of $P(t)$ induced by a large value of q may give rise to the so-called "blow-up" phenomenon, see [1], which occurs when, after a long quiescence interval with poorly informative regressors, there are sudden jumps in the gain producing undesired oscillations in the estimates.

5   Central H_∞-filter

The mean-square performance of the central H_∞-filter can be studied similarly to what done for the Kalman filter. Indeed, letting $r = \left(1 - \gamma^{-2}\right)^{-1}$, Lemma 5 can be used to derive an upper bound for the solution of the Riccati equation (20). Then, a performance bound is given in the following result, which is similar to Theorem 2.

Theorem 6. *[5] Assume that ηI is an upper bound for the solution $P(\cdot)$ of (20) on the interval $[0, T - 1]$, namely $P(t) \leq \eta I, \forall t \in [0, T - 1]$. Moreover, suppose that $\{\varphi(\cdot)\}$ is UPE. Then, the mean-square performance of the tuned central H_∞-filter with $\gamma > 1$ satisfies the inequality*

$$J_2 \leq \left(1 + \frac{1}{\eta\beta}\right)^{-1} \tag{27}$$

As a corollary, it is clear that $J_2 \leq 1$ for any central H_∞-filter with $\gamma > 1$. On the other hand, it will be shown later in this section that the central H_∞-filter with $\gamma = 1$ approaches the simplistic estimator (22), whose H_2-performance is $J_2 = 1$. Then, in view of the above theorem, for any (finite or infinite) time horizon, the mean-square performance of the central H_∞-filter ranges from $J_2 = 1$ (achieved for $\gamma = 1$, simplistic estimator) to the H_2-performance of the Kalman filter (corresponding to $\gamma = \infty$), which is the minimum achievable one.

Note that, when the bound η is computed according to (25) with $r = \left(1 - \gamma^{-2}\right)^{-1}$, then the upper bound (27) is monotonically decreasing with γ. It would be tempting to argue that the actual mean-square performance (and not only its bound) of the central H_∞-filter is monotonic as well. As a matter of fact, such a property can be proven in the scalar parameter case ($n = 1$) or in the case of constant regressor vector, but its general validity is still an open issue.

We now consider the analysis of the H_∞-performance. As expected, easy but cumbersome computations show that if $\gamma = \bar{\gamma}$ and $q = \bar{q}$, the solution $\Pi(\cdot)$ of (15) coincides with the solution $P(\cdot)$ of (20). Moreover, the feasibility condition (16) reduces to

$$P(t)^{-1} + \varphi(t)\varphi(t)'(1 - \frac{1}{\gamma^2}) > 0, \quad \forall t \in [0, T-1] \tag{28}$$

Since the above condition is trivially satisfied for $\bar{\gamma} \geq 1$, the following result is straightforward.

Theorem 7. *For any $\bar{\gamma} \geq 1$, the tuned H_∞-filter guarantees $J_\infty(\bar{\gamma}) \leq 0$ over an arbitrarily long horizon T, irrespective of the regressor sequence $\{\varphi(\cdot)\}$.*

In other words, an H_∞-filter with a certain design parameter $\gamma \geq 1$ ensures an attenuation level $\bar{\gamma} = \gamma$. As a matter of fact, the next theorem shows that using $\gamma > 1$ actually guarantees an easy-to-compute attenuation level $\bar{\gamma}$ strictly less than γ.

Theorem 8. *[5], [6] For any time horizon $[0, T-1]$, the central H_∞-filter with $\gamma \geq 1$, guarantees an attenuation level less than*

$$\bar{\gamma} = \frac{\underline{r} + \sqrt{\underline{r}(1 - \gamma^{-2}) + \gamma^{-2}}}{\underline{r} + \gamma^{-2}} \leq \frac{2}{1 + \gamma^{-2}} \tag{29}$$

where $\underline{r} = \inf_t (1 + \varphi(t)'P(t)\varphi(t))$, and $P(t)$ is the solution of the Riccati equation (20).

Note that, since $1 + \gamma^{-2} \geq 2\gamma^{-1}$, then $\bar{\gamma} \leq \gamma$, with strict inequality for $\gamma > 1$. Consequently, the attenuation level guaranteed by a central H_∞-filter is always less than the design parameter γ. Note also that, as γ tends to infinity, the H_∞-filter tends to the Kalman one, and Theorem 4 is recovered. Although the bound (29) is monotonically increasing as a function of γ, this does not imply that the actual H_∞-norm is monotonic as well. For the time being, such a property can only be conjectured.

For what concerns the asymptotic behaviour as T tends to infinity, when the regressor sequence is UPE it is possible to show that the central H_∞-filter with $\gamma > 1$ is exponentially stable. As a consequence the parameter estimation error $\vartheta(t) - \hat{\vartheta}(t-1)$ exponentially converges to zero when the disturbances $w(\cdot)$ and $v(\cdot)$ are either zero or square summable. Moreover, when the disturbances are stationary stochastic processes, the parameter estimation error has bounded variance.

Turning now to the case $\bar{\gamma} < 1$, note that the fulfillment of condition (28) depends on the sequence of regressors $\{\varphi(\cdot)\}$. Moreover, from (20) it is apparent that $P(\cdot)$ is monotonically increasing with time, so that condition (28) is eventually violated unless the regressors are poorly exciting. In this respect, an interesting result is stated in the following theorem.

Theorem 9. *[4] For any* $\bar{\gamma} < 1$ *and any* ϵ*-exciting sequence* $\{\varphi(\cdot)\}$*, there exists a sufficiently long interval* $[0, T-1]$ *such that the tuned* H_∞*-filter does not guarantee* $J_\infty(\bar{\gamma}) \leq 0$.

In view of Theorems 7 and 9, it can be argued that, under suitable assumptions on regressor excitation, $\bar{\gamma} = 1$ is the minimum achievable attenuation level when considering an indefinitely long horizon. In particular, when $\bar{\gamma} = 1$, it results $P(t) = P_0 + \bar{q}tI$, $\forall t \geq 0$, and the tuned H_∞-filter yields the estimate

$$\hat{z}(t) = \frac{\varphi(t)'}{1 + \varphi(t)'P_0\varphi(t) + \bar{q}t\varphi(t)'\varphi(t)}\hat{\vartheta}(t-1)$$

$$+ \frac{\varphi(t)'\,P_0\,\varphi(t) + \bar{q}t\varphi(t)'\varphi(t)}{1 + \varphi(t)'P_0\varphi(t) + \bar{q}t\varphi(t)'\varphi(t)}y(t)$$

Under the assumption that $\varphi(t) \neq 0, \forall t$ and $\lim_{t\to\infty} t\varphi(t)'\varphi(t) = \infty$, it follows that

$$\lim_{t\to\infty} (\hat{z}(t) - y(t)) = 0$$

so that $\hat{z}(t)$ asymptotically approaches the simplistic estimator (22). This observation highlights the inadequacy of robustness against disturbances as the sole criterion in the design of adaptive filters. It seems more sensible to pursue a design approach which jointly takes into account the worst-case and mean-square performance. This idea is further discussed in the next section.

6   On mixed H_2/H_∞ design

When facing an adaptive filtering problem, the designer can follow three main approaches: H_2-design, H_∞-design, mixed H_2/H_∞-design. If the minimization of the mean-square performance is pursued, the solution is provided by the Kalman filter. As pointed out in a previous section, such a design, which completely disregards H_∞-robustness, nevertheless leads to an estimator guaranteeing an attenuation level less than 2 (Theorem 4).

In the opposite, when the accent is put on H_∞-robustness alone, striving for the minimum achievable attenuation level ($\bar{\gamma} = 1$) leads to a trivial result, since the central H_∞-filter with $\gamma = 1$ converges to the simplistic estimator (22), which is the worst in the H_2-sense ($J_2 = 1$).

Consequently, the third approach, namely the mixed H_2/H_∞-design, appears to be the unique viable alternative to the Kalman filter, if an attenuation level less than 2 is required. Usually, the mixed H_2/H_∞-design aims at minimizing the H_2-performance (or an upper bound for the H_2-performance) subject to an H_∞-constraint [2]. However, most results available in the literature cannot be applied in the present context, because the regressor sequence $\{\varphi(\cdot)\}$ is time-varying and the synthesis of the estimator cannot rely on future values of $\varphi(\cdot)$, due to the real-time constraint.

A first practical way to address the mixed design consists of properly detuning the Kalman filter in order to achieve a satisfactory compromise (see Theorem 5).

Another reasonable suboptimal solution is the use of the central H_∞-filter with the design parameter γ acting as a tuning knob. In general, an increase of γ is expected to improve the mean-square performance at the cost of decreased robustness. Hence, one could use the largest value of the design parameter γ compatible with the prescribed attenuation level $\bar{\gamma}$, $1 < \bar{\gamma} < 2$. Obviously, the tuned H_∞-filter ($\gamma = \bar{\gamma}$) complies with the attenuation constraint, but Theorem 8 suggests a more clever choice, that is

$$\gamma = \sqrt{\frac{\bar{\gamma}}{2 - \bar{\gamma}}} > \bar{\gamma}$$

Such a recalibration is likely to improve on the H_2-performance of the tuned filter, without violating the robustness constraint.

Remark 3. Observe that the central H_∞-filter contains two tuning knobs, namely q and γ. As for their effects, either increasing q or decreasing γ, enhances the promptness of the estimator. Indeed, both actions tends to inflate the matrix $P(t)$, and hence the gain $K(t)$. A remarkable difference between these two strategies is that, while the parameter q produces a uniform inflation of the diagonal entries of $P(t)$, changes of γ influence the "information matrix" $P(t)^{-1}$ only in the subspace spanned by the regressor vector $\varphi(t)$. This directional feature bears some analogy with RLS algorithms equipped with directional forgetting [15], and helps alleviating the blow-up phenomenon. Indeed γ does not contribute to the growth of $P(t)$ in the directions that are not excited by the regressors during a quiescence period. As a consequence, the tuning of γ instead of q could be suggested as a way to improve robustness and promptness without exacerbating the blow-up problem.

An alternative way to improve the mean-square performance of a central H_∞-filter relies on the extension of the post-optimization procedure proposed in [7] for the mixed H_2/H_∞ control problem. It consists of correcting the filter gain according to the following equation

$$K(t) = (1 - \sigma)K_c(t) + \sigma \frac{P_2(t)\varphi(t)}{1 + \varphi(t)'P_2(t)\varphi(t)}$$

where σ is a scalar parameter, $K_c(t)$ is the gain of the tuned central H_∞-filter, and $P_2(t)$ is the solution of (13), still depending on $K(t)$. Note that $P_2(t)$ can be recursively computed from

$$P_2(t + 1) = A_c(t)P_2(t)A_c(t)' + \bar{q}I + K_c(t)K_c(t)'$$

$$-\frac{\left(2\sigma - \sigma^2\right)\left(A_c(t)P_2(t)\varphi(t) - K_c(t)\right)\left(A_c(t)P_2(t)\varphi(t) - K_c(t)\right)'}{1 + \varphi(t)'P_2(t)\varphi(t)}$$

$$A_c(t) = I - K_c(t)\varphi(t)'$$

As apparent, if $\sigma = 0$, $K(t)$ coincides with the gain of the central H_∞-filter, while $\sigma = 1$ corresponds to the Kalman filter. Moreover, it can be shown that the H_2-performance is a monotonically nonincreasing function of $\sigma \in [0, 1]$. Then the ideal strategy would be to select the maximum value of $\sigma \in [0, 1]$ such that feasibility is preserved. Note however, that it is not possible to check such a feasibility condition without knowing the regressor sequence in advance.

7 An illustrative example

In order to compare the H_2 and H_∞ performances of different filtering algorithms, consider the signal model

$$y(t) = \varphi_1(t)\vartheta_1(t) + \varphi_2(t)\vartheta_2(t) + v(t), \quad t = 0, 1, \cdots, T-1$$

where $\vartheta(t) = \left[\vartheta_1(t)\ \vartheta_2(t)\right]'$ satisfies (2) and $\varphi_1(t)$, $\varphi_2(t)$, $t = 0, 1, \cdots, T-1$, are independent random variables taking the values 0 and 1 with probability 0.7 and 0.3, respectively. A sample of length $T = 100$ was generated, and the weights $P_0 = 10^{-6}I$, $\bar{q} = 10^{-2}$ were selected.

For a given filter characterized by the gain $K(\cdot)$, eqs. (10), (11) can be used to derive the matrix H linking the (weighted) disturbances to the estimation error, i.e.

$$\begin{bmatrix} e(0) \\ e(1) \\ e(2) \\ \vdots \\ e(T-1) \end{bmatrix} = H \begin{bmatrix} P_0^{-1/2}x(0) \\ \tilde{d}(0) \\ \tilde{d}(1) \\ \vdots \\ \tilde{d}(T-1) \end{bmatrix}$$

where $\tilde{d}(t) = \left[\bar{q}^{-1/2}w(t)'\ v(t)\right]'$. The H_2 performance is measured by

$$N_2 = \left(\frac{1}{T}trace(HH')\right)^{1/2}$$

which coincides with the square-root of J_2 defined in (12). Consistently with (14), the H_∞-performance N_∞ is measured by the maximum singular value of H, namely

$$N_\infty = (\lambda_{max}(HH'))^{1/2}$$

The associated unit-norm singular vector provides the (unit-norm) worst-case disturbance vector.

The adaptive filters discussed in the paper have been designed with different values of the tuning parameters and the corresponding indices N_∞ and N_2 have been evaluated. In particular, the central H_∞-filter with $q = \bar{q}$ and γ ranging from 1 to ∞ was considered. As for the Kalman filter, the tuning parameter q varied in the interval $[0.005, 0.2]$. Finally, the NLMS learning rate μ was selected in $[0.1, 1]$.

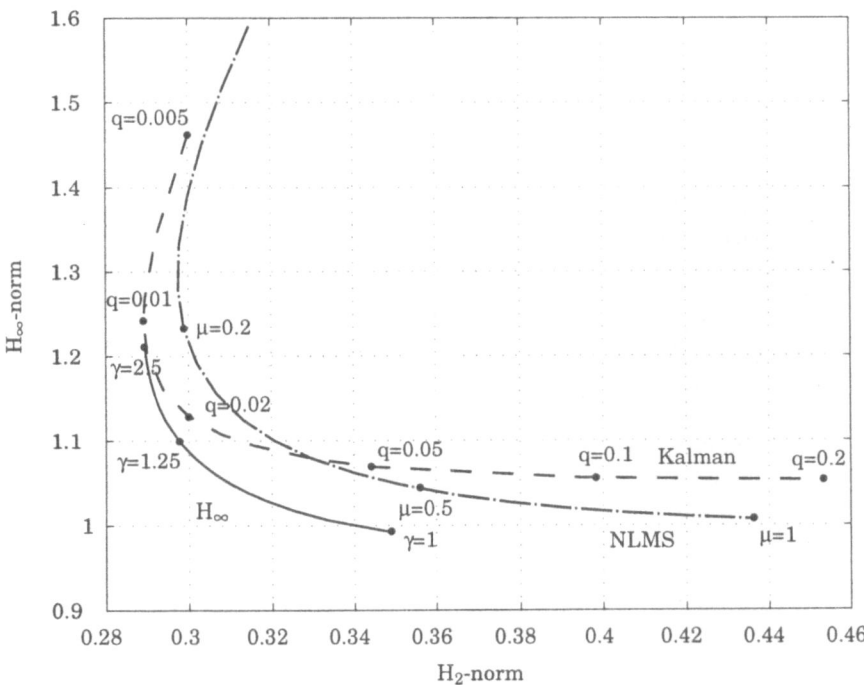

Fig. 1. H_∞ versus H_2 performance of the filters: NLMS (dash-dot), Kalman (dash), central H_∞ (solid).

The results are shown in Fig.1. It is apparent that the central H_∞-filter dominates (in a Pareto sense) the other two filters. This implies that it is superior with respect to any cost function defined as a linear combination of N_2 and N_∞. As expected, for $\gamma \to \infty$, the H_∞-filter converges to the tuned Kalman filter yielding $N_2 = 0.289$ and $N_\infty = 1.2423$. Note that the minimum achievable H_∞-norm is less than 1 due to the finiteness of the time horizon. A nice feature of the H_∞-filter is the small sensitivity of its performance with respect to the design parameter γ. Conversely, trying to robustify either the Kalman filter or the NMLS one by increasing q or μ leads to substantial deterioration of the H_2-performance. Also, using $q < \bar{q} = 0.01$ in the design of the Kalman filter impairs both the H_2 and H_∞

performances. By comparison, the performances of the simplistic estimator are $N_2 = 0.721$, $N_\infty = 1$.

Acknowledgements

The paper was partially supported by CESTIA, CIB and ICE of the Italian National Research Council (CNR) and by MURST project "Identification and control of industrial systems".

References

1. Åström, K.J. (1980) Design principles for self-tuning regulators. In *Methods and Applications in Adaptive Control*, U.Unbenhauen Ed., Springer, Berlin, 1-20.
2. Bernstein, D.S., Haddad, W.M. (1989) Steady-state Kalman filtering with an H_∞ error bound. *Systems and Control Letters*, **12**, 9-16.
3. Bolzern, P., Colaneri, P., De Nicolao, G. (1997) Robustness in adaptive filtering: how much is enough? *Proc. 36th Conf. on Decision and Control*, San Diego, USA, 4677-4679.
4. Bolzern, P., Colaneri, P., De Nicolao, G. (1999) H_∞ robustness of adaptive filters against measurement noise and parameter drift. To appear in *Automatica*.
5. Bolzern, P., Colaneri, P., De Nicolao, G. (1999) Tradeoff between mean-square and worst-case performances in adaptive filtering. Submitted to *European J. Control*.
6. Bolzern, P., Colaneri, P., De Nicolao, G. (1999) γ-recalibration of a class of H_∞-filters. *IEEE American Control Conference*, San Diego, USA.
7. Colaneri, P., Geromel, J.C., Locatelli, A. (1997) *Control Theory and Design - An RH_2 and RH_∞ viewpoint*. Academic Press, San Diego, USA.
8. Didinsky, G., Pan, Z., Basar, T. (1995) Parameter identification for uncertain plants using H_∞ methods. *Automatica*, **31**, N.9, 1227-1250.
9. Grimble, M.J., Hashim, R., Shaked, U. (1992) Identification algorithms based on H_∞ state-space filtering techniques. *Proc. 31st Conference on Decision and Control*, Tucson, USA, 2287-2292.
10. Guo, L., Ljung, L. (1995) Performance analysis of general tracking algorithms, *IEEE Trans. Automatic Control*, **AC-40**, N.8, 1388-1402.
11. Guo, L., Xia, L., Moore, J.B. (1991) Tracking randomly varying parameters - Analysis of a standard algorithm. *Mathematics of Control, Signal and Systems*, **4**, N.1, 1-16.
12. Hassibi, B. (1996) Indefinite metric spaces in estimation, control and adaptive filtering. *PhD Thesis*, Stanford University.
13. Hassibi, B., Sayed, A.H., Kailath, T. (1996) H_∞ optimality of the LMS algorithm. *IEEE Trans. Signal Processing*, **AC-44**, 267-280.
14. Hassibi, B., Sayed, A.H., Kailath, T. (1996) Linear estimation in Krein spaces - Part II: Applications. *IEEE Trans. Automatic Control*, **AC-41**, N.1, 34-49.
15. Kulhavy, R., Karny, M. (1984) Tracking of slowly varying parameters by directional forgetting. *Proc. 9th Triennal IFAC World Congress*, Budapest, Hungary, 79-83.

16. Ljung, L., Gunnarsson, S. (1990) Adaptation and tracking in system identification - a survey. *Automatica*, **26**, N.1, 7-21.

17. Nagpal, K.M., Khargonekar, P.P. (1991) Filtering and smoothing in an H_∞ setting. *IEEE Trans. Automatic Control*, **AC-36**, N.2, 152-166.

18. Rupp, M., Sayed, A.H. (1996) A time-domain feedback analysis of filtered-error adaptive gradient algorithms. *IEEE Trans. Signal Processing*, **AC-44**, N.6, 1428-1439.

19. Sayed, A.H., Kailath, T. (1994) A state-space approach to adaptive RLS filtering. *IEEE Signal Processing Magazine*, **11**, 18-60.

20. Sayed, A.H., Rupp, M. (1997) An l_2-stable feedback structure for nonlinear adaptive filtering and identification. *Automatica*, **33**, N.1, 13-30.

21. Theodor, Y., Shaked, U., Berman, N. (1996) Time-domain H_∞ identification. *IEEE Trans. Automatic Control*, **AC-41**, 1019-1024.

22. Yaesh, I., Shaked, U. (1991) H_∞-optimal estimation - the discrete-time case. *Proc. Symp. MTNS*, Kobe, Japan, 261-267.

23. Yaesh, I., Shaked, U. (1991) A transfer function approach to the problems of discrete-time systems: H_∞-optimal linear control and filtering. *IEEE Trans. Automatic Control*, **AC-36**, N.11, 1264-1271.

Nonlinear Identification Based on Unreliable Priors and Data, with Application to Robot Localization

Michel Kieffer[1], Luc Jaulin[1,2], Eric Walter[1]*, and Dominique Meizel[3]

[1] Laboratoire des Signaux et Systèmes,
   CNRS-Supélec, 91192 Gif-sur-Yvette, France
[2] On leave from Laboratoire d'Ingénierie des Systèmes Automatisés
   Université d'Angers, 49045 Angers Cedex, France
[3] HEUDIASYC, UMR CNRS 6599
   Université de Technologie de Compiègne, B.P. 20529 Compiègne Cedex, France

Abstract. This paper deals with characterizing all feasible values of a parameter vector. Feasibility is defined by a finite number of tests based on experimental data and priors. It is assumed that some of these tests may be unreliable, because of the approximate nature of the priors and of the presence of outliers in the data. The methodology presented, which can be applied to models nonlinear in their parameters, makes it possible to compute a set guaranteed to contain all values of the parameter vector that would have been obtained if all tests had been reliable, provided that an upper bound on the number of faulty tests is available. This remains true even if a majority of the tests turns out to be faulty. The methodology is applied to the localization of a robot from distance measurements.

1 Introduction and Problem Statement

Let $\mathbf{y} \in \mathbb{R}^n$ be a vector of experimental data collected on a system to be modeled. Let $\mathbf{p} \in \mathcal{P} \subset \mathbb{R}^m$ be a vector of unknown parameters of a mathematical model, to be estimated from these data so as to make the model output $\mathbf{y}_m(\mathbf{p})$ resemble the data as much as possible. In general, $n > m$, and the model is only an approximation of the system generating the data, so there is no $\widehat{\mathbf{p}} \in \mathcal{P}$ such that $\mathbf{y}_m(\widehat{\mathbf{p}}) = \mathbf{y}$. The usual statistical approach is then to estimate $\mathbf{p}$ by solving an optimization problem

$$\widehat{\mathbf{p}} = \arg\min_{\mathbf{p} \in \mathcal{P}} J(\mathbf{p}), \qquad (1)$$

---

* Corresponding author

where $J(.)$ is some cost function quantifying the difference in behavior of the actual system and its model, deduced from prior knowledge (or hypotheses) on the statistical nature of the noise corrupting the data and sometimes on some prior distribution for the parameters. If, for instance, it is assumed that the data have been generated by

$$\mathbf{y} = \mathbf{y}_m(\mathbf{p}^*) + \mathbf{b}^*, \tag{2}$$

where $\mathbf{p}^*$ is the (unknown) true value of the parameters and $\mathbf{b}^*$ is a realization of a Gaussian random vector with zero mean and known covariance Σ, then the maximum-likelihood estimate of $\mathbf{p}^*$ is obtained as

$$\widehat{\mathbf{p}} = \arg \min_{\mathbf{p} \in \mathcal{P}} \; [\mathbf{y} - \mathbf{y}_m(\mathbf{p})]^T \Sigma^{-1} [\mathbf{y} - \mathbf{y}_m(\mathbf{p})]. \tag{3}$$

Often, however, such priors on the nature of the noise corrupting the data are at best doubtful. The main contribution to the error may for instance be deterministic and due to the fact that the model is only an approximation of the actual system. Even when the errors are random, nothing may be known about their desirable statistical distribution. An attractive alternative is then *parameter bounding* (or *set-membership estimation*), in which one characterizes the set S of all values of $\mathbf{p}$ in $\mathcal{P}$ such that all components $y_m(i, \mathbf{p})$ of the model output $\mathbf{y}_m(\mathbf{p})$ satisfy

$$e_{\min}(i) \leq y_i - y_m(i, \mathbf{p}) \leq e_{\max}(i), \qquad i = 1, \cdots, n, \tag{4}$$

where the bounds $e_{\min}(i)$ and $e_{\max}(i)$ are assumed available *a priori*. A more general formulation is

$$\mathbf{y}_m(\mathbf{p}) \in \mathcal{Y}, \tag{5}$$

where $\mathcal{Y}$ is the predefined set of all acceptable model outputs, given the data $\mathbf{y}$ and priors about the various sources of error. See, *e.g.*, [1], [2], [3], [4], [5] and the references therein for an extensive bibliography on parameter bounding, and the methods available for linear and nonlinear models.

These methods make it possible to enclose S in ellipsoids, polytopes or unions of such sets, without making any other assumption on the distribution of the errors, thereby providing information on the nominal value of the parameter vector as well as on the uncertainty on this value. The price to be paid is that if the priors and data upon which the definition of $\mathcal{Y}$ is based are not reliable, then S may be unrealistically small or even empty. The choice of $\mathcal{Y}$ is therefore crucial. When priors are doubtful and/or abnormal data points are to be feared, one is left with an alternative, the two terms of which are equally dissatisfying. Either one increases the size of $\mathcal{Y}$ enough to enclose the largest error that could be attributed to the wildest possible outlier, in which case S might become so large that no information is gained on the value of the parameters of interest. Or one keeps a set $\mathcal{Y}$

that is compatible with regular errors, at the risk of obtaining a void estimate when outliers are present. The classical set-membership estimators are therefore not robust to such outliers.

In this paper, guaranteed parameter bounding from unreliable priors and data is considered. The (posterior) feasible set S for $\mathbf{p}$ is assumed to be defined by a finite number of elementary tests $t_i(\mathbf{p})$, which must all hold true for $\mathbf{p}$ to belong to S. An example of such test would be

$$\begin{cases} t_i(\mathbf{p}) = \text{true if } e_{\min}(i) \leq y_i - y_{im}(\mathbf{p}) \leq e_{\max}(i), \\ t_i(\mathbf{p}) = \text{false otherwise.} \end{cases} \tag{6}$$

S is then defined as

$$S = \{\mathbf{p} \in \mathcal{P} \mid t(\mathbf{p}) \text{ holds true}\}, \tag{7}$$

where

$$t(\mathbf{p}) = \bigwedge_{i=1}^{n} t_i(\mathbf{p}), \tag{8}$$

with $\bigwedge$ the AND operator.

Remark 1. In this example, the number of elementary tests is equal to the dimension of $\mathbf{y}$, and we shall continue to denote the number of elementary tests by n, but this is of course a special case. Note also that more complicated logical expressions than (8) could be tailored to the need of specific applications, which could benefit from a treatment similar to that presented here.

We assume in this paper that some of the elementary tests $t_i(.)$ may be erroneous. The methodology to be presented will nevertheless make it possible to compute a set guaranteed to contain all acceptable values of the parameter vector, provided that an upper bound on the number of tests at fault is available. The model output may be nonlinear in the parameters, and even a majority of faulty tests can be dealt with.

This methodology will be described in Section 2, before applying it in Section 3 to the localization of a robot in a partially known environment from distance measurements provided by a belt of onboard ultrasonic sensors.

2   Guaranteed Robust Nonlinear Estimator

The estimator to be presented is a guaranteed generalization of the *Outlier Minimum Number Estimator (OMNE)* [6], [7], defined as

$$\widehat{\mathbf{p}} = \arg \min_{\mathbf{p} \in \mathcal{P}} (\text{number of error bounds violated}). \tag{9}$$

Data points for which the error bounds are violated are considered as outliers by the estimator, hence its name. The resulting estimate is a set, equal to S when there are no outliers. When $y_m(p)$ is linear in p and the number of data points tends to infinity, it has been shown [8] that the breakdown point of OMNE tends to the theoretical limit of 50% outliers organized so as to fool the estimator. When the outliers are not designed with this purpose in mind, OMNE can even handle a majority of outliers. The initial implementation of OMNE combined global optimization by adaptive random search to get a vector belonging to the feasible set for the parameters and exploration of cost contours to characterize its boundary. Although the method could be used to characterize sets consisting of several disconnected parts [9], no guarantee could be given that the result provided would indeed contain the entire set defined by (9). Such a guarantee can be obtained by taking advantage of the possibilities offered by interval computation.

When p is a vector, $t(p)$ is either true or false, but this is no longer true for interval vectors (or *boxes*) $[p]$, because the test may be true on one part of $[p]$ and false on the other. To implement the resulting three-valued logic, define a Boolean interval as an element of $\mathbb{IB} = \{0, [0, 1], 1\}$, where 0 stands for *false*, 1 for *true*, and $[0, 1]$ for *indeterminate*. Let $\mathbb{IR}^m$ be the set of all m-dimensional real boxes. An *inclusion test* for the *test* $t : \mathbb{R}^m \to \{0, 1\}$ is a function $t_{[]} : \mathbb{IR}^m \to \mathbb{IB}$ such that for any $[p]$ in $\mathbb{IR}^m$

$$t_{[]}([p]) = 1 \Rightarrow \forall p \in [p], t(p) = 1,$$
$$t_{[]}([p]) = 0 \Rightarrow \forall p \in [p], t(p) = 0. \tag{10}$$

When, as in (8), the test $t(.)$ is obtained by combining elementary tests $t_i(.)$, $t_{[]}(.)$ may be obtained by replacing each occurrence of any operator in the expression of $t(.)$ by its interval counterpart, and each elementary test $t_i(.)$ by $t_{i[]}(.)$. By analogy with the classical terminology for real functions [10], we call the result a *natural interval extension* of the initial test.

Assume, for the sake of simplicity, that the prior feasible set $\mathcal{P}$ for the parameters is some (possibly very large) box $[p_0]$ in parameter space. S as defined by (7) can then be rewritten as

$$S = \{p \in [p_0] \mid t(p) = 1\} = t_{[p_0]}^{-1}(1). \tag{11}$$

Characterizing S is thus cast in the framework of *set-inversion*, and can be solved using the SIVIA algorithm [11], [12]. This algorithm starts from $[p_0]$ and splits it into subboxes $[p]$, until it becomes able to classify them as inside S (because $t_{[]}([p]) = 1$), outside S (because $t_{[]}([p]) = 0$) or still ambiguous but too small to be considered any further. An outer approximation $\hat{S}$ of S is thus obtained as the union of the set of all subboxes that have been proved inside with the set of all ambiguous subboxes. For any box $[p]$ such that $t_{i[]}([p])$ is equal to 0 or 1, but not to $[0, 1]$, the result of this test will remain so for any subbox of [**p**]. It is therefore useless to reevaluate it provided that the results already obtained are stored in a mask. This masked version of SIVIA turns out to be much more efficient than the previous one.

So far, it was implicitly assumed that all elementary tests $t_i(.)$ could be relied upon. As already mentioned, this is sometimes quite unrealistic, hence the interest of a robust version of this set estimator.

Let q and n be positive integers, with $q < n$; by definition,

$$\bigotimes_q(t_1, \ldots, t_n) = \bigotimes_{i=1}^n {}_q(t_i) \tag{12}$$

holds true if and only if at least $n - q$ of the Booleans t_i $(i = 1, \cdots, n)$ are true. The $\bigotimes_q$ operator will be called a q-relaxed AND. To evaluate it, it suffices to compute the sum s (in the usual arithmetic sense) of the t_i's. $\bigotimes_q$ will be equal to 1 if and only if $s \geq n - q$. Otherwise it will be equal to 0. The usual rules are employed to obtain the interval counterpart $\bigotimes_{q[]}$ of the q-relaxed AND.

Remark 2. One could also define a weighted q-relaxed AND, by replacing s by a weighted sum of the t_i's. This would allow one to take into account prior information on the relative reliability of these elementary tests.

To tolerate up to q faulty tests out of n, it suffices to use SIVIA with

$$t_{[]}([\mathbf{p}], q) = \bigotimes_{i=1}^n {}_{q[]} \left(t_{i[]}([\mathbf{p}]) \right) \tag{13}$$

instead of $t_{[]}([\mathbf{p}])$. The resulting outer approximation will be denoted by $\widehat{S}_q$. Various policies may then be considered to obtain a robust estimate of **p**. The first one starts with $q = 0$, and increases q until $\widehat{S}_q$ as computed by SIVIA becomes nonempty [13]. With this policy, some faulty tests may escape elimination. To guard against such a risk, one may choose to increase q by some prespecified positive integer. Alternatively, one may choose q at the outset, as the total number of faulty tests to be guarded against. Remember that q could even be chosen larger than $n/2$. Increasing the value of q improves robustness at the cost of potentially enlarging $\widehat{S}_q$, so a compromise must of course be struck. It is important to note that the algorithm does not proceed by examining all possible combinations of $n - q$ tests, and thus escapes combinatorial explosion of the complexity when n and q become large. Note also that the tests discarded may vary within $\widehat{S}_q$.

3 Application to Robot Localization

The robot considered is represented on Fig. 1. It is equipped with a belt of 24 ultrasonic sensors (or sonars), nine of which are visible on the picture.

Fig. 1. Robot, with its belt of sonars

Let $\mathcal{R}$ be a frame tied to the robot, and $\mathcal{W}$ be a frame tied to the world in which the robot lives (Fig. 2). The parameters to be estimated are the coordinates x_c and y_c of the origin of $\mathcal{R}$ in $\mathcal{W}$ and the heading angle θ of robot. They form the *configuration vector* $\mathbf{p} = (x_c, y_c, \theta)^{\mathrm{T}}$.

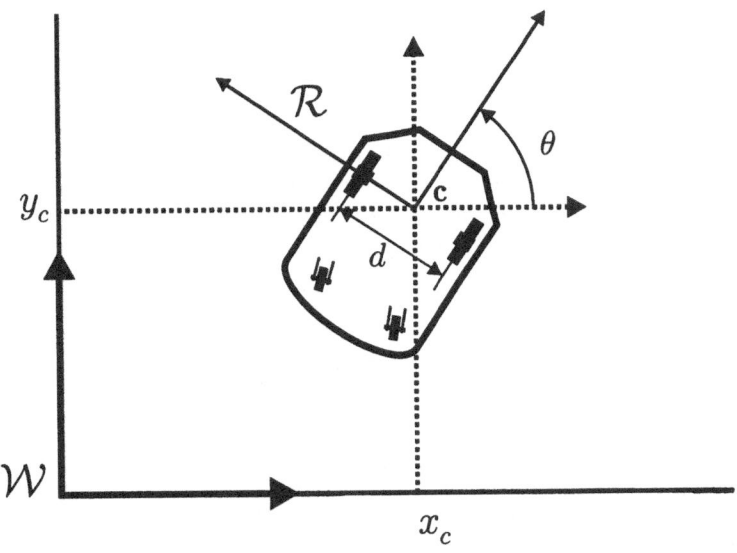

Fig. 2. World frame $\mathcal{W}$ and robot frame $\mathcal{R}$

The data **y** to be used consist of the 24 distances to obstacles measured by the sonars ($n = 24$). The prior information available to the robot comprises three parts. The first one is a map (Fig. 3) consisting of oriented segments $[\mathbf{a}_j, \mathbf{b}_j]$ which describe the landmarks (walls, doors, pillars, etc.)

$$\mathcal{M} = \{[\mathbf{a}_j, \mathbf{b}_j] \,|\, j = 1, ..., n_w\}. \tag{14}$$

The second one is a model $\mathbf{y}_m(\mathbf{p})$ of the distances that should be obtained

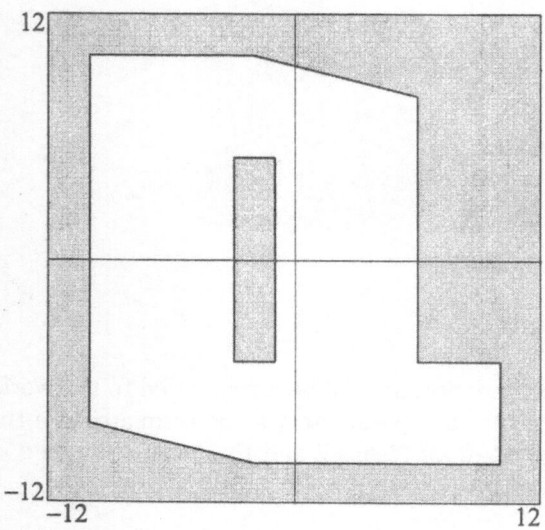

Fig. 3. Map of the room available to the robot

by the 24 sonars in a given configuration **p**. To keep this model simple, it is assumed that the ith sonar reports the distance d_i to the nearest portion of the intersection of a segment of the map with its emission cone (see Fig. 4). This corresponds to the notion of remoteness of a sensor from the environment, to be explained below. The third part is a model of the inaccuracy attached to each measurement. Based on prior laboratory experiments, it is assumed that the actual distance belongs to an interval $[d_i] = [(1 - \alpha_i)d_i, (1 + \alpha_i)d_i]$, with $\alpha_i = 0.02$. The half-aperture angle $\tilde{\gamma}_i$ is taken equal to 0.2 rad.

To any data vector **y**, one can then attach an emission diagram for the 24 sensors (Fig. 5). For each of them, an obstacle should be located at least partly between the two arcs of circles indicated, and no other obstacle closer to the sonar should intercept its beam. Under these idealized hypotheses, the configuration of the robot should be such that segments of the map would be responsible for each of the distances reported.

One of the main difficulties encountered by localization methods is deciding which landmark is responsible for which distance measurement. When

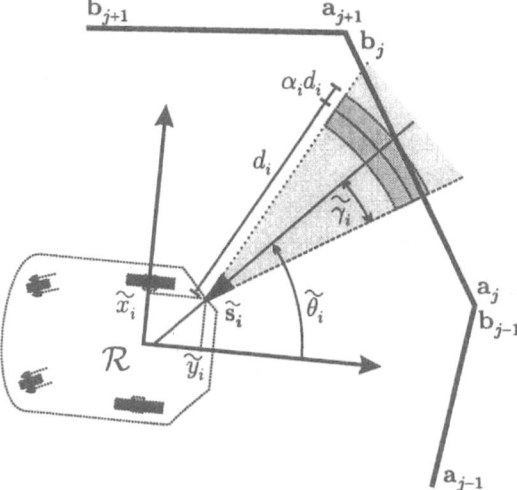

Fig. 4. Emission cone of the ith sensor

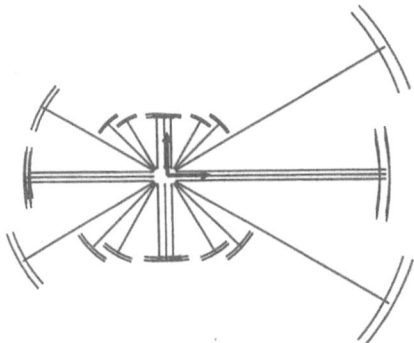

Fig. 5. Emission diagram. At least one obstacle should lie somewhere between the two arcs of circles associated with any given sensor, and no obstacle closer to the sensor should intercept the beam.

the configuration of the robot is at least approximately known, it may be possible to limit the possibilities to a few cases. When this is not so, data-association methods can be used to enumerate all segments of the map that may correspond to the distances measured. Combinatorial complexity may however become tremendous (see, *e.g.*, [14], [15], [16], [17]). For each of the data-association hypotheses generated, one may proceed to estimate the configuration by minimizing some cost function, *e.g.*, the sum of the squares of the deviations between the distances $\mathbf{y}$ measured by the sensors and those $\mathbf{y_m}(\mathbf{p})$ predicted by a model. When $\mathbf{y_m}$ is nonlinear in $\mathbf{p}$, the algorithm employed is usually local, with no guarantee of convergence to the global optimum. When $\mathbf{y_m}$ is linearized with respect to $\mathbf{p}$, least-squares provide an analytic expression for $\hat{p}$, but the validity of this expression is also local. The various approaches based on employing ellipsoidal outerbounding algorithms in the context of set-membership estimation (see, *e.g.*, [18]) suffer from the same limitations. In all cases, outliers may ruin the results, either because they are the major contributors in the quadratic criterion or because the set of all $\mathbf{p}$ consistent with the data and error bounds turns out to be empty. By contrast, the method advocated here only assumes that the actual configuration of the robot belongs to some possibly very large box $[\mathbf{p_0}]$ in configuration space, and looks for the set of *all* configurations that are consistent with the data and priors, assuming knowledge of some upper bound on the number of faulty tests.

For the sake of simplicity, we shall only consider one type of elementary test here. Let $r_{ij}(\mathbf{p})$ be the *remoteness* of the ith sonar from the jth segment of the map, defined as follows: if the jth segment intersects the emission cone of the ith sonar, then $r_{ij}(\mathbf{p})$ is equal to the minimum distance between the sensor and the intersection between the segment and the emission cone. Else, the distance is infinite. The remoteness $r_i(\mathbf{p})$ of the environment from the ith sonar is then the minimum of $r_{ij}(\mathbf{p})$ over all segments j of the map. This remoteness is easy to compute for any configuration, and is taken as the ith entry of $\mathbf{y_m}(\mathbf{p})$. The configuration $\mathbf{p}$ will thus be considered as consistent with the measurement provided by the ith sonar if $r_i(\mathbf{p}) \in [d_i]$. The elementary test $t_i(.)$ is then defined as

$$\begin{cases} t_i(\mathbf{p}) = 1 \text{ if } r_i(\mathbf{p}) \in [d_i], \\ t_i(\mathbf{p}) = 0 \text{ otherwise.} \end{cases} \tag{15}$$

Note that this is a rather crude model. It would be easy to refine, *e.g.*, by taking into account the fact that no reflection will take place if the angle between the incident beam and the normal to the segment is too large. Other tests accelerating localization could also be defined. See [19] or [20] for more details.

In an interval context, the test (15) becomes

$$\begin{cases} t_{i[]}([\mathbf{p}]) = 1 & \text{if } r_{i[]}([\mathbf{p}]) \subset [d_i], \\ t_{i[]}([\mathbf{p}]) = 0 & \text{if } r_{i[]}([\mathbf{p}]) \cap [d_i] = \emptyset, \\ t_{i[]}([\mathbf{p}]) = [0, 1] \text{ otherwise.} \end{cases} \tag{16}$$

If all our assumptions were true, $\mathcal{S}$ as defined by (11) with $t(\mathbf{p})$ given by (8), and thus its outer approximation $\hat{\mathcal{S}}$ provided by SIVIA, would be guaranteed to contain the actual configuration of the robot.

There are many reasons, however, why not all the elementary tests $t_i(.)$ can be relied upon. The map may be partly out of date, some sensors may be faulty, persons or pieces of furniture may have intercepted some beams, multiple reflections may have taken place, the model $\mathbf{y_m}(.)$ is a rough approximation of reality... The robust version of the set estimator based on (13) can therefore not be dispensed with.

In the example to be considered now, the map available to the robot is that of Fig. 3. As we shall see later, this map is partly out of date. A set of simulated data is obtained using a more sophisticated model than the one used for localization. This simulation model takes into account the fact that if the beam hits a landmark with an angle of incidence greater than some predefined angle taken here as $\frac{\pi}{4}$, then the wave is not reflected or refracted back to the sensor.

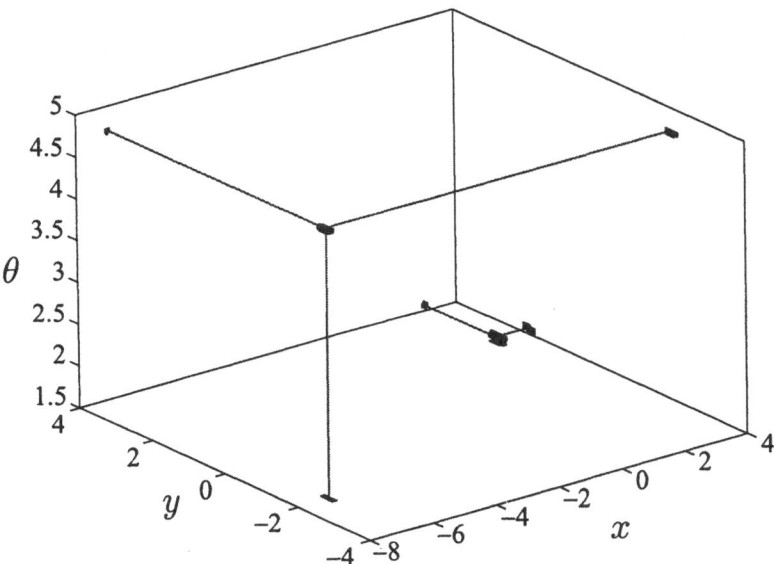

Fig. 6. The two disconnected components of $\widehat{S}_6$, with their projections.

The emission diagram obtained with this simulation model for a correct map of the room corresponds to Fig. 5. It goes without saying that the actual map used for the simulation is not communicated to the localization algorithm. The localization procedure is then applied, starting with $q = 0$. The configuration search box is $[\mathbf{p}_0] = [-12 \text{ m}, 12 \text{ m}] \times [-12 \text{ m}, 12 \text{ m}] \times [0, 2\pi]$. The feasible set is found to be empty as long as $q < 6$. For $q = 6$, Fig. 6 shows the outer approximation $\widehat{S}_q$ computed by SIVIA. It consists of two disconnected subsets $\widehat{S}'$ and $\widehat{S}''$, which correspond to radically different associations of distances to landmarks. The ambiguity is due to the local symmetry of the room. $\widehat{S}_6$ is guaranteed to contain the actual robot configuration if there are no more than $q = 6$ outliers. Two possible configurations (one in $\widehat{S}'$ and the other in $\widehat{S}''$) are displayed on Fig. 7. The true configuration used for the simulation $\mathbf{p}^* = (3, 2, \pi/2)^{\mathrm{T}}$ does belong to one of them. SIVIA was implemented in C++. Using various accelerating tests discussed in [20], the entire search takes 268 s on a Pentium 233MMX, to be compared with 1065 s for a basic SIVIA. Most of the acceleration is due to the use of the mask shortly described in Section 2. The actual configuration is pre-

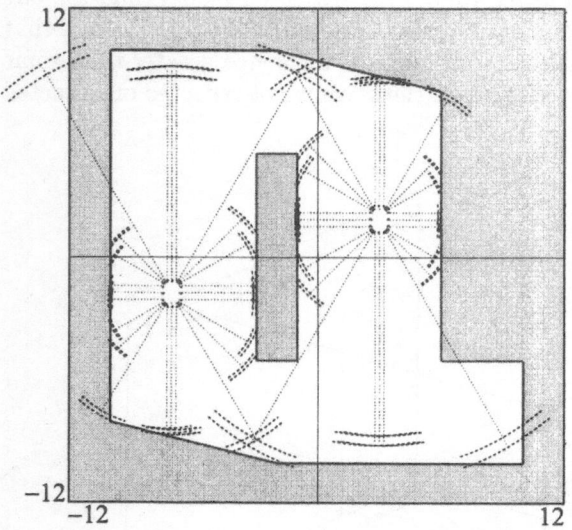

Fig. 7. Two of the configurations in $\widehat{S}_q$, together with the map available to the robot

sented in Fig. 8 on a correct map of the room. A cupboard, not mentioned on the map available to the robot and indicated by a hatched box on the figure, has caused three outliers. The three others, correspond to beams for which the angle of incidence on the closest landmark was too large.

If a number $q > 6$ of outliers is tolerated, a larger is obtained. For $q = 7$, $\widehat{S}_6$ consists of six disconnected components, shown on Fig. 9. One of them again contains $\mathbf{p}^*$.

4   Conclusions and Perspectives

In the context of parameter bounding, priors and data are used to define a number of conditions (or tests) that the parameter vector must satisfy to be consistent with the experimental results. When the priors are doubtful and the data may contain outliers, one would like to ensure some robustness of the estimated set of feasible values for the parameter vector. The method proposed here is based on accepting any value of the parameter vector such that at most q of the tests hold false. As long as there are indeed no more than q faulty tests, it is guaranteed to provide an outer approximation for the set that would have been obtained if priors were always true and there were no outliers. To the best of our knowledge, no other method can make similar claims.

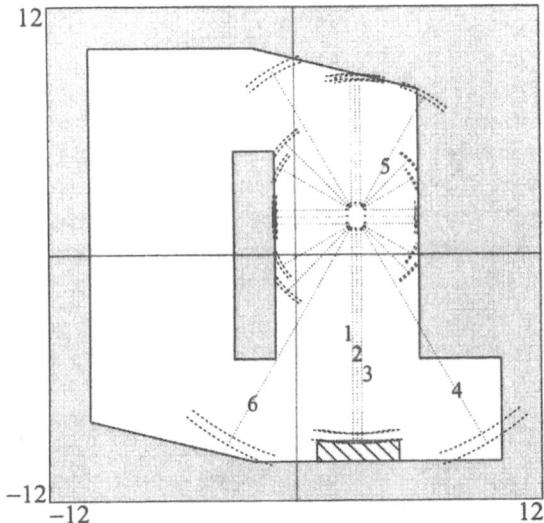

Fig. 8. Actual configuration of the robot in the room. The elementary tests associated with beams 1 to 3 are faulty because of a cupboard not reported on the map available to the robot and represented by a hatched box. the elementary tests associated with beams 4 to 6 are faulty because their angles of incidence on the closest landmarks are too large.

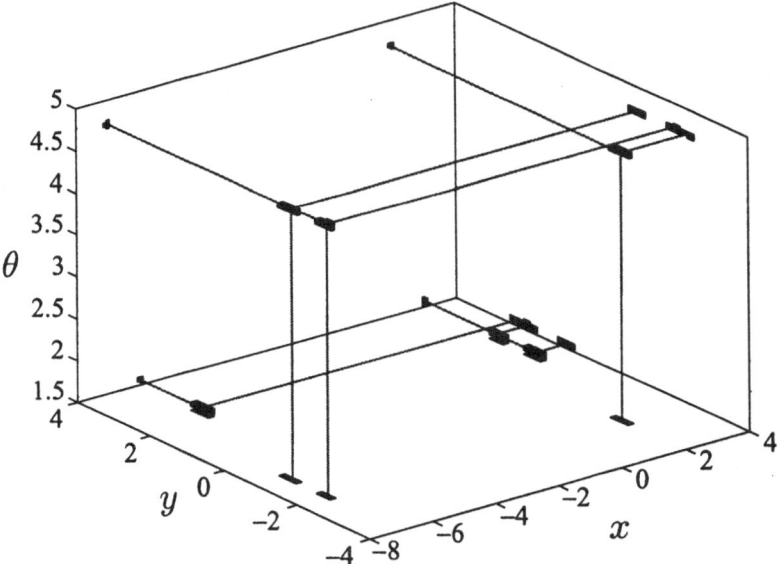

Fig. 9. The six disconnected components of $\hat{\mathcal{S}}_7$, with their projections.

Although other types of applications could obviously be considered, autonomous robot localization from onboard distance measurements by sonars is a case in point for the interest of this method, because the number of parameters is small enough for a branch-and-bound technique to remain tractable and because the presence of outliers can hardly be avoided. On this problem, the method presented here has definite advantages over conventional numerical methods. The combinatorics of associating segments to distances and of considering all possible configurations of q faulty tests are avoided, and the association of the distances with the segments of the map or with outliers is obtained as a by-product of the procedure. A majority of faulty tests poses no problem. The results obtained are global, and ambiguities due, e.g., to the symmetries of the landscape are taken into account. The elementary tests considered here were fairly crude, and additional physical insight could easily be incorporated, as well as tests facilitating elimination of boxes in configuration space. The method can also be extended to tracking , thus providing a guaranteed and robust nonlinear state estimator [21].

References

1. E. Walter (Ed.), "Special issue on parameter identifications with error bounds," *Mathematics and Computers in Simulation*, vol. 32, no. 5&6, pp. 447–607, 1990.
2. J. P. Norton (Ed.), "Special issue on bounded-error estimation: Issue 1," *Int. J. of Adaptive Control and Signal Processing*, vol. 8, no. 1, pp. 1–118, 1994.
3. J. P. Norton (Ed.), "Special issue on bounded-error estimation: Issue 2," *Int. J. of Adaptive Control and Signal Processing*, vol. 9, no. 1, pp. 1–132, 1995.
4. M. Milanese, J. Norton, H. Piet-Lahanier, and E. Walter (Eds), *Bounding Approaches to System Identification*. New York: Plenum Press, 1996.
5. E. Walter and L. Pronzato, *Identification of Parametric Models from Experimental Data*. London: Springer-Verlag, 1997.
6. H. Lahanier, E. Walter, and R. Gomeni, "OMNE: a new robust membership-set estimator for the parameters of nonlinear models," *J. of Pharmacokinetics and Biopharmaceutics*, vol. 15, pp. 203–219, 1987.
7. E. Walter and H. Piet-Lahanier, "Estimation of the parameter uncertainty resulting from bounded-error data," *Math. Biosci.*, vol. 92, pp. 55–74, 1988.
8. L. Pronzato and E. Walter, "Robustness to outliers of bounded-error estimators and consequences on experiment design," in *Bounding Approaches to System Identification* (M. Milanese, J. Norton, H. Piet-Lahanier, and E. Walter, eds.), (New York), pp. 199–212, Plenum Press, 1996.
9. H. Piet-Lahanier and E. Walter, "Characterization of non-connected parameter uncertainty regions," *Math. and Comput. in Simulation*, vol. 32, pp. 553–560, 1990.
10. R. E. Moore, *Methods and Applications of Interval Analysis*. Philadelphia, Pennsylvania: SIAM Publ., 1979.
11. L. Jaulin and E. Walter, "Set inversion via interval analysis for nonlinear bounded-error estimation," *Automatica*, vol. 29, no. 4, pp. 1053–1064, 1993.

12. L. Jaulin and E. Walter, "Guaranteed nonlinear parameter estimation from bounded-error data via interval analysis," *Math. and Comput. in Simulation*, vol. 35, pp. 1923–1937, 1993.
13. L. Jaulin, E. Walter, and O. Didrit, "Guaranteed robust nonlinear parameter bounding," *CESA '96 IMACS Multiconference (Symposium on Modelling, Analysis and Simulation)*, pp. 1156–1161, 1996.
14. M. Drumheller, "Mobile robot localization using sonar," *IEEE Trans. on Pattern Analysis and Machine Intelligence*, vol. 9, no. 2, pp. 325–332, 1987.
15. W. E. Grimson and T. Lozano-Pérez, "Localizing overlapping parts by searching the interpretation tree," *IEEE Trans. on Pattern Analysis and Machine Intelligence*, vol. 9, no. 4, pp. 469–482, 1987.
16. J. J. Leonard and H. F. Durrant-Whyte, "Mobile robot localization by tracking geometric beacons," *IEEE Trans. on Robotics and Automation*, vol. 7, no. 3, pp. 376–382, 1991.
17. E. Halbwachs and D. Meizel, "Multiple hypothesis management for mobile vehicle localization," in *CD Rom of the European Control Conference*, (Louvain), 1997.
18. D. Meizel, A. Preciado-Ruiz, and E. Halbwachs, "Estimation of mobile robot localization: geometric approaches," in *Bounding Approaches to System Identification* (M. Milanese, J. Norton, H. Piet-Lahanier, and E. Walter, eds.), (New York), pp. 463–489, Plenum Press, 1996.
19. M. Kieffer, L. Jaulin, E. Walter, and D. Meizel, "Robust autonomous robot localization using interval analysis." To appear in *Reliable Computing*, 1999.
20. M. Kieffer, *Estimation ensembliste par analyse par intervalles, application à la localisation d'un véhicule.* Phd thesis, Université Paris-Sud, Orsay, 1999.
21. M. Kieffer, L. Jaulin, and E. Walter. Guaranteed recursive nonlinear state estimation using interval analysis. In *Proc. 37th IEEE Conference on Decision and Control*, pages 3966–3971, Tampa, Florida, 16-18 december 1998.

Part II

Robust Control

Robust Model Predictive Control: A Survey

Alberto Bemporad and Manfred Morari

Automatic Control Laboratory, Swiss Federal Institute of Technology (ETH),
Physikstrasse 3, CH-8092 Zürich, Switzerland,
bemporad,morari@aut.ee.ethz.ch, http://control.ethz.ch/

Abstract. This paper gives an overview of robustness in *Model Predictive Control* (MPC). After reviewing the basic concepts of MPC, we survey the uncertainty descriptions considered in the MPC literature, and the techniques proposed for robust constraint handling, stability, and performance. The key concept of "closed-loop prediction" is discussed at length. The paper concludes with some comments on future research directions.

1 Introduction

Model Predictive Control (MPC), also referred to as *Receding Horizon Control* and *Moving Horizon Optimal Control*, has been widely adopted in industry as an effective means to deal with multivariable constrained control problems [42,57]. The ideas of receding horizon control and model predictive control can be traced back to the 1960s [28], but interest in this field started to surge only in the 1980s after publication of the first papers on IDCOM [59] and *Dynamic Matrix Control* (DMC) [25,26], and the first comprehensive exposition of *Generalized Predictive Control* (GPC) [22,23]. Although at first sight the ideas underlying the DMC and GPC are similar, DMC was conceived for multivariable constrained control, while GPC is primarily suited for single variable, and possibly adaptive control.

The conceptual structure of MPC is depicted in Fig. 1. The name MPC stems from the idea of employing an explicit *model* of the plant to be controlled which is used to *predict* the future output behavior. This prediction capability allows solving optimal control problems on line, where tracking error, namely the difference between the predicted output and the desired reference, is minimized over a future horizon, possibly subject to constraints on the manipulated inputs and outputs. When the model is linear, then the optimization problem is quadratic if the performance index is expressed through the ℓ_2-norm, or linear if expressed through the ℓ_1/ℓ_∞-norm. The result of the optimization is applied according to a *receding horizon* philosophy: At time t only the first input of the optimal command sequence is

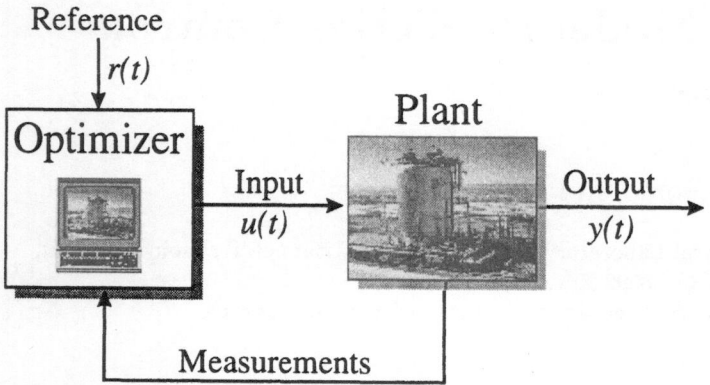

Fig. 1. Basic structure of Model Predictive Control

actually applied to the plant. The remaining optimal inputs are discarded, and a new optimal control problem is solved at time $t + 1$. This idea is illustrated in Fig. 2. As new measurements are collected from the plant at each time t, the receding horizon mechanism provides the controller with the desired feedback characteristics.

The issues of feasibility of the on-line optimization, stability and performance are largely understood for systems described by linear models, as testified by several books [14,65,46,24,12,17] and hundreds of papers [36][1]. Much progress has been made on these issues for nonlinear systems [48], but for practical applications many questions remain, including the reliability and efficiency of the on-line computation scheme. Recently, application of MPC to hybrid systems integrating dynamic equations, switching, discrete variables, logic conditions, heuristic descriptions, and constraint prioritizations have been addressed in [9]. They expanded the problem formulation to include integer variables, yielding a Mixed-Integer Quadratic or Linear Program for which efficient solution techniques are becoming available.

A fundamental question about MPC is its *robustness* to model uncertainty and noise. When we say that a control system is robust we mean that stability is maintained and that the performance specifications are met for a specified range of model variations and a class of noise signals (uncertainty range). To be meaningful, any statement about "robustness" of a particular control algorithm must make reference to a specific uncertainty range as well as specific stability and performance criteria. Although a rich theory has been developed for the robust control of linear systems, very little is known about the robust control of linear systems with constraints. Recently, this type of problem has been addressed in the context of MPC. This paper will give an overview of these attempts to endow MPC with some robustness

---

[1] Morari [51] reports that a simple database search for "predictive control" generated 128 references for the years 1991-1993. A similar search for the years 1991-1998 generated 2802 references.

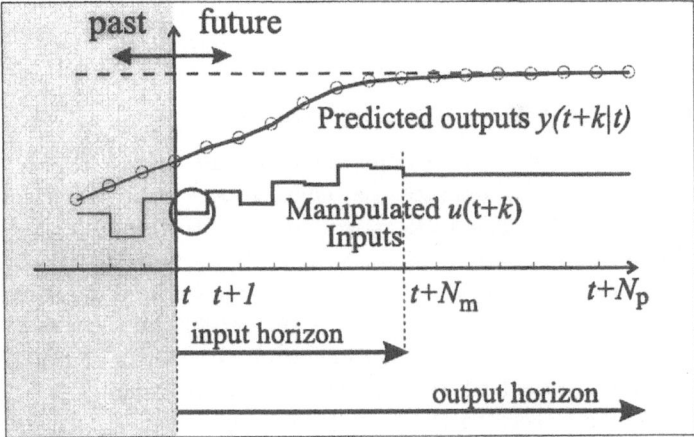

Fig. 2. Receding horizon strategy: only the first one of the computed moves $u(t)$ is implemented

guarantees. The discussion is limited to linear time invariant (LTI) systems with constraints. While the use of MPC has also been proposed for LTI systems without constraints, MPC does not have any practical advantage in this case. Many other methods are available which are at least equally suitable.

2 MPC Formulation

In the research literature MPC is formulated almost always in the state space. Let the model Σ of the plant to be controlled be described by the linear discrete-time difference equations

$$\Sigma : \begin{cases} x(t+1) = Ax(t) + Bu(t), & x(0) = x_0, \\ y(t) = Cx(t) \end{cases} \tag{1}$$

where $x(t) \in \mathbb{R}^n$, $u(t) \in \mathbb{R}^m$, $y(t) \in \mathbb{R}^p$ denote the state, control input, and output respectively. Let $x(t + k, x(t), \Sigma)$ or, in short, $x(t + k|t)$ denote the prediction obtained by iterating model (1) k times from the current state $x(t)$.

A receding horizon implementation is typically based on the solution of the following open-loop optimization problem:

$$\min_{\mathbf{U} \triangleq \{u(t+k|t)\}_{k=t}^{t+N_m-1}} J(\mathbf{U}, x(t), N_p, N_m) = x^T(N_p)P_0 x(N_p)$$

$$+ \sum_{k=0}^{N_p-1} x'(t+k|t)Qx(t+k|t) + \sum_{k=0}^{N_m-1} u'(t+k|t)Ru(t+k|t) \tag{2a}$$

$$\text{subject to} \qquad \begin{aligned} F_1 u(t+k|t) &\le G_1 \\ E_2 x(t+k|t) + F_2 u(t+k|t) &\le G_2 \end{aligned} \qquad (2b)$$

$$\text{and} \qquad \text{``stability constraints''} \qquad (2c)$$

where, as shown in Fig. 2, N_p denotes the length of the *prediction hori-zon* or *output horizon*, and N_m denotes the length of the *control horizon* or *input horizon* ($N_m \le N_p$). When $N_p = \infty$, we refer to this as the *in-finite horizon problem*, and similarly, when N_p is finite, as a *finite horizon* problem. For the problem to be meaningful we assume that the polyhedron $\{(x,u): F_1 u \le G_1, E_2 x + F_2 u \le G_2\}$ contains the origin ($x = 0, u = 0$). The constraints (2c) are inserted in the optimization problem in order to guarantee closed-loop stability, and will be discussed in the sequel.

The basic MPC law is described by the following algorithm:

Algorithm 1:

1. Get the new state $x(t)$
2. Solve the optimization problem (2)
3. Apply only $u(t) = u(t+0|t)$
4. $t \leftarrow t+1$. Go to 1.

2.1   Some Important Issues

Feasibility Feasibility of the optimization problem (2) at each time t must be ensured. Typically one assumes feasibility at time $t = 0$ and chooses the cost function (2a) and the stability constraints (2c) such that feasibility is preserved at the following time steps. This can be done, for instance, by ensuring that the shifted optimal sequence $\{u(t+1|t), \ldots, u(t+N_p|t), 0\}$ is feasible at time $t+1$. Also, typically the constraints in (2b) which involve state components are treated as *soft* constraints, for instance by adding the slack variable ϵ

$$E_2 x + F_2 u \le G_2 + \epsilon \begin{bmatrix} 1 \\ \vdots \\ 1 \end{bmatrix}, \qquad (3)$$

while pure input constraints $F_1 u \le G_1$ are maintained as *hard*. Relaxing the state constraints removes the feasibility problem at least for stable systems. Keeping the state constraints tight does not make sense from a practical point of view because of the presence of noise, disturbances, and numerical errors. As the inputs are generated by the optimization procedure, the input constraints can always be regarded as hard.

Stability In the MPC formulation (2) we have not specified the stability constraints (2c). Below we review some of the popular techniques used in the literature to "enforce" stability. They can be divided into two main classes. The first uses the value $V(t) = J(\mathbf{U}^*, x(t), N_p, N_m)$ attained for the minimizer $\mathbf{U}^* \triangleq \{u^*(t+1|t), \ldots, u^*(t+N_m|t)\}$ of (2) at each time t as a Lyapunov function. The second explicitly requires that the state $x(t)$ is shrinking in some norm.

- *End (Terminal) Constraint* [38,39]. The stability constraint (2c) is

$$x(t + N_p|t) = 0 \qquad (4)$$

This renders the sequence $\mathbf{U}_1 \triangleq \{u^*(t+1|t), \ldots, u^*(t+N_m|t), 0\}$ feasible at time $t+1$, and therefore $V(t+1) \leq J(\mathbf{U}_1, x(t+1), N_p, N_m) \leq J(\mathbf{U}^*, x(t), N_p, N_m) = V(t)$ is a Lyapunov function of the system [34,10]. The main drawback of using terminal constraints is that the control effort required to steer the state to the origin can be large, especially for short N_p, and therefore feasibility is more critical because of (2b). The domain of attraction of the closed-loop (MPC+plant) is limited to the set of initial states x_0 that can be steered to 0 in N_p steps while satisfying (2b), which can be considerably smaller then the set of initial states steerable to the origin in an arbitrary number of steps. Also, performance can be negatively affected because of the artificial terminal constraint. A variation of the terminal constraint idea has been proposed where only the unstable modes are forced to zero at the end of the horizon [58]. This mitigates some of the mentioned problems.
- *Infinite Output Prediction Horizon* [34,58,71]. For asymptotically stable systems, no stability constraint is required if $N_p = +\infty$. The proof is again based on a similar Lyapunov argument.
- *Terminal Weighting Matrix* [37,40]. By choosing the terminal weighting matrix P_0 in (2a) as the solution of a Riccati inequality, stability can be guaranteed without the addition of stability constraints.
- *Invariant terminal set* [64]. The idea is to relax the terminal constraint (4) into the set-membership constraint

$$x(t + N_p|t) \in \Omega \qquad (5)$$

and set $u(t + k|t) = F_{LQ}x(t + k|t)$, $\forall k \geq N_m$, where F_{LQ} is the LQ feedback gain. The set Ω is invariant under LQ regulation and such that the constraints are fulfilled inside Ω. Again, stability can be proved via Lyapunov arguments.
- *Contraction Constraint* [53,73]. Rather then relying on the optimal cost $V(t)$ as a Lyapunov function, the idea is to require explicitly that the state $x(t)$ is decreasing in some norm

$$\|x(t + 1|t)\| \leq \alpha \|x(t)\|, \ \alpha < 1 \qquad (6)$$

Following this idea, Bemporad [4] proposed a technique where stability is guaranteed by synthesizing a quadratic Lyapunov function for the system, and by requiring that the terminal state lies within a level set of the Lyapunov function, similar to (5).

Computation The complexity of the solver for the optimization problem (2) depends on the choice of the performance index and the stability constraint (2c). When $N_p = +\infty$, or the stability constraint has the form (4), or the form (5) and Ω is a polytope, the optimization problem (2) is a *Quadratic Program* (QP). Alternatively, one obtains a *Linear Program* (LP) by formulating the performance index (2a) in $\| \cdot \|_1$ or $\| \cdot \|_\infty$ [19]. The constraint (6) is convex, and is quadratic or linear depending if $\| \cdot \|_2$ or $\| \cdot \|_1 / \| \cdot \|_\infty$ is chosen. When $\| \cdot \|_2$ is used, second-order cone programming algorithms [44] can be adopted conveniently.

3   Robust MPC — Problem Definition

The basic MPC algorithm described in the previous section assumes that the plant Σ_0 to be controlled and the model Σ used for prediction and optimization are the same, and no unmeasured disturbance is acting on the system. In order to talk about robustness issues, we have to relax these hypotheses and assume that (i) the true plant $\Sigma_0 \in S$, where S is a given family of LTI systems, and/or (ii) an unmeasured noise $w(k)$ enters the system, namely

$$\Sigma : \begin{cases} x(t+1) = Ax(t) + Bu(t) + Hw(t), & x(0) = x_0, \\ y(t) = Cx(t) + Kw(t) \end{cases} \tag{7}$$

where $w(t) \in \mathcal{W}$ and $\mathcal{W}$ is a given set (usually a polytope).

We will refer to *robust stability*, *robust constraint fulfillment*, and *robust performance* of the MPC law if the respective property is guaranteed for all possible $\Sigma_0 \in S$, $w(t) \in \mathcal{W}$.

As part of the modelling effort it is necessary to arrive at an appropriate description of the uncertainty, i.e. the sets S and $\mathcal{W}$. This is difficult because there is very little experience and no systematic procedures are available. On one hand, the uncertainty description should be "tight", i.e. it should not include "extra" plants which do not exist in the real situation. On the other hand, there is a trade-off between realism and the resulting computational complexity of the analysis and controller synthesis. In other words, the uncertainty description should lead to a simple (non-conservative) *analysis* procedure to determine if a particular system with controller is stable and meets the performance requirements in the presence of the specified uncertainty. Alternatively, a computationally tractable *synthesis* procedure should exist to design a controller which is robustly stable and satisfies the robust performance specifications.

At present all the proposed uncertainty descriptions and associated analysis/synthesis procedures do little more than provide different handle to the engineer to detect and avoid sensitivity problems. They do not address the trade-off alluded to above, in a systematic manner.

For example, for simplicity some procedures consider only the uncertainty introduced by the set of unmeasured bounded inputs. There is the implicit assumptions that the other model uncertainty is in some way covered in this manner. There has been no rigorous analysis, however, to determine the exact relationship between the input set $\mathcal{W}$ and the covered set $\mathcal{S}$ — if such a relationship does indeed exist.

In the remaining part of the paper we will describe the different uncertainty descriptions which have been used in robust MPC, comment on the robustness analysis of standard (no uncertainty description) MPC, and give an overview of the problems associated with the synthesis of robust MPC control laws.

4   Uncertainty Descriptions

Different uncertainty sets $\mathcal{S}$, $\mathcal{W}$ have been proposed in the literature in the context of MPC, and are mostly based on *time-domain* representations. Frequency-domain descriptions of uncertainty are not suitable for the formulation of robust MPC because MPC is primarily a time-domain technique.

4.1   Impulse/Step-Response

Uncertainties on the impulse-response or step-response coefficients provide a practical description in many applications, as they can be easily determined from experimental tests, and allow a reasonably simple way to compute robust predictions. Uncertainty is described as range intervals over the coefficients of the impulse- and/or step-response. In the simplest SISO (single-input single-output) case, this corresponds to set

$$\Sigma: \; y(t) = \sum_{k=0}^{N} h(t)u(t-k) \tag{8}$$

and

$$\mathcal{S} = \{\Sigma: \; h_t^- \le h(t) \le h_t^+\}, \; t = 0,\ldots,N \tag{9}$$

where $[h_t^-, h_t^+]$ are given intervals. For $N < \infty$, $\mathcal{S}$ is a set of FIR models.

A similar type of description can be used for step-response models

$$y(t) = \sum_{k=0}^{N} s(t)[u(t-k) - u(t-k-1)], \; s(t) \in [s_t^-, s_t^+] \tag{10}$$

Impulse- and step-response descriptions are only equivalent when there is no uncertainty. If there is uncertainty they behave rather differently [8]. In order to arrive at a tight uncertainty description both may have to be used simultaneously and further constraints may have to be imposed on the coefficient variations as we will explain.

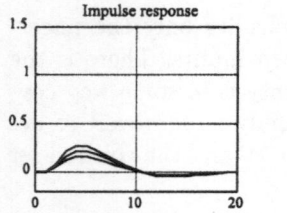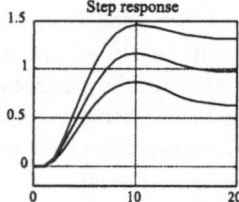

Fig. 3. Step-response interval ranges (right) arising from an impulse-response description (left)

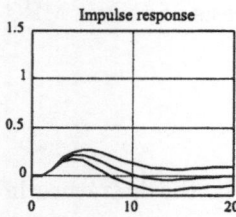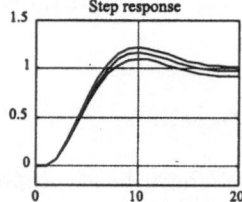

Fig. 4. Impulse-response interval ranges (left) arising from a step-response description (right)

Consider Fig. 3, which depicts perturbations expressed only in terms of the impulse response. The resulting step-response uncertainty is very large as $t \to \infty$. This may not be a good description of the real situation. Conversely, as depicted in Fig. 4, uncertainty expressed only in terms of the step response could lead to nonzero impulse-response samples at large values of t, for instance because the DC-gain from u to y is uncertain. Hence any a priori information about asymptotic stability properties would not be exploited.

Also, the proposed bounds would allow the step response to be highly oscillatory, though the process may be known to be overdamped. Similar comments apply to the impulse response. Thus this description may introduce high frequency model uncertainty artificially and may lead to a conservative design. This deficiency can be alleviated by imposing a correlation between neighboring uncertain coefficients as proposed by Zheng [73].

Another subtle point is that the uncertain FIR model (8) is usually unsuitable if the coefficients must be assumed to be time varying in the analysis or synthesis. In this case, the model would predict output variations even when the input is constant, which is usually undesirable. Writing the model in the form

$$\Sigma: \quad y(t) = y(t-1) + \sum_{k=0}^{N} h(t)[u(t-k) - u(t-k-1)] \tag{11}$$

removes this problem.

In conclusion, simply allowing the step- or impulse-response coefficients to vary within intervals is rarely a useful description of model uncertainty

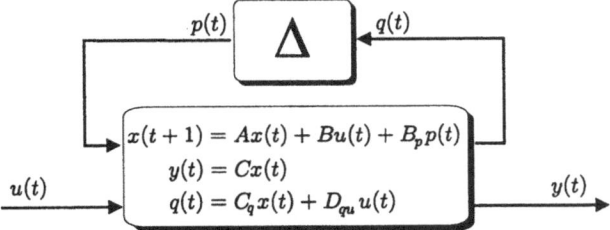

Fig. 5. Structured feedback uncertainty

unless additional precautions are taken. Nevertheless, compared to other descriptions, it leads to computationally simpler algorithms when adopted in robust MPC design, as will be discussed in Sect. 9

4.2 Structured Feedback Uncertainty

A common paradigm for robust control consists of a linear time-invariant system with uncertainties in the feedback loop, as depicted in Fig. 5 [35]. The operator Δ is block-diagonal, $\Delta = \text{diag}\{\Delta_1, \ldots, \Delta_r\}$, where each block Δ_i represents either a memoryless time-varying matrix with $\|\Delta_i(t)\|_2 = \bar{\sigma}(\Delta_i(t)) \leq 1$, $\forall i = 1, \ldots, r$, $t \geq 0$; or a convolution operator (e.g. a stable LTI system) with the operator norm induced by the truncated ℓ_2-norm less than 1, namely $\sum_{j=0}^t p'(j)p(j) \leq \sum_{j=0}^t q'(j)q(j)$, $\forall t \geq 0$. When Δ_i are stable LTI systems, this corresponds to the frequency domain specification on the z-transform $\hat{\Delta}_i(z)$ $\|\hat{\Delta}(z)\|_{\mathcal{H}_\infty} < 1$.

4.3 Multi-Plant

We refer to a *multi-plant* description when model uncertainty is parameterized by a finite list of possible plants [3]

$$\Sigma \in \{\Sigma_1, \ldots, \Sigma_n\} \tag{12}$$

When we allow the real system to vary within the convex hull defined by the list of possible plants we obtain the so called polytopic uncertainty.

4.4 Polytopic Uncertainty

The set of models S is described as

$$x(t+1) = A(t)x(t) + B(t)u(t)$$
$$y(t) = Cx(t)$$
$$[A(t)\ B(t)] \in \Omega$$

and $\Omega = \text{Co}\{[A_1\ B_1], \ldots, [A_M\ B_M]\}$, the convex hull of the "extreme" models $[A_i\ B_i]$ is a polytope. As remarked by Kothare *et al.* [35], polytopic uncertainty is a conservative approach to model a nonlinear system $x(t + 1) = f(x(k), u(k), k)$ when the Jacobian $[\frac{\partial f}{\partial x} \ \frac{\partial f}{\partial u}]$ is known to lie in the polytope Ω.

4.5 Bounded Input Disturbances

The uncertainty is limited to the unknown disturbance $w \in \mathcal{W}$ in (7), the plant Σ_0 is assumed to be known ($\mathcal{S} = \{\Sigma_0\}$). Also, one assumes that bounds on the disturbance are known, i.e. $\mathcal{W}$ is a given set. Although the assumption of knowing model Σ_0 might seem restrictive, the description of uncertainty by additive terms $w(t)$ that are known to be bounded in some norm is a reasonable choice, as shown in the recent literature on robust control and identification [50,45].

5 Robustness Analysis

We distinguish robustness *analysis*, i.e. analysis of the robustness properties of *standard* MPC designed for a nominal model without taking into account uncertainty, and *synthesis* of MPC algorithms which are robust by construction.

The robustness analysis of MPC control loops is more difficult than the synthesis, where the controller is designed in such a way that *it is* robustly stabilizing. This is not unlike the situation in the nominal case where the stability analysis of a closed loop MIMO system with multiple constraints is essentially impossible. On the other hand, the MPC technology leads naturally to a controller such that the closed loop system is guaranteed to be stable. There is a need for analysis tools, however, because standard MPC algorithms typically require less on-line computations, which is desirable for implementation.

Indeed, there are very few analysis methods discussed in the literature. By using a contraction mapping theorem, Zafiriou [68] derives a set of sufficient conditions for nominal and robust stability of MPC. Because the conditions are difficult to check he also states some necessary conditions for these sufficient conditions.

Genceli and Nikolaou [29] give sufficient conditions for robust closed-loop stability and investigate robust performance of dynamic matrix control (DMC) systems with hard input/soft output constraints. The authors consider an ℓ_1-norm performance index, a terminal state condition as a stability constraint, an impulse-response model with bounds on the variations of the coefficients. They derive a robustness test in terms of simple inequalities to be satisfied. This simplicity is largely lost in the extension to the MIMO case.

Primbs and Nevistíc [56] provide an off-line robustness analysis test of constrained finite receding horizon control which requires the solution of a set of linear matrix inequalities (LMIs). The test is based on the so called S-procedure and provides a (conservative) sufficient condition for $V(t)$ to be decreasing for all $\Sigma \in \mathcal{S}$, $\forall w(t) \in \mathcal{W}$. Both polytopic and structured uncertainty descriptions are considered. The authors also extend the idea to develop a robust synthesis method. It requires the solution of bilinear matrix inequalities (BMIs) and is computationally demanding.

More recently, Primbs [55] presented a new formulation of the analysis technique which is less conservative. The idea is to express the (optimal) input $u(t)$ obtained by the MPC law through the Lagrangian multipliers λ associated with the optimization problem (2a), and then to write the S-procedure in the $[x, u, \lambda]$-space.

6 Robust MPC Synthesis

In light of the discussion in Section 2.1, one has the following alternatives when synthesizing robust MPC laws:

1. Optimize performance of the nominal model or robust performance ?
2. Enforce state constraints on the nominal model or robustly ?
3. Adopt an *open-loop* or a *closed-loop prediction* scheme ?
4. How to guarantee robust stability ?

In the remaining part of the section we will discuss these questions.

6.1 Nominal vs. Robust Performance

The performance index (2a) depends on one particular model Σ and disturbance realization $w(t)$. In an uncertainty framework, two strategies are possible: (i) define a nominal model $\hat{\Sigma}$ and nominal disturbance $\hat{w}(t) = 0$, and optimize nominal performance; or (ii) solve the min-max problem to optimize robust performance

$$\min_{\mathbf{U}} \quad \max_{\substack{\Sigma \in \mathcal{S} \\ \{w(k+t)\}_{k=0}^{N_p-1} \subseteq \mathcal{W}}} \quad J(\mathbf{U}, x(t), \Sigma, w(\cdot)) \tag{13}$$

Min-max robust MPC was first proposed by Campo and Morari [18], and further developed in [2] and [69] for SISO FIR plants. Kothare *et al.* [35] optimize robust performance for polytopic/multi-model and structured feedback uncertainty, Scokaert and Mayne [63] for input disturbances only, and Lee and Yu [41] for linear time-varying and time-invariant state-space models depending on a vector of parameters $\theta \in \Theta$, where Θ is either an ellipsoid or a polyhedron. However it has two possible drawbacks. The first one is computational: Solving the problem (13) is computationally much more demanding than solving (2a) for a nominal model $\hat{\Sigma}$, $w(t) = 0$. However, under slightly restrict assumptions on the uncertainty, quite efficient algorithms are possible [73]. The second one is that the control action may be excessively conservative.

6.2   Input and State Constraints

In the presence of uncertainty, the constraints on the states variables (2b) can be enforced for all plant $\Sigma \in S$ (robust constraint fulfillment) or for a nominal system $\hat{\Sigma}$ only. One also has to distinguish between hard and soft state constraints, although the latter are preferable for the reasons discussed in Section 2.1. As command inputs are directly generated by the optimizer, input constraints do not present any additional difficulty relative to the nominal MPC case.

For uncertainty described in terms of $w(t) \in \mathcal{W}$ only, when the set $\mathcal{W}$ is a polyhedron, state constraints can be tackled through the theory of *maximal output admissible sets* MOAS developed in [31,32]. The theory provides tools to enforce hard constraints on states despite the presence of input disturbances, by computing the minimum output prediction horizon N_p which guarantees robust constraint fulfillment.

In [47,63] tools from MOAS theory are used to synthesize robust minimum-time control on line. The technique is based on the computation of the level sets of the value function, and deals with hard input/state constraints.

Bemporad and Garulli [7] also consider the effect of the worst input disturbance over the prediction horizon, and enforce constraint fulfillment for all possible disturbance realizations (output prediction horizons are again computed through algorithms inspired by MOAS theory). In addition, the authors consider the case when full state information is not available. They use the so-called *set-membership* (SM) state estimation   [62,13], through recursive algorithms based on parallelotopic approximation of the state uncertainty set [66,21].

When impulse-response descriptions are adopted, output constraints can be easily related to the uncertainty intervals of the impulse-response coefficients. For embedding input and state constraint into LMIs, the reader is referred to [35].

Robust fulfillment of state constraints can result in a very conservative behavior. Such an undesirable effect can be mitigated by using closed-loop prediction (see Sect. 8). Alternatively, when violations of the constraints are allowed, it can be more convenient to impose constraint satisfaction on the nominal plant $\hat{\Sigma}$ only.

Although unconstrained MPC for uncertain systems has been investigated, we do not review this literature here, because many superior linear robust control techniques are available.

7   Robust Stability

The minimum closed-loop requirement is robust stability, i.e., stability in the presence of uncertainty. In MPC the various design procedures achieve robust stability in two different ways: indirectly by specifying the performance objective and uncertainty description in such a way that the optimal control computations lead to robust stability; or directly by enforcing a type of robust contraction constraint which guarantees that the state will shrink for all plants in the uncertainty set.

7.1 Min-max performance optimization

While the generalization (13) of nominal MPC to the robust case appears natural, it is not without pitfalls. The min-max formulation as proposed in [18] alone does not guarantee robust stability as was demonstrated by Zheng [73] through a counterexample. To ensure robust stability the uncertainty must be assumed to be time varying. This added conservativeness may be prohibitive for demanding applications.

7.2 Robust contraction constraint

For stable plants, Zheng [73] introduces the stability constraint

$$\|x(t+1|t)\|_P \leq \lambda \|x(t)\|_P, \lambda < 1. \tag{14}$$

which forces the state to contract. When $P \succ 0$ is chosen as the solution of the Lyapunov equation $A'PA - P = -Q$, $Q \succ 0$, then this constraint can always be met for some u ($u(t+k) = 0$ satisfies this constraint and any constraint on u). Zheng [73] achieves robust stability by requiring the state to contract for all plants in $\mathcal{S}$. For the uncertain case constraint (14) is generalized by maximizing $\|x(t+1|t)\|_P$ over $\Sigma \in \mathcal{S}$.

For the multi-plant description, Badgwell [3] proposes a robust MPC algorithm for stable, constrained, linear plants that is a direct generalization of the nominally stabilizing regulator presented by Rawlings and Muske [58]. By using Lyapunov arguments, robust stability can be proved when the following stability constraint is imposed for each plant in the set.

$$J(\mathbf{U}, x(t), \Sigma_i) \leq J(\mathbf{U}_1^*, x(t), \Sigma_i) \tag{15}$$

This can be seen as a special case of the contraction constraint, where $J(\mathbf{U}, x(t), \Sigma_i)$ is the cost associated with the prediction model Σ_i for a fixed pair (N_p, N_m), and $\mathbf{U}_1 \triangleq \{u^*(t|t-1), \ldots, u^*(t-1+N_m|t-1), 0\}$ is the shifted optimal sequence computed at time $t-1$. Note that the stability constraints (15) are quadratic.

7.3 Robustly Invariant Terminal Sets

Invariant ellipsoidal terminal sets have been proposed recently in the nominal context as relaxations of the terminal equality constraint mentioned in Section 2.1 (see for instance [4] and references therein). Such techniques can be extended to robust MPC formulations, for instance by using the LMI techniques developed in [35]. Invariant terminal ellipsoid inevitably lead to Quadratically Constrained Quadratic Programs (QCQP), which can be solved through interior-point methods [44]. Alternatively, one can determine polyhedral robustly terminal invariant sets [16], which would lead to linear constraints, and therefore quadratic programming (QP), which is computationally cheaper than QCQP, at least for small/medium size problems.

8   Closed-Loop Prediction

Let us consider the design of a predictive controller which guarantees that hard state constraints are met in the presence of input disturbances $w(t)$. In order to achieve this task for every possible disturbance realization $w(t) \in \mathcal{W}$, the control action must be chosen safe enough to cope with the effect of the worst disturbance realization [30]. This effect is typically evaluated by predicting the *open-loop* evolution of the system driven by such a worst-case disturbance. This can be very conservative because in actual operation the disturbance effect is mitigated by feedback.

Lee and Yu [41] show that this problem can be addressed rigorously via Bellman's principle of optimality but that this is impractical for all but the simplest cases. As a remedy they introduce the concept of *closed-loop prediction*. For closed-loop prediction a feedback term $F_k x(t + k|t)$ is included in the expression for $u(t + k|t)$,

$$u(t + k|t) = F_k x(t + k|t) + v(k), \tag{16}$$

and the MPC controller optimizes with respect to both F_k and $v(k)$.

The benefit of this feedback formulation is discussed in [5] and is briefly reviewed here. In open-loop prediction the disturbance effect is passively suffered, while *closed-loop prediction* attempts to reduce the effect of disturbances. In open-loop schemes the uncertainty produced by the disturbances grows over the prediction horizon.

As an example, consider a hard output constraint $y_{min} \leq y(t) \leq y_{max}$. The output evolution due to (16) from initial state $x(t)$ for $F_k \equiv F$ is

$$y(t + k|t) = C(A + BF)^k x(t) + \sum_{k=0}^{t-1} C(A + BF)^k v(k) +$$

$$+ \sum_{k=0}^{t-1} C(A + BF)^k Hw(t - 1 - k) + Kw(t) \tag{17}$$

It is clear that F offers some degrees of freedom to counteract the effect of $w(t)$ by modifying the multiplicative term $(A + BF)^k$. For instance, if F renders $(A + BF)$ nilpotent, $y(t + k|t)$ is only affected by the last n disturbance inputs $w(k - n + 1), \ldots, w(k)$, and consequently no uncertainty accumulation occurs. On the other hand, if F is set to 0 (open-loop prediction) and A has eigenvalues close to the unit circle, the disturbance action leads to

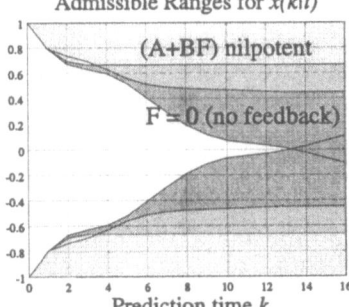

Fig. 6. Benefits of closed-loop prediction: Admissible ranges for the output $y(t + k|t)$ for different feedback LQ gains F (input weight $\rho = 0, 1, +\infty$)

very conservative constraints, and consequently to poor performance. Fig. 6 shows this effect for different gains F, selected by solving LQ problems with unit output weight and input weights $\rho = 0$, $\rho = 1$, and $\rho = +\infty$. The last one corresponds to open-loop prediction ($F = 0$).

For a wide range of uncertainty models, Kothare *et al.* [35] design, at each time step, a state-feedback control law that minimizes a 'worst-case' infinite horizon objective function, subject to input and output constraints. The authors transform the problem of minimizing an upper bound on the worst-case objective function to a convex optimization involving linear matrix inequalities (LMIs). A robustly stable MPC algorithm results. On one hand the closed-loop formulation reduces the conservativeness. On the other hand, the algorithm requires the uncertainty to be time-varying which may be conservative for some applications.

9 Computation

In the previous sections we discussed the formulation of various robust MPC algorithms, which differed with respect to the uncertainty descriptions, the performance criteria, and the type of stability constraints. In practice the choice is often dictated by computational considerations.

Uncertainty descriptions involving impulse/step-response coefficients or bounded input disturbances are easier to deal with, as the optimization problem can often be recast as an LP.

In [35] the authors solve optimal control problems with state-space uncertainty descriptions through LMIs. For the technique proposed in [33], where a worst case quadratic performance criterion is minimized over a finite set of models subject to input/state constraints, the authors report that problems with more than 1000 variables and 5000 constraints can be solved in a few minutes on a workstation by using interior-point methods.

For impulse and step response uncertainty, Bemporad and Mosca [8] propose a computationally efficient approach based on the *reference gover-*

nor [32,6]. The main idea is to separate the stabilization problem from the robust constraint fulfillment problem. The first is left to a conventional linear robust controller. Constraints are enforced by manipulating the desired set-points at a higher level (basically the reference trajectory is smoothed out when abrupt set-point changes would lead to constraint violations). The advantages of this scheme are that typically only one scalar degree of freedom suffices, as reported in [8], where the on-line optimization is reduced to a small number of LPs.

10   Conclusions and Research Directions

While this review is not complete it reflects the state of the art. It is apparent that none of the methods presented is suitable for use in industry except maybe in very special situations. The techniques are hardly an alternative to ad hoc MPC tuning based on exhaustive simulations for ranges of operating conditions. Choosing the right robust MPC technique for a particular application is an art and much experience is necessary to make it work — even on a simulation case study. Much research remains to be done, but the problems are difficult. Some topics for investigation are suggested next.

Contraction constraints have been shown to be successful tools to get stability guarantees, but typically performance suffers. By forcing the state to decrease in a somewhat arbitrary manner, the evolution is driven away from optimality as measured by the performance index. The contraction constraints which are in effect Lyapunov functions are only sufficient for stability. In principle, less restrictive criteria could be found. *Integral Quadratic Constraints* [49] could be embedded in robust MPC in order to deviate as little as possible from optimal performance but still guarantee robust stability.

Robustly invariant terminal sets can be adopted as an alternative to contraction constraints. As mentioned in Sect. 7, ellipsoids and polyhedra can be determined off-line, by utilizing tools from robustly invariant set theories [16].

The benefits of closed-loop prediction were addressed in Sect. 8. However very little research has been done toward the development of computationally efficient MPC algorithms.

Finally, the algorithms should be linked to appropriate identification procedures for obtaining the models and the associated uncertainty descriptions.

Acknowledgments

The authors thank Dr. James A. Primbs and Prof. Alexandre Megretski for useful discussions. Alberto Bemporad was supported by the Swiss National Science Foundation.

References

1. Allwright, J. C. (1994). On min-max model-based predictive control. In: *Advances in Model-Based Predictive Control.* pp. 415–426. Oxford Press Inc.,N. Y.. New York.

2. Allwright, J.C. and G.C. Papavasiliou (1992). On linear programming and robust model-predictive control using impulse-responses. *Systems & Control Letters* **18**, 159–164.

3. Badgwell, T. A. (1997). Robust model predictive control of stable linear systems. *Int. J. Control* **68**(4), 797–818.

4. Bemporad, A. (1998*a*). A predictive controller with artificial Lyapunov function for linear systems with input/state constraints. *Automatica* **34**(10), 1255–1260.

5. Bemporad, A. (1998*b*). Reducing conservativeness in predictive control of constrained systems with disturbances. In: *Proc. 37th IEEE Conf. on Decision and Control.* Tampa, FL. pp. 1384–1391.

6. Bemporad, A., A. Casavola and E. Mosca (1997). Nonlinear control of constrained linear systems via predictive reference management. *IEEE Trans. Automatic Control* **AC-42**(3), 340–349.

7. Bemporad, A. and A. Garulli (1997). Predictive control via set-membership state estimation for constrained linear systems with disturbances. In: *Proc. European Control Conf..* Bruxelles, Belgium.

8. Bemporad, A. and E. Mosca (1998). Fulfilling hard constraints in uncertain linear systems by reference managing. *Automatica* **34**(4), 451–461.

9. Bemporad, A. and M. Morari (1999). Control of systems integrating logic, dynamics, and constraints. *Automatica* **35**(3), 407–427. `ftp://control.ethz.ch/pub/reports/postscript/AUT98-04.ps`.

10. Bemporad, A., L. Chisci and E. Mosca (1994). On the stabilizing property of the zero terminal state receding horizon regulation. *Automatica* **30**(12), 2013–2015.

11. Benvenuti, L. and L. Farina (1998). Constrained control for uncertain discrete-time linear systems. *Int. J. Robust Nonlinear Control* **8**, 555–565.

12. Berber, R., Ed.) (1995). *Methods of Model Based Process Control.* Vol. 293 of *NATO ASI Series E: Applied Sciences.* Kluwer Academic Publications. Dortrecht, Netherlands.

13. Bertsekas, D.P. and I.B. Rhodes (1971). Recursive state estimation for a set-membership description of uncertainty. *IEEE Trans. Automatic Control* **16**, 117–128.

14. Bitmead, R. R., M. Gevers and V. Wertz (1990). *Adaptive Optimal Control. The Thinking Man's GPC.* International Series in Systems and Control Engineering. Prentice Hall.

15. Blanchini, F. (1990). Control synthesis for discrete time systems with control and state bounds in the presence of disturbances. *J. of Optimization Theory and Applications* **65**(1), 29–40.

16. Blanchini, F. (1999). Set invariance in control — a survey. *Automatica.* In press.

17. Camacho, E.F. and C. Bordons (1995). *Model Predictive Control in the Process Industry.* Advances in Industrial Control. Springer Verlag.

18. Campo, P.J. and M. Morari (1987). Robust model predictive control. In: *Proc. American Contr. Conf.*. Vol. 2. pp. 1021–1026.
19. Campo, P.J. and M. Morari (1989). Model predictive optimal averaging level control. *AIChE Journal* 35(4), 579–591.
20. Chen, H., C. W. Scherer and F. Allgöwer (1997). A game theoretic approach to nonlinear robust receding horizon control of constrained systems. In: *Proc. American Contr. Conf.*. Vol. 5. pp. 3073–3077.
21. Chisci, L., A. Garulli and G. Zappa (1996). Recursive state bounding by parallelotopes. *Automatica* 32(7), 1049–1056.
22. Clarke, D. W., C. Mohtadi and P. S. Tuffs (1987a). Generalized predictive control–I. The basic algorithm. *Automatica* 23, 137–148.
23. Clarke, D. W., C. Mohtadi and P. S. Tuffs (1987b). Generalized predictive control–II. Extensions and interpretations. *Automatica* 23, 149–160.
24. Clarke, D.W., Ed.) (1994). *Advances in Model-Based Predictive Control*. Oxford University Press.
25. Cutler, C. R. and B. L. Ramaker (1979). Dynamic matrix control– A computer control algorithm. In: *AIChE 86th National Meeting*. Houston, TX.
26. Cutler, C. R. and B. L. Ramaker (1980). Dynamic matrix control– A computer control algorithm. In: *Joint Automatic Control Conf.*. San Francisco, California.
27. De Nicolao, G., L. Magni and R. Scattolini (1996). Robust predictive control of systems with uncertain impulse response. *Automatica* 32(10), 1475–1479.
28. Garcia, C.E., D.M. Prett and M. Morari (1989). Model predictive control: Theory and practice – a survey. *Automatica*.
29. Genceli, H. and M. Nikolaou (1993). Robust stability analysis of constrained ℓ_1-norm model predictive control. *AIChE J.* 39(12), 1954–1965.
30. Gilbert, E.G. and I. Kolmanovsky (1995). Discrete-time reference governors for systems with state and control constraints and disturbance inputs. In: *Proc. 34th IEEE Conf. on Decision and Control*. pp. 1189–1194.
31. Gilbert, E.G. and K. Tin Tan (1991). Linear systems with state and control constraints: the theory and applications of maximal output admissible sets. *IEEE Trans. Automatic Control* 36, 1008–1020.
32. Gilbert, E.G., I. Kolmanovsky and K. Tin Tan (1995). Discrete-time reference governors and the nonlinear control of systems with state and control constraints. *Int. J. Robust Nonlinear Control* 5(5), 487–504.
33. Hansson, A. and S. Boyd (1998). Robust optimal control of linear discrete time systems using primal-dual interior-point methods. In: *Proc. American Contr. Conf.*. Vol. 1. pp. 183–187.
34. Keerthi, S.S. and E.G. Gilbert (1988). Optimal infinite-horizon feedback control laws for a general class of constrained discrete-time systems: stability and moving-horizon approximations. *J. Opt. Theory and Applications* 57, 265–293.
35. Kothare, M.V., V. Balakrishnan and M. Morari (1996). Robust constrained model predictive control using linear matrix inequalities. *Automatica* 32(10), 1361–1379.
36. Kwon, W. H. (1994). Advances in predictive control: Theory and application. In: *1st Asian Control Conf.*. Tokyo. (updated in October, 1995).
37. Kwon, W.H., A.M. Bruckstein and T. Kailath (1983). Stabilizing state-feedback design via the moving horizon method. *Int. J. Control* 37(3), 631–643.

38. Kwon, W.H. and A.E. Pearson (1977). A modified quadratic cost problem and feedback stabilization of a linear system. *IEEE Trans. Automatic Control* **22**(5), 838–842.

39. Kwon, W.H. and A.E. Pearson (1978). On feedback stabilization of time-varying discrete linear systems. *IEEE Trans. Automatic Control* **23**, 479–481.

40. Kwon, W.H. and D. G. Byun (1989). Receding horizon tracking control as a predictive control and its stability properties. *Int. J. Control* **50**(5), 1807–1824.

41. Lee, J. H. and Z. Yu (1997). Worst-case formulations of model predictive control for systems with bounded parameters. *Automatica* **33**(5), 763–781.

42. Lee, J.H. and B. Cooley (1997). Recent advances in model predictive control. In: *Chemical Process Control - V*. Vol. 93, no. 316. pp. 201–216b. AIChe Symposium Series - American Institute of Chemical Engineers.

43. Lee, K. H., W. H. Kwon and J. H. Lee (1996). Robust receding-horizon control for linear systems with model uncertainties. In: *Proc. 35th IEEE Conf. on Decision and Control*. pp. 4002–4007.

44. Lobo, M., L. Vandenberghe and S. Boyd (1997). Software for second-order cone programming. user's guide. http://www-isl.stanford.edu/ boyd/SOCP.html.

45. Mäkilä, P. M., J. R. Partington and T. K. Gustafsson (1995). Worst-case control-relevant identification. *Automatica* **31**, 1799–1819.

46. Martín Sánchez, J.M. and J. Rodellar (1996). *Adaptive Predictive Control*. International Series in Systems and Control Engineering. Prentice Hall.

47. Mayne, D. Q. and W. R. Schroeder (1997). Robust time-optimal control of constrained linear systems. *Automatica* **33**(12), 2103–2118.

48. Mayne, D.Q. (1997). Nonlinear model predictive control: an assessment. In: *Chemical Process Control - V*. Vol. 93, no. 316. pp. 217–231. AIChe Symposium Series - American Institute of Chemical Engineers.

49. Megretski, A. and A. Rantzer (1997). System analysis via integral quadratic constraints. *IEEE Trans. Automatic Control* **42**(6), 819–830.

50. Milanese, M. and A. Vicino (1993). Information-based complexity and non-parametric worst-case system identification. *Journal of Complexity* **9**, 427–446.

51. Morari, M. (1994). Model predictive control: Multivariable control technique of choice in the 1990s ?. In: *Advances in Model-Based Predictive Control*. pp. 22–37. Oxford University Press Inc.. New York.

52. Noh, S. B., Y. H. Kim, Y. I. Lee and W. H. Kwon (1996). Robust generalised predictive control with terminal output weightings. *J. Process Control* **6**(2/3), 137–144.

53. Polak, E. and T.H. Yang (1993a). Moving horizon control of linear systems with input saturation and plant uncertainty–part 1. robustness. *Int. J. Control* **58**(3), 613–638.

54. Polak, E. and T.H. Yang (1993b). Moving horizon control of linear systems with input saturation and plant uncertainty–part 2. disturbance rejection and tracking. *Int. J. Control* **58**(3), 639–663.

55. Primbs, J.A. (1999). The analysis of optimization based controllers. In: *Proc. American Contr. Conf.*. San Diego, CA.

56. Primbs, J.A. and V. Nevistíc (1998). A framework for robustness analysis of constrained finite receding horizon control. In: *Proc. American Contr. Conf.*. pp. 2718–2722.

57. Qin, S.J. and T.A. Badgewell (1997). An overview of industrial model predictive control technology. In: *Chemical Process Control - V*. Vol. 93, no. 316. pp. 232–256. AIChe Symposium Series - American Institute of Chemical Engineers.

58. Rawlings, J.B. and K.R. Muske (1993). The stability of constrained receding-horizon control. *IEEE Trans. Automatic Control* **38**, 1512–1516.

59. Richalet, J., A. Rault, J.L. Testud and J. Papon (1978). Model predictive heuristic control: applications to industrial processes. *Automatica* **14**(5), 413–428.

60. Santis, E. De (1994). On positively invariant sets for discrete-time linear systems with disturbance: an application of maximal disturbance sets. *IEEE Trans. Automatic Control* **39**(1), 245–249.

61. Santos, L. O. and L. T. Biegler (1998). A tool to analyze robust stability for model predictive controllers. *J. Process Control*.

62. Schweppe, F.C. (1968). Recursive state estimation: unknown but bounded errors and system inputs. *IEEE Trans. Automatic Control* **13**, 22–28.

63. Scokaert, P.O.M. and D.Q. Mayne (1998). Min-max feedback model predictive control for constrained linear systems. *IEEE Trans. Automatic Control* **43**(8), 1136–1142.

64. Scokaert, P.O.M. and J.B. Rawlings (1996). Infinite horizon linear quadratic control with constraints. In: *Proc. IFAC*. Vol. 7a-04 1. San Francisco, USA. pp. 109–114.

65. Soeterboek, R. (1992). *Predictive Control - A Unified Approach*. International Series in Systems and Control Engineering. Prentice Hall.

66. Vicino, A. and G. Zappa (1996). Sequential approximation of feasible parameter sets for identification with set membership uncertainty. *IEEE Trans. Automatic Control* **41**, 774–785.

67. Yang, T.H. and E. Polak (1993). Moving horizon control of nonlinear systems with input saturation, disturbances and plant uncertainty. *Int. J. Control* **58**, 875–903.

68. Zafiriou, E. (1990). Robust model predictive control of processes with hard constraints. *Computers & Chemical Engineering* **14**(4/5), 359–371.

69. Zheng, A. and M. Morari (1993). Robust stability of constrained model predictive control. In: *Proc. American Contr. Conf.*. Vol. 1. San Francisco, CA. pp. 379–383.

70. Zheng, A. and M. Morari (1994). Robust control of linear time-varying systems with constraints. In: *Proc. American Contr. Conf.*. Vol. 3. pp. 2416–2420.

71. Zheng, A. and M. Morari (1995). Stability of model predictive control with mixed constraints. *IEEE Trans. Automatic Control* **40**, 1818–1823.

72. Zheng, A. and M. Morari (1998). Robust control of lineary systems with constraints. *Unpublished report*.

73. Zheng, Z. Q. (1995). Robust Control of Systems Subject to Constraints. Ph.D. dissertation. California Institute of Technology. Pasadena, CA, U.S.A.

Intrinsic Performance Limits of Linear Feedback Control*

Jie Chen

Department of Electrical Engineering
University of California
Riverside, CA 92521
Tel: (909)787-3688   Fax: (909)787-3188
Email: jchen@ee.ucr.edu

Abstract. Nevanlinna-Pick interpolation techniques are employed in this paper to derive for multivariable systems exact expressions and bounds for the best attainable $\mathcal{H}_\infty$ norms of the sensitivity and complementary sensitivity functions. These results improve the previously known performance bounds and provide intrinsic limits on the performance of feedback systems irreducible via compensator design, leading to new insights toward the understanding of fundamental limitations in feedback control. It becomes clear that in a multivariable system the best achievable performance is constrained by plant nonminimum phase zeros and unstable poles, and additionally, is dependent on the mutual orientation of the zero and pole directions.

1 Introduction

Control system design is dictated by many performance considerations and physical constraints. An important step in the design process is to analyze how plant characteristics may impose constraints upon design and thus may limit the level of achievable performance, and accordingly, to determine the best achievable performance under the design constraints. This task has become even more tangible nowadays, as current control design theory and practice relies heavily on optimization-based numerical routines and tools. A performance analysis in this spirit can aid control design in several aspects. First, it may yield an intrinsic limit on the best performance attainable irrespective of compensator design. Not only does this give the optimal performance *a priori*, but also it serves as a guiding benchmark in the design process. Secondly, it may provide a clear indication on what and how plant properties may inherently conflict and thus undermine performance objectives, and hence valuable insight regarding the inherent limitation in achieving performance, and more generally, the difficulty to

---

* This research was supported in part by the NSF under Grant ECS-9623228.

control the plant. This in turn will provide analytical justifications to long held classical heuristic observations, and indeed, to extend such heuristics much further beyond.

For both its intrinsic appeal and fundamental implication, performance limitation of feedback control has been a topic of enduring interest, in both classical and recent control literature [20]. Most notably, the pioneering work by Bode [3] has had a profound impact in this area of study and has inspired renewed research effort dated most recently, leading to a variety of extensions and new results which seek to quantify design constraints and performance limitations by logarithmic integral relations of Bode and Poisson type [10,4,17,5,6]. On the other hand, study of best achievable performance is an integrable issue in optimal control and it has lended bounds on optimal performance indices defined under various criteria. These include bounds on the minimal $\mathcal{H}_\infty$ norm of closed loop transfer functions typically employed to represent performance objectives in feedback design, such as the sensitivity and complementary sensitivity functions [25,4,15]. Bounds and exact expressions have also been obtained for LQR and $\mathcal{H}_2$ related optimal performance indices, concerning, e.g., cheap regulator problem [16], servomechanism problem [19], and optimal tracking [18,9]. These developments have been substantial and are continuing to branch to different problems and different system categories; recent extensions are found for, e.g., filtering problems [12], and sampled data systems [11]. The understanding in the limitations of feedback control has also been compelling. While the results in question may differ in forms and contexts, they all unequivocally point to the fact that such plant characteristics as right half plane zeros and poles will inherently constrain a system's achievable performance, never reducible no matter how the compensator may be designed.

This work continues the aforementioned studies on performance limitations in linear feedback control systems. We consider multivariable, continuous-time and discrete-time systems, and we derive exact expressions and bounds for the best achievable $\mathcal{H}_\infty$ norms of the sensitivity and complementary sensitivity functions. Unlike in the previous work, which relies on either Bode and Poisson type integrals [10,4,17,5] or properties of analytic functions, notably, the maximum modulus principle [25,4], we use a technique based upon analytic function interpolation theory. Specifically, the optimal performance problems are recast as one of Nevanlinna-Pick interpolation with a minimal $\mathcal{H}_\infty$ norm. This formulation furnishes a unified framework for studying intrinsic performance issues, and it leads to new and more general results, particularly for multivariable systems. Specifically, our derivation results in exact performance limits expressed in terms of plant nonminimum phase zeros and unstable poles. The expressions characterize explicitly the dependence of the performance measures, quantified by the sensitivity and complementary sensitivity magnitudes, on the zeros and poles, as well as the alignment between their directions.

2   Preliminaries

We begin by introducing the basic notation in this paper. Let $\mathbb{D} := \{z : |z| < 1\}$, $\overline{\mathbb{D}} := \{z : |z| \leq 1\}$, and $\mathbb{D}^c := \{z : |z| > 1\}$ denote the open unit disc, the closed unit disc, and its exterior, respectively. Similarly, let $\mathbb{C}_+ := \{z : \text{Re}(z) > 0\}$ denote the open right half plane and $\overline{\mathbb{C}}_+$ the closed right half plane. For any complex number z, we denote its conjugate by $\overline{z}$. For any complex vector x, we denote its conjugate transpose by x^H, and its Euclidean norm by $\|x\|$. Similarly, for any complex matrix A, A^H denotes its conjugate transpose. The largest singular value of a matrix A will be written as $\overline{\sigma}(A)$. If A is a Hermitian matrix, we denote by $\overline{\lambda}(A)$ its largest eigenvalue. In particular, we write $A \geq 0$ if A is nonnegative definite, and $A > 0$ if it is positive definite. For any unitary vectors u, $v \in \mathbb{C}^n$, we denote by $\angle(u, v)$ the *principal angle* [2] between the two one-dimensional subspaces, also known as the directions, spanned by u and v:

$$\cos \angle(u, v) := |u^H v|.$$

Following the usual convention, we denote the transfer function matrix of a continuous-time system by $G(s)$, and that of a discrete-time system by $G(z)$. If $G(s)$ is stable, we may define its $\mathcal{H}_\infty$ norm by

$$\|G(s)\|_\infty := \sup_{\text{Re}(s) > 0} \overline{\sigma}(G(s)).$$

Similarly, the $\mathcal{H}_\infty$ norm of a stable $G(z)$ is defined as

$$\|G(z)\|_\infty := \sup_{|z| < 1} \overline{\sigma}(G(z)).$$

Throughout this paper all transfer function matrices are assumed to satisfy conjugate symmetry property.

We consider the feedback system depicted in Figure 1, which is assumed to be linear, time-invariant. For a continuous-time system, we represent the plant and compensator transfer function matrices by $P(s)$ and $F(s)$, and for a discrete-time system we write $P(z)$ and $F(z)$. The open loop transfer function, the sensitivity function, and the complementary sensitivity function are defined respectively by

$$L = PF, \quad S = (I + L)^{-1}, \quad T = L(I + L)^{-1}, \tag{1}$$

with an appropriate argument s or z.

For a left-invertible transfer function matrix P, it is well-known that any $z \in \mathbb{C}$ is a zero of P if and only if $P(z)\eta = 0$ for some unitary vector η, where η is called an input zero direction vector associated with z. Similarly, if P is right-invertible, then $z \in \mathbb{C}$ is a zero of P if and only if $w^H P(z) = 0$

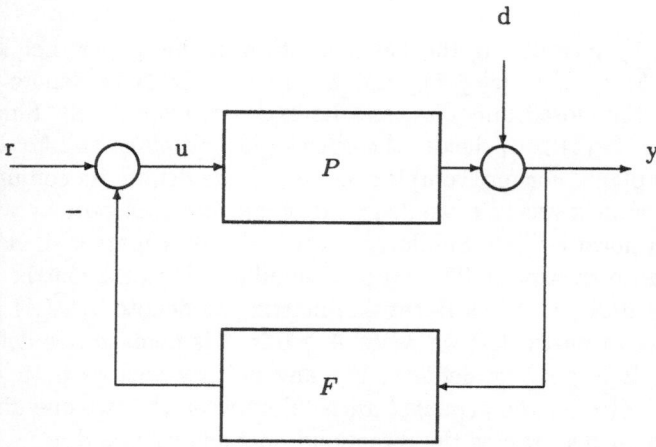

Fig. 1. The feedback system

for some unitary vector w, and w is called an output zero direction vector associated with z. Additionally, for a square and invertible P, it is well-known that any $p \in \mathbb{C}$ is a pole of P if and only if it is a zero of P^{-1}, and as such one may define analogously the pole direction vectors associated with the pole p via $P^{-1}(p)\eta = 0$ and $w^H P^{-1}(p) = 0$. Note that an alternative, equivalent definition of zeros and poles can be made by resorting to system representations via coprime factorizations. The latter can be found in [4].

For a continuous-time system, a pole in $\overline{\mathbb{C}}_+$ is called an unstable pole and a zero in $\overline{\mathbb{C}}_+$ is called a nonminimum phase zero. A transfer function matrix is said to be minimum phase if it has no zero in $\overline{\mathbb{C}}_+$, and otherwise nonminimum phase. Similarly, for a discrete-time system, unstable poles and nonminimum phase zeros lie outside $\mathbb{D}$, and a transfer function matrix is minimum phase if all its zeros are in $\mathbb{D}$ and nonminimum phase otherwise. For any continuous-time or discrete-time system, in the sequel, we shall assume that no unstable pole and nonminimum phase zero will coincide.

The following result is well-known, and can be found in [4] and [5].

Lemma 2.1 *Let L, S, and T be defined by (2.1). Then, $p \in \mathbb{C}_+$ (resp. $\mathbb{D}^c$) is a zero of S with zero direction vector η if and only if it is a pole of L with pole direction vector η, and $z \in \mathbb{C}_+$ (resp. $\mathbb{D}^c$) is a zero of T with zero direction vector w if and only if it is a zero of L with zero direction vector w. Suppose that $p \in \mathbb{C}_+$ (resp. $\mathbb{D}^c$) is a pole of L with input pole direction vector η, and $z \in \mathbb{C}_+$ (resp. $\mathbb{D}^c$) is a zero of L with output zero direction vector w. Then, the following conditions hold:*

$$S(p)\eta = 0, \qquad T(p)\eta = \eta,$$

$$w^H S(z) = w^H, \qquad w^H T(z) = 0.$$

3 Main Results

We now derive a number of exact expressions and bounds for the best achievable $\mathcal{H}_\infty$ norms of the sensitivity and complementary sensitivity functions. These results provide *a priori* intrinsic performance limits independent of compensator design, and are useful in several aspects. First, they improve the earlier performance bounds, derived mainly for SISO systems [15,25]. Secondly, both the exact expressions and bounds are shown to be directly dependent on the zero and pole direction vectors, demonstrating explicitly how the relative orientation between zero and pole directions may impose limits on the best achievable performance. Finally, when weighted sensitivity and complementary sensitivity functions are of interest, they shed light on how the weighting functions should be selected to reflect the effects of plant nonminimum phase zeros and unstable poles. Our development draws upon the recognition that a sensitivity minimization problem, in essence, may be posed as one of the classical Nevanlinna-Pick interpolation; this fact is well-known and has been used in the early formulations of $\mathcal{H}_\infty$ control problems (see, e.g., [25]).

3.1 Continuous-Time Systems

We first develop results for continuous-time systems. The following preliminary lemma, which gives a necessary and sufficient condition to a two-sided tangential Nevanlinna-Pick interpolation problem and is taken from [1], plays a key role in our derivation.

Lemma 3.1 *Consider two sets of distinct points $z_i \in \mathbb{C}_+$, $i = 1, \cdots, l$ and $p_i \in \mathbb{C}_+$, $i = 1, \cdots, k$. Assume that $z_i \neq p_j$ for any i and j. Then, there exists a matrix function $H(s)$ such that (i) $H(s)$ is analytic in $\mathbb{C}_+$, (ii) $\|H(s)\|_\infty \leq 1$, and (iii) $H(s)$ satisfies the conditions*

$$x_i^H H(z_i) = y_i^H, \qquad i = 1, \cdots, l, \tag{1}$$
$$H(p_i)u_i = v_i, \qquad i = 1, \cdots, k \tag{2}$$

for some vector sequences x_i, y_i, $i = 1, \cdots, l$ and u_i, v_i, $i = 1, \cdots, k$, of compatible dimensions, if and only if

$$Q := \begin{bmatrix} Q_1 & Q_{12}^H \\ Q_{12} & Q_2 \end{bmatrix} \geq 0,$$

where $Q_1 \in \mathbb{C}^{l \times l}$, $Q_2 \in \mathbb{C}^{k \times k}$, and $Q_{12} \in \mathbb{C}^{k \times l}$ are defined as

$$Q_1 := \left[\frac{x_i^H x_j - y_i^H y_j}{z_i + \overline{z}_j} \right],$$

$$Q_2 := \left[\frac{u_i^H u_j - v_i^H v_j}{\overline{p}_i + p_j} \right],$$

$$Q_{12} := \left[\frac{y_i^H u_j - x_i^H v_j}{z_i - p_j} \right].$$

Furthermore, $H(s)$ is nonunique if and only if $Q > 0$.

Suppose now that the plant transfer function matrix $P(s)$ satisfies

Assumption 3.1 $P(s)$ has neither finite zero nor pole on the imaginary axis.

Then, an application of Lemma 3.1 gives rise to the following best achievable $\mathcal{H}_\infty$ norms of the sensitivity and complementary sensitivity functions.

Theorem 3.1 *Let $z_i \in \mathbb{C}_+$, $i = 1, \cdots, l$, be the zeros of $P(s)$ with output direction vectors $\tilde{w}_i$, and $p_i \in \mathbb{C}_+$, $i = 1, \cdots, k$, be the poles of $P(s)$ with input direction vectors $\tilde{\eta}_i$. Furthermore, assume that z_i and p_i are all distinct. Define*

$$\gamma_{\min}^S := \inf \{ \|S(s)\|_\infty : F(s) \text{ stabilizes } P(s) \},$$
$$\gamma_{\min}^T := \inf \{ \|T(s)\|_\infty : F(s) \text{ stabilizes } P(s) \}.$$

Then,

$$\gamma_{\min}^S = \gamma_{\min}^T = \sqrt{1 + \overline{\sigma}^2 \left(Q_p^{-1/2} Q_{zp} Q_z^{-1/2} \right)}, \tag{3}$$

where $Q_z \in \mathbb{C}^{l \times l}$, $Q_p \in \mathbb{C}^{k \times k}$, and $Q_{zp} \in \mathbb{C}^{k \times l}$ are given by

$$Q_z := \left[\frac{\tilde{w}_i^H \tilde{w}_j}{z_i + \overline{z}_j} \right], \qquad Q_p := \left[\frac{\tilde{\eta}_i^H \tilde{\eta}_j}{\overline{p}_i + p_j} \right], \qquad Q_{zp} := \left[\frac{\tilde{w}_i^H \tilde{\eta}_j}{z_i - p_j} \right].$$

Proof. We shall prove (3) for $\gamma_{\min}^S$, but the proof for $\gamma_{\min}^T$ follows analogously. Let $H(s) := S(s)/\gamma$. It follows that

$$\tilde{w}_i^H H(z_i) = \frac{\tilde{w}_i^H}{\gamma}, \qquad i = 1, \cdots, l,$$
$$H(p_i)\tilde{\eta}_i = 0, \qquad i = 1, \cdots, k.$$

In light of Lemma 3.1, in order for the closed loop system to be stable and $\|S(s)\|_\infty \leq \gamma$, it is necessary and sufficient that

$$\left[\begin{matrix} \left(1 - \frac{1}{\gamma^2}\right) Q_z & \frac{1}{\gamma} Q_{zp}^H \\ \frac{1}{\gamma} Q_{zp} & Q_p \end{matrix} \right] \geq 0. \tag{4}$$

We claim that $Q_p > 0$. This is clear by noting that $H_1(s) :\equiv 0$ and

$$H_2(s) := \left(\prod_{i=1}^n \frac{s - p_i}{s + \overline{p}_i} \right) I$$

satisfy the interpolation constraints $H_1(p_i)\tilde{\eta}_i = 0$, and $H_2(p_i)\tilde{\eta}_i = 0$. Since $Q_p > 0$, we may invoke Schur complement (see, e.g., [14], pp. 472), arriving at the equivalent condition

$$\left(1 - \frac{1}{\gamma^2}\right) Q_z - \frac{1}{\gamma} Q_{zp}^H Q_p^{-1} \frac{1}{\gamma} Q_{zp} \geq 0,$$

or alternatively,

$$Q_z - \frac{1}{\gamma^2} \left(Q_z + Q_{zp}^H Q_p^{-1} Q_{zp}\right) \geq 0.$$

In a similar manner, it is easy to recognize that $Q_z > 0$. By pre- and post-multiplying $Q_z^{-1/2}$, the above condition can be further written as

$$\gamma^2 I \geq I + \left(Q_p^{-1/2} Q_{zp} Q_z^{-1/2}\right)^H \left(Q_p^{-1/2} Q_{zp} Q_z^{-1/2}\right),$$

which is equivalent to

$$\gamma \geq \sqrt{1 + \bar{\sigma}^2 \left(Q_p^{-1/2} Q_{zp} Q_z^{-1/2}\right)}.$$

The proof may now be completed by noting that

$$\gamma_{\min}^S = \inf \left\{\gamma : \quad \text{The closed loop system is stable and } \|S(s)\|_\infty \leq \gamma\right\}$$

$$= \inf \left\{\gamma : \quad \gamma \geq \sqrt{1 + \bar{\sigma}^2 \left(Q_p^{-1/2} Q_{zp} Q_z^{-1/2}\right)}\right\}$$

$$= \sqrt{1 + \bar{\sigma}^2 \left(Q_p^{-1/2} Q_{zp} Q_z^{-1/2}\right)}.$$

∎

A distinct advantage of (3) is that it exhibits how the plant unstable poles and nonminimum phase zeros may interact to impose an irreducible limit to the peak sensitivity and complementary sensitivity magnitudes, independently of compensator design. This is manifested through the dependence of $\gamma_{\min}^S$ and $\gamma_{\min}^T$ on the inner products $\tilde{w}_i^H \tilde{w}_j$, $\tilde{\eta}_i^H \tilde{\eta}_j$, and $\tilde{w}_i^H \tilde{\eta}_j$, as seen in the construction of the matrices Q_z, Q_p, and Q_{zp}. A clear message here is that the plant unstable poles and nonminimum phase zeros affect a system's performance in a rather complex manner: not only the zeros or poles alone can couple to have a negative effect, but also any zero and pole may be aligned in ways leading to a undesirable consequence. The latter statement may be seen explicitly from the following corollary.

Corollary 3.1 *Suppose that $P(s)$ has only one nonminimum phase zero $z \in \mathbb{C}_+$, with an output direction vector $\tilde{w}$, and one unstable pole $p \in \mathbb{C}_+$, with an input direction vector $\tilde{\eta}$. Furthermore, assume that $z \neq p$. Then,*

$$\gamma_{\min}^S = \gamma_{\min}^T = \sqrt{\sin^2 \angle(\tilde{w}, \tilde{\eta}) + \left|\frac{p + \bar{z}}{p - z}\right|^2 \cos^2 \angle(\tilde{w}, \tilde{\eta})}. \tag{5}$$

Proof. The result follows by noting that

$$Q_z = \frac{1}{2\mathrm{Re}(z)}, \qquad Q_p = \frac{1}{2\mathrm{Re}(p)}, \qquad Q_{zp} = \frac{\tilde{w}^H \tilde{\eta}}{z - p},$$

and by use of the identity $|p + \bar{z}|^2 - |p - z|^2 = 4\mathrm{Re}(z)\mathrm{Re}(p)$. ∎

Remark 3.1 In [15], Khargonekar and Tannenbaum obtained a similar expression for SISO systems. Corollary 3.1 generalizes their result to MIMO systems. Moreover, Corollary 5.1 can be extended to yield the bound

$$\gamma_{\min}^S = \gamma_{\min}^T \geq \sqrt{\sin^2 \angle(\tilde{w}_i, \tilde{\eta}_j) + \left|\frac{p_j + \bar{z}_i}{p_j - z_i}\right|^2 \cos^2 \angle(\tilde{w}_i, \tilde{\eta}_j)},$$

which was obtained previously in [5]. Indeed, since a necessary condition for (4) to hold is that

$$\begin{bmatrix} \left(1 - \frac{1}{\gamma^2}\right) \frac{1}{2\mathrm{Re}(z_i)} & \frac{1}{\gamma} \frac{\tilde{\eta}_j^H \tilde{w}_i}{\bar{z}_i - \bar{p}_j} \\ \frac{1}{\gamma} \frac{\tilde{w}_i^H \tilde{\eta}_j}{z_i - p_j} & \frac{1}{2\mathrm{Re}(p_j)} \end{bmatrix} \geq 0,$$

or equivalently,

$$\gamma^2 \geq 1 + \frac{4\mathrm{Re}(z_i)\mathrm{Re}(p_j)}{|z_i - p_j|^2} |\tilde{w}_i^H \tilde{\eta}_j|^2,$$

we have

$$\gamma_{\min}^S \geq \inf \left\{ \gamma : \ \gamma^2 \geq 1 + \frac{4\mathrm{Re}(z_i)\mathrm{Re}(p_j)}{|z_i - p_j|^2} |\tilde{w}_i^H \tilde{\eta}_j|^2 \right\}$$

$$= \sqrt{\sin^2 \angle(\tilde{w}_i, \tilde{\eta}_j) + \left|\frac{p_j + \bar{z}_i}{p_j - z_i}\right|^2 \cos^2 \angle(\tilde{w}_i, \tilde{\eta}_j)}.$$

More generally, note that $\gamma_{\min}^S$ and $\gamma_{\min}^T$ are strictly greater than one, unless every zero direction is orthogonal to all pole directions; in the latter case, $Q_{zp} = 0$.

Remark 3.2 The main restriction on Theorem 3.1 is that all the zeros and poles must be distinct. This assumption, however, is imposed mainly for simplicity. When the plant transfer function matrix has multiple zeros or poles in $\mathbb{C}_+$, one can resort to a more general interpolation condition and hence obtain similar bounds. The expression of these bounds, understandably, will be more complicated. Moreover, Assumption 3.1–that $P(s)$ has neither imaginary zero nor imaginary pole–can be made without loss of generality; in fact, such zeros and poles have no effect on $\gamma_{\min}^S$ and $\gamma_{\min}^T$. Indeed, by a limiting argument, imaginary zeros or poles will result in infinitely large diagonal elements in the matrices Q_z and Q_p, and whenever

$\gamma > 1$, the positivity condition (4) is not altered. The condition thus reduces to one involving only zeros and poles in the open right half plane.

For SISO systems, Corollary 3.1 can be further generalized, leading to the following result which also extends that of [15].

Corollary 3.2 *Let $P(s)$ be a SISO transfer function.*
(i) Suppose that $P(s)$ has distinct unstable poles $p_i \in \mathbb{C}_+$, $i = 1, \cdots, k$, but only one nonminimum phase zero $z \in \mathbb{C}_+$. Assume that $z \neq p_i$ for all $i = 1, \cdots, k$. Then,

$$\gamma_{\min}^S = \gamma_{\min}^T = \prod_{i=1}^{k} \left| \frac{z + \overline{p}_i}{z - p_i} \right|. \tag{6}$$

(ii) Suppose that $P(s)$ has distinct nonminimum phase zeros $z_i \in \mathbb{C}_+$, $i = 1, \cdots, l$, but only one unstable pole $p \in \mathbb{C}_+$. Assume that $p \neq z_i$ for all $i = 1, \cdots, l$. Then,

$$\gamma_{\min}^S = \gamma_{\min}^T = \prod_{i=1}^{l} \left| \frac{p + \overline{z}_i}{p - z_i} \right|. \tag{7}$$

Proof. We shall prove (6), and (7) can be established analogously. Toward this end, we first note that

$$Q_z = \frac{1}{2\mathrm{Re}(z)}, \qquad Q_p = \begin{bmatrix} \frac{1}{p_1 + \overline{p}_1} & \cdots & \frac{1}{p_1 + \overline{p}_k} \\ \vdots & \cdots & \vdots \\ \frac{1}{p_k + \overline{p}_1} & \cdots & \frac{1}{p_k + \overline{p}_k} \end{bmatrix}, \qquad Q_{zp} = \begin{bmatrix} \frac{1}{z - p_1} \\ \vdots \\ \frac{1}{z - p_k} \end{bmatrix}.$$

From Theorem 3.1, it follows that

$$\begin{aligned} \gamma_{\min}^S = \gamma_{\min}^T &= \sqrt{1 + \overline{\lambda}\left(Q_z^{-1/2} Q_{zp}^H Q_p^{-1} Q_{zp} Q_z^{-1/2}\right)} \\ &= \sqrt{1 + 2\mathrm{Re}(z) Q_{zp}^H Q_p^{-1} Q_{zp}}. \end{aligned}$$

Next, consider the function

$$f(s) = \frac{1}{z - s} \prod_{i=1}^{k} \frac{s - p_i}{s + \overline{p}_i}.$$

It can be expanded via partial fraction as

$$f(s) = \sum_{i=1}^{k} \frac{2\mathrm{Re}(p_i)}{z + \overline{p}_i} f_i \frac{1}{s + \overline{p}_i} + \frac{f_0}{z - s},$$

where

$$f_0 = \prod_{i=1}^{k} \frac{z - p_i}{z + \bar{p}_i}, \qquad f_i = \prod_{\substack{j=1 \\ j \neq i}}^{k} \frac{p_j + \bar{p}_i}{\bar{p}_j - \bar{p}_i}, \quad i = 1, \cdots, k.$$

By evaluating $f(p_i)$, $i = 1, \cdots, k$, we obtain

$$Q_p \begin{bmatrix} \frac{2\mathrm{Re}(p_1)}{z+\bar{p}_1} f_1 \\ \vdots \\ \frac{2\mathrm{Re}(p_k)}{z+\bar{p}_k} f_k \end{bmatrix} + f_0 Q_{zp} = 0,$$

which in turn leads to

$$Q_{zp}^H Q_p^{-1} Q_{zp} = -\frac{1}{f_0} \sum_{i=1}^{k} \frac{2\mathrm{Re}(p_i) f_i}{z + \bar{p}_i} \frac{1}{\bar{z} - \bar{p}_i} = \frac{1}{f_0} \left(f(-\bar{z}) - \frac{f_0}{2\mathrm{Re}(z)} \right).$$

Since

$$f(-\bar{z}) = \frac{1}{2\mathrm{Re}(z)} \prod_{i=1}^{k} \frac{\bar{z} + p_i}{\bar{z} - \bar{p}_i} = \frac{1}{2\mathrm{Re}(z)\overline{f_0}},$$

we have

$$Q_{zp}^H Q_p^{-1} Q_{zp} = \frac{1}{2\mathrm{Re}(z)} \left(\frac{1}{|f_0|^2} - 1 \right).$$

Consequently,

$$\gamma_{\min}^S = \gamma_{\min}^T = \frac{1}{|f_0|} = \prod_{i=1}^{k} \left| \frac{z + \bar{p}_i}{z - p_i} \right|.$$

This completes the proof. ∎

Next, we extend Theorem 3.1 to weighted sensitivity and complementary sensitivity functions. With no loss of generality, we assume that the weighting transfer function matrices $R_1(s)$ and $R_2(s)$ are stable and have stable inverses, and we consider the weighted functions $R_1(s)S(s)R_2(s)$ and $R_1(s)T(s)R_2(s)$. Let us define analogously

$$Q_{z1} := \left[\frac{\tilde{w}_i^H R_1^{-1}(z_i) R_1^{-H}(z_j) \tilde{w}_j}{z_i + \bar{z}_j} \right],$$

$$Q_{z2} := \left[\frac{\tilde{w}_i^H R_2(z_i) R_2^H(z_j) \tilde{w}_j}{z_i + \bar{z}_j} \right],$$

$$Q_{p1} := \left[\frac{\tilde{\eta}_i^H R_1^H(p_i) R_1(p_j) \tilde{\eta}_j}{\bar{p}_i + p_j} \right],$$

$$Q_{p2} := \left[\frac{\tilde{\eta}_i^H R_2^{-H}(p_i) R_2^{-1}(p_j) \tilde{\eta}_j}{\bar{p}_i + p_j} \right],$$

$$Q_{zp1} := \left[\frac{\tilde{w}_i^H R_1^{-1}(z_i) R_1(p_j) \tilde{\eta}_j}{z_i - p_j} \right],$$

$$Q_{zp2} := \left[\frac{\tilde{w}_i^H R_2(z_i) R_2^{-1}(p_j) \tilde{\eta}_j}{z_i - p_j} \right].$$

Theorem 3.2 *Let the assumptions in Theorem 3.1 hold. Define*

$$\hat{\gamma}_{min}^S := \inf \{ \|R_1(s)S(s)R_2(s)\|_\infty : F(s) \text{ stabilizes } P(s) \},$$
$$\hat{\gamma}_{min}^T := \inf \{ \|R_1(s)T(s)R_2(s)\|_\infty : F(s) \text{ stabilizes } P(s) \}.$$

Then,

$$\hat{\gamma}_{min}^S = \bar{\lambda}^{1/2} \left(Q_{z1}^{-1/2} \left(Q_{z2} + Q_{zp2}^H Q_{p2}^{-1} Q_{zp2} \right) Q_{z1}^{-1/2} \right), \tag{8}$$

$$\hat{\gamma}_{min}^T = \bar{\lambda}^{1/2} \left(Q_{p2}^{-1/2} \left(Q_{p1} + Q_{zp1} Q_{z1}^{-1} Q_{zp1}^H \right) Q_{p2}^{-1/2} \right). \tag{9}$$

Proof. The proof is similar to that for Theorem 3.1. To establish (8), simply define $H(s) := R_1(s)S(s)R_2(s)/\gamma$ and note the interpolation constraints

$$\tilde{w}_i^H R_1^{-1}(z_i) H(z_i) = \frac{1}{\gamma} \tilde{w}_i^H R_2(z_i), \qquad i = 1, \cdots, l,$$

$$H(p_i) R_2^{-1}(p_i) \tilde{\eta}_i = 0, \qquad i = 1, \cdots, k.$$

The proof can then be completed as in that for Theorem 3.1. ∎

Remark 3.3 Following the idea suggested in Remark 3.1, it is straightforward to find that

$$\hat{\gamma}_{min}^S \geq \frac{\|\tilde{w}_i^H R_2(z_i)\|}{\|\tilde{w}_i^H R_1^{-1}(z_i)\|} \sqrt{\sin^2 \phi_{ij} + \left| \frac{p_j + \bar{z}_i}{p_j - z_i} \right|^2 \cos^2 \phi_{ij}}, \tag{10}$$

and

$$\hat{\gamma}_{min}^T \geq \frac{\|R_1(p_j)\tilde{\eta}_j\|}{\|R_2^{-1}(p_j)\tilde{\eta}_j\|} \sqrt{\sin^2 \psi_{ij} + \left| \frac{p_j + \bar{z}_i}{p_j - z_i} \right|^2 \cos^2 \psi_{ij}}, \tag{11}$$

where

$$\cos \phi_{ij} = \frac{\left| \tilde{w}_i^H R_2(z_i) R_2^{-1}(p_j) \tilde{\eta}_j \right|}{\|\tilde{w}_i^H R_2(z_i)\| \, \|R_2^{-1}(p_j)\tilde{\eta}_j\|},$$

$$\cos \psi_{ij} = \frac{\left| \tilde{w}_i^H R_1^{-1}(z_i) R_1(p_j) \tilde{\eta}_j \right|}{\|\tilde{w}_i^H R_1^{-1}(z_i)\| \, \|R_1(p_j)\tilde{\eta}_j\|}.$$

From these expressions, it becomes clear that the weighting functions $R_1(s)$ and $R_2(s)$ may be appropriately selected to counter the effects of plant unstable poles and nonminimum phase zeros, so as to prevent the lower bounds in (10-11) from being large; more specifically, they must be selected carefully to reflect the directional properties of the zeros and poles, taking into account the zero and pole directions. On the other hand, undesirable spatial properties of these weighting functions may exacerbate the negative effects incurred by the zeros and poles which may be otherwise not so critical. Note that both bounds will become exact when $P(s)$ has a single nonminimum phase zero and a single unstable pole.

3.2   Discrete-Time Systems

The discrete-time counterparts to the above results can be derived in parallel in a straightforward manner. The main technical foundation, similar to Lemma 3.1, is the following result pertaining to interpolation over the unit disc, also found in [1].

Lemma 3.2 *Consider two sets of distinct points $s_i \in \mathbb{D}$, $i = 1, \cdots, m$ and $p_i \in \mathbb{D}$, $i = 1, \cdots, n$. Assume that $s_i \neq p_j$ for any i and j. Then, there exists a matrix function $H(z)$ such that (i) $H(z)$ is analytic in $\mathbb{D}$, (ii) $\|H(z)\|_\infty \leq 1$, and (iii) $H(z)$ satisfies the conditions*

$$x_i^H H(s_i) = y_i^H, \qquad i = 1, \cdots, m,$$
$$H(p_i)u_i = v_i, \qquad i = 1, \cdots, n$$

for some vector sequences x_i, y_i, $i = 1, \cdots, m$ and u_i, v_i, $i = 1, \cdots, n$, of compatible dimensions, if and only if

$$Q := \begin{bmatrix} Q_1 & Q_{12}^H \\ Q_{12} & Q_2 \end{bmatrix} \geq 0,$$

where $Q_1 \in \mathbb{C}^{m \times m}$, $Q_2 \in \mathbb{C}^{n \times n}$, and $Q_{12} \in \mathbb{C}^{n \times m}$ are defined as

$$Q_1 := \left[\frac{x_i^H x_j - y_i^H y_j}{1 - s_i \bar{s}_j} \right],$$
$$Q_2 := \left[\frac{u_i^H u_j - v_i^H v_j}{1 - \bar{p}_i p_j} \right],$$
$$Q_{12} := \left[\frac{y_i^H u_j - x_i^H v_j}{s_i - p_j} \right].$$

Furthermore, $H(z)$ is nonunique if and only if $Q > 0$.

Similarly, we assume that the plant transfer function matrix $P(z)$ satisfies

Assumption 3.2 $P(z)$ has neither zero nor pole on the unit circle.

Based upon Lemma 3.2, we state below a number of results parallel to those in Section 3.1. The proofs follow analogously, and similar remarks can be drawn as well.

Theorem 3.3 *Let $s_i \in \mathbb{D}^c$, $i = 1, \cdots, m$, be the nonminimum phase zeros of $P(z)$ with output direction vectors $\tilde{w}_i$, and $p_i \in \mathbb{D}^c$, $i = 1, \cdots, n$, be the*

unstable poles of $P(z)$ with input direction vectors $\tilde{\eta}_i$. Furthermore, assume that s_i and p_i are all distinct, and that $s_i \neq p_j$. Define

$$\gamma_{\min}^S := \inf \{\|S(z)\|_\infty : F(z) \text{ stabilizes } P(z)\},$$
$$\gamma_{\min}^T := \inf \{\|T(z)\|_\infty : F(z) \text{ stabilizes } P(z)\}.$$

Then,

$$\gamma_{\min}^S = \gamma_{\min}^T = \sqrt{1 + \bar{\sigma}^2 \left(Q_p^{-1/2} Q_{sp} Q_s^{-1/2}\right)}, \tag{12}$$

where $Q_s \in \mathbb{C}^{m \times m}$, $Q_p \in \mathbb{C}^{n \times n}$, and $Q_{sp} \in \mathbb{C}^{n \times m}$ are given by

$$Q_s := \left[\frac{s_i \bar{s}_j}{s_i \bar{s}_j - 1} \tilde{w}_i^H \tilde{w}_j\right],$$

$$Q_p := \left[\frac{\bar{p}_i p_j}{\bar{p}_i p_j - 1} \tilde{\eta}_i^H \tilde{\eta}_j\right],$$

$$Q_{sp} := \left[\frac{s_i p_j}{p_j - s_i} \tilde{w}_i^H \tilde{\eta}_j\right].$$

Corollary 3.3 *Suppose that the assumptions in Theorem 3.3 hold. Then,*

$$\gamma_{\min}^S = \gamma_{\min}^T \geq \sqrt{\sin^2 \angle(\tilde{w}_i, \ \tilde{\eta}_j) + \cos^2 \angle(\tilde{w}_i, \ \tilde{\eta}_j) \left|\frac{1 - p_j \bar{s}_i}{p_j - s_i}\right|^2}. \tag{13}$$

Furthermore, if $P(z)$ has only one nonminimum phase zero $s \in \mathbb{D}^c$, with an output direction vector $\tilde{w}$, and one unstable pole $p \in \mathbb{D}^c$, with an input direction vector $\tilde{\eta}$, and $s \neq p$, Then,

$$\gamma_{\min}^S = \gamma_{\min}^T = \sqrt{\sin^2 \angle(\tilde{w}, \ \tilde{\eta}) + \cos^2 \angle(\tilde{w}, \ \tilde{\eta}) \left|\frac{1 - p\bar{s}}{p - s}\right|^2}. \tag{14}$$

Corollary 3.4 *Let $P(z)$ be a SISO transfer function.*
(i) Suppose that $P(z)$ has distinct unstable poles $p_i \in \mathbb{D}^c$, $i = 1, \cdots, m$, but only one nonminimum phase zero $s \in \mathbb{D}^c$. Assume that $s \neq p_i$ for all $i = 1, \cdots, n$. Then,

$$\gamma_{\min}^S = \gamma_{\min}^T = \prod_{i=1}^{n} \left|\frac{1 - s\bar{p}_i}{s - p_i}\right|. \tag{15}$$

(ii) Suppose that $P(z)$ has distinct nonminimum phase zeros $s_i \in \mathbb{D}^c$, $i = 1, \cdots, m$, but only one unstable pole $p \in \mathbb{D}^c$. Assume that $p \neq s_i$ for all $i = 1, \cdots, m$. Then,

$$\gamma_{\min}^S = \gamma_{\min}^T = \prod_{i=1}^{m} \left|\frac{1 - p\bar{s}_i}{p - s_i}\right|. \tag{16}$$

4 Conclusion

While similar bounds have been developed in the past for MIMO systems based upon Bode and Poisson type integrals and properties of analytic functions, in this paper we derived *exact* performance limits on the sensitivity and complementary sensitivity magnitudes independent of compensator design. These expressions characterize the best achievable $\mathcal{H}_\infty$ norms representing typical feedback design objectives, and show how the latter may depend on the plant nonminimum phase zeros and unstable poles. A particularly noteworthy feature throughout our results is that they display a clear dependence of the best achievable $\mathcal{H}_\infty$ performance on the pole and zero direction vectors, which in turn shows that mutual orientation of pole and zero directions plays a crucial role.

References

1. J.A. Ball, I. Gohberg, and L. Rodman, *Interpolation of Rational Matrix Functions*, Operator Theory: Advances and Applications, vol. 45, Basel: Birkhäuser, 1990.
2. A. Bjorck and G.H. Golub, "Numerical methods for computing angles between linear subspaces," *Math. of Computation*, vol. 27, no. 123, pp. 579-594, July 1973.
3. H.W. Bode, *Network Analysis and Feedback Amplifier Design*, Princeton, NJ: Van Nostrand, 1945.
4. S. Boyd and C.A. Desoer, "Subharmonic functions and performance bounds in linear time-invariant feedback systems," *IMA J. Math. Contr. and Info.*, vol. 2, pp. 153-170, 1985.
5. J. Chen, "Sensitivity integral relations and design tradeoffs in linear multivariable feedback systems," *IEEE Trans. Auto. Contr.*, vol. AC-40, no. 10, pp. 1700-1716, Oct. 1995.
6. J. Chen, "Multivariable gain-phase and sensitivity integral relations and design tradeoffs," *IEEE Trans. Auto. Contr.*, vol. AC-43, no. 3, pp. 373-385, March 1998.
7. J. Chen, "On logarithmic integrals and performance bounds for MIMO systems, part I: sensitivity integral inequalities," *Proc. 36th IEEE Conf. Decision Contr.*, San Diego, CA, Dec. 1997, pp. 3620-3625.
8. J. Chen and C.N. Nett, "Sensitivity integrals for multivariable discrete-time systems," *Automatica*, vol. 31, no. 8, pp. 1113-1124, Aug. 1995.
9. J. Chen, O. Toker, and L. Qiu, "Limitations on maximal tracking accuracy, part 1: tracking step signals," *Proc. 35th IEEE Conf. Decision Contr.*, Kobe, Japan, Dec. 1996, pp. 726-731.
10. J.S. Freudenberg and D.P. Looze, "Right half plane zeros and poles and design tradeoffs in feedback systems," *IEEE Trans. Auto. Contr.*, vol. AC-30, no. 6, pp. 555-565, June 1985.
11. J.S. Freudenberg, R.H. Middleton, and J.H. Braslavsky, "Inherent design limitations for linear sampled-data feedback systems," *Int. J. Contr.*, vol. 35, no. 6, pp. 1387-1421, 1994.

12. G.C. Goodwin, D. Mayne, and J. Shim, "Tradeoffs in linear filter design," *Automatica*, vol. 31, no. 10, pp. 1367-1376, Oct. 1995.

13. S. Hara and H.K. Sung, "Constraints on sensitivity characteristics in linear multivariable discrete-time control systems," *Linear Algebra and its Applications*, vol. 122/123/124, 1989, pp. 889-919.

14. R.A. Horn and C.R. Johnson, *Matrix Analysis*, Cambridge, UK: Cambridge Univ. Press, 1985.

15. P.P. Khargonekar and A. Tannenbaum, "Non-Euclidean metrics and the robust stabilization of systems with parameter uncertainty," *IEEE Trans. Auto. Contr.*, vol. AC-30, no. 10, pp. 1005-1013, Oct. 1985.

16. H. Kwakernaak and R. Sivan, "The maximally achievable accuracy of linear optimal regulators and linear optimal filters," *IEEE Trans. Auto. Contr.*, vol. AC-17, no. 1, pp. 79-86, Feb. 1972.

17. R.H. Middleton, "Trade-offs in linear control system design," *Automatica*, vol. 27, no. 2, pp. 281-292, Feb. 1991.

18. M. Morari and E. Zafiriou, *Robust Process Control*, Englewood Cliffs, NJ: Prentice Hall, 1989.

19. L. Qiu and E.J. Davison, "Performance limitations of non-minimum phase systems in the servomechanism problem," *Automatica*, vol. 29, no. 2, pp. 337-349, 1993.

20. M.M. Seron, J.H. Braslavsky, and G.C. Goodwin, *Fundamental Limitations in Filtering and Control*, London: Springer-Verlag, 1997.

21. M.M. Seron, J.H. Braslavsky, P.V. Kokotovic, and D.Q. Mayne, "Feedback limitations in nonlinear systems: from Bode integrals to cheap control," Tech. Rept. CCEC97-0304, University of California, Santa Barbara, CA, March 1997.

22. H.K. Sung and S. Hara, "Properties of sensitivity and complementary sensitivity functions in single-input single-output digital control systems," *Int. J. Contr.*, vol. 48, no. 6, pp. 2429-2439, 1988.

23. H.K. Sung and S. Hara, "Properties of complementary sensitivity function in SISO digital control systems," *Int. J. Contr.*, vol. 48, no. 4, pp. 1283-1295, 1989.

24. O. Toker, J. Chen, and L. Qiu, "Tracking performance limitations for multivariable discrete-time systems," *Proc. 1997 Amer. Contr. Conf.*, Albuquerque, NM, June 1996.

25. G. Zames and B.A. Francis, "Feedback, minimax sensitivity, and optimal robustness," *IEEE Trans. Auto. Contr.*, vol. AC-28, no. 5, pp. 585-600, May 1985.

Puzzles in Systems and Control

Pertti M. Mäkilä

Automation and Control Institute, Tampere University of Technology, P.O. Box 692, FIN-33101 Tampere, FINLAND

Abstract. This tour of four puzzles is to highlight fundamental and surprising open problems in linear systems and control theory, ranging from system identification to robust control. The first puzzle deals with complications caused by fragility of poles and zeros in standard parameterizations of systems due to unavoidable parameter inaccuracies. The second and the third puzzle are concerned with difficulties in generalizing robust H_∞ control theory to realistic persistent signal setups. Surprisingly it seems very difficult to generalize the theory to such setups and simultaneously keep the terminology robust H_∞ control intact. These puzzles have implications also to model validation. The fourth puzzle, due to Georgiou and Smith, deals with difficulties in doubly-infinite time axis formulations of input-output stabilization theory.

1 Introduction

Linear systems have been claimed to be, at regular time intervals, a mature subject that can not yield new big surprises. There are, however, many misunderstandings and messy fundamental issues that need to be cleaned up. These are often known only to a small group of specialists, so the purpose of this paper is to discuss four such neglected topics, misunderstandings, or puzzles, in a tutorial fashion.

The 1970s witnessed the growth of stochastic system identification theory. The text booxs [13], [3], [25] summarize many of the developments obtained during this very dynamic phase that lasted to the mid 1980s. Developments in robust control dominated the 1980s. The text books [26], [9], [31] summarize many of the developments during the very dynamic phase of robust control research that lasted to the first part of the 1990s.

System identification and model validation for robust control design have been two popular research themes in the 1990s (see e.g. the special issues, IEEE TAC, July 1992; Automatica, December 1995; MMOS, January 1997, and the monograph [22]). So what could be more appropriate than to start our tour with an often overlooked puzzle that has deep implications for control, systems and system identification. This is the puzzle of the wandering poles and zeros. That is, the exponential explosion of the accuracy required, as a function of system order, in many standard system models and representations to reproduce system poles and zeros at least in a qualitatively correct manner. The sensitivity of poles and zeros is an old finding but we shall try to put the puzzle in a more quantitative form.

We are all familiar with least squares and prediction error identification of AR, ARX, and ARMAX models [13]. Such models are popular in pole placement and minimum variance type control design procedures [2]. The examples we provide should shake the reader's confidence in black box methodology and in such control desing methods that rely on accurate knowledge of system poles and zeros. Fortunately, modern robust control theory provides a partial rescue of black box identification methodology.

Recently several difficulties have surfaced in H_∞ control and systems theory. In the present work we shall discuss and study three such puzzles. One of them is the Georgiou-Smith puzzle in doubly-infinite time axis input-output H_∞ stabilization theory [8]. Georgiou and Smith [8] conclude their study by stating that in such a setting linear time-invariant systems with right-half plane poles can not be considered to be both causal and stabilizable. As this puzzle is very technical we shall leave it as the last puzzle of our tour. The second puzzle of our tour deals with difficulties with standard bounded power signal set formulations of H_∞ control [6],[31] due to e.g. lack of closedness with respect to addition of signals in such sets. The third puzzle has to do with the difficulty to define frequency domain concepts for general H_∞ functions and H_∞ uncertainty models, in contrast to what is assumed in many papers and books. These have also consequences for H_∞ model validation [23].

2   Puzzle A : Wandering Poles and Zeros

The sensitivity, or fragility, of the roots of polynomials to small coefficient perturbations is a well-recognized issue in numerical mathematics. This problem is sometimes discussed briefly in the connection of implementation issues for controllers [2]. This is, however, only a part of the puzzle of poles and zeros.

Let us study two examples. We have used in all computations the numerical and symbolic mathematics package MAPLE^{TM}. The computations have been directly latex documented within the Scientific Workplace^{TM} package using its MAPLE facility, so as to minimize the risk for typographical errors in reproducing long expressions.

Example A.1 : System 1

Let us study the system

$$G_n(q^{-1}) = \frac{0.30003^n q^{-1}}{(1 + 0.69997 q^{-1})^n} = \frac{0.30003^n q^{n-1}}{(q + 0.69997)^n}, \ n \geq 1. \quad (1)$$

Here q and q^{-1} denote the forward and the backward time shift operator, respectively (so that $(qu)(t) = u(t+1)$, $(q^{-1}u)(t) = u(t-1)$). This system is stable and its poles are concentrated at -0.69997. (We have normalized G_n to have unit supremum norm on the unit circle for any $n \geq 1$.) It is convenient to define the spread of the system poles as $\max_{i,j} |p_i - p_j|$, where $\{p_i\}$ denote the system poles. Hence the spread of the system poles is zero. The poles of the system are determined from $(q + 0.69997)^n = 0$. In black box identification and in control design the nominator and denominator of the system transfer function are usually given in expanded form. To illustrate the expanded form of the denominator, it is given below for $n = 8$.

$$(q + 0.69997)^8 = q^8 + 5.\,59976q^7 + 13.\,71882\,40252q^6 +$$
$$19.\,20553\,05058\,38488q^5 + 16.\,80411\,89852\,14708\,0567q^4 +$$
$$9.\,40990\,33328\,64591\,35875\,86392q^3 +$$
$$3.\,29332\,50179\,52614\,00669\,51423\,40412q^2 +$$
$$.\,65863\,67750\,90368\,92179\,03996\,52576\,62504q +$$
$$.0\,57628\,24793\,25006\,91773\,20325\,56017\,57528\,6561 \qquad (2)$$

This long expression explains why we have included several digits in the pole of the system as otherwise there would be no digits to round off in the higher degree terms in q.

Let us see what happens with the poles when we round off the expanded form of the denominator to 5 significant digits. The approximate denominator is then given by

$(z + 0.69997)^8 \approx z^8 + 5.\,5998z^7 + 13.\,719z^6 + 19.\,206z^5 + 16.\,804z^4 +$
$9.\,4099z^3 + 3.\,2933z^2 + .\,65864z + 5.\,7628 \times 10^{-2},$

The roots of the approximate expression are :
$$\begin{bmatrix} -1.\,1064 \\ -.\,91455 - .\,30897i \\ -.\,91455 + .\,30897i \\ -.\,62584 - .\,30426i \\ -.\,62584 + .\,30426i \\ -.\,48387 - .\,15917i \\ -.\,48387 + .\,15917i \\ -.\,44484 \end{bmatrix}$$

Note that the spread of the poles is now about 0.66 ! Furthermore, one of the poles is now outside the unit circle, that is the rounded-off system is unstable. Repeating the computation with a 6 significant digit approximation of the expanded denominator gives poles that are all inside the unit circle. (The same is true if we use more than 6 significant digits). That is, 5 significant digits is the largest number of significant digits for which the rounded off system is unstable.

We have determined in a similar manner the required number, $d(n)$, of significant digits at which the rounded off system changes from unstable to stable as a function of the order n of the system. To our surprise it was very difficult to complete these computations for $n > 20$ even if more than 200 digits were used in the computations. The MAPLE root solver complained ill-conditioning and even using an eigenvalue problem formulation became computationally very demanding at about $n = 20$. We list some $d(n)$ values : $d(8) = 5$, $d(10) = 6$, $d(12) = 8$, $d(15) = 10$, and $d(20) = 13$. From these values it seems reasonable to claim that the required number of significant digits, $d(n)$, to predict from the rounded-off expanded form that the poles of the system are stable, grows about linearly as a function of the order n of the system. This is a devastating observation.

Example A.2 : System 2

Let us study the system

$$H_n(q^{-1}) = \frac{q^{n-1}}{[(q + 0.513)^2 + 0.703^2]^{n/2}}, \ n = 2, 4, 6, \ldots \tag{3}$$

Here the system poles are within the radius 0.7574, that is well inside the unit circle. The spread of the system poles is 2*0.703=1.406, so one would expect a reduced sensitivity of the poles against round-off.

The quantity $d(n)$, as defined in the previous Example, was computed for this system for a number of n values : $d(10) = 5$, $d(12) = 6$, $d(16) = 8$, $d(20) = 11$, $d(24) = 13$, and $d(26) = 14$. Due to the large spread of the system poles it was possible to compute $d(n)$ for somewhat larger n values than in the previous example. This list indicates that the required number of significant digits increases again about linearly in n.

Some consequences of observations like the above are as follows. Do not implement controllers in standard expanded transfer function form contrary to what is studied in [10] ! The paradigm of black box identification needs to be re-examined. If the required accuracy to predict even qualitative behaviour of the system correctly can increase so rapidly with system order then the magic of the claim *let the data speak*, seems partly removed as the required experiment lengths can become astronomical even for fairly small n. It is very important to understand what system properties can be reliably identified for reasonable experiment lengths. Accuracy requirements can become very important at about order 4-9. By the way how many significant digits would you expect that is needed for all the roots of the rounded-off expansion of the denominator to be within 1 % of the system G_8 ($n = 8$) pole at -0.69997? The answer is 19 digits.

Observations such as the above indicate that controller design methods that rely heavily on poles and zeros, such as pole placement and minimum variance control, loose partly their justification. It is surprising that black box identification-based ARX, ARMAX etc other expanded transfer function form models are presented in standard control text books as convenient to use in connection with such controller design methods, when in fact the same black box identification theory implies that the information needed to apply such methods in a reliable manner simply can not be obtained even in the ideal world of linear time invariant (LTI) systems.

To convince you that the fragility of poles and zeros in the expanded transfer function form is not just a very rare phenomenon, consider the following readily proved result.

Proposition 6. *Let a sequence of polynomials $\{P_n(z) = \prod_{k=1}^{n}(z+z_{n,k})\}_{n \geq 1}$ be given, such that $\sup_{k,n} |z_{n,k}| < 1$ and $\inf_n |\sum_{k=1}^{n} z_{n,k}| > 0$. Furthermore, let there exist a complex number $|c| > 1$ such that $\sup_{k,n} |c + z_{n,k}| < 1$. Denote $P_n(z) = z^n + p_{n,1}z^{n-1} + \ldots + p_{n,n}$. Let $P_n(\epsilon, z) = P_n(z) + \epsilon p_{n,1}z^{n-1}$. Then $P_n(\epsilon, z)$ has a pole at $z = c$ if*

$$\epsilon = -\frac{\prod_{k=1}^{n}(c + z_{n,k})}{\left(\sum_{k=1}^{n} z_{n,k}\right) c^{n-1}}. \tag{4}$$

That is, the relative size, ϵ, of a perturbation in the coefficient $p_{n,1}$ of the z^{n-1} term in $P_n(z)$ to move a pole of the perturbed $P_n(\epsilon, z)$ out of the unit circle tends to zero exponentially as a function of n.

Note that here the polynomial coefficients $p_{n,k}$ (and hence possibly also ϵ) are allowed to be complex numbers.

Here are some difficult questions : Is there a useful way to describe those LTI systems whose expanded transfer functions do not exhibit wildly wandering poles and zeros? Is there some fundamental way to describe those representations of systems and controllers that do not exhibit high pole-zero sensitivity to small perturbations in the representation parameters? This is a difficult issue and has been studied in connection with issues of implementation of digital controllers [21],[12]. Keel and Bhattacharyya [10] neglected this issue completely in their fragility critique of modern robust and optimal control as explained in [14]. How can appropriate *a priori* information be included in system identification so as to develop a methodology that works also when black-box methodology based expanded transfer function like models fail? Set membership identification, approximation theory, metric complexity theory and information based complexity theory provide some non-conventional concepts and tools for undertaking such a study [29,20,18,22].

3   Puzzle B : Bounded Power Signals and Robustness

During recent years it has become more common to motivate robust H_∞ control via persistent rather than transient signals [6], [31]. This has many reasons, such as measurement noise being persistent and e.g. load disturbances being more naturally described as persistent signals. Furthermore, frequency domain methods are classically described via stationary responses to certain periodic signals, that is via simple persistent signals. Certain bounded power signal spaces analysed by Wiener have been instrumental in this extension of H_∞ control [6], [31], [19]. Along the way it was somehow forgotten that these Wiener spaces are not closed under addition, that is they are not linear vector spaces.

There are other technical problems as well, so surprisingly there is some confusion about how to extend H_∞ control to persistent signals, especially in the doubly infinite time axis case. This is all the more surprising as even in the seminal paper of Zames [30], the new H_∞ theory is motivated via spectral properties of signals, that is via persistent signals.

There is a very elegant theory of robust control for persistent signals, namely robust L_1/ℓ_1 control [27,5,4], which does not seem to have similar problems. (In fact, there is an elegant symmetry in the applicability of the theory for finite and infinite dimensional systems alike, in both the singly-infinite and the doubly-infinite time axis setups [15,16].)

Let us start by defining two signal seminorms and a norm in the doubly-infinite time axis case for discrete systems. Let $\{u_k\}$ denote a real or complex sequence. Introduce the Wiener space seminorm

$$\|u\|_A = \left(\lim_{n \to \infty} \frac{1}{2n+1} \sum_{k=-n}^{n} |u_k|^2 \right)^{1/2} \tag{5}$$

and the Marcinkiewicz space seminorm

$$\|u\|_B = \left(\limsup_{n \to \infty} \frac{1}{2n+1} \sum_{k=-n}^{n} |u_k|^2 \right)^{1/2} \tag{6}$$

and the Besicovitch/Beurling space norm

$$\|u\|_{BP} = \left(\sup_{n} \frac{1}{2n+1} \sum_{k=-n}^{n} |u_k|^2 \right)^{1/2}. \tag{7}$$

Analogous definitions apply for the one-sided sequences case and for continuous time systems, see [19] and the references therein. The problem with the first seminorm is that it defines signal spaces that are not closed under addition, leading to big technical problems especially if used to analyse robustness etc for non-LTI systems. Furthermore, it excludes unnecessarily many natural signals that do not possess the limit indicated. In any case this seminorm has been used in certain bounded power generalizations of H_∞ control [6,31].

BP norm like conditions have been used by Wiener [28] in his generalized harmonic analysis. The B seminorm and the BP norm have been used in [19] (see also [17]) to solve the one-sided time axis case of bounded power H_∞ control theory. In these setups there is a beautiful symmetry with the Hardy space H_∞ of bounded analytic functions and bounded power stability. It is a very unfortunate fact that this symmetry breaks down in the doubly-infinite time axis case [19].

A problem with the BP norm is that it is signal translation sensitive. So let e^{hs} denote the Laplace transform of a continuous-time translation operator (a delay operator if $h < 0$). Unfortunately, the induced norm of this operator is infinite in the space of signals defined by boundedness of the BP norm for any $h > 0$ in the one-sided time axis case and for any $h \neq 0$ in the double-sided time axis case.

The B seminorm is probably a better alternative, but boundedness of the B seminorm defines a signal space that results in an induced system stability requirement which is even stronger than stability in the sense of the robust L_1/ℓ_1 control theory. This is also true for the nonlinear signal space defined by boundedness of the A seminorm. If one includes some extra conditions on signals then one typically looses completeness of the resulting space of bounded induced norm LTI systems. We are presently aware of only one persistent signal norm which automatically guarantees that certain pathological situations do no occur whilst resulting in H_∞ control techniques, namely the seminorm studied in [19]

$$\|u\|_C = \left(\limsup_{n \to \infty} \left[\sup_{t_0} \frac{1}{2n+1} \sum_{-n+t_0}^{n+t_0} |u_k|^2 \right] \right)^{1/2}. \tag{8}$$

(An analogous definition applies in the continuous-time case.) A conclusion from the work done so far in generalizing H_∞ control to doubly-infinite time axis persistent signal spaces is that several concepts from transient signal H_∞ control break down ! In fact, a generic H_∞ transfer function which is stable in the transient signal H_∞ theory turns out to be unstable in the generalized theory. By this we mean that in the generalized theory the unit impulse response of the system should be (at least) absolutely summable (or representable by a bounded variation (convolution) measure in the continuous-time case) for stability, and the space of such systems is separable, whilst the Hardy space H_∞ is a (much larger) non-separable space.

Hence H_∞ space techniques are not needed in the generalized robust control theory, as stable systems have now trivial transfer functions from the point of view of true Hardy space techniques ! Unfortunately, this means also that many definitions of robust stability and uncertainty models need to be modified. In retrospect, it is most unfortunate that the terminology robust H_∞ control has been taken into use as we see that the close relationship between the mathematics of H_∞ spaces and robust control must be abandoned in order for the persistent signal theory to deserve the terminology robust in any reasonable sense.

4   Puzzle C : Non-Unique Frequency Responses ?

In H_∞ control and systems theory a certain boundary function of the system transfer function is identified as the system frequency response. Is this function the frequency response of the H_∞ system to simple periodic inputs? Unfortunately, as will be discussed shortly, frequency domain control and modelling concepts become quite puzzling for general H_∞ transfer functions.

To provide an example as to why general H_∞ functions often appear in studies of robust control, consider the standard additive uncertainty model for stable systems

$$\{G = G_n + W\Delta \mid \|\Delta\|_\infty \leq 1\}, \tag{9}$$

where G_n is a stable (often rational) nominal LTI model, W is a stable (often rational) frequency weighting function, and Δ is an arbitrary stable perturbation in the unit ball of the appropriate H_∞ space. In fact, in some papers it is stated that real systems are so complex that in the linear case it is best to model them as balls of H_∞ transfer functions. This is one reason why general H_∞ functions are often believed to be important in applications of robust control.

Consider now the discrete time case. Let G be the transfer function of a general H_∞ system defined by

$$G(z) = \sum_{k \geq 0} g_k z^k, \ |z| < 1, \tag{10}$$

where $\{g_k\}_{k \geq 0}$ are the unit impulse response coefficients of the LTI system. Note that $G(z^{-1})$ is the usual Z-transform of the system unit impulse response. Recall that the H_∞ norm of G is defined by

$$\|G\|_\infty = \sup_{|z|<1} |G(z)|. \tag{11}$$

Let $F_G(e^{i\omega})$ denote the Fourier series associated with G, that is

$$F_G(e^{i\omega}) = \sum_{k \geq 0} g_k e^{ik\omega}. \tag{12}$$

There is a third quantity of interest, the so-called boundary function of G, defined by

$$G_b(e^{i\omega}) = \lim_{r \uparrow 1} G(re^{i\omega}). \tag{13}$$

Note that the radial limit exists almost everywhere on the unit circle and defines an L_∞ function on the unit circle, and whose Fourier series is equal to $F_G(e^{i\omega})$. It is also known that $\|G_b\|_{L_\infty} = \text{ess sup}_\omega |G_b(e^{i\omega})| = \|G\|_\infty$.

Green and Limebeer [9] suggest that one should drop the b notation, writing $G(e^{i\omega})$ instead of $G_b(e^{i\omega})$. Now this choice is done in mathematics in the Poisson integral analysis of H_∞ [7], but is it the correct choice for the frequency response of G? We have three candidates for the frequency response of G, namely $G(e^{i\omega})$, $F_G(e^{i\omega})$, and $G_b(e^{i\omega})$? Note that these choices are not equivalent for a general H_∞ function. To make this more concrete let us consider a simple example.

Example C.1 Let the system have the transfer function

$$G(z) = e^{-\frac{1+z}{1-z}}, \; |z| < 1. \tag{14}$$

Clearly G is an H_∞ function and $\|G\|_\infty = 1$. Furthermore, G has an essential singularity at $z = 1$. Note that the Taylor series defining G has thus radius of convergence $= 1$. But clearly we can try to use the function expression for G even on the unit circle. We see that $G(1)$ is undefined by this expression, but that it defines an analytic function for all $z \neq 1$. Furthermore, we can not define $G(1)$ so that $G(z)$ would become continuous in the closed unit circle. Thus the function $G(e^{i\omega})$ can not be made continuous at $\omega = 0$. Note also that $|G(e^{i\omega})| = 1$ for $\omega \in (0, 2\pi)$. So clearly $G(e^{i\omega})$ is a square integrable function on the unit circle.

It is easy to verify that $G_b(e^{i\omega}) = G(e^{i\omega})$ for $\omega \in (0, 2\pi)$. It is also easy to check that $G_b(1) = 0$. What about the Fourier series F_G of G. This is trickier to evaluate. Due to a result of Carleson on Fourier series convergence for square integrable functions, we know that F_G converges to $G(e^{i\omega})$ almost everywhere on the unit circle. In fact, as $G(e^{i\omega})$ is an absolutely integrable function, Dini's test gives that $F_G(e^{i\omega}) = G(e^{i\omega})$ for $\omega \in (0, 2\pi)$. What can be said about $F_G(1)$? By a result of Abel we know that if the Fourier series of G converges at $\omega = 0$, then its value must be given by the radial limit of G at that point, that is if $F_G(1)$ is well-defined then $F_G(1) = 0$. However, to prove that the Fourier series actually converges at $\omega = 0$ appears tricky.

Consider now the output response y of the system G to an input u, assuming that the output is defined solely by the input u. That is write

$$y(t) = \sum_{k \geq 0} g_k u(t - k). \tag{15}$$

Consider an input which is periodic for $t > 0$. Specifically, let $u(t) = e^{i\omega t}$ for $t \geq 0$, and $u(t) = 0$ for $t < 0$. Then $y(t) = [\sum_{k=0}^{t} g_k e^{-i\omega k}] e^{i\omega t}$ for $t \geq 0$. Hence the convergence of $y(t)$ when $t \to \infty$ is determined by the convergence of the partial sums of the Fourier series F_G of G. In particular take $u(t) = 0$ for $t < 0$ and $u(t) = 1$, that is $\omega = 0$, for $t \geq 0$. Then $y(t) = \sum_{k=0}^{t} g_k$, $t \geq 0$. We see that the unit step response of G tends to a definite limit when $t \to \infty$ if the Fourier series of G converges at $\omega = 0$, that is if $F_G(1)$ exists. If we would use $G(1)$ as the steady state step response value, then we would have to conclude that the step response does not tend to any definite value. And if we would use $G_b(1)$ we would have to conclude that the steady state value is 0. But does the steady state value exist? That is, does $F_G(1)$ exist?

Note that the unit impulse response coefficients $\{g_k\}_{k\geq 0}$ of G are not absolutely summable (obviously $\lim_{k\to\infty} g_k = 0$). Hence this system gives an infinite output value $y(0)$, say for the binary input $u(-k) = \text{sign}(g_k)$, $k \geq 0$ (put $u(-k)$ equal to, say, $+1$ if $g_k = 0$). So would you call this system stable, even if it does not amplify the one-sided ℓ_2 norm of any ℓ_2 signal passing through it?

Let us return to the general case. It is known [11] that there is an H_∞ function G such that $\|G\|_\infty \leq 1$, yet the Fouries series of G diverges unboundedly at $\omega = 0$. So even if G_b is the correct choice for the frequency response of G when one-sided ℓ_2 signals are considered, this choice can fail badly if true frequency responses for periodic signals are aimed at, as in classical control. G_b smooths out the frequency response in a way that may make robust control theory appear less than *robust* !

Consequences for H_∞ model validation

General H_∞ uncertainty models are routinely used in H_∞ model validation theory [23] under the assumption that the system starts at zero initial conditions. Unfortunately, as the above discussion shows, such model validation problems become ill-conditioned when reasonable non-zero initial conditions are allowed.

It is also important to note that the H_∞ norm has other properties that make it difficult for model validation purposes in the general case. Let $\{g_k\}_{k\geq 0}$ denote the H_∞ function mentioned earlier, whose Fourier series diverges unboundedly at $\omega = 0$. Hence, for any large real number $C \gg 1$ and any integer $N \geq 1$, there exists an integer $n_N \geq N$ such that $|\sum_{k=0}^{k=n_N} g_k| \geq C$. Let H denote the system with unit impulse response coefficients $\{h_k\}_{k\geq 0}$ given by $h_k = g_k$ for $k = 0, 1, \ldots, n_N$, and $h_k = 0$ otherwise. Let $Z_N = \{u(t), y(t)\}_{t=0}^{N}$ denote a set of $N+1$ input-output pairs for the system H.

Let us say that we wish to test the assumption that the data Z_N was generated by a system F with $\|F\|_\infty \leq 1$. Indeed, such an assumption would pass the model validation test against the data Z_N (whatever the input u is). Present H_∞ model validation theory [23] is therefore potentially rather misleading (the real system could have a very large H_∞ norm as the system H above). The theory should at least say something relevant about how the test should be made so that it gives reasonably reliable results.

More on Boundary Functions

Let us consider an example that illustrates the smoothing effects of taking radial limits in a surprising manner. So consider the well-known analytic function in $|z| < 1$

$$G(z) = (1 - z) \exp\left[-\exp\left(\frac{1+z}{1-z}\right)\right]. \tag{16}$$

It turns out that the H_∞ norm of G is infinite, that is $\sup_{|z|<1} |G(z)| = \infty$. Hence $G \notin H_\infty$. However, by defining $G(1) = 0$, we see that the above expression for $G(z)$ defines a continuous function on the unit circle. In fact,

$$G(e^{i\omega}) = (1 - e^{i\omega}) \exp[-\exp(i \cot \frac{\omega}{2})], \tag{17}$$

and $\lim_{\omega \to 0} G(e^{i\omega}) = 0$. So is $G(e^{i\omega})$ the frequency response of the system G? By taking radial limits we see that

$$G_b(e^{i\omega}) = \lim_{r \uparrow 1} G(re^{i\omega}) = G(e^{i\omega}). \tag{18}$$

Hence G_b is a continuous function on the unit circle, yet $G \notin H_\infty$. That is, both $G_b(e^{i\omega})$ and $G(e^{i\omega})$ describe a very much smoothed out behaviour of the system G. So let us consider the 3rd candidate for the frequency response of G, namely the Fourier series associated with $G(e^{i\omega})$. Assume that $G \in H_p$ for some $p \geq 1$, where H_p denotes standard Hardy spaces [7]. But then by a standard Hardy space result, the Fourier coefficients of $G_b(e^{i\omega})$ (and of $G(e^{i\omega})$) are the Taylor series coefficients of $G(z)$. But this contradicts the fact that $G \notin H_\infty$. So in particular, $G \notin H_2$ and as both $G(e^{i\omega})$ and G_b are square integrable on the unit circle, it follows that the Taylor series coefficients of $G(z)$ are not given by the Fourier coefficients of $G(e^{i\omega})$ (or of $G_b(e^{i\omega})$).

Denote, as before, $F_G(e^{i\omega}) = \sum_{k \geq 0} g_k e^{ik\omega}$, where $\{g_k\}$ are the Taylor series coefficients of $G(z)$. (Equivalently, they are the unit impulse response coefficients of G). It follows by the above reasoning that $\{g_k\} \notin \ell_2$, that is the unit impulse response coefficients of the system G are not square integrable. It is concluded that in this example both $G(e^{i\omega})$ and $G_b(e^{i\omega})$ are bad candidates for the frequency response of the system G. F_G is the true frequency response, but it is difficult to analyse this quantity for the present example.

In conclusion certain standard notions in robust H_∞ control become problematic if we want to keep the symmetry between classical frequency domain ideas and H_∞ control. So it appears that puzzles B and C demand a re-examination of standard uncertainty models and several other concepts in robust control. It is seen that the terminology robust H_∞ control is misleading in several ways. The theory is neither a frequency domain theory nor a robust theory in the classical sense if general H_∞ functions are considered.

5   Puzzle D : The Doubly-Infinite Axis Stabilization Problem

Georgiou and Smith [8] arrived at the surprising conclusion that the input/output stabilization problem for the doubly-infinite L_2 signal setup contains an intrinsic difficulty. To put this differently, their very counter-intuitive conclusion was that only stable systems are stabilizable in the considered setup.

Let us discuss this fundamental puzzle using the causal system example in [8], namely

$$y(t) = \int_{-\infty}^{\infty} h(t - \tau)u(\tau) \, d\tau = h * u, \tag{19}$$

where $h(t) = e^t$ for $t \geq 0$ and zero otherwise. Denote the convolution operator above as C_h, that is $C_h u = h * u$.

Let $u \in L_2$, that is let u be square integrable on the real axis. The *graph*, $\mathcal{G}_h$, of the system h is defined as the set of all input-output pairs $(u, h * u)$, such that both u and $h * u$ are in L_2. It is well-known that the graph of a system is a fundamental notion in stabilization theory. Georgiou and Smith [8] show that the graph $\mathcal{G}_h$ is not a closed set, and hence they argue that the system h can not be stabilized in the doubly-infinite axis L_2 setup.

Georgiou and Smith [8] discuss this puzzle relying heavily on causality and stabilizability arguments. Let us now point out another possibility. Let us solve, implicitly, the convolution equation (19) for a smooth enough u. This gives

$$y(t) = e^{t-t_0}y(t_0) + \int_{t_0}^{t} e^{t-\tau}u(\tau) \, d\tau, \; t > t_0. \tag{20}$$

This integral equation has an associated differential equation, for smooth enough u, namely

$$\frac{dy}{dt} = y + u. \tag{21}$$

Consider, as in [8], the input-output pair $u(t) = e^{-t}$ for $t \geq 0$ and zero otherwise, $y(t) = -e^{-|t|}/2$. Note that this pair (u, y) can not be in the graph $\mathcal{G}_h$ as y is non-zero for $t < 0$. Georgiou and Smith [8] essentially provided a Cauchy sequence in $\mathcal{G}_h$ that converges to the above pair (u, y), hence showing that $\mathcal{G}_h$ is not a closed set in the product topology, implying non-stabilizability of h. Interestingly, however, y solves (20) for any real t and (21) for any $t \neq 0$. That is, y solves the associated differential equation in an L_2 sense. So we arrive at the conclusion that there is a reasonable case for considering the pair (u, y) as being in the definition set of some more general definition of the convolution operator !

Add to $u(t)$ as given above a small pulse $p_i(t)$ for $t < 0$. It is easy to check that one can define a sequence of pulses $\{p_i(t)\}$ so that $\|p_i\|_2 \to 0$, that is the pulse L_2 norm tends to zero, and the center of the pulse tends to $-\infty$, whilst the solution of $u + p_i$ tends to y in an L_2 sense. It seems unavoidable to arrive at the following conclusion : as there is no physical way to make a distinction between u and $u + p_i$ for large enough i, any theory of physical importance should not arrive at essentially different conclusions depending on whether the signal u or $u + p_i$ is used in the analysis. Hence it appears natural to look at extensions of the convolution operator.

So we suggest the following possible resolution of the Georgiou and Smith puzzle. Let us for simplicity just state this for causal convolution operators having a rational, strictly proper, transfer function. Let h be the kernel of any such convolution operator C_h, so that $C_h u = h * u$. Let $D_h \subset L_2$ denote the usual domain of definition of the convolution operator C_h. Add to D_h all the elements $u \in L_2$, which are limits of convergent sequences $u_i \in D_h$ and which generate convergent image sequences $h * u_i$, and put $\overline{C}_h u = \lim_{i \to \infty} h * u_i$, where the extended convolution operator is denoted as $\overline{C}_h$. Similarly, the extended definition set is denoted as $\overline{D}_h$. The operator $\overline{C}_h$ is a *closed extension* of C_h [1]. The smallest closed extension (when one exists), which is contained in every closed extension of C_h, is called the *minimal closed extension*, or simply the *closure*, of the convolution operator C_h. *Does C_h always have a closure?* What if C_h is infinite dimensional?

It is important to understand that the operation of closing an operator, in general, is not the same as closing the graph of the operator. Note that closing the convolution operator C_h in the example considered in [8] means that the graph of the extended convolution operator now includes those input-output pairs that caused problems, recovering the possibility to stabilize this simple system, as one would hope.

Let us consider again the example studied in [8]. Consider the input-output relationships

$$y = h * u + w \tag{22}$$
$$u = -ky, \tag{23}$$

where w is an output disturbance and $k > 0$ is the gain of the proportional controller $u = -ky$. Let $\lim_{t \to -\infty} d(t) = \lim_{t \to -\infty} y(t) = 0$. The output y is given by

$$y(t) = w(t) - k \int_0^\infty e^{-(k-1)\tau} w(t - \tau) d\tau \tag{24}$$

when w is such that the above integral is well-defined. In particular, the integral is well-defined for $w(t) = Ce^t$ for any (real) t, where C is an arbitrary constant. Computing y gives $y(t) = 0$ for all real t. Hence $u(t) = 0$ for any t, i.e. $(h * u)(t) = -Ce^t$ for any t for a zero input. Thus the unavoidable conclusion is that the system h can exhibit a non-zero response to a zero input when the system h is in a closed-loop configuration. Note that the above computations are valid for any $k > 0$, that is also for *certainly* non-stabilizing $0 < k < 1$. This means that closing the loop also implies that we need to look at an extended definition of the convolution operator as argued earlier from a different point of view. This conclusion is independent of whether the closed loop is stable or not !

It was assumed in [8] that the closed-loop expressions must have, for a stabilizing controller, unique well-defined solutions for any $L_2(-\infty, \infty)$ external signals without extending the operator. But we saw that this assumption can not be valid unless we extend the convolution operator ! Note that the remedy suggested here does not destroy causality of the closed-loop system.

To summarize, the argumentation in [8] was incomplete. In an input-output $L_2(-\infty, \infty)$ setting operator extensions appear to be a key concept in stabilization studies.

6   Conclusions

We have discussed four fundamental puzzles in systems and control theory. The first puzzle highlights the need to gain a greater understanding of the limitations of standard model and controller parametrizations, and of limits of standard black-box identification methodology. The second and the third puzzle point out important limitations in standard formulations of robust H_∞ control theory when robustness against persistent signals is desired. These limitations raise doubt over whether one should at all use the terminology H_∞ control in the context of robust control. The fourth puzzle deals with intrinsic difficulties in doubly-infinite time axis input-output stabilization theory as pointed out by Georgiou and Smith [8]. This is a messy problem that needs to be cleaned up. It is clear that the conclusions in [8] were based on an incomplete analysis.

Acknowledgements

Financial support to the author from the Academy of Finland is gratefully acknowledged.

References

1. N.I. Akhiezer and I.M. Glazman. *Theory of Linear Operators in Hilbert Space*, New York : Frederick Ungar Publ., 1961.
2. K.J. Aström and B. Wittenmark. *Computer-Controlled Systems*, Englewood Cliffs, NJ : Prentice-Hall, 1990.
3. P.E. Caines. *Linear Stochastic Systems*, New York : Wiley, 1988.
4. M.A. Dahleh and I.J. Diaz-Bobillo. *Control of Uncertain Systems : A Linear Programming Approach*, Englewood Cliffs, NJ : Prentice-Hall, 1995.
5. M.A. Dahleh and J.B. Pearson. ℓ^1 optimal feedback controllers for MIMO discrete-time systems. *IEEE Trans. Automat. Control*, vol. AC-32, pp. 314-322, 1987.
6. J.C. Doyle, B.A. Francis and A.R. Tannenbaum. *Feedback Control Theory*, New York : MacMillan, 1992.

7. P.L. Duren. *Theory of H^P Spaces*, New York : Academic Press, 1970.
8. T.T. Georgiou and M.C. Smith. Intrinsic difficulties in using the doubly-infinite time axis for input-output control theory. *IEEE Trans. Automat. Control*, vol. 40, pp. 516-518, 1995.
9. M. Green and D.J.N. Limebeer. *Linear Robust Control*, Englewood Cliffs, NJ : Prentice-Hall, 1995.
10. L.H. Keel and S.P. Bhattacharyya. Robust, Fragile, or Optimal? *IEEE Trans. Automat. Control*, vol. 42, pp. 1098-1105, 1997.
11. E. Landau. *Darstellung und Begründung einiger neuerer Ergebnisse der Funktionentheorie*. Berlin : Springer-Verlag, 1929.
12. G. Li. On the structure of digital controllers with finite word length consideration. *IEEE Trans. Automat. Control*, vol. 43, pp. 689-693, 1998.
13. L. Ljung. *System Identification*, Englewood Cliffs, NJ : Prentice-Hall, 1987.
14. P.M. Mäkilä. Comments on "Robust, Fragile, or Optimal?". *IEEE Trans. Automat. Control*, vol. 43, pp. 1265-1267, 1998.
15. P.M. Mäkilä and J.R. Partington. Robust stabilization − BIBO stability, distance notions and robustness optimization. *Automatica*, vol. 23, pp. 681-693, 1993.
16. P.M. Mäkilä and J.R. Partington. On bounded-error identification of feedback systems. *Int. J. Adapt. Control Signal Proc.*, vol. 9, pp. 47-61, 1995.
17. P.M. Mäkilä and J.R. Partington. Lethargy results in LTI system modelling. *Automatica*, vol. 34, pp. 1061-1070, 1998.
18. P.M. Mäkilä, J.R. Partington and T.K. Gustafsson. Worst-case control-relevant identification. *Automatica*, vol. 31, pp. 1799-1819, 1995.
19. P.M. Mäkilä, J.R. Partington and T. Norlander. Bounded power signal spaces for robust control and modelling. *SIAM J. Control & Optim.*, vol. 00, pp. 00-00, 1999. In print.
20. M. Milanese and A. Vicino. Information based complexity and nonparametric worst-case system identification. *J. Complexity*, vol. 2, pp. 78-94, 1993.
21. P. Moroney. *Issues in the Implementation of Digital Compensators*, Cambridge, U.S.A : The MIT Press, 1983.
22. J.R. Partington. *Interpolation, Identification and Sampling*, Oxford : Oxford Univ. Press, 1997.
23. K. Poolla, P. Khargonekar, A. Tikku, J. Krause and K. Nagpal. A time-domain approach to model validation. *IEEE Trans. Automat. Control*, vol AC-39, 951-959, 1994.
24. W. Rudin. *Real and Complex Analysis*, New York : McGraw-Hill, 1987.
25. T. Söderström and P. Stoica. *System Identification*, Englewood Cliffs, NJ : Prentice-Hall, 1989.
26. M. Vidyasagar. *Control System Synthesis*, MA : MIT Press, 1985.
27. M. Vidyasagar. Optimal rejection of persistent bounded disturbances. *IEEE Trans. Automat. Control*, vol. 31, pp. 527-534, 1986.
28. N. Wiener. *The Fourier Integral*, Cambridge Univ. Press, 1933.
29. G. Zames. On the metric complexity of causal linear systems : ϵ-entropy and ϵ-dimension for continuous-time. *IEEE Trans. Automat. Control*, vol. 24, pp. 222-230, 1979.
30. G. Zames. Feedback and optimal sensitivity : Model reference transformations, multiplicative seminorms, and approximate inverses. *IEEE Trans. Automat. Control*, vol. AC-26, pp. 301-320, 1981.

31. K. Zhou, J.C. Doyle and K. Glover. *Robust and Optimal Control*, Englewood Cliffs, NJ : Prentice-Hall, 1996.

Robust Regional Pole Placement: An Affine Approximation

Vladimír Kučera[1,2] and František J. Kraus[3]

[1] Academy of Sciences, Institute of Information Theory and Automation, CZ–182 08 Praha, Czech Republic
[2] Czech Technical University, Trnka Laboratory for Automatic Control, CZ–166 27 Praha, Czech Republic
[3] Swiss Federal Institute of Technology, Automatic Control Laboratory, CH–8092 Zürich, Switzerland

Abstract. Assignment of the coefficients of factors of the characteristic polynomial of an uncertain multi–input multi–output system to polytopic regions is considered. Owing to a special upper triangular form of the desired closed–loop polynomial matrix, this formulation leads to an affine design problem. This design is closer to the ultimate pole placement than the assignment of the characteristic polynomial direct. As a result, one obtains a set of affine inequalities with respect to the controller coefficients. The solution (if it exists) of these inequalities then defines all admissible controllers.

1 Introduction

Pole placement is a common approach to designing closed–loop controllers in order to meet desired control specifications. For convenience, the objective of assigning closed–loop poles is often replaced by that of assigning characteristic polynomials. These two formulations are equivalent (up to numerical considerations) assuming that the models used in the design are accurate and that the desired pole locations are given exactly. The design is usually based on polynomial system description and leads to the solution of a diophantine equation [1], [7]. A state–space solution is also available, often as a modification of an optimal control problem [11], [12].

Supported by the Ministry of Education of the Czech Republic under Project VS97–034, Grant Agency of the Czech Republic under Project 102/97/0861, and Swiss National Science Foundation under Project 7TRPJ 038631.

Preliminary versions of this paper were presented at the European Control Conference, Brussels 1997 and the Workshop on Robustness in Identification and Control, Siena 1998.

A practical design based on this technique usually seeks to guarantee that the closed–loop poles are placed within a specified subset of the complex plane, rather than at exact positions. The solution of this problem is addressed in [5], [6] using state–space techniques. A polynomial solution for single–input single–output systems is presented in [8], a generalization to the case of multi–input multi–output systems can be found in [4].

Another important practical issue is that of model uncertainty. If the model uncertainty is relatively small, then it is possible to use sensitivity-based methods to calculate variations of the poles due to uncertainty of the model. If the model uncertainty (described as variation of some model parameters) is bounded but large the sensitivity techniques, being valid only in an infinitesimal neighbourhood of the nominal values, may lead to misleading results. A robust formulation of the problem is then appropriate.

In each case mentioned above, i.e. plant description uncertainty and/or unsharp pole positions, the design problem of pole placement is different from that of characteristic polynomial assignment . The former corresponds to the primary design goal while the latter is easier to handle, especially when the desired region of characteristic polynomial coefficients is convex. The pole assignment problem is addressed in [3], [10] whereas the characteristic polynomial assignment is discussed in [9].

This substitution is sometimes considered as the main drawback of the approach. Indeed, for a given pole region the corresponding region in the space of characteristic polynomial coefficients is not necessarily convex. In general, an adequate convex approximation of this region is not evident. Any such approximation, naturally, rules out some of the valid combinations of the closed–loop poles.

Despite these difficulties, we adopt the approach of assigning polynomials rather than poles. The design problem we consider in the paper is formulated for a multi–input multi–output uncertain plant with polytopic uncertainty. The plant family is described as a "fraction" of polytopic polynomial matrices, i.e. polynomial matrices whose coefficients live in a polytope region. This uncertainty model is easy to handle and it overbounds the true uncertainty of the plant.

It is quite natural then to formulate the design goal as the assignment of the closed–loop polynomial "denominator" matrix to a polytopic region. As unimodular transformations leave the determinant of this matrix unchanged, it is convenient to consider upper triangular denominator matrices which contain factors of the characteristic polynomial on the diagonal. We then propose to assign these factors rather than the characteristic polynomial itself.

This formulation leads to an affine design problem, even for multi–input multi–output plants. In contrast, the problem of assigning the characteristic polynomial directly is in general a multiaffine one, being affine in the single–input single–output case. We also emphasize that the assignment of polytopic regions for factors of the characteristic polynomial is closer to the ultimate goal – pole placement – than the assignment of polytopic regions for the characteristic polynomial. The more even is the distribution of the degrees among the factors the better. Factors of degree one or two actually imply the true pole placement within a polytopic region.

The design procedure involves some degrees of freedom. Firstly, only a partial assignment of the closed–loop polynomial matrix is necessary. Thanks to the special upper triangular structure of this matrix, we can formulate the assignment problem just for its lower triangular part. Secondly, the diagonal entries of the matrix can be chosen in many different ways to yield the same characteristic polynomial. The two sources of freedom can be used to satisfy additional specifications or to enlarge the family of admissible controllers.

It is clear that, depending on the description of plant uncertainty and the desired region of pole positions, the problem is either solvable or not. The first step is therefore to parametrize the set of all controllers that satisfy the given specifications. If it is not empty, the parametrization can be used in a second step to meet additional design specifications.

2   Single–input Single–output Plants

Let us first review the results available for single–input single–output plants. Consider a plant described by real–rational strictly proper transfer function

$$S(s) = \frac{B(s)}{A(s)} \tag{1}$$

where A and B are polynomials. Now for a real–rational proper controller of the form

$$R(s) = -\frac{Q(s)}{P(s)} \tag{2}$$

where P and Q are polynomials, the closed–loop system poles are given by

$$C(s) = P(s)A(s) + Q(s)B(s). \tag{3}$$

Thus by setting C as the polynomial for the closed–loop poles we can, under certain assumptions, solve equation (3) for the polynomials P and Q. These assumptions are as folows:

(α) A and B are coprime polynomials, so that we may choose the coefficients of C at will;

(β) C has high enough a degree, at least twice the degree of A less one, so that we obtain a proper controller for any A and B.

Furthermore, if the state–space realizations of S and R are both controllable and observable, the polynomial C defined by (3) is the characteristic polynomial of the closed–loop system.

Now suppose that A has degree n, C has degree $2n - 1$ (the least degree indicated in β) so that P has degree $n - 1$, and let

$$a = [a_0, a_1, ..., a_n]^T$$
$$b = [b_0, b_1, ..., b_{n-1}]^T$$
$$p = [p_0, p_1, ..., p_{n-1}]^T$$
$$q = [q_0, q_1, ..., q_{n-1}]^T$$

be polynomial coefficients of A, B, P, Q, respectively, and

$$c = [c_0, c_1, ..., c_{2n-1}]^T$$

be those of C. Define

$$g = \begin{bmatrix} a \\ b \end{bmatrix}$$

as the plant vector,

$$x = \begin{bmatrix} p \\ q \end{bmatrix}$$

as the controller vector and two Sylvester matrices,

$$G = \begin{bmatrix}
a_0 & & 0 & b_0 & & 0 \\
a_1 & \ddots & \vdots & b_1 & \ddots & \vdots \\
\vdots & & a_0 & \vdots & & b_0 \\
\vdots & & a_1 & b_{n-1} & & b_1 \\
a_n & & \vdots & \vdots & \ddots & \vdots \\
\vdots & \ddots & \vdots & \vdots & & b_{n-1} \\
0 & & a_n & 0 & & 0
\end{bmatrix}$$

of size $2n \times 2n$ and

$$X = \begin{bmatrix}
p_0 & & 0 & q_0 & & 0 \\
p_1 & \ddots & \vdots & q_1 & \ddots & \vdots \\
\vdots & & p_0 & \vdots & & q_0 \\
\vdots & & p_1 & \vdots & & q_1 \\
p_{n-1} & & \vdots & q_{n-1} & & \vdots \\
\vdots & \ddots & \vdots & \vdots & \ddots & q_{n-1} \\
0 & & p_{n-1} & 0 & & 0
\end{bmatrix}$$

of size $2n \times (2n + 1)$. Then it is straightforward to rewrite (3) as either $Gx = c$ or $Xg = c$.

We shall now assume that the plant is uncertain in the sense that A and B are not fixed but are known to lie within polytope regions. This means that the plant vector g lies in a polytope

$$P_G = \text{ convex hull } \{g^1, ..., g^t\}$$

where g^i, $i = 1, 2, ..., t$ are vertices of the convex polytope. For example, if the plant parameters are intervals then g^i, $i = 1, 2, ..., 2^{2n+1}$ are vertices of the rectangular solid.

Further, we assume that there is some flexibility in the desired pole locations and that the region desired is bounded and defined by a convex polytope

$$P_C = \{c \in \mathbb{R}^{2n} : Hc \geq 0\}$$

where H is a real matrix; this is a dual definiton of a convex polytope.

For a fixed controller, we denote the corresponding set of all possible closed–loop polynomial coefficients as

$$P_X = \{c \in \mathbb{R}^{2n} : c = Xg, \; g \in P_G\}.$$

Then P_X is a polytope,

$$P_X = \text{ convex hull } \{Xg^1, ..., Xg^t\}$$

and if G^i is the Sylvester matrix corresponding to the plant vector g^i, we have

$$P_X = \text{ convex hull } \{G^1 x, ..., G^t x\}.$$

Thus any controller R will take any and every plant vector g from P_G and map it into a closed–loop polynomial vector c in P_X. Then the problem of robust pole assignment seeks to find a controller vector x such that P_X is included in the prescribed polytope P_C.

Since both P_X and P_C are convex polytopes, we have [10] that $P_X \subset P_C$ if and only if

$$HG^i x \geq 0, \quad i = 1, 2, ..., t. \tag{4}$$

Any such x defines a robust controller. The set (4) is an intersection of convex polytopes; denoting its vertices as $x^1, ..., x^d$ we obtain a neat parametrization of all robust controllers R as

$$x = \lambda_1 x^1 + ... + \lambda_d x^d$$

where λ_i, $i = 1, 2, ..., d$ are non–negative reals whose sum is one. It may be a costly task to identify the vertices $x^1, x^2, ..., x^d$. The number d of vertices is an exponential in t. From the computational point of view, it is easier to characterize the set of robust controllers using (4) and apply a convex optimization tool to search for a feasible solution, if any.

When the degree of C exceeds $2n - 1$, the McMillan degree of R exceeds $n - 1$ and x lives in a space of dimension higher than $2n$. On the other hand, if some coefficients of A and B (or linear combinations thereof) are known, the number of constraints t gets smaller thus reducing the complexity of the problem.

Finally, we point out that the desired closed-loop polynomials C are specified here by its coefficient regions rather than by pole regions. The relationship between these two is complicated and known only in special cases.

3   Multi–input Multi–output Plants

Let us now consider the problem for multi–input multi–output plants . We suppose our plant gives rise to a real–rational strictly proper $l \times m$ transfer function matrix

$$S(s) = B(s)A^{-1}(s) \tag{5}$$

where A and B are polynomial matrices, respectively $m \times m$ and $l \times m$. Now for a real–rational proper $m \times l$ controller transfer matrix

$$R(s) = -P^{-1}(s)Q(s) \tag{6}$$

where P and Q are polynomial matrices, respectively $m \times m$ and $m \times l$, the closed–loop system poles are given by the determinant of the matrix

$$C(s) = P(s)A(s) + Q(s)B(s). \tag{7}$$

Thus by setting C as the polynomial matrix for the closed–loop poles we can, under certain assumptions, solve equation (7) for the polynomial matrices P and Q. These assumptions are as follows:

(a) A and B are right coprime polynomial matrices, so that we may choose the coefficients of C at will;
(b) C has high enough degrees, so that we obtain a proper controller.

Furthermore, if the state–space realizations of S and R are both controllable and observable, the invariant factors of the matrix C defined by (7) are the invariant polynomials of the closed–loop system. In particular, the determinant of C is the characteristic polynomial of the closed–loop system.

The polynomial factorization (5) is unique up to the right multiplication by a unimodular matrix. This freedom can be used to normalize A. We shall suppose that A has been chosen

(c) column reduced, so that the determinant of A has degree equal to the sum of the column degrees, $\alpha_1, \alpha_2, ..., \alpha_m$, of A;

(d) decreasingly column–degree ordered, so that $\alpha_1 \geq \alpha_2 \geq ... \geq \alpha_m$;

(e) having a diagonal highest–column–degree coefficient matrix.

The polynomial factorization (6) is unique up to the left multiplication by a unimodular matrix. This freedom can be used to normalize C, since the unimodular factor is absorbed in P and Q when solving equation (7), thus producing the same R. We shall suppose that C has the following form:

(f) C is upper triangular and has the identity highest–column–degree coefficient matrix;

(g) the diagonal elements C_{ii} of C have degrees equal to $\pi_i + \alpha_i$ while the off–diagonal elements C_{ij}, $i < j$ have degrees less that $\pi_i + \alpha_j$, where

$$\alpha_1 \geq \alpha_2 \geq ... \geq \alpha_m$$

are the column degrees of A and

$$\pi_1 \leq \pi_2 \leq ... \leq \pi_m$$

is a list of non–negative integers.

This special form of C is motivated by our design goal. We wish to describe the flexibility one has in choosing pole locations by using polytopes. In the multi–input multi–output systems, however, we do not know how to assign the pole polynomial to a polytope region for a general C, since the coefficients of the determinant of C depend on those of C_{ij} in a multi–affine manner. So we take C upper triangular, make the diagonal polynomials C_{ii} appear as factors of the pole polynomial, and assign these factors independently of each other to polytope regions. This formulation puts us closer to the original problem of assigning the closed–loop poles. Indeed, the lower the degree of a polynomial, the simpler relation exists between pole regions and coefficient polytopes.

It is important to realize that assumptions (f) and (g) impose no restrictions on the set of controllers that result from equation (7). The requirement (f) means that C is taken in the upper Hermite form; this can certainly be achieved by applying left unimodular transformations to any square non–singular polynomial matrix C. The requirement (g) reflects the special way in which C is constructed using (7) and it makes it possible to determine *a priori* the degrees of the solution matrices P and Q, if they exist, as follows:

(h) P is row reduced with row degrees $\pi_1, \pi_2, ..., \pi_m$ and with a diagonal highest–row–degree coefficient matrix;

(i) Q has row degrees less than or equal to the corresponding degrees of P.

As a result, whenever P and Q with properties (h) and (i) exist and satisfy equation (7), the controller transfer matrix (6) is proper.

We observe that the McMillan degree of S is given by

$$\delta S = \alpha_1 + \alpha_2 + ... + \alpha_m$$

and that of R by

$$\delta R = \pi_1 + \pi_2 + ... + \pi_m. \tag{8}$$

This provides an interpretation for the list of integers $\pi_1, \pi_2, ..., \pi_m$ involved in assumption (g). This provides also a clue how large these integers should be chosen to actually guarantee the existence of X and Y in (7) having the properties (h) and (i); see assumption (b). In contrast to the single–input single–output case, where a simple estimate $\delta R = n - 1$ exists, no general estimates for $\pi_1, \pi_2, ..., \pi_m$ are available in the multi–input multi–output case. It is helpful to know [2] that $\delta R = \alpha_1 - 1$ will generically do. Then (8) is used to generate a guess for the list $\pi_1, \pi_2, ..., \pi_m$.

Finally we observe that assigning the pole polynomial factor–wise fits well the format of equation (7). When the polynomial matrix C is set to define a desired pole polynomial, it defines at the same time a factorization of this polynomial into invariant factors. One could opt to assign these particular factors to polytopes. Their degrees, however, are not well distributed. Indeed, the invariant factors of C are $1, 1, ..., 1, \psi_m$ (= determinant of C) in the generic case. The specification of factors via (f) and (g) provides a better alternative: the designer retains some control over the degrees of the factors and can achieve a more even split. That is why the two lists, namely $\alpha_1, \alpha_2, ..., \alpha_m$ and $\pi_1, \pi_2, ..., \pi_m$ are ordered in the opposite way.

Now we suppose that a suitable list $\pi_1, \pi_2, ..., \pi_m$ has been found and proceed to define the polytopes which accommodate the plant uncertainty and the available flexibility in the desired pole locations.

Let A_{ij} be the polynomial in position i, j in A and let $a_{ij,k}$ be the coefficient of s^k in A_{ij}; similarly for B, C and P, Q. Define

$$a_{ij} = \begin{cases} [\, a_{jj,0}, ..., a_{jj,\alpha_j} \,]^T, & i = j \\ [\, a_{ij,0}, ..., a_{ij,\alpha_j-1} \,]^T, & i \neq j \end{cases}$$

$$b_{ij} = [\, b_{ij,0}, ..., b_{ij,\alpha_j-1} \,]^T,$$

$$p_{ij} = \begin{cases} [\, p_{ii,0}, ..., p_{ii,\pi_i} \,]^T, & i = j \\ [\, p_{ij,0}, ..., p_{ij,\pi_i-1} \,]^T, & i \neq j \end{cases}$$

$$q_{ij} = [\, q_{ij,0}, ..., q_{ij,\pi_i} \,]^T.$$

The plant vector g is defined column–wise,

$$g = \begin{bmatrix} g_1 \\ \vdots \\ g_m \end{bmatrix}, \quad g_j = \begin{bmatrix} a_j \\ b_j \end{bmatrix}, \quad j = 1, 2, ..., m$$

where

$$a_j = \begin{bmatrix} a_{1j} \\ \vdots \\ a_{mj} \end{bmatrix}, \quad b_j = \begin{bmatrix} b_{1j} \\ \vdots \\ b_{lj} \end{bmatrix}$$

and the dimension of g is $(l+m)\,\delta S + m$. The controller vector x is defined row–wise,

$$x = \begin{bmatrix} x_1 \\ \vdots \\ x_m \end{bmatrix}, \quad x_i = \begin{bmatrix} p_i \\ q_i \end{bmatrix}, \quad i = 1, 2, ..., m$$

where

$$p_i = \begin{bmatrix} p_{i1} \\ \vdots \\ p_{im} \end{bmatrix}, \quad g_i = \begin{bmatrix} q_{i1} \\ \vdots \\ q_{il} \end{bmatrix}$$

and the dimension of x is $(l+m)\,\delta R + lm + m$. Further let

$$c_{ii} = [\, c_{ii,0}, ..., c_{ii,\pi_i + \alpha_i}\,]^T$$
$$c_{ij} = 0 \in \mathbb{R}^{\pi_i + \alpha_j}, \quad i > j$$
$$c_{ij} \text{ not specified}, \quad i < j$$

and define the vector

$$c = \begin{bmatrix} c_1 \\ \vdots \\ c_m \end{bmatrix}, \quad c_i = \begin{bmatrix} c_{i1} \\ \vdots \\ c_{im} \end{bmatrix}, \quad i = 1, 2, ..., m$$

of dimension $m(\delta R + \delta S + 1)$.

Then it is straightforward to rewrite (7) as either $Gx = c$ or $Xg = c$, where G and X are two block Sylvester matrices, each having a different structure and size.

We assume that the plant vector g lies in a polytope

$$P_G = \text{convex hull } \{\, g^1, ..., g^t\,\}$$

where g^i, $i = 1, 2, ..., t$ are vertices of the convex polytope. Further, we assume that the vector c belongs to a convex polytope

$$P_C = \{\, c \in \mathbb{R}^{m(\delta R + \delta S + 1)} : Hc \geq 0\,\}$$

where H is a real matrix. Consistent with our goal, we set $c_{ij} = 0$ for $i > j$ and do not impose any constraint on c_{ij}, $i < j$.

For a fixed controller, we denote the corresponding set of all possible closed–loop polynomial matrix coefficients as

$$P_X = \{c \in I\!R^{m(\delta R + \delta S + 1)} : c = Xg, \quad g \in P_G\}.$$

Then P_X is a polytope,

$$P_X = \text{convex hull } \{X g^1, ..., X g^t\}$$

and if G^i is the block Sylvester matrix corresponding to the plant vector g^i, we have

$$P_X = \text{convex hull } \{G^1 x, ..., G^t x\}.$$

The single–input single–output case is then routinely followed to conclude that $P_X \subset P_C$ if and only if

$$HG^i x \geq 0, \quad i = 1, 2, ..., t. \tag{9}$$

Hence a controller is robust if and only if the vector x belongs to the set

$$\{x \in I\!R^{(l+m)\delta R + lm + m} : HG^i x \geq 0, \quad i = 1, 2, ..., t\}.$$

This set is an intersection of convex polytopes; denoting its vertices as $x^1, ..., x^d$ we obtain a neat parametrization of all robust controllers R as

$$x = \lambda_1 x^1 + ... + \lambda_d x^d$$

where λ_i, $i = 1, 2, ..., d$ are non–negative reals whose sum is one.

4   Example

We present a numerical example to illustrate the proposed design method.
Let us consider a multi–input multi–output plant giving rise to the transfer function matrix

$$S(s) = \frac{1}{s+1} \begin{bmatrix} b & 1 \\ 0 & 1 \end{bmatrix}$$

with uncertainty in the parameter b,

$$b \in [\frac{1}{2}, \frac{3}{2}].$$

An associated coprime factorization is

$$S(s) = \begin{bmatrix} b & 1 \\ 0 & 1 \end{bmatrix} \begin{bmatrix} s+1 & 0 \\ 0 & s+1 \end{bmatrix}^{-1} = B(s)A^{-1}(s).$$

We are looking for a proper, first order controller. One possible structure of such a controller is

$$R(s) = - \begin{bmatrix} s + x_1 & x_2 \\ 0 & 1 \end{bmatrix}^{-1} \begin{bmatrix} y_1 s + y_2 & y_3 s + y_4 \\ 0 & y_5 \end{bmatrix}$$

where x_i and y_j are parameters to be designed.

The design specification is to obtain a closed–loop polynomial matrix of the form

$$C(s) = \begin{bmatrix} s^2 + \alpha s + \beta & \delta \\ 0 & s + \gamma \end{bmatrix}$$

whose coefficients live in a convex polytopic region, defined by the hyper-planes

$$\begin{bmatrix} -14 & 1 & 0 \\ 16 & -2 & 0 \\ -2 & 1 & 0 \\ \hline 0 & 0 & -1 \\ 0 & 0 & 1 \end{bmatrix} \begin{bmatrix} \alpha \\ \beta \\ \gamma \end{bmatrix} + \begin{bmatrix} 196 \\ -56 \\ 4 \\ \hline -2 \\ 14 \end{bmatrix} > 0 \tag{10}$$

and where δ is an uspecified real number.

This design specification results in all poles of the closed–loop system lying within the disk

$$|s + 8| < 6.$$

The design equation (7) reads

$$\begin{bmatrix} s + x_1 & x_2 \\ 0 & 1 \end{bmatrix} \begin{bmatrix} s + 1 & 0 \\ 0 & s + 1 \end{bmatrix} + \begin{bmatrix} y_1 s + y_2 & y_3 s + y_4 \\ 0 & y_5 \end{bmatrix} \begin{bmatrix} b & 1 \\ 0 & 1 \end{bmatrix} =$$

$$\begin{bmatrix} s^2 + \alpha s + \beta & \delta \\ 0 & s + \gamma \end{bmatrix}.$$

From the lower triangular part of this matrix we obtain

$$\begin{bmatrix} 1 & b & 0 & 0 \\ 1 & 0 & b & 0 \\ \hline 0 & 0 & 0 & 1 \end{bmatrix} \begin{bmatrix} x_1 \\ \hline y_1 \\ y_2 \\ y_5 \end{bmatrix} + \begin{bmatrix} 1 \\ 0 \\ \hline 1 \end{bmatrix} = \begin{bmatrix} \alpha \\ \beta \\ \hline \gamma \end{bmatrix} \tag{11}$$

The upper triangular part yields

$$x_2 + y_1 + y_3 = 0, \quad x_2 + y_2 + y_4 = \delta.$$

Using (11) in (10) results in

$$\left(\begin{bmatrix} -14 & 1 & | & 0 \\ 16 & -2 & | & 0 \\ -2 & 1 & | & 0 \\ \hline 0 & 0 & | & -1 \\ 0 & 0 & | & 1 \end{bmatrix} \begin{bmatrix} 1 & | & b & 0 & 0 \\ 1 & | & 0 & b & 0 \\ \hline 1 & | & 0 & 0 & 1 \end{bmatrix} \begin{bmatrix} x_1 \\ - \\ y_1 \\ y_2 \\ y_5 \end{bmatrix} + \begin{bmatrix} -14 + 196 \\ 16 - 56 \\ -2 + 4 \\ \hline -1 - 2 \\ 1 + 14 \end{bmatrix}\right) > 0$$

For $b \in [\frac{1}{2}, \frac{3}{2}]$ we finally obtain the inequalities

$$-15 < y_5 < -3$$

and

$$\left(\begin{bmatrix} -13 & -7 & 0.5 \\ 14 & 8 & -1 \\ -1 & -1 & 0.5 \\ \hline -13 & -21 & 1.5 \\ 14 & 24 & -3 \\ -1 & -3 & 1.5 \end{bmatrix} \begin{bmatrix} x_1 \\ y_1 \\ y_2 \end{bmatrix} + \begin{bmatrix} 182 \\ -40 \\ 2 \\ \hline 182 \\ -40 \\ 2 \end{bmatrix}\right) > 0 .$$

These inequalities characterize all admissible robust controllers. Explicit parametrization of the controller set is as follows:

$$\begin{bmatrix} x_1 \\ y_1 \\ y_2 \end{bmatrix} = \sum_{i=1}^{9} \lambda_i V_i, \qquad \begin{matrix} -15 < y_5 < -3 \\ x_2 = -y_1 - y_3, \quad y_3, y_4 \text{ arbitrary} \end{matrix}$$

where V_i denote the vertices of a convex polytope given by the columns of the matrix (suitably rounded)

$$V = \begin{bmatrix} 18.3 & 14 & 14 & 14 & 2.9 & 2.9 & 2 & 2 & -2.3 \\ -6.6 & 26 & 2 & 8.7 & 0.3 & 16.1 & 8.7 & 2 & 11.5 \\ 19.4 & 364 & 28 & 121.3 & 2.3 & 128.8 & 17.3 & 4 & 20.2 \end{bmatrix}$$

and where λ_i are free real positive parameters such that

$$\sum_{i=1}^{9} \lambda_i = 1.$$

This parametrization can be used to meet additional design specifications.

5   Concluding Remarks

A solution has been presented of a design problem for multi–input multi–output systems which is close to the ultimate goal of robust pole placement within a desired region. The design has been formulated as an assignment of polynomial factors of the characteristic polynomial for a special upper triangular structure of the desired closed–loop polynomial matrix. This leads to an affine formulation of and provides all admissible solutions to the design problem.

Future research should focus on the effect of the unimodular transformations needed to bring the closed–loop polynomial matrix to the special triangular form as well as on the freedom in factoring the desired characteristic polynomial.

References

1. Åström, K. J., Wittenmark, B. (1990) Computer–controlled Systems: Theory and Design (2nd ed.). Prentice–Hall, Englewood Cliffs, NJ
2. Brasch, F. M., Pearson, J. B. (1970) Pole placement using dynamic compensators. IEEE Trans. Automat. Control, 15, 34–43
3. Figueroa, J. L., Romagnoli, J. A. (1994) An algorithm for robust pole assignment via polynomial approach. EEE Trans. Automat. Control, 39, 831–835
4. Jetto, L. (1996) Assigning invariant polynomials over polytopes. IEEE Trans. Automat. Control, 41, 144–148
5. Kawasaki, N., Shimemura, E. (1983) Determining quadratic weighting matrices to locate poles in a specified region. Automatica, 19, 558–560
6. Kim, S. B., Furuta, K. (1988) Regulator design with poles in a specified region. Int. J. Control, 47, 143–160
7. Kučera, V. (1994) Pole placement equation: A survey. Kybernetika, 30, 578–584.
8. Kučera, V., Kraus, F. J. (1995) Regional pole placement. Kybernetika, 31, 541–546
9. Rotstein, H., Sanchez Peña, R., Bandoni, J., Desages, A. Romagnoli, J. A. (1991) Robust characteristic polynomial assignment. Automatica, 27, 711–715
10. Soh, Y. C., Evans, R. J., Petersen, I. R., Betz, R. E. (1987) Robust pole assignment. Automatica, 23, 601–610
11. Solheim, O. A. (1972) Design of optimal control systems with prescribed eigenvalues. Int. J. Control, 15, 143–160
12. Wittenmark, B., Evans, R. J., Soh, Y. C. (1987) Constrained pole–placement using transformation and LQ–design. Automatica, 23, 767–769

On Achieving L_p (ℓ_p) Performance with Global Internal Stability for Linear Plants with Saturating Actuators

Anton A. Stoorvogel[1], Ali Saberi[2], and Guoyong Shi[2]

[1] Department of Mathematics and Computing Science, Eindhoven Univ. of Technology, P.O. Box 513, 5600 MB Eindhoven, The Netherlands
[2] School of Electrical Engineering and Computer Science, Washington State University, Pullman, WA 99164-2752, U.S.A.

Abstract. In this paper we show that for both continuous-time and discrete-time critically unstable linear systems with saturating actuators one can not simultaneously achieve the global internal stabilization and the global finite-gain L_p (ℓ_p) performance whenever the external input (disturbance) is not input-additive. However, one can achieve the global internal stabilization and global L_p (ℓ_p) stabilization (without finite-gain) for any $p \in [1, \infty)$.

1 Introduction

The problem of internal stabilization while simultaneously achieving L_p (ℓ_p) performance in the presence of external input (disturbance), in global and/or semiglobal setting, of linear systems with saturating actuators has been the subject of quite a few recent papers, see [1–6,8]. Most of the existing

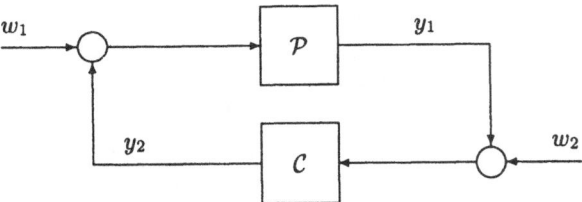

Fig. 1. Input-additive configuration.

papers in the literature that deal with this problem in global or semiglobal setting are restricted to a special case when the external input (disturbance) is input-additive as shown in Fig. 1. This particular case is labeled as the simultaneous internal and external stabilization problem (see [5] and [3]). Moreover, for this special case the majority of the results reported in the open literature deal only with the state feedback case, i.e. assuming $y_1 = x$ and $w_2 = 0$ in Fig. 1. It has been established that for continuous-time systems which are open-loop *critically unstable* the simultaneous global internal stabilization and global finite-gain L_p stabilization is achieveable for all $p \in [1, \infty]$, moreover one can make the L_p-induced norm for the closed-loop system arbitrarily small, i.e. almost disturbance decoupling can be achieved [8]. For discrete-time system the analogous global result is not available. However, for the special case that the discrete-time open-loop system is *critically stable*, this result, without possibility of achieving arbitrarily small ℓ_p-induced norm for the closed-loop system, has been established recently for $p \in (1, \infty]$ [1]. (There is yet no result available for the case $p = 1$.) For the simultaneous internal and external stabilization problems in the semiglobal setting, the interested readers are referred to [3].

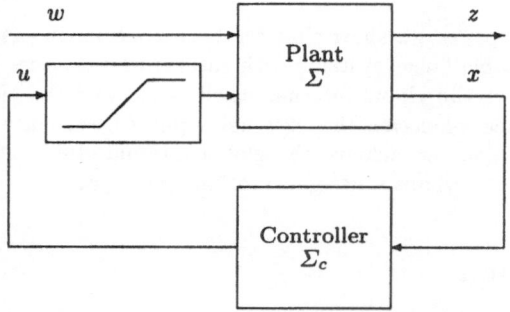

Fig. 2. Not input-additive configuration.

In this paper we would like to examine the problem of simultaneously achieving global internal stabilization (in the absence of disturbance) and global L_p (ℓ_p) performance in the presence of disturbance in a general setting where the external input (disturbance) is *not* input-additive as shown in Fig. 2. Here we provide three fundamental results. The first one is a negative result. It shows that, for both continuous-time and discrete-time *critically unstable* systems, whenever the external input (disturbance) is *not* input-additive, the problem of global internal stabilization (in the absence of disturbance) and simultaneous global L_p (ℓ_p) performance with *finite-gain* in the presence of disturbance is intrinsically not solvable. The second one is a positive result. It shows that for the same class of systems, whenever the external input (disturbance) is *not* input-additive, the simultaneous global internal stabilization (in the absence of disturbance) and global L_p (ℓ_p) performance *without finite-gain* in the presence of disturbance is possible for

$p \in [1, \infty)$. It is trivial to see that for $p = \infty$ such a problem is impossible to solve. Our third result is this. For both continuous-time and discrete-time *critically unstable* systems, whenever the external input (disturbance) is *not* input-additive, the problem of internal stabilization (in the absence of disturbance) and simultaneous L_2 (ℓ_2) performance with *finite-gain* in the presence of disturbance is solvable in a semi-global framework rather than in the global framework. As such, our results here point out that for linear systems with saturating actuators, whenever the external input (disturbance) is *not* input-additive, for internal stabilization (in the absence of disturbance) and for simultaneous L_p (ℓ_p) performance with *finite-gain* in the presence of disturbance, one should limit oneself to semi-global framework rather than using the global framework.

2   Preliminary Notations and Statements of Problems

For $x \in \mathbb{R}^n$, define $\|x\|_p^p := \sum_{i=1}^{n} |x|_i^p$ and $\|x\|_\infty := \max\{x_i : i = 1, \cdots, n\}$. The prime $(\cdot)'$ denotes the transpose of a vector or a matrix. I denotes the identity matrix of appropriate dimension. $\lambda_i(A)$ denotes any eigenvalue of A. For any $p \in [1, \infty]$, L_p^n denotes the space of all measurable functions $x(\cdot) : [0, \infty) \to \mathbb{R}^n$ such that

$$\|x\|_{L_p} := \left(\int_0^\infty \|x(t)\|_p^p \, dt \right)^{1/p} < \infty$$

if $p < \infty$ and

$$\|x\|_{L_\infty} := \text{ess sup}_{t \in [0, \infty)} \|x(t)\|_\infty < \infty.$$

Similarly, for any $p \in [1, \infty]$, ℓ_p^n denotes the space of all measurable sequences $x(\cdot) : \mathbb{Z}_+ \to \mathbb{R}^n$ such that

$$\|x\|_{\ell_p} := \left(\sum_{k=0}^{\infty} \|x(k)\|_p^p \right)^{1/p} < \infty$$

if $p < \infty$ and

$$\|x\|_{\ell_\infty} := \sup_k \|x(k)\|_\infty < \infty.$$

Further, for any $p \in [1, \infty]$, $L_p^s(D) := \{w \in L_p^s : \|w\|_{L_p} \leq D\}$. $\ell_p^s(D)$ is defined similarly. For any operator $\Upsilon : L_p^s \to L_p^r, w \mapsto z$, the L_p induced norm is defined as

$$\|\Upsilon\|_{i, L_p} := \sup_{\substack{w \in L_p^s \\ w \neq 0}} \frac{\|\Upsilon w\|_{L_p}}{\|w\|_{L_p}}$$

while $\|\Upsilon\|_{i,\ell_p}$ is defined similarly. Now we introduce the class of systems we consider in this paper. Consider the general system (for both discrete-time and continuous-time) described by

$$\Sigma : \begin{cases} \rho x = Ax + B\sigma(u) + Ew, & x \in \mathbb{R}^n,\ u \in \mathbb{R}^m, \\ y = C_1 x + D_1 w, & y \in \mathbb{R}^p,\ w \in \mathbb{R}^s, \\ z = C_2 x + D_2 \sigma(u), & z \in \mathbb{R}^r \end{cases} \tag{1}$$

where ρ is an operator indicating the time derivative $\frac{d}{dt}$ for continuous-time systems and a forward unit time shift for discrete-time systems. x, y and z are the state, the measurement and the controlled output vectors, respectively. u is the control input, and w is the external disturbance. $\sigma(\cdot)$ is a standard saturation function which is defined below.

Definition 1. (Saturation function) A function $\sigma(\cdot) : \mathbb{R}^m \to \mathbb{R}^m$ is called a saturation function if

1. $\sigma(\cdot)$ is decentralized, i.e.

$$\sigma(u) = [\sigma_1(u_1)\ \sigma_2(u_2)\ \cdots\ \sigma_m(u_m)]';$$

2. σ_i is globally Lipschitz, i.e. for some $\delta > 0$,

$$|\sigma_i(s_1) - \sigma_i(s_2)| \leq \delta |s_1 - s_2|;$$

3. $s\sigma_i(s) > 0$ whenever $s \neq 0$ and $\sigma_i(0) = 0$;
4. $\liminf\limits_{s \to 0} \frac{\sigma_i(s)}{s} > 0$
5. $\liminf\limits_{s \to \pm\infty} |\sigma_i(s)| > 0$.
6. σ_i is bounded.

Remark 3. Graphically, the assumption on $\sigma(\cdot)$ is that each $\sigma_i(\cdot)$ is in the first and the third quadrants and there exist $\Delta > 0$, $M > 0$ and $\delta > k > 0$ such that the nonlinearity lies in the nonlinear sector between the graph $(s, \delta\,\mathrm{sat}_{M/\delta}(s))$ and the graph $(s, k\,\mathrm{sat}_\Delta(s))$, where

$$\mathrm{sat}_a(s) = \mathrm{sgn}(s) \min\{|s|, a\}$$

for some $a > 0$. In other words,

$$ks\,\mathrm{sat}_\Delta(s) \leq s\sigma_i(s) \leq \delta s\,\mathrm{sat}_{M/\delta}(s).$$

For simplicity we assume that $k = 1$ which can always be achieved by rescaling.

Next we state the problems which are the focus of this paper. They are stated for the continuous-time case. The discrete-time case can be stated in a similar way. Each problem can be studied for state feedback where u is a causal but not necessarily linear function of the state, for full-information feedback where u is a causal but not necessarily linear function of state and disturbance and for measurement feedback where u is a causal function of the measurements y. The feedback operator will be denoted by Υ.

Problem 1. Given any $p \in [1, \infty]$, the *simultaneous global asymptotic stabilization and global finite-gain L_p performance problem* is to find a feedback Υ, possibly nonlinear, such that the following properties hold:

1. In the absence of the external input signal w, the equilibrium point $x = 0$ of the closed-loop system is globally asymptotically stable.
2. In the presence of the external input signal w, there exists a positive constant γ_p such that with $x(0) = 0$ the following inequality holds

$$\|z\|_{L_p} \leq \gamma_p \|w\|_{L_p}, \quad \text{for all } w \in L_p^s.$$

Next we formulate the above problem in a semiglobal framework for L_p performance. It seems reasonable to expect that given the saturation we can only guarantee performance (in the sense of a finite gain) for all disturbances in some compact set.

Problem 2. Given any $p \in [1, \infty]$, the *simultaneous global asymptotic stabilization and semiglobal finite-gain L_p performance problem* is for any $D > 0$ to find a feedback Υ_D, possibly nonlinear, such that the following properties hold:

1. In the absence of the external input signal w, the equilibrium point $x = 0$ of the closed-loop system is globally asymptotically stable.
2. There exists a positive constant $\gamma_p = \gamma_p(D)$ such that with $x(0) = 0$ and for any external input signal $w \in L_p^s(D)$, the following inequality holds

$$\|z\|_{L_p} \leq \gamma_p \|w\|_{L_p}.$$

As we will see it is in most cases impossible to keep the gain bounded if the size of the disturbances increases. Therefore a reasonable question is whether in a global setting we are at least able to guarantee that the output remains a signal in L_p. This question is formulated in the following problem.

Problem 3. Given any $p \in [1, \infty]$, the *simultaneous global asymptotic stabilization and global L_p performance problem* is to find a feedback Υ, possibly nonlinear, such that the following properties hold:

1. In the absence of the external input signal w, the equilibrium point $x = 0$ of the closed-loop system is globally asymptotically stable.
2. For any $w \in L_p^s$ and $x(0) = 0$, we have $z \in L_p^r$.

3   Main Results

The first question is when it is possible to achieve global internal stability and a global finite gain from disturbance w to output z by either full-information, state or measurement feedback as formulated in detail in problem 1. We present a necessary condition which is extremely restrictive and shows that this problem is in most cases not solvable.

We present and prove this theorem for continuous time systems but it is obvious that the corresponding discrete time result is equally valid.

Theorem 1. *Consider the system Σ in (1). The problem of simultaneous global asymptotic stabilization and global finite-gain L_p performance via either state, full-information, or measurement feedback is solvable only if the auxiliary system Σ_a defined as*

$$\Sigma_a : \begin{cases} \rho x = Ax + Ew \\ z = C_2 x \end{cases} \tag{2}$$

is globally L_p finite-gain externally stable, i.e. there exists $\gamma_p > 0$ such that

$$\|z\|_{L_p} < \gamma_p \|w\|_{L_p}.$$

for zero initial conditions.

Remark 4. In the case of state feedback and if (C_2, A) is detectable, the condition is also sufficient in the sense that the smallest global L_p gain we can achieve for Σ under the constraint of global asymptotic stability is equal to the L_p gain of the auxiliary system Σ_a. We design a stabilizing controller for the unstable part. Since Σ_a is input-output stable and detectable, the unstable internal dynamics must be uncontrollable from w. But then for zero initial conditions and any disturbance w we have $u = 0$ and we immediately obtain this result. In the measurement feedback this argument does not work because in that case the disturbance can affect the unstable dynamics via the observer which we have to use to estimate the state of the unstable dynamics.

Note that even when the open-loop system is critically stable, Problem 1 is not solvable in general. Another interesting aspect of Theorem 1 is that one can not improve the L_p performance of a stable linear system with saturating actuators via any feedback law. This indicates that, whenever the disturbance is not input-additive, the global framework for L_p performance is not meaningful for linear systems with saturating actuators. Hence one needs to work with the notion of L_p performance with the disturbance being restricted to a bounded set, (i.e. semiglobal L_p performance). A very interesting open question in this regard is how to identify the set of all possible disturbances in L_p^s for which L_p performance can be achieved with some a priori specified finite gain.

Proof. Assume that we have zero initial conditions. We note that $z = z_{u,0} + z_{0,w}$ where $z_{u,0}$ is the output of the system for input u and zero disturbances while $z_{0,w}$ is the output of the system for zero input and disturbance w. Since the saturation function is bounded, there exists for any given $T > 0$, a constant $M > 0$ such that $\|z_{u,0}\|_{[0,T]} < M$ for any input u. Let $w \in L_p$. Then

$$\|z_{u,w}\|_{[0,T]} \geq \|z_{0,w}\|_{[0,T]} - \|z_{u,0}\|_{[0,T]}.$$

It follows that

$$\|z_{u,\lambda w}\|_{[0,T]} \geq \lambda \|z_{0,w}\|_{[0,T]} - M$$

for any λ. This yields

$$\frac{\|z_{u,\lambda w}\|}{\|\lambda w\|} \geq \frac{\|z_{0,w}\|_{[0,T]}}{\|w\|} - \frac{M}{\lambda \|w\|}.$$

This is true for any disturbance w and therefore by letting $\lambda \to \infty$ we find

$$\sup_w \frac{\|z_{u,w}\|_{[0,T]}}{\|w\|_{[0,T]}} \geq \sup_w \frac{\|z_{0,w}\|_{[0,T]}}{\|w\|_{[0,T]}} \tag{3}$$

and hence

$$\inf_{\Sigma_c} \sup_w \frac{\|z_{u,w}\|_{[0,T]}}{\|w\|_{[0,T]}} \geq \sup_w \frac{\|z_{0,w}\|_{[0,T]}}{\|w\|_{[0,T]}}. \tag{4}$$

Suppose that Σ_a is not finite-gain L_p stable. Then for any $\gamma > 0$ and any controller, there exist $w_* \in L_p$, $(w_* \neq 0)$ and $T > 0$ such that

$$\|z_{0,w_*}\|_{[0,T]} > \gamma \|w_*\|$$

Hence

$$\sup_w \frac{\|z_{0,w}\|_{[0,T]}}{\|w\|_{[0,T]}} \geq \frac{\|z_{0,w_*}\|_{[0,T]}}{\|w_*\|_{[0,T]}} \geq \frac{\|z_{0,w_*}\|_{[0,T]}}{\|w_*\|} > \gamma.$$

By use of (4) we get

$$\inf_{\Sigma_c} \sup_w \frac{\|z_{u,w}\|}{\|w\|} \geq \inf_{\Sigma_c} \sup_w \left\{ \frac{\|z_{u,w}\|}{\|w\|} : w(t) = 0 \text{ for } t > T \right\}$$

$$\geq \inf_{\Sigma_c} \sup_w \frac{\|z_{u,w}\|_{[0,T]}}{\|w\|_{[0,T]}} > \gamma.$$

Since this last inequality is satisfied for all $\gamma > 0$, we see that we can never obtain a finite gain.

Before stating our second result, we recall the scheduled low-gain design for continuous time from [8] and for discrete case from [3].

Lemma 1. *Let* $Q : [0,1] \to \mathbb{R}^{n \times n}$ *be a continuously differentiable matrix-valued function such that* $Q(0) = 0$, *and* $\dot{Q}(\varepsilon) > 0$ *for any* $\varepsilon \in (0,1]$. *Assume* (A,B) *is stabilizable and the eigenvalues of* A *are in the closed left half plane. Then the continuous-time algebraic Riccati equation defined as*

$$PA + A'P - PBB'P + Q(\varepsilon) = 0 \tag{5}$$

has a unique positive definite solution $P(\varepsilon)$ *for any* $\varepsilon \in (0,1]$. *Moreover, this positive definite solution* $P(\varepsilon)$ *has the following properties:*

1. For any $\varepsilon \in (0,1]$, $P(\varepsilon)$ is such that $A - BB'P(\varepsilon)$ is Hurwitz-stable.
2. $\lim_{\varepsilon \to 0} P(\varepsilon) = 0$.
3. $P(\varepsilon)$ is continuously differentiable and $\dot{P}(\varepsilon) > 0$, for any $\varepsilon \in (0,1]$.

Lemma 2. Let $Q(\cdot) : (0,1] \to \mathbb{R}^{n \times n}$ be a continuously differentiable matrix-valued function such that $Q(\varepsilon) > 0$ and $\dot{Q}(\varepsilon) > 0$ for any $\varepsilon \in (0,1]$. Assume (A,B) is stabilizable and the eigenvalues of A are in the closed unit disc. Then the discrete-time algebraic Riccati equation

$$P = A'PA + Q(\varepsilon) - A'PB(B'PB + I)^{-1}B'PA \tag{6}$$

has a unique positive definite solution $P(\varepsilon) > 0$ for any $\varepsilon \in (0,1]$. Moreover, this positive definite solution $P(\varepsilon)$ has the following properties:

1. For any $\varepsilon \in (0,1]$, $A - BF(\varepsilon)$ is Schur-stable, where

$$F(\varepsilon) = (I + B'P(\varepsilon)B)^{-1}B'P(\varepsilon)A.$$

2. $\lim_{\varepsilon \to 0} P(\varepsilon) = 0$.
3. For any $\varepsilon \in (0,1]$, there exists $M_p > 0$ such that

$$\|P(\varepsilon)^{1/2}AP(\varepsilon)^{-1/2}\| \leq M_p.$$

4. For any $\varepsilon \in (0,1]$, $\lambda_i(Q(\varepsilon)) \leq \lambda_i(P(\varepsilon))$. The strict inequality holds whenever A is nonsingular.
5. $P(\varepsilon)$ is continuously differentiable and $\dot{P}(\varepsilon) > 0$ for any $\varepsilon \in (0,1]$.

Definition 2. (Scheduled low-gain) The scheduled low-gain parameter ε, denoted as $\varepsilon_s(x)$, for the discrete-time systems is given by

$$\varepsilon_s(x) = \max\{\, r \in (0,\bar{\varepsilon}_s] \mid (x'P(r)x)\,\mathrm{tr}(P(r)) \leq \tfrac{\Delta^2}{M_p^2\|B\|^2} \,\},$$

where M_p is the constant in Lemma 2, and $\bar{\varepsilon}_s \in (0,1]$ satisfies

$$(\delta - 1)\lambda_{max}(B'P(\bar{\varepsilon}_s)B) \leq 1$$

with δ the Lipschitz constant of Definition 1.

For continuous-time systems, the scheduled low-gain parameter ε, denoted as $\varepsilon_s(x)$, is given by

$$\varepsilon_s(x) = \max\{\, r \in (0,1] \mid (x'P(r)x)\,\mathrm{tr}\,P(r) \leq \tfrac{\Delta^2}{\|B\|^2} \,\}, \tag{7}$$

Now we are ready to state our second main result.

Theorem 2. Given the discrete-time or continuous-time system Σ in (1). Assume that (A, B) is stabilizable, (C_1, A) detectable and the eigenvalues of A are in the closed left half plane (continuous time) or in the closed unit disc (discrete time). Then the problem of simultaneous global asymptotic stabilization and global L_p performance, i.e. Problem 3, is solvable for any $p \in [1, \infty)$ via both dynamic state feedback and measurement feedback.

Proof. First note that since we are only interested in getting the output z in L_p, we can ignore the part of the state space which is uncontrollable by u. Because of stabilizability, it must be exponentially stable and it is affected by disturbances in L_p and therefore the state of the uncontrollable part will also be in L_p. Therefore, whether the output is determined only by the controllable part of the state space or not, we can without loss of generality assume that (A, B) is controllable. We split the proof of this theorem in two parts: discrete-time and continuous-time systems.

Part I: We first prove the result under state feedback, then we extend the result to measurement feedback. Consider the linear system:

$$\xi(k + 1) = A\xi(k) + B\mu(k), \qquad\qquad \xi(0) = \xi_0. \qquad\qquad (8)$$

Since (A, B) is controllable, there exist for every ξ_0, an input $\{\mu(i) : i = 0, \cdots, n - 1\}$ such that $\xi(n) = 0$. It can be easily seen that this implies that there exist F_i and G_i such that

$$\mu(i) = H_i\xi_0, \quad \xi(i) = G_i\xi_0 \qquad\qquad\qquad (9)$$

for $i = 0, \ldots, n - 1$ satisfy (8) and result in $\xi(n) = 0$ for any ξ_0. Note that $G_0 = I$ and $G_n = 0$. Substituting (9) into (8) yields

$$G_{i+1} = AG_i + BH_i, \qquad i = 0, \cdots, n - 1. \qquad\qquad (10)$$

Our control law consists of a scheduled low-gain state feedback part and an additional term, which is employed for decoupling the disturbance term in the original open-loop system.

$$u(k) = -\frac{1}{2}\sigma(F(v(k))v(k)) + f(k), \qquad\qquad\qquad (11)$$

where $F(x)$ is the discrete-time scheduled low-gain feedback for the pair $A, \tilde{B}$ with $\tilde{B} = B/2$, i.e.

$$F(v) = (I + \tilde{B}'P(\varepsilon_s(v))\tilde{B})^{-1}\tilde{B}'P(\varepsilon_s(v))A,$$

where P is the solution of the Riccati equation (6) with B replaced by $\tilde{B}$ and

$$f(k) = \sum_{i=0}^{n-1} BH_iEw(k - i - 1)$$

$$= \sum_{i=0}^{n-1} BH_i\left(x(k - i) - Ax(k - i - 1) - Bu(k - i - 1)\right) \qquad (12)$$

$$v(k) = x(k) - \sum_{i=0}^{n-1} G_iEw(k - i - 1)$$

$$= x(k) - \sum_{i=0}^{n-1} G_i\left(x(k - i) - Ax(k - i - 1) - Bu(k - i - 1)\right) \qquad (13)$$

It is easy to see that with the above definition we have a well-defined dynamic state feedback controller.

If $w(k) \equiv 0$, then the control law (11) reduces to a scheduled low-gain design and it has been shown in [3] that this control law guarantees the global asymptotic stability. It remains to show that for any given $w \in \ell_p$ we have $x \in \ell_p$. Let $w \in \ell_p$ $(p < \infty)$. Since $w(t) \to 0$ as $t \to \infty$, there exists $T_1 > 0$ such that

$$\left\| \sum_{i=0}^{n-1} BH_i Ew(k - i - 1) \right\| < \frac{\Delta}{2}$$

for all $k > T_1$. It follows that for $k > T_1$ we have $\|u(k)\| \leq \Delta$; that is, $u(k)$ does not saturate for sufficiently large k. By the definition of $v(k)$ in (13), it is easily verified that for $k > T_1$

$$v(k + 1) = Av(k) - \tilde{B}\sigma(F(k)v(k)). \tag{14}$$

Indeed, noticing that $G_0 = I$, $G_n = 0$, $u(k)$ does not saturate for $k > T_1$, and using (10) we have

$$v(k + 1) = x(k + 1) - \sum_{i=0}^{n-1} G_i Ew(k - i)$$

$$= Ax(k) + B\sigma(u(k)) - \sum_{i=0}^{n-1} G_{i+1} Ew(k - i - 1)$$

$$= Av(k) - \tilde{B}\sigma(F(v(k))v(k)).$$

Since the dynamics of (14) does not depend on w, the global asymptotic stability ensures that $v(t) \to 0$ as $t \to \infty$. That is, there exists $T_2 > T_1$ such that (14) does not saturate for $k > T_2$. Therefore, for $k > T_2$ (14) reduces to

$$v(k + 1) = [A - \tilde{B}(I + \tilde{B}'P(\varepsilon_s(x(k)))\tilde{B})^{-1}\tilde{B}'P(\varepsilon_s(x(k)))A]v(k) \tag{15}$$

Since the dynamics of $v(k)$ in (15) is exponentially stable (by Lemma 2), we see that the tail of $v(k)$ is exponentially decaying, hence $v \in \ell_p$. By (13) it follows that $x \in \ell_p$ for any $w \in \ell_p$.

Now we extend the result proved above to measurement feedback. The construction of a scheduled low-gain measurement feedback controller is based upon a linear observer architecture given below.

$$\Sigma_c : \begin{cases} \hat{x}(k + 1) = A\hat{x}(k) + B\sigma(u(k)) + K(y(k) - C_1\hat{x}(k)) \\ u(k) = -\frac{1}{2}\sigma(F(\hat{v}(k))\hat{v}(k)) + \sum_{i=0}^{n-1} BH_i s(k - i - 1) \end{cases}$$

where

$$\hat{v}(k) = \hat{x}(k) - \sum_{i=0}^{n-1} G_i KD_1 w(k - i - 1)$$

$$s(k) = Ky(k) - KC_1\hat{x}(k)$$

and K is chosen such that $A - KC$ is Schur-stable. If we define

$$e(k) = \hat{x}(k) - x(k)$$

where e is the estimation error, then the error dynamics becomes

$$e(k+1) = (A - KC)e(k) + (KD_1 - E)w(k).$$

Since $A - KC$ is Schur-stable, it follows that $e \in \ell_p$ and therefore $s \in \ell_p$ since

$$s(k) = -KC_1 e(k) + KD_1 w(k)$$

On the other hand, it is easy to verify that the closed-loop system represented in $\hat{v}(k)$ becomes (as derived in the state feedback case)

$$\hat{v}(k+1) = A\hat{v}(k) + B(\sigma(-F(\hat{v}(k))\hat{v}(k))) \tag{16}$$

for $k > T_1$ with T_1 such that

$$\left\| \sum_{i=0}^{n-1} BH_i s(k - i - 1) \right\| < \frac{\Delta}{2}$$

for all $k > T_1$. Then the same arguments as in the state feedback case allows us to conclude that $\hat{v} \in \ell_p$ which yields $x \in \ell_p$

Part II: For the continuous-time system Σ, the proof essentially follows the same line of Part I. So we only give out the steps that might be different from those in Part I and omit the others.

First consider the linear system:

$$\dot{\xi}(t) = A\xi(t) + B\mu(t), \qquad \xi(0) = \xi_0. \tag{17}$$

where w_0 is any arbitrary vector in $\mathbb{R}^s$. The solution of (17) is

$$\xi(t) = e^{At}\xi_0 + \int_0^t e^{A(t-\tau)} B\mu(\tau)d\tau.$$

Since (A, B) is controllable, for any w_0 there exists t_1 and $\mu(t)$ such that

$$0 = \xi(t_1) = e^{At_1}\xi_0 + \int_0^{t_1} e^{A(t_1-\tau)} B\mu(\tau)d\tau.$$

Specifically, $\mu(t)$ is chosen as

$$\mu(t) = -B'e^{A'(t_1-t)} W^{-1}(t_1)e^{At_1}\xi_0,$$

with $t < t_1$ and where

$$W(t_1) = \int_0^{t_1} e^{A(t_1-\tau)} BB'e^{A'(t_1-\tau)}d\tau > 0.$$

Therefore, there exist time-varying matrices $G(t)$ and $H(t)$ with $t \in [0, t_1]$ such that

$$\mu(t) = H(t)\xi_0, \qquad \xi(t) = G(t)\xi_0. \tag{18}$$

Substituting (18) into (17) yields

$$\dot{G}(t) = AG(t) + BH(t). \tag{19}$$

Now, define

$$v(t) = x(t) - \int_0^{t_1} G(\tau)Ew(t - \tau)d\tau. \tag{20}$$

Then, using (19) and noticing $G(0) = I$, one can easily verify that

$$\dot{v}(t) = Av(t) + B\sigma(u(t)) - \int_0^{t_1} BH(\tau)Ew(t - \tau)d\tau. \tag{21}$$

Since $F(t)$ is bounded for $t \in [0, t_1]$ and $w(t) \in L_p$, $p \in [1, \infty)$, there exists T_1 such that

$$\left\| \int_0^{t_1} BH(\tau)Ew(t - \tau)d\tau \right\| < \frac{\Delta}{2}.$$

for $t > T_1$. All the remaining steps follow the same lines as in the proof of Part I. This completes the proof.

Remark 5. In the case of discrete-time systems, Theorem 2 implies that for any $w \in \ell_p$ and for all $p \in [1, \infty)$ one can design a measurement feedback law such that the controlled output goes to zero asymptotically. In other words, the closed-loop system remains globally attractive in the presence of any $w \in \ell_p$.

Remark 6. The control laws that achieve global L_p (ℓ_p) performance with global internal stability as given in the proof of Theorem 2 are nonlinear. However, it is straightforward to show that if the plant is open-loop *critically stable*, then one can modify the control laws to be a linear one by removing the scheduling of low-gain parameters.

Next we focus on the problem of simultaneous global internal stabilization (in the absence of the disturbance) and semiglobal finite-gain L_p (ℓ_p) performance (in the presence of the disturbance). This problem as formulated in Problem 2 has already been solved for the case $p = 2$ in [7]. But we conjecture that this problem is also solvable for $p \neq 2$. For the sake of completeness and in light of the fact that the reference [7] does not have the proof, here we provide the next theorem.

Theorem 3. *Given the discrete-time or continuous-time system Σ in (1). Assume that (A, B) is stabilizable, (C_1, A) detectable and the eigenvalues of A are in the closed left half plane (continuous time) or in the closed unit disc (discrete time). Then the problem of simultaneous global asymptotic stabilization and semiglobal finite-gain L_2 performance, i.e. Problem 2, is solvable via both state and measurement feedback.*

Remark 7. Assume the auxiliary system (2) is L_2 finite-gain externally stable with an external finite gain. It can then be shown that the minimal achievable semiglobal finite gain for the system Σ under the requirement of global asymptotic stability which is obviously increasing in D will have a finite upper bound.

If the auxiliary system (2) is *not* L_2 finite-gain externally stable and the upper bound D on the disturbance w goes to infinity then the minimal achievable finite gain for the system Σ under the requirement of global asymptotic stability also converges to ∞.

Proof. Due to space limitations we only prove the continuous time result.

First we need to adapt the scheduled low gain feedback. Consider the following Riccati equation:

$$PA + A'P - PBB'P + \gamma^{-2}PEE'P + Q(\varepsilon) = 0. \tag{22}$$

It is well known that there exists γ and ε_1^* such that this Riccati equation has a positive definite solution $P(\varepsilon)$ for all $\varepsilon \in [0, \varepsilon_1^*]$. We know that $P(\varepsilon)$ converges to zero as ε goes to zero. We use the same scheduling as before, i.e. ε_s is defined by (7) and the following scheduled low-gain feedback:

$$u = g(x) := B'P(\varepsilon_s(x))x \tag{23}$$

It is easy to check that this controller will never saturate. It is known that the scheduled low-gain controller $u = -B'P(\varepsilon_s(x))x$ achieves global asymptotic stability. It remains to show that it achieves a semiglobal finite gain when the disturbances w are bounded in L_2 norm by D. First consider:

$$\dot{V} - \gamma^2\|w\|^2 \tag{24}$$

This is obviously negative if V is decreasing. On the other hand if V is increasing then we have:

$$\dot{V} - \gamma^2\|w\|^2 \le \dot{x}'Px + x'P\dot{x} + x'\dot{P}x + x'Qx + \|u\|^2 - \gamma^2\|w\|^2.$$

Since the scheduling of ε_s guarantees that if V is increasing then P is non-increasing, a simple completion-of-the-squares argument then yields

$$\dot{V} - \gamma^2\|w\|^2$$
$$\le -x'Qx - \|u\|^2 + \|u - B'P(\varepsilon_s(x))x\|^2 - \gamma^2\|w - E'P(\varepsilon_s(x))x\|^2$$

which is obviously negative when $u = B'P(\varepsilon_s(x))x$. Therefore (24) is always less than or equal to zero and after integrating (24) from 0 to t, we obtain:

$$V(x(t)) \leq \gamma^2 \int_0^t \|w(t)\|^2 \, dt \leq D^2$$

In other words the state remains bounded. This implies that there exists ε_2^* such that $\varepsilon_s(x(t)) \geq \varepsilon_2^*$ for all disturbances $w \in L_2(D)$ and all $t > 0$.

Next we consider:

$$\dot{V} + x'Qx + \|u\|^2 - \gamma^2\|w\|^2. \tag{25}$$

If $\varepsilon_s = \varepsilon_1^*$ then $\dot{P} = 0$. Also, if $\varepsilon_s < \varepsilon_1^*$ and V is increasing then $\dot{P} \leq 0$. In both of these cases we find:

$$\dot{V} + x'Qx + \|u\|^2 - \gamma^2\|w\|^2 \leq -\gamma^2\|w - E'P(\varepsilon_s(x))x\|^2 \leq 0$$

via the same completion-of-the-squares argument as used before.

The remaining case is $\varepsilon_s < \varepsilon_1^*$ and V decreasing. This implies that P is increasing and we find:

$$\operatorname{tr} \dot{P} \leq -\frac{\dot{V}}{V} \operatorname{tr} P \leq -M\dot{V}$$

with

$$M = n \max_{\varepsilon \in [\varepsilon_2^*, \varepsilon_1^*]} \frac{\sigma_{\max}(P)}{\sigma_{\min}(P)}$$

We consider the following expression:

$$(1 + M)\dot{V} + x'Qx + \|u\|^2 - \gamma^2\|w\|^2 \tag{26}$$

The same completion of the squares argument combined with the upper bound for $x'\dot{P}x$ then shows that the expression in (26) is less than or equal to zero. Since we know that either (25) or (26) is less than or equal to zero we find that:

$$\dot{V} + \tfrac{1}{1+M}x'Qx + \tfrac{1}{1+M}\|u\|^2 - \gamma^2\|w\|^2 \leq 0 \tag{27}$$

which is valid in all situations. Then integrating from 0 to ∞ and using the zero initial conditions we obtain:

$$\tfrac{\delta}{1+M}\|x\|_{L_2}^2 + \tfrac{1}{1+M}\|u\|_{L_2}^2 \leq \gamma^2\|w\|_{L_2}^2 \tag{28}$$

where δ is such that $Q(\varepsilon_2^*) > \delta I$. It is then obvious that the system has a finite gain.

For the measurement feedback case we use the following controller:

$$\Sigma_c : \begin{cases} \dot{\hat{x}} = A\hat{x} + B\sigma(u) + K(y - C_1\hat{x}) \\ u = g(\hat{x}) \end{cases} \tag{29}$$

where g is the nonlinear state feedback we have just constructed in (23) but with E replaced by $(KD_1 \quad -KC_1)$. Then we obtain

$$\dot{e} = (A - KC_1)e + (KD_1 - E)w. \tag{30}$$

Note that from (30) we can conclude that

$$\|e\|_{L_2} \leq \gamma_1 \|w\|_{L_2},$$

for some $\gamma_1 > 0$ and from (29) we can conclude that

$$\|\hat{x}\|_{L_2} + \|u\|_{L_2} \leq \gamma_2 (\|e\|_{L_2} + \|w\|_{L_2}).$$

for some $\gamma_2 > 0$. Therefore, we obtain that

$$\begin{aligned} \|x\|_{L_2} + \|u\|_{L_2} &\leq \|\hat{x}\|_{L_2} + \|u\|_{L_2} + \|e\|_{L_2} \\ &\leq \gamma_2 \|w\|_{L_2} + (\gamma_2 + 1)\|e\|_{L_2} \\ &\leq (\gamma_2 + \gamma_1(\gamma_2 + 1))\|w\|_{L_2} \end{aligned}$$

i.e. we have a finite gain.

Remark 8. The controller constructed in the above proof has the very nice feature that it does not depend on the size D of the disturbances. Therefore, it is easy to see that this controller also solves the problem of simultaneous global asymptotic stabilization and global L_p performance problem (of course without a finite-gain). This controller has the advantage over the controller constructed in the proof of theorem 2, which is a static state feedback or a finite-dimensional measurement feedback controller.

4 Conclusion

In this paper we have shown that for both continuous time and discrete time systems the problem of simultaneously achieving global internal stabilization (in the absence of disturbance) and global finite-gain L_p (ℓ_p) performance (in the presence of the disturbance) is intrinsically not solvable whenever the external input (disturbance) is not input-additive. However, for the same class of systems, whenever the external input (disturbance) is not input-additive, the simultaneous global internal stabilization (in the absence of disturbance) and global L_p (ℓ_p) performance *without finite-gain* is possible. These results essentially point out that for this class of systems the control problems involving L_p (ℓ_p) performance in a global framework are meaningless. These control problems must be considered in the semiglobal setting where the external input (disturbance) resides in an a priori given compact set in L_p (ℓ_p) space.

References

1. X. Bao, Z. Lin, and E. D. Sontag. Finite gain stabilization of discrete-time linear systems subject to actuator saturation. Submitted for publication, 1998.
2. P. Hou, A. Saberi, and Z. Lin. On ℓ_p-stabilization of strictly unstable discrete-time linear systems with saturating actuators. *Int. J. Robust & Nonlinear Control*, 8:1227–1236, 1998.
3. P. Hou, A. Saberi, Z. Lin, and P. Sannuti. Simultaneously external and internal stabilization for continuous and discrete-time critically unstable systems with saturating actuators. *Automatica*, 34(12):1547–1557, 1998.
4. Z. Lin, A. Saberi, and A. Teel. The almost disturbance decoupling problem with internal stability for linear systems subject to input saturation – state feedback case. *Automatica*, 32:619–624, 1996.
5. Z. Lin, A. Saberi, and A.R. Teel. Simultanuous l_p-stabilization and internal stabilization of linear systems subject to input saturation — state feedback case. *Syst. & Contr. Letters*, 25:219–226, 1995.
6. W. Liu, Y. Chitour, and E.D. Sontag. On finite-gain stabilizability of linear systems subject to input saturation. *SIAM J. Contr. & Opt.*, 34(4):1190–1219, 1996.
7. A. Megretski. L_2 BIBO output feedback stabilization with saturated control. In *Proc. 13th IFAC world congress*, volume D, pages 435–440, San Francisco, 1996.
8. Ali Saberi, Ping Hou, and Anton A. Stoorvogel. On simultaneous global external and global internal stabilization of critically unstable linear systems with saturating actuators. Available from ftp://ftp.win.tue.nl/pub/techreports/wscoas/global.ps, 1997.

Control Under Structural Constraints: An Input-Output Approach *

Petros G. Voulgaris

Coordinated Science Laboratory
University of Illinois at Urbana-Champaign

Abstract. In this paper we present an input-output point of view of certain optimal control problems with constraints on the processing of the measurement data. In particular, considering linear controllers and plant dynamics, we present solutions to the ℓ^1, $\mathcal{H}^\infty$ and $\mathcal{H}^2$ optimal control problems under the so-called one-step delay observation sharing pattern. Extensions to other decentralized structures are also possible under certain conditions on the plant. The main message from this unified input-output approach is that, structural constraints on the controller appear as linear constraints of the same type on the Youla parameter that parametrizes all controllers, as long as the part of the plant that relates controls to measurements possesses the same off-diagonal structure required in the controller. Under this condition, ℓ^1, $\mathcal{H}^\infty$ and $\mathcal{H}^2$ optimization transform to nonstandard, yet convex problems. Their solution can be obtained by suitably utilizing the Duality, Nehari and Projection theorems respectively.

1 Introduction

Optimal control under decentralized information structures is a topic that, although it has been studied extensively over the last forty years or so, still remains a challenge to the control community. The early encounters with the problem date back in the fifties and early sixties under the framework of team theory (e.g., [11,13].) Soon it was realized that, in general, optimal decision making is very difficult to obtain when decision makers have access to private information, but do not exchange their information [25]. Nonetheless, under particular decentralized information schemes such as the partially nested information structures [9] certain optimal control problems admit trackable solutions. Several results where LQG and linear exponential-quadratic Gaussian (LEQG) critria are considered exist when exchange of information is allowed with a one-step time delay (which is a special case of the partially nested information structure.) e.g., [1,15,2,18–21].

---

* This work is supported by ONR grant N00014-95-1-0948/N00014-97-1-0153 and National Science Foundation Grant ECS-9308481

In this paper, in contrast to the state-space view-point of the works previously referenced, we undertake an input-output approach to optimal control under the quasiclassical information scheme known as the one-step delay observation sharing pattern (e.g., [1]). Under this pattern measurement information can be exchanged between the decision makers with a delay of one time step. In the paper we define and present solutions to three optimal control problems: ℓ^1, $\mathcal{H}^\infty$ and $\mathcal{H}^2$ (or LQG) optimal disturbance rejection. The key ingredient in this approach is the transformation of the decentralization constraints on the controller to *linear* constraints on the Youla parameter used to characterize all controllers. Hence, the resulting problems in the input-output setting are, although nonstandard, convex. These problems resemble the ones appearing in optimal control of periodic systems when lifting techniques are employed [6,23], and can be solved by suitably utilizing the Duality, Nehari and Projection theorems respectively. Other structured control problems can also be dealt similarly *provided that the part of the plant that relates controls to measurements possesses the same off-diagonal structure required in the controller.* This condition is crucial in transforming linearly structural constraints. If it is satisfied, problems with n-step delay observation sharing patterns where $n > 1$, or with fully decentralized operation can be solved in a similar fashion.

Some generic notation that is used throughout is as follows: $\rho(A)$, $\overline{\sigma}[A]$ denote the spectral radius and maximum singular value of a matrix A respectively; $\hat{H}(\lambda)$ denotes the λ-transform of a real sequence $H = \{H(k)\}_{k=-\infty}^\infty$ defined as $\hat{H}(\lambda) = \sum_{k=-\infty}^\infty H(k)\lambda^k$; X^* denotes the dual space of the normed linear space X: BX is the closed unit ball of X; $^\perp S$ denotes the left annihilator of $S \subset X^*$; $S^\perp$ denotes the right annihilator of $S \subset X$.

2 Problem Definition

The standard block diagram for the disturbance rejection problem is depicted in Figure 1. In this figure, P denotes some fixed linear, discrete-time, causal plant, C denotes the compensator, and the signals w, z, y, and u are defined as follows: w, exogenous disturbance; z, signals to be regulated; y, measured plant output; and u, control inputs to the plant. P can be thought as a four block matrix each block being a linear causal system. In what follows we will assume that both P and C are linear, time-invariant (LTI) systems; we comment on this restriction on C later. Furthermore, we assume that there is a predefined information structure that the controller C has to respect when operating on the measurement signal y. The particular information structure is precisely defined in the sequel.

2.1 The one-step delay observation sharing pattern

To simplify our analysis we will consider the case where the control input u and plant output y are partitioned into two (possibly vector) components

u_1, u_2 and y_1, y_2 respectively, i.e., $u = (u_1 \; u_2)^T$ and $y = (y_1 \; y_2)^T$. Let $Y_k :=$ $\{y_1(0), y_2(0), \ldots, y_1(k), y_2(k)\}$ represent the measurement set at time k. The controllers that we are considering (henceforth, admissible controllers) are such that $u_1(k)$ is a function of the data $\{Y_{k-1}, y_1(k)\}$ and $u_2(k)$ is a function of the data $\{Y_{k-1}, y_2(k)\}$. We refer to this particular information processing structure imposed on the controller as the *one-step delay observation sharing pattern*. Alternatively, partitioning the controller C accordingly as

$$C = \begin{pmatrix} C_{11} & C_{12} \\ C_{12} & C_{22} \end{pmatrix},$$

we require that both C_{12} and C_{21} be strictly causal operators. Let now

$$S := \{C \text{ stabilizing and LTI}: \; C_{12}, \; C_{21} \text{ strictly causal}\}$$

and let T_{zw} represent the resulting map from w to z for a given compensator $C \in S$. The problems of interest are as follows.

The first two problems are deterministic: w is assumed to be any ℓ^α disturbance with $\alpha = \infty, 2$ and we are interested in minimizing the worst case ℓ^α norm of z. Namely, our objective can be stated as

(OBJ$_\alpha$): Find C such that the resulting closed loop system is stable and also the induced norm $\|T_{zw}\|$ over ℓ^α for $\alpha = \infty, 2$ is minimized.

The third problem we want to solve is stochastic: we assume that w is a stationary zero mean Gaussian white noise with $E[ww^T] = I$ and we seek to minimize the average noise power in z. This is nothing else but a LQG problem. So our objective is stated as

(OBJ$_{LQG}$): Find C such that the resulting closed loop system is stable and also

$$\lim_{M \to \infty} (1/2M) \sum_{k=-M}^{M-1} \text{trace}(E[z(k)z^T(k)])$$

is minimized.

To solve the above problems the following assumptions are introduced. Let $P = \begin{pmatrix} P_{11} & P_{12} \\ P_{21} & P_{22} \end{pmatrix}$ then,

Assumption 1 *P is finite dimensional and stabilizable.*

Assumption 1 means that P has a state space description

$$P \sim (A, (B_1 \; B_2), \begin{pmatrix} C_1 \\ C_2 \end{pmatrix}, \begin{pmatrix} D_{11} & D_{12} \\ D_{12} & D_{22} \end{pmatrix})$$

with the pairs (A, B_2) and (A, C_2) being stabilizable and detectable respectively. In addition we assume that

Assumption 2 *The subsystem P_{22} is strictly causal, i.e., $D_{22} = 0$.*

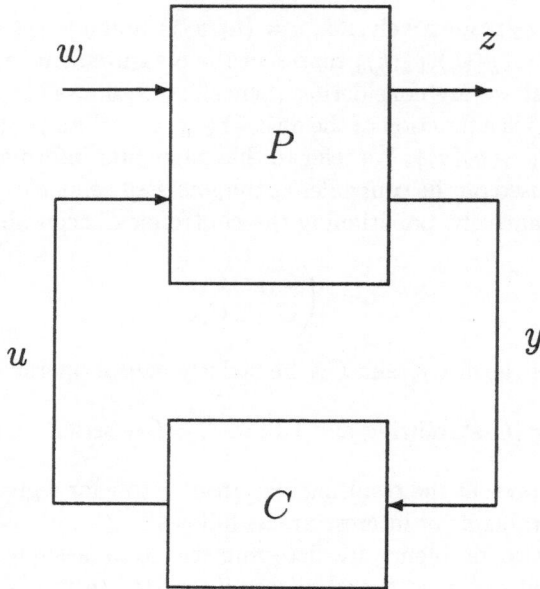

Fig. 1. Block Diagram for Disturbance Rejection.

This assumption has the implication that the system of Figure 1 is well-posed [8,7]. More important than this, however, is the fact that it allows for a convenient characterization of the structural constraints on the controller as we shall see in the following section.

3   Problem Solution

The problems defined in the previous section can be related to problems in periodic systems where additional constraints that ensure causality appear in the so-called lifted system [6,23]. These constraints are of similar nature as with the problems at hand. The methods of solutions we develop herein are along the same lines with [6,23]. A common step in the solution of all of the problems defined earlier is the convenient characterization of all controllers that are in $\mathcal{S}$. This is done in the sequel.

3.1   Parametrization of all stabilizing controllers and feasible maps

Since we have assumed that P is finite dimensional with a stabilizable and detectable state space description we can obtain a doubly coprime factorization (dcf) of P_{22} using standard formulas (e.g., [8,24]) i.e., having P_{22}

associated with the state space description $P_{22} \sim (A, B_2, C_2, D_{22})$ the coprime factorization such as in [8,24] is $P_{22} = N_l D_l^{-1} = D_r^{-1} N_r$ with

$$\begin{pmatrix} X_r & -Y_r \\ -N_r & D_r \end{pmatrix} \begin{pmatrix} D_l & Y_l \\ N_l & X_l \end{pmatrix} = I$$

where

$$N_l \sim (A_K, B_2, C_K, D_{22}), \quad D_l \sim (A_K, B_2, K, I)$$

$$N_r \sim (A_M, B_M, C_2, D_{22}), \quad D_r \sim (A_M, M, C_2, I)$$

$$X_l \sim (A_K, -M, C_K, I), \quad Y_l \sim (A_K, B_2, -M, K, 0)$$

$$X_r \sim (A_M, -B_M, K, I), \quad Y_r \sim (A_M, -M, K, 0)$$

with K, M selected such that $A_K = A + B_2 K$, $A_M = A + MC_2$ are stable (eigenvalues in the open unit disk) and $B_M = B_2 + MD_{22}$, $C_K = C_2 + D_{22}K$. Note that the above formulas indicate that the coprime factors of P_{22} have as feedforward terms the matrices D_{22} or I or 0 which are all block diagonal. Denoting by the space of time invariant operators that are ℓ^α-stable by $\mathcal{B}_{TI}(\ell^\alpha)$ where $\alpha = \infty, 2$ we have the following well-known result (e.g.,[8,24]):

Fact 1 *All ℓ^α-stabilizing LTI controllers C (possibly not in $\mathcal{S}$) of P are given by*

$$C = (Y_l - D_l Q)(X_l - N_l Q)^{-1} = (X_r - QN_r)^{-1}(Y_r - QD_r).$$

where $Q \in \mathcal{B}_{TI}(\ell^\alpha)$.

The above fact characterizes the set of all stabilizing controllers in terms of the so-called Youla parameter Q. The set $\mathcal{S}$ of interest is clearly a subset of the set implied by Fact 1 and is characterized by the constraint that the feedforward term of C should be block diagonal i.e.,

$$C(0) = \begin{pmatrix} C_{11}(0) & 0 \\ 0 & C_{22}(0) \end{pmatrix}.$$

However, a simple characterization is possible as the following lemma indicates

Lemma 6. *All ℓ^α-stabilizing controllers C in $\mathcal{S}$ of P are given by*

$$C = (Y_l - D_l Q)(X_l - N_l Q)^{-1} = (X_r - QN_r)^{-1}(Y_r - QD_r).$$

where $Q \in \mathcal{B}_{TI}(\ell^\alpha)$ and $Q(0)$ is block diagonal.

Proof. It follows from the particular structure of the doubly coprime factors of P_{22} since $C(0) = -Q(0)(I - D_{22}Q(0))^{-1}$ with D_{22} block diagonal (in fact equal to zero) and hence $C(0)$ is block diagonal if and only if $Q(0)$ is block diagonal.

Using the above lemma it is easy to show that all the feasible closed-loop maps are given as $T_{zw} = H - UQV$ where $H, U, V \in \mathcal{B}_{TI}(\ell^\alpha)$ and $Q \in \mathcal{B}_{TI}(\ell^\alpha)$ with $Q(0)$ block diagonal. Moreover, H, U, V are determined by P. Hence, we obtain in a straightforward manner the following lemma which shows how the objectives defined earlier transform to distance problems.

Lemma 7. *The objective (OBJ_α) with $\alpha = \infty$ or 2 is equivalent to the problem (OPT_α):*

$$\inf_{Q \in \mathcal{B}_{TI}(\ell^\alpha)} \|H - UQV\|_{\mathcal{B}_{TI}(\ell^\alpha)}$$

subject to $Q(0)$ is block diagonal. The objective (OBJ_{LQG}) is equivalent to the problem (OPT_{LQG}):

$$\inf_{Q \in \mathcal{H}^\infty} \|H - UQV\|_{\mathcal{H}^2}$$

subject to $Q(0)$ is block diagonal. Moreover, if Q_o is an optimal solution to any of the above problems then the corresponding optimal compensator is given as

$$C_o = (Y_l - D_l Q_o)(X_l - N_l Q_o)^{-1} = (X_r - Q_o N_r)^{-1}(Y_r - Q_o D_r).$$

It should be noted that all of the problems in Lemma 7 are, although infinite dimensional, minimizations of convex functionals over convex domains. It is also important to note that Assumption 2 plays a central role in transforming the structural constraints on C to *linear* constraints on Q as indicated in Lemma 6. In fact, as it can be seen from the proof of Lemma 6, the constraint on Q remains unchanged even if we relax Assumption 2 to requiring a block diagonal D_{22} instead of $D_{22} = 0$ (which is of course block diagonal.) More generally, as long as P_{22} has the same off-diagonal structure as the one required on C, then the Youla parameter Q will have to have the same structure. Hence, structural constraints on C transform to linear structural constraints on Q via the parametrization of Fact 1 provided P_{22} satisfies the same structural constraints as C. If on the other hand Assumption 2 is completely relaxed allowing for fully populated D_{22}, then, the constraints on Q will no longer be linear or convex and hence the resulting optimization problem is hard to solve.

3.2   Equivalent problem formulation

We start solving the problems stated in Lemma 7 by first trying to transform the constraints on $Q(0)$ to constraints on the closed loop. As a first step, we perform an inner outer factorization [8] for U, V to obtain

$$U = U_i U_o, \quad V = V_o V_i$$

where the subscript i stands for "inner" and o for "outer"; i.e., $\hat{U}_i^T(\lambda^{-1})\hat{U}_i(\lambda) = I$ and $\hat{V}_i(\lambda^{-1})\hat{V}_i^T(\lambda) = I$. We will also make the simplifying technical assumption that $\hat{U}(\lambda), \hat{V}(\lambda)$ do not lose rank on the unit circle and hence U_o, V_o have stable right and left inverses respectively. Note that the various factors in the inner outer factorization do not possess necessarily the block diagonal structure at $\lambda = 0$. Then we proceed by reflecting the constraints of $Q(0)$ on U_oQV_o. Towards this end let $Z = U_oQV_o$; the following proposition shows how Z is affected due to the constraints on Q.

Proposition 7. *Let $Z \in \mathcal{B}_{TI}(\ell^\alpha)$ then*

$$\exists Q \in \mathcal{B}_{TI}(\ell^\alpha) \quad with \quad Q(0) \quad block\ diagonal\ and \quad Z = U_oQV_o$$

if and only if

$$Z(0) \in S_A = \{U_o(0)AV_o(0) : A \quad block\ diagonal\ matrix\}.$$

Proof. The "if" direction goes as follows: Let U_{or}, V_{ol} denote any right and left stable inverses of U_o, V_o respectively. Then $U_{or}, V_{ol} \in \mathcal{B}_{TI}(\ell^\alpha)$. Let A be a block diagonal matrix such that $Z(0) = U_o(0)AV_o(0)$; define $Q_A = \{A, 0, 0, \ldots\}$ and let $\tilde{Z} = Z - U_oQ_AV_o$ then $\tilde{Z} \in \mathcal{B}_{TI}(\ell^\alpha)$ and $\tilde{Z}(0) = 0$. Define now Q as $Q = U_{or}\tilde{Z}V_{ol} + Q_A$. It then follows that $Q \in \mathcal{B}_{TI}(\ell^\alpha)$, $Q(0)$ is block diagonal and $Z = U_oQV_o$.

The "only if" direction is immediate.

Proposition 7 shows that only $Z(0)$ is constrained to lie in a certain finite dimensional subspace (i.e., S_A) otherwise Z can be arbitrary in $\mathcal{B}_{TI}(\ell^\alpha)$. Note that the characterization of this subspace is independent of the choice of right and left inverses for U_o, V_o respectively; hence it is exact. One can easily find a basis for this subspace by considering $S_A^\perp$ and finding a basis for this subspace. This is done as follows: For each element j of $Q(0)$ with indices (l_j, m_j) that has to equal 0 (i.e., the elements that are not in the block diagonal portion of $Q(0)$) we associate a matrix R_j with the same dimension as $Q(0)$ that has all its entries but one equal to 0. The nonzero entry is taken to equal 1 and its indices are precisely the ones that correspond to j i.e., (l_j, m_j). If r is the number of the elements in $Q(0)$ that are necessarily equal to 0 then we have the following proposition.

Proposition 8. *Let*

$$S_B := \{B : U_o^T(0)BV_o^T(0) \in span(\{R_j\}_{j=1}^r)\}.$$

Then,

$$S_A^\perp = S_B.$$

Proof. Let A be a block diagonal matrix and $B \in S_A^\perp$; then since

$$\langle U_o(0)AV_o(0), B\rangle = \langle A, U_o^T(0)BV_o^T(0)\rangle \quad \forall A$$

it follows that $B \in S_B$. Conversely, it also follows that if $B \in S_B$ then $\langle U_o(0)AV_o(0), B \rangle = 0$. Hence,

$$S_A{}^\perp = S_B$$

or equivalently

$$S_B{}^\perp = S_A$$

which proves the proposition.

A basis $\{B_1, B_2 \dots, B_{j_B}\}$ for this subspace can be computed in a routine fashion. When U_o, V_o are square the computation of the basis is immediate. Namely,

$$B_j = U_o^{-T}(0) R_j V_o^{-T}(0) \quad j = 1, \dots r.$$

In view of the previous developments we have

$$Z(0) \in S_A \text{ if and only if } \langle Z(0), B_j \rangle = 0 \quad \forall j = 1, \dots j_B.$$

Summarizing, the optimization problems become

$$(\text{OPT}_\alpha): \quad \inf_{Z \in \mathcal{B}_{TI}(\ell^\alpha), Z(0) \in S_A} \|H - U_i Z V_i\|_{\mathcal{B}_{TI}(\ell^\alpha)}, \quad \alpha = \infty, 2$$

and

$$(\text{OPT}_{\text{LQG}}): \quad \inf_{Z \in \mathcal{H}^\infty, Z(0) \in S_A} \|H - U_i Z V_i\|_{\mathcal{H}^2}.$$

where S_A is characterized in terms of the basis $\{B_1, B_2 \dots, B_{j_B}\}$ of $S_A{}^\perp = S_B$ of Proposition 8

3.3   Optimal ℓ^1 control

In this subsection we present the solution to the problem of optimal rejection of bounded persistent disturbances. In the previous section we defined precisely this problem i.e., (OBJ_∞) and demonstrated that this problem transforms to (OPT_∞) or, equivalently

$$(\text{OPT}_\infty): \quad \inf_{Z \in \ell^1, \, Z(0) \in S_A} \|H - U_i Z V_i\|_{\ell^1}$$

The unconstrained problem i.e., when $Q(0)$ and hence $Z(0)$ is not constrained is solved in [4,5]. In [4,5] the problem is transformed to a tractable linear programming problem, via duality theory. In this subsection, we show that the same approach can be extended to yield the optimal solution for the constrained problem. In particular, we show that the constraints on $Q(0)$ can be transformed as linear constraints on the closed loop map of interest i.e., T_{zw}. Once this is done, duality theory can be used to provide the solution.

First, we consider the 1-block case by assuming that $\hat{U}(\lambda), \hat{V}(\lambda)$ have full row and column rank, respectively, for almost all λ; we will come back to the general case later on. Also assume that $\hat{U}(\lambda), \hat{V}(\lambda)$ have no zeros on the unit circle. Let now $\{P_n\}_{n=1}^{N_s}$ be as in [4,5] the basis for the functionals in c^0 that annihilate the space

$$S_s = \{K : K = UQV, Q \in \ell^1\}$$

i.e.,

$$\langle UQV, P_i\rangle = 0 \quad \forall i = 1, 2, \ldots, N_s, \quad Q \in \ell^1.$$

These functionals are attributed to the unstable zeros of U and V. Suppose now that we are able to find functionals $\{X_j\}_{j=1}^{J}$ in c^0 having the following property (PROP):
if $K \in S_s$ then

$$\langle K, X_j\rangle = 0 \quad \forall j = 1, 2, \ldots, J$$

if and only if

$$\exists Q \in \ell^1 \quad \text{with} \quad K = UQV \quad \text{and} \quad Q(0) \quad \text{block diagonal}.$$

Next, define S as

$$S = \{K : K = UQV, Q \in \ell^1, Q(0) \quad \text{block diagonal}\}.$$

The following lemma, given without proof, stems from standard results in functional analysis (for example [10]):

Lemma 8. *Let $\{P_n\}_{n=1}^{N_s} \in c^0$ as above and let $\{X_j\}_{j=1}^{J}$ in c^0 satisfy (PROP) as above. Then the annihilator subspace $^\perp S$ of S can be characterized as*

$$^\perp S = \text{span}(\{P_n\}_{n=1}^{N_s} \cup \{X_j\}_{j=1}^{J})$$

All that the above lemma says is that the functionals $\{X_j\}_{j=1}^{J}$ add the extra constraints of causality of $Q(0)$ required to solve (OPT$_\infty$) by enlarging the subspace $^\perp S_s$ to $^\perp S$. Since we now have a complete characterization of $^\perp S$ we can proceed exactly as in [4,5] to solve (OPT$_\infty$). Namely, using duality we can transform (OPT$_\infty$) to a maximization problem inside $B(^\perp S)$: Since $(c^0)^* = \ell^1$ and M is the subspace in c^0 defined as $M = \text{span}(\{P_n\}_{n=1}^{N_s} \cup \{X_j\}_{j=1}^{J})$ then from the definitions of $\{X_j\}_{j=1}^{J}$ and $\{P_n\}_{n=1}^{N_s}$ it is easy to verify as in [4,5] that $M^\perp = S$ which implies that S is weak * closed. Hence

$$\inf_{K \in S} \|H - K\| = \min_{K \in (^\perp S)^\perp} \|H - K\| = \sup_{G \in B(^\perp S)} \langle G, H\rangle$$

but since $^\perp S$ is finite dimensional

$$\sup_{G \in B(^\perp S)} \langle G, H\rangle = \max_{G \in B(^\perp S)} \langle G, H\rangle$$

therefore

$$\inf_{K \in S} \|H - K\| = \max_{G \in B(^\perp S)} \langle G, H \rangle.$$

The right-hand side of the above equality can be turned into a finite dimensional linear programming problem [4,5] and hence we obtain the optimal $G = G_0$. The optimal K_0 is found by using the alignment conditions [4,5]

$$\langle G_0, H - K_0 \rangle = \|H - K_0\|.$$

In the sequel we show how to obtain these $\{X_j\}_{j=1}^J$. Towards this end define the following functionals in c^0:

$$R_{z_j} = \{B_j, 0, 0, \ldots\} \quad \forall j = 1, 2, \ldots, j_B.$$

where $\{B_j\}_{j=1}^{j_B}$ be a basis for the finite-dimensional Euclidean space

$$S_A^\perp = S_B = \{B : U_o^T(0) B V_o^T(0) \in (\{R_j(0)\}_{j=1}^r)\}.$$

Let $Z \in \ell^1$ then clearly

$$Z(0) \in S_A$$

if and only if

$$\langle Z, R_{z_j} \rangle = 0 \quad \forall j = 1, 2, \ldots, j_B.$$

In view of the above, (OPT_∞) can be stated as

$$\inf_Z \|H - U_i Z V_i\|$$

with

$$Z \in \ell^1, \langle Z, R_{z_j} \rangle = 0 \quad \forall j = 1, \ldots, j_B.$$

We now show how to obtain the functionals $\{X_j\}_{j=1}^J$ that have the property (PROP) mentioned in the beginning of this subsection.

Theorem 1. *The functionals*

$$X_j = U_i R_{z_j} V_i \quad j = 1, 2, \ldots j_B$$

satisfy (PROP).

Proof. Consider the bounded operators T_{U_i}, T_{V_i} on ℓ^1 defined as

$$(T_{U_i} X)(t) = \sum_{\tau=0}^t U_i(\tau) X(t - \tau)$$

$$(T_{V_i} X)(t) = \sum_{\tau=0}^t X(\tau) V_i(t - \tau)$$

where $X \in \ell^1$. Their (weak*) adjoints $T_{U_i}^*, T_{V_i}^*$ on c^0 which are given by

$$(T_{U_i}^* Y)(t) = \sum_{\tau=0}^{\infty} U_i^T(\tau) Y(\tau + t)$$

$$(T_{V_i}^* Y)(t) = \sum_{\tau=0}^{\infty} Y(\tau + t) V_i^T(\tau)$$

where $Y \in c^0$

Notice, that since U_i, V_i inner then $\hat{U}_i^T(\lambda^{-1})\hat{U}_i(\lambda) = I$ and $\hat{V}_i(\lambda^{-1})\hat{V}_i^T(\lambda) = I$. Note also that $T_{U_i}^*, T_{U_i}$ represent multiplication from the right whereas $T_{V_i}^*, T_{V_i}$ represent multiplication from the left. Hence, it follows that $T_{U_i}^* T_{U_i} = I$ and $T_{V_i}^* T_{V_i} = I$

Interpreting

$$U_i Z V_i = T_{U_i}(T_{V_i}(Z))$$

and

$$U_i R_z V_i = T_{U_i}(T_{V_i}(R_z))$$

with

$$R_z \in \text{span}(\{R_{z_j}\}_{j=1}^{j_B})$$

we can verify that

$$\langle U_i Z V_i, U_i R_z V_i \rangle = \langle Z, R_z \rangle$$

Hence, if $X = U_i R_z V_i$ then

$$\langle UQV, X \rangle = 0 \text{ if and only if } \langle Z, R_z \rangle = 0$$

which completes the proof.

Hence, the additional (finitely many) functionals due to the structural constraints are completely characterized in Theorem 1 and the solution to the 1-block problem follows the duality approach described earlier. Note that since the additional functionals X_j of Theorem 1 are in c^0 (in fact, in ℓ^1), the optimal solution has a finite impulse response (FIR) as in the unconstrained case.

So far in this subsection we assumed that $\hat{U}(\lambda), \hat{V}(\lambda)$ have full row and column rank respectively. However, there is no loss of generality since in the "bad" rank case, or the 4-block problem, (i.e., when the above assumption does not hold [4,5,12]) it is shown in [12] that in order to solve the unconstrained problem it is necessary to satisfy the feasibility conditions of a square subproblem. In particular, we can partition U, V as

$$U = \begin{pmatrix} \bar{U} \\ U_2 \end{pmatrix}, V = (\bar{V} \quad V_2)$$

where $\bar{U}, \bar{V}$ are square and invertible. Let $K = UQV$ then

$$K = \begin{pmatrix} \bar{K} & K_{12} \\ K_{21} & K_{22} \end{pmatrix}$$

A necessary condition for the existence of a solution is that $\bar{K}$ interpolates $\bar{U}, \bar{V}$ which is the aforementioned subproblem. For the constrained problem, in addition to the interpolation conditions on K (e.g., [3]) we can embed the constraints on $Q(0)$ in $\bar{K}$. This embedding relates to the 1-block subproblem and can be done as before and thus the additional constraints on the closed loop can be completely characterized. Thus the standard 4-block procedures [3] can be applied to solve the problem.

3.4   Optimal $\mathcal{H}^\infty$ control

In this subsection we present the solution to the problem of optimal rejection of energy bounded disturbances i.e.,

$$(\text{OPT}_2): \quad \inf_{Z \in \mathcal{H}^\infty, Z(0) \in S_A} \|H - U_i Z V_i\|_{\mathcal{H}^\infty}.$$

This is a constrained $\mathcal{H}^\infty$ problem. We solve this $\mathcal{H}^\infty$ problem, by suitably modifying the standard Nehari's approach [8] in order to account for the additional constraint on the parameter Q (and hence on Z.) This modification yields to a finite dimensional convex optimization problem over a convex set that needs to be solved before applying the standard solution to the Nehari problem. The solution to the above convex finite dimensional problem can be obtained using standard programming techniques. Once this is done, we obtain the optimal LTI controller by solving a standard Nehari's problem.

First we assume that U_i, V_i are square. We will come back to the general 4-block problem later. The solution to the 1-block case is as follows: Let $R = U_i^* H V_i^*$ where $\hat{U}_i^*(\lambda) = \hat{U}_i^T(\lambda^{-1}), \hat{V}_i^*(\lambda) = \hat{V}_i^T(\lambda^{-1})$ and define for each $J \in S_A$ the system R_J as:

$$\hat{R}_J(\lambda) = \lambda^{-1}(\hat{R}(\lambda) - J).$$

For each $J \in S_A$, let Γ_{R_J} represent the Hankel operator [8] with symbol R_J. Before we present the solution we will need to compute $\|\Gamma_{R_J}\|$ using state space formulae. In particular we are going to compute the controllability and observability grammians [8] associated with Γ_{R_J}. Towards this end let R correspond via the Fourier transform to the double-sided (since R is not necessarily causal) sequence $(R(i))_{i=-\infty}^\infty$ then R_J will correspond to $(R_J(i))_{i=-\infty}^\infty$ with $R_J(i) = R(i+1)$ $\forall i \neq -1$ and $R_J(-1) = R(0) - J$. Let now $\overline{G}$ represent the stable (causal) system associated with the pulse response $\{0, R(-1), R(-2), \ldots\}$ and let $(\overline{A}, \overline{B}, \overline{C}, 0)$ be a minimal state space description of it. Let also G represent the stable system associated with the pulse response $\{0, R_J(-1), R_J(-2), \ldots\}$ i.e., G is the anticausal part of R_J

but viewed as a causal (one-sided) system. Then it easy to check that G has the state space description $(A, B, C, 0)$ with

$$A = \begin{pmatrix} \overline{A} & \overline{B} \\ 0 & 0 \end{pmatrix}, \quad B = \begin{pmatrix} 0 \\ I \end{pmatrix}, \quad C = (\overline{C} \quad \overline{J})$$

where $\overline{J} = R(0) - J$. Finally, let W_c, W_o be the controllability and observability grammians for G i.e.,

$$W_c = \sum_{k=0}^{\infty} A^k BB^T (A^T)^k$$

$$W_o = \sum_{k=0}^{\infty} (A^T)^k C^T C A^k.$$

Then W_c and W_o are the solutions to the Lyapunov equations :

$$W_c - AW_c A^T = BB^T, \quad W_o - A^T W_o A = C^T C.$$

Similarly, let $\overline{W}_c, \overline{W}_o$ be the controllability and observability grammians for $\overline{G}$.

Following [8] we have that $\|\Gamma_{R_J}\| = \rho^{1/2}(W_c^{1/2} W_o W_c^{1/2})$. Using the state space description we compute

$$W_c = \begin{pmatrix} \overline{W}_c & 0 \\ 0 & I \end{pmatrix}$$

$$W_o = \begin{pmatrix} \overline{C}^T \overline{C} & \overline{C}^T \overline{J} \\ \overline{J}^T \overline{C} & \overline{J}^T \overline{J} \end{pmatrix} + K$$

where

$$K = \sum_{k=1}^{\infty} (A^T)^k C^T C A^k = \begin{pmatrix} \overline{W}_o - \overline{C}^T \overline{C} & \overline{A}^T \overline{W}_o \overline{B} \\ \overline{B}^T \overline{W}_o \overline{A} & \overline{B}^T \overline{W}_o \overline{B} \end{pmatrix}.$$

Note that K does not depend on J. Also, since $K \geq 0$ then $K = K^{1/2} K^{1/2}$ with $K^{1/2} \geq 0$. Now, proceeding with the computations and rearranging certain terms we obtain:

$$W_c^{1/2} W_o W_c^{1/2} = \begin{pmatrix} I & 0 \\ 0 & \overline{J}^T \end{pmatrix} M^T M \begin{pmatrix} I & 0 \\ 0 & \overline{J} \end{pmatrix} + L^T L$$

where

$$M = \begin{pmatrix} \overline{C} \overline{W}_c^{1/2} & I \\ 0 & 0 \end{pmatrix}, \quad L = K^{1/2} W_c^{1/2}.$$

Hence

$$\|\Gamma_{R_J}\| = \rho^{1/2}(W_c^{1/2} W_o W_c^{1/2}) = \overline{\sigma}[\begin{pmatrix} M\overline{H} \\ L \end{pmatrix}]$$

with $\overline{H} = \begin{pmatrix} I & 0 \\ 0 & \overline{J} \end{pmatrix}$. The following lemma shows that $\|\Gamma_{R_J}\|$ is convex in J

Lemma 9. $\bar{\mu} = \inf_{J \in S_A} \|\Gamma_{R_J}\|$ *is a finite dimensional optimization of a convex and continuous functional on a convex closed set.*

Proof. From the preceding discussion we have

$$\inf_{J \in S_A} \|\Gamma_{R_J}\| = \inf_{\overline{H} \in \overline{S}} \overline{\sigma}[\begin{pmatrix} M\overline{H} \\ L \end{pmatrix}].$$

where

$$\overline{S} = \{\begin{pmatrix} I & 0 \\ 0 & \overline{J} \end{pmatrix} : \quad \overline{J} = R(0) + J, \quad J \in S_A\}.$$

Clearly, since S_A is a subspace then $\overline{S}$ is a convex set. Moreover if $\overline{H}_1, \overline{H}_2 \in \overline{S}$, and $t \in [0, 1]$ we have

$$\overline{\sigma}[\begin{pmatrix} M(t\overline{H}_1 + (1-t)\overline{H}_2) \\ L \end{pmatrix}] = \overline{\sigma}[\begin{pmatrix} tM\overline{H}_1 \\ tL \end{pmatrix} + \begin{pmatrix} (1-t)M\overline{H}_2 \\ (1-t)L \end{pmatrix}]$$

or

$$\overline{\sigma}[\begin{pmatrix} M(t\overline{H}_1 + (1-t)\overline{H}_2) \\ L \end{pmatrix}] \leq t\overline{\sigma}[\begin{pmatrix} M\overline{H}_1 \\ L \end{pmatrix}] + (1-t)\overline{\sigma}[\begin{pmatrix} M\overline{H}_2 \\ L \end{pmatrix}]$$

which shows that $\overline{\sigma}[\begin{pmatrix} M\overline{H} \\ L \end{pmatrix}]$ is convex in $\overline{H}$ and consequently in J. Also, continuity of $\overline{\sigma}[\begin{pmatrix} M\overline{H} \\ L \end{pmatrix}]$ with respect to J is apparent and therefore our claim is proved.

We are now ready to show how to obtain the optimal solution to

$$\mu_{\mathcal{H}^\infty} = \inf_{Z \in \mathcal{H}^\infty, Z(0) \in S_A} \|H - U_i Z V_i\|_{\mathcal{H}^\infty}.$$

Theorem 2. *The following hold:*

1. $\mu_{\mathcal{H}^\infty} = \inf_{Z \in \mathcal{H}^\infty, Z(0) \in S_A} \|R - Z\|_{\mathcal{H}^\infty} = \bar{\mu}$,
2. *A minimizer J_o of the preceding convex programming problem of Lemma 9 always exists. Moreover, if X_o is the solution to the standard Nehari problem*

$$\inf_{X \in \mathcal{H}^\infty} \|R_{J_o} - X\|$$

then the optimal solution Z_o is given by

$$\hat{Z}_o(\lambda) = J_o + \lambda \hat{X}_o(\lambda).$$

Proof. For the first part note that since $\hat{U}_i(\lambda), \hat{V}_i(\lambda), \lambda I$ are inner then

$$\|H - U_i Z V_i\| = \|U_i^* H V_i^* - Z\| = \|R - Z\|$$

Writing $\hat{Z}(\lambda) = Z(0) + \lambda \hat{\bar{Z}}(\lambda)$ with $\bar{Z}$ arbitrary in $\mathcal{H}^\infty$ we have

$$\|R - Z\| = \left\| \hat{R}(\lambda) - Z(0) - \lambda \hat{\bar{Z}}(\lambda) \right\| = \left\| \lambda^{-1}(\hat{R}(\lambda) - Z(0)) - \hat{\bar{Z}}(\lambda) \right\|.$$

Now, if $J \in S_A$ then from Nehari's theorem we have:

$$\inf_{\bar{Z} \in \mathcal{H}^\infty} \|R_J - \bar{Z}\| = \|\Gamma_{R_J}\|$$

hence the first part of the proof follows.

The second part of the theorem is immediate given that a bounded minimizer J_o of the convex minimization in Lemma 9 can be found in S_A (note S_A is unbounded). In fact, this is always the case and the proof of it follows from the fact that the optimal solution Z has to be bounded.

The previous theorem indicates what is the additional convex minimization problem that has to be solved in order to account for the constraint on $Q(0)$. The following corollary is a direct consequence from the proof of the previous analysis.

Corollary 1. *The convex minimization problem of Lemma 9 is*

$$\mu_{\mathcal{H}^\infty} = \bar{\mu} = \min_{J \in \bar{S}_A} \rho^{1/2}(W_c^{1/2} W_o W_c^{1/2})$$

with

$$W_c = \begin{pmatrix} \bar{W}_c & 0 \\ 0 & I \end{pmatrix}, \quad W_o = \begin{pmatrix} 0 & \bar{C}^T \bar{J} \\ \bar{J}^T \bar{C} & \bar{J}^T \bar{J} \end{pmatrix} + \begin{pmatrix} \bar{W}_o & \bar{A}^T \bar{W}_o \bar{B} \\ \bar{B}^T \bar{W}_o \bar{A} & \bar{B}^T \bar{W}_o \bar{B} \end{pmatrix},$$

$\bar{J} = R(0) + J$ *and* $\bar{S}_A = \{J \in S_A : \quad \bar{\sigma}[J] \leq 2\|H\|\}.$

The above convex programming problem can be solved with descent algorithms. In [17] the authors treating a problem of $\mathcal{H}^\infty$ optimization with time domain constraints arrive at a similar finite dimensional convex programming problem.

The full 4-block problem i.e., when U_i and/or V_i are not square is treated analogously as in the standard Nehari approach [8] with the so-called γ-iterations. In particular, using exactly the same arguments as in [8] the same iterative procedure can be established where at each iteration step a 1-block (square) problem with the additional causality constraints on the free parameter Q needs to be solved. Hence, the aforementioned procedure of solving the $\mathcal{H}^\infty$ constrained problem is complete.

3.5   Optimal $\mathcal{H}^2$ control

The problem of interest is

$$(\text{OPT}_{\text{LQG}}): \quad \inf_{Z \in \mathcal{H}^\infty, Z(0) \in S_A} \|H - U_i Z V_i\|_{\mathcal{H}^2}.$$

The solution to this nonstandard $\mathcal{H}^2$ problem is obtained by utilizing the Projection theorem as follows: Let again $R = U_i^* H V_i^*$ and let $Y = \{Y(0), Y(1), Y(2), \ldots\}$ represent the projection of R onto $\mathcal{H}^2$ i.e., $Y = \Pi_{\mathcal{H}^2}(R)$. We note that U_i and V_i need not be square. Consider now the finite dimensional Euclidean space E of real matrices with dimensions equal to those of $Y(0)$ and let Π_{S_A} represent the projection operator onto the subspace S_A of E. Then

Theorem 3. *The optimal solution Z_o for the problem*

$$\mu_{\mathcal{H}^2} = \inf_{Z \in \mathcal{H}^\infty, Z(0) \in S_A} \|H - U_i Z V_i\|_{\mathcal{H}^2}$$

is given by

$$Z_o = \{\Pi_{S_A}(Y(0)), Y(1), Y(2), \ldots\}.$$

Proof. The proof follows from a direct application of the Projection theorem in the Hilbert space $\mathcal{L}^2$. Let $\mathcal{H}_S = \{Z : Z \in \mathcal{H}^2, Z(0) \in S_A\}$ then $\mathcal{H}_S$ is a closed subspace of $\mathcal{L}^2$. Also, let $\langle \bullet, \bullet \rangle$ denote the inner product in $\mathcal{L}^2$. Viewing U_i and V_i as operators on $\mathcal{L}^2$ we have that Z_o is the optimal solution if and only if

$$\langle H - U_i Z_o V_i, U_i Z V_i \rangle = 0 \quad \forall Z \in \mathcal{H}_S$$

or equivalently

$$\langle U_i^* H V_i^* - Z_o, Z \rangle = 0 \quad \forall Z \in \mathcal{H}_S$$

or equivalently

$$\Pi_{\mathcal{H}_S}(U_i^* H V_i^* - Z_o) = 0.$$

But

$$\Pi_{\mathcal{H}_S}(U_i^* H V_i^*) = \{\Pi_{S_A}(Y(0)), Y(1), Y(2), \ldots\} \in \mathcal{H}^\infty$$

which completes the proof.

The above theorem states that only the first component of the classical solution Y is affected. The computation of $\Pi_{S_A}(Y(0))$ is routine; for example having an orthonormal basis $\{B_j\}_{j=1}^r$ for $S_A^\perp$ we have that

$$\Pi_{S_A}(Y(0)) = Y(0) - \sum_{j=1}^r \langle Y(0), B_j \rangle B_j.$$

4   Concluding Remarks

In this paper we presented the solutions to the optimal ℓ^1, $\mathcal{H}^\infty$ and $\mathcal{H}^2$ disturbance rejection problems in the case of a one-step delay observation sharing pattern. We took an input-output point of view that enabled us to convert structural constraints on the controller to linear constraints on the Youla parameter characterizing all possible controllers. In the optimal ℓ^1 disturbance rejection problem, the key observation was that we can obtain a finite number of linear constraints (functionals) to account for the constraint on the Youla parameter. These functionals combined with the functionals of the unconstrained problem can be used exactly as in the standard ℓ^1 problem to yield a tractable linear programming problem. The $\mathcal{H}^\infty$ problem was solved using the Nehari's theorem whereas in the $\mathcal{H}^2$ problem the solution was obtained using the Projection theorem. In particular, the $\mathcal{H}^\infty$ problem was solved by modifying the standard Nehari's approach in order to account for the additional constraint on the compensator. This modification yielded a finite dimensional convex optimization problem over a convex set that needs to be solved before applying the standard solution to the Nehari problem. The solution to the above convex finite dimensional problem can be obtained easily using standard programming techniques. In the $\mathcal{H}^2$ case the solution was obtained from the optimal (standard) unconstrained problem by projecting only the feedforward term of the standard solution to the allowable subspace.

It should be realized that the key element in obtaining convex problems through the Youla parametrization approach was the assumption that, the part of plant that connects controls to measurements, i.e., P_{22} in Figure 1, has the same structure as the one that is required in the controller, i.e., a block diagonal feedthrough term. This is what makes the key Lemma 6 work. Note also that the other parts of the plant, i.e., P_{11}, P_{12} and P_{21} can have arbitrary structure. More generally, if the off-diagonal structure of P_{22} is the same as the structure required on the controller C then, the same methods presented herein are applicable.

In the development herein we assumed that the admissible controllers were LTI. This may or may not be restrictive depending on the particular measure of interest. For the $\mathcal{H}^2$, or more precisely LQG, problem for example, it is well known (e.g., [9]) that it admits a linear optimal solution. A similar result has not established so far for the $\mathcal{H}^\infty$ case with state-space methods. Using input-output averaging arguments however [16], it can be shown that the optimal controller is indeed linear. For the ℓ^1 problem it can be shown that, in the unconstrained case, nonlinear controllers may [22] outperform linear ones. Thus, it seems likely that this will be the case for the constrained problem as well. Nonetheless, in both the $\mathcal{H}^\infty$ and ℓ^1 problems one can show using the exact same arguments as in [16] that linear time varying controllers do not outperform LTI ones.

References

1. T. Başar. "Two-criteria LQG decision problems with one-step delay observation sharing pattern," *Information and Control*, vol. 38, pp. 21-50, 1978.

2. T. Başar and R. Srikant. " Decentralized control of stochastic systems using risk-sensitive criterion," *Advances in Communications and Control*, UNLV Publication, pp. 332-343, 1993.

3. M.A. Dahleh and I.J. Diaz-Bobillo. *Control of Uncertain Systems: A Linear Programming approach*, Prentice Hall, 1995.

4. M.A. Dahleh and J.B. Pearson. "l^1 optimal feedback controllers for MIMO discrete-time systems," *IEEE Trans. A-C*, Vol AC-32, April 1987.

5. M.A. Dahleh and J.B. Pearson. "Optimal rejection of persistent disturbances, robust stability and mixed sensitivity minimization," *IEEE Trans. Automat. Contr.*, Vol AC-33, pp. 722-731, August 1988.

6. M. A. Dahleh, P.G. Voulgaris, and L. Valavani, "Optimal and robust controllers for periodic and multirate systems," *IEEE Trans. Automat. Control*, vol. AC-37, pp. 90–99, January 1992.

7. C.A. Desoer and M. Vidyasagar. *Feedback Systems: Input-Output Properties*, 1975, Academic Press, Inc, N.Y.

8. B.A. Francis. *A Course in H_∞ Control Theory*, Springer-Verlag, 1987.

9. Y.C. Ho and K.C. Chu. "Team decision theory and information structures in optimal control problems-parts I and II," *IEEE Trans. A-C*, Vol AC-17, 15-22, 22-28, 1972.

10. D.G. Luenberger. *Optimization by Vector Space Methods*, New York: Wiley, 1969.

11. J. Marschak and R. Rander. "The firm as a team," *Econometrica*, 22, 1954

12. J.S. McDonald and J.B. Pearson. "Constrained optimal control using the ℓ^1 norm", *Automatica*, vol. 27, March 1991.

13. R. Rander. "Team decision problems," *Ann. Math. Statist.*, vol 33, pp. 857-881, 1962

14. H. Rotstein and A. Sideris, "$\mathcal{H}^\infty$ optimization with time domain constraints," *IEEE Trans. A-C*, Vol AC-39,pp. 762-779, 1994.

15. N. Sandell and M. Athans. "Solution of some nonclassical LQG stochastic decision problems," *IEEE Trans. A-C*, Vol AC-19, pp. 108-116, 1974.

16. J.S. Shamma and M.A. Dahleh. "Time varying vs. time invariant compensation for rejection of persistent bounded disturbances and robust stability," *IEEE Trans. A-C*, Vol AC-36, July 1991.

17. A. Sideris and H. Rotstein. "$\mathcal{H}^\infty$ optimization with time domain constraints over a finite horizon," *Proceedings of the 29th CDC*, Honolulu, Hawaii, December 1990.

18. J.L. Speyer, S.I. Marcus and J.C. Krainak. "A decentralized team decision problem with an exponential cost criterion," *IEEE Trans. A-C*, Vol AC-25,pp. 919-924, 1980.

19. J.C. Krainak, F. W. Machel, S.I. Marcus and J.L. Speyer. " The dynamic linear exponential Gaussian team problem" *IEEE Trans. A-C*, Vol AC-27,pp. 860-869, 1982.

20. C. Fan, J.L. Speyer and C. Jaensch. "Centralized and decentralized solutions to the linear exponential-Gaussian problem," *IEEE Trans. A-C*, Vol AC-39, pp. 1986-2003, 1994.

21. R. Srikant, "Relationships between decentralized controllers design using $\mathcal{H}^{\infty}$ and stochastic risk-averse criteria," *IEEE Trans. A-C*, Vol AC-39,pp. 861-864, 1994.

22. A.A. Stoorvogel. Nonlinear $\mathcal{L}_1$ optimal controllers for linear systems. *IEEE Transactions on Automatic Control*, AC–40(4):694-696, 1995.

23. P.G. Voulgaris, M.A. Dahleh and L.S. Valavani, "$\mathcal{H}^{\infty}$ and $\mathcal{H}^2$ optimal controllers for periodic and multirate systems," *Automatica*, vol. 30, no. 2, pp. 252-263, 1994.

24. M. Vidyasagar. *Control Systems Synthesis: A Factorization Approach*, MIT press, 1985.

25. H.S. Witsenhausen, "A countrexample in stochastic optimal control," *SIAM J. Contr.*, vol 6, pp. 131-147, 1968.

Multi-objective MIMO Optimal Control Design without Zero Interpolation*

Murti V. Salapaka and Mustafa Khammash

Electrical and Computer Engineering,
Iowa State University,
Ames, IA 50011

Abstract. In this article optimal controller designs which address the concerns of the $\mathcal{H}_2$ and the ℓ_1 norms is studied. In the first problem a positive combination of the $\mathcal{H}_2$ and the ℓ_1 norms of the closed loop is minimized. In the second the ℓ_1 norm of a transfer function is minimized while restraining the $\mathcal{H}_2$ norm below a prespecified level.

Converging upper and lower bounds are obtained. Relation to the pure ℓ_1 problem is established. The solution methodology does not involve zero interpolation to characterize the achievable closed-loop maps. Thus obtaining the controller from the optimal closed-loop map is straightforward.

1 Introduction

It has been recognized that controllers which optimize a particular measure (particularly the $\mathcal{H}_2, \mathcal{H}_\infty$ or the ℓ_1 norm) might be unacceptable because their performance might be poor with respect to an alternate measure. Indeed it has been shown that in many cases optimal performance with respect to a certain measure can compromise the performance with respect to some other measure. This has led to a number of results which address the design of controllers that incorporate the objectives of two or more measures.

An important class of controllers which address the concerns of the $\mathcal{H}_2$ criterion and relevant time-domain criteria have been the recent focus of attention. In [5,8,9] it is shown that such problems can be solved via finite-dimensional convex-optimization. In this work, controller design to optimize a performance index which reflect the objectives of both the ℓ_1 and the $\mathcal{H}_2$ norm is addressed. However, the referred work is limited to single-input single-output problems.

In [2,6] the interaction of the $\mathcal{H}_2$ and the ℓ_1 norms for the multi-input multi-output case was studied. In [6] it was shown that approximating solutions within any *a priori* given tolerance to the optimal can be obtained via

---

* This work was supported by NSF grants ECS-9733802 and ECS-9457485.

quadratic programming problems. The approach was based on the Delay-Augmentation method which was first introduced for solving the pure ℓ_1 problem [1]. In [2] a method based on positive cones was used to minimize an $\mathcal{H}_2$ measure of the closed-loop map subject to an ℓ_1 constraint.

Most methods which address the ℓ_1 ($\ell_1 - \mathcal{H}_2$) (e.g. [1,2,6]) problem involve the following steps; parametrization of all stable closed-loop maps which are achievable via stabilizing controllers (using Youla parametrization), casting the ℓ_1 ($\ell_1 - \mathcal{H}_2$) problem as an infinite dimensional linear (convex) programming problem and establishing ways to obtain upper and lower bounds to the optimal via finite dimensional linear (convex) programming. Also, the methods mentioned cast the ℓ_1 ($\ell_1 - \mathcal{H}_2$) problem as an optimization problem with the elements of the impulse response of the closed-loop map as the optimization variables. The achievability of such an impulse-response sequence via stabilizing controllers is enforced by a linear constraint on the sequence. The characterization of this linear constraint from the Youla parametrization involves computation of zeros and zero directions of transfer matrices (see [1]), which can get computationally intensive.

Also in these schemes the optimization procedure yields the optimal closed-loop map's impulse response sequence as its solution. Thus the task of retrieving the optimal controller from the optimal closed-loop map needs to be performed. This involves inversion of certain transfer matrices and it becomes essential to ascertain which unstable zeros get interpolated by the optimal solution. This can be particularly difficult due to the errors caused by the finite precision inherent in any numerical scheme.

Recent research has focussed on alternate methods of solving the pure ℓ_1 problem which avoid the difficulties mentioned [2,3]. A scheme for the pure ℓ_1 problem which does not use the zero-interpolation method was introduced in [3]. This method is particularly attractive because obtaining the controller from the optimal solution is straightforward and the method can be generalized to solve multi-objective problems.

In this article we present two problems. The first problem considers a positive combination of the ℓ_1 and the $\mathcal{H}_2$ norms. It is shown that approximating solutions which converge in the ℓ_2 norm to the optimal can be obtained. Conditions are given under which the optimal is unique. Furthermore this problem reduces to the pure ℓ_1 problem when the $\mathcal{H}_2$ part is excluded. These relate well to the problem studied in [3]. In the second problem the ℓ_1 norm of a transfer function is minimized while keeping the two norm of another transfer function below a prespecified level.

The solution method adopted in this article includes the Youla parameter Q as an optimization variable. Thus obtaining the optimal controller from the optimal solution is immediate. The difficulties associated with the inversion of maps and the correct interpolation of zeros are therefore avoided. Also, for the method presented there is no need to characterize achiveable closed-loop map via a linear constraint on the closed-loop's impulse-response sequence. Thus the computationally intensive task of finding the zeros and the zero directions is eliminated.

This article is organized as follows: In Section 2 we present the combination problem, in Section 3 we present the $\ell_1/\mathcal{H}_2$ problem and in Section 4 we present the conclusions. Section 5 is the Appendix which can be reffered to by the reader for the notation and the mathematical results used in the article.

2   Combination problem

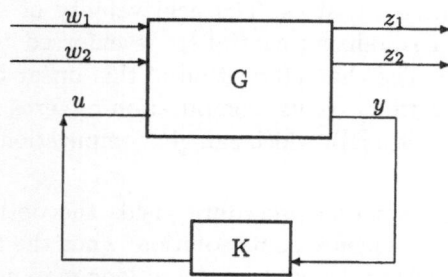

Fig. 1. Closed Loop System

Consider the system described by Figure 1, where $w := (w_1 \ w_2)'$ is the exogenous disturbance, $z := (z_1 \ z_2)'$ is the regulated output, u is the control input and y is the measured output. In feedback control design the objective is to design a controller K such that with $u = Ky$ the resulting closed loop map Φ_{zw} from w to z is stable (see Figure 1) and satisfies certain performance criteria. In [10] a nice parameterization of all closed loop maps which are achievable via stabilizing controllers was first derived. A good treatment of the issues involved in the parameterization is presented in [1]. Following the notation used in [1] we denote by n_u, n_w, n_z and n_y the number of control inputs, exogenous inputs, regulated outputs and measured outputs respectively of the plant G. We represent by Θ, the set of closed loop maps of the plant G which are achievable through stabilizing controllers. H in $\ell_1^{n_z \times n_w}$, U in $\ell_1^{n_z \times n_u}$ and V in $\ell_1^{n_y \times n_w}$ characterize the Youla parameterization of the plant [10]. The following theorem follows from Youla parameterization.

Theorem 1. $\Theta = \{\Phi \in \ell_1^{n_z \times n_w} :$ there exists a $Q \in \ell_1^{n_u \times n_y}$ with $\hat{\Phi} = \hat{H} - \hat{U}\hat{Q}\hat{V}\}$, where $\hat{f}$ denotes the λ transform of f (see [1]).

If Φ is in Θ we say that Φ is an *achievable* closed-loop map. We assume throughout the article that $\hat{U}$ has normal rank n_u and $\hat{V}$ has normal rank n_y. There is no loss of generality in making this assumption [1].

The performance criteria required on the closed-loop map might be best described in terms of two or more measures. For example, in Figure 1, the ℓ_∞ induced norm between w_1 and z_1 might be of concern, whereas the variance of the regulated output z_2 for a white noise input w_2 might be relevant. This would imply that the ℓ_1 norm of the map between w_1 and z_1, and the $\mathcal{H}_2$ norm of the map between w_2 and z_2 are the appropriate norms to employ. Note that these objectives will not be reflected by the ℓ_1 or the $\mathcal{H}_2$ norm alone. It is also true that optimizing the $\mathcal{H}_2$ (ℓ_1) norm might lead to a compromise in the performance when quantified using the ℓ_1 ($\mathcal{H}_2$) norm. With this motivation we study two problems which combine the objectives of the $\mathcal{H}_2$ and the ℓ_1 norms.

In this section we study the combination problem where a positive combination of the ℓ_1 norm of a transfer function and the square of the $\mathcal{H}_2$ norm of another transfer function are minimized. An auxiliary problem (which is more tractable) is defined and the relationship between the auxiliary and the main problem is established. An efficient method is developed to design controllers which guarantee performance within any given tolerance of the optimal value.

2.1   Problem Statement

Given a plant G, let the Youla parameterization be characterized by the transfer function matrices H, U, and V so that a closed loop map is achievable via a stabilizing controller if and only if it can be written as $H - U * Q * V$ for some Q in $\ell_1^{n_u \times n_y}$. Further let H, U, and V be partitioned according to the following equation;

$$H - U * Q * V = \begin{pmatrix} H^{11} & H^{12} \\ H^{21} & H^{22} \end{pmatrix} - \begin{pmatrix} U^1 \\ U^2 \end{pmatrix} * Q * \begin{pmatrix} V^1 & V^2 \end{pmatrix}. \tag{1}$$

The *combination* problem statement is: Given a plant G, positive constants c_1 and c_2 solve the following problem,

$$\inf_{Q \in \ell_1^{n_u \times n_y}} c_1 \|H^{11} - U^1 * Q * V^1\|_1 + c_2 \|H^{22} - U^2 * Q * V^2\|_2^2. \tag{2}$$

Now we define an associated problem which is more tractable. The *auxiliary combination* problem statement is: Given a plant G, positive constants c_1 and c_2 solve the following problem,

$$\nu = \inf_{\|Q\|_1 \leq \alpha} c_1 \|H^{11} - U^1 * Q * V^1\|_1 + c_2 \|H^{22} - U^2 * Q * V^2\|_2^2. \tag{3}$$

In the next part of the article we will study the relationship between the combination problem and the associated auxiliary problem.

2.2   Relation between the auxiliary and the main problem

It is clear that the optimization in (2) can be restricted to the set

$$\{Q : c_1 \|H^{11} - U^1 * Q * V^1\|_1 \leq c_1 \|H^{11}\|_1 + c_2 \|H^{22}\|_2^2\}.$$

This implies that for any relevant Q in the optimization in (2)

$$\|U * Q * V\|_1 \leq 2\|H^{11}\|_1 + \frac{c_2}{c_1}\|H^{22}\|_2^2.$$

Suppose, U^1 has more rows than columns and V^1 has more columns than rows and both have full normal rank. Thus the left inverse of U^1 exists (given by $(U^1)^{-l}$) and the right inverse of V^1 exists (given by $(V^1)^{-r}$). Further suppose that U^1 and V^1 have no zeros on the unit circle (note that in most techniques which are used to solve the ℓ_1 or $\ell_1 - \mathcal{H}_2$ problem and which employ the zero-interpolation method, similar conditions are imposed on U and V). It can be shown that there exists a β (which depends only on $(U^1)^{-l}$ and $(V^1)^{-r}$) such that $\|Q\|_1 \leq \beta$. Thus if in the auxiliary problem we choose $\alpha = \beta$ then the constraint $\|Q\|_1 \leq \alpha$ is redundant in the problem statement of ν and the solutions of (2) and (3) will be identical. The extra constraint in the problem statement of ν is useful because it regularizes the problem (as will be seen).

Now we state a result which shows that problem (3) does not have any anamolous behavior with respect to the constraint level α on the one norm of the Youla parameter Q.

Theorem 2. *Define* $\nu : [0, \infty) \to R$ *by*

$$\nu(\alpha) := \inf_{\|Q\|_1 \leq \alpha} c_1 \|H^{11} - U^1 * Q * V^1\|_1 + c_2 \|H^{22} - U^2 * Q * V^2\|_2^2,$$

where α *is a non-negative real number. Then* $\nu(\alpha)$ *is continuous with respect to* α *on* $(0, \infty)$.

Note that the difference between the combination and its associated auxiliary problem is the extra constraint on the one norm of the parameter Q in the auxiliary problem's definition. It is possible that the plant G is such that there is no Q in $\ell_1^{n_u \times n_v}$ which achieves the minimum for the combination problem and the optimization proceeds towards smaller and smaller objective value with the one norm of Q going unbounded. It can be argued that for such cases it is reasonable to impose a constraint on the one norm of Q so that the problem admits a solution and so that the solution can be used to build stabilizing controllers. With this motivation in the rest of this section we will focus on the solution of the auxiliary combination problem given by (3).

2.3 Converging lower bounds

Let

$$\nu_n := \inf_{\|Q\|_1 \leq \alpha} c_1 \|P_n(H^{11} - U^1 * Q * V^1)\|_1 + c_2 \|P_n(H^{22} - U^2 * Q * V^2)\|_2^2. \quad (4)$$

The above problem can be reduced to a finite dimensional quadratic programming problem. Thus every such approximate problem has a solution. It is also clear that $\nu_n \leq \nu_m \leq \nu$ for all $n < m$.

Theorem 3. *There exists a solution Q^o to problem (3). Also, $\nu_n \nearrow \nu$. Furthermore, if*

$$(\Phi^{11,n}, \Phi^{22,n}) := (P_n(H^{11} - U^1 * Q^n * V^1), P_n(H^{22} - U^2 * Q^n * V^2)),$$
$$(\Phi^{11,o}, \Phi^{22,o}) := (H^{11} - U^1 * Q^o * V^1, H^{22} - U^2 * Q^o * V^2) \text{ and}$$
$$\|(\Phi^{11,o})_p\|_1 = \|\Phi^{11,o}\|_1,$$

where Q^n is a solution to (4), P_n denotes the truncation operator and $(\Phi^{11})_p$ denotes the pth row of Φ^{11}, then there exists a subsequence $\{(\Phi^{11,n_m}, \Phi^{22,n_m})\}$ such that

$$c_1 \|(\Phi^{11,n_m})_p - (\Phi^{11,o})_p\|_1 + c_2 \|\Phi^{22,n_m} - \Phi^{22,o}\|_2^2 \text{ converges to 0.}$$

Proof: Let Q^n denote a solution to (4). As the sequence $\{Q^n\}$ is uniformly bounded by α in $\ell_1^{n_u \times n_y}$ we know that from Banach-Alaoglu theorem (see [4]) that there exists a subsequence $\{Q^{n_m}\}$ of $\{Q^n\}$ and Q^0 in $\ell_1^{n_u \times n_y}$ such that $Q_{ij}^{n_m}$ converges to Q_{ij}^0 in the $W(c_0^*, c_0)$ topology. This implies that $Q^{n_m}(t)$ converges to $Q^0(t)$ for all $t = 0, 1, \ldots$. Therefore for all n, $P_n(U * Q^{n_m} * V)$ converges to $P_n(U * Q^0 * V)$.

Thus for any $n > 0$ and for any $n_m > n$,

$$c_1 \|P_n(H^{11} - U^1 * Q^{n_m} * V^1)\|_1 + c_2 \|P_n(H^{22} - U^2 * Q^{n_m} * V^2)\|_2^2 \leq \nu.$$

This implies that

$$c_1 \|P_n(H^{11} - U^1 * Q^o * V^1)\|_1 + c_2 \|P_n(H^{22} - U^2 * Q^o * V^2)\|_2^2 \leq \nu.$$

Since n is arbitrary, it follows that

$$c_1 \|H^{11} - U^1 * Q^o * V^1\|_1 + c_2 \|H^{22} - U^2 * Q^o * V^2\|_2^2 \leq \nu.$$

It follows that Q^o is an optimal solution for (3). To prove that $\nu_n \nearrow \nu$, we note that

$$c_1 \|P_n(H^{11} - U^1 * Q^{n_m} * V^1)\|_1 + c_2 \|P_n(H^{22} - U^2 * Q^{n_m} * V^2)\|_2^2 \leq$$
$$c_1 \|P_{n_m}(H^{11} - U^1 * Q^{n_m} * V^1)\|_1 + c_2 \|P_{n_m}(H^{22} - U^2 * Q^{n_m} * V^2)\|_2^2 =$$
ν_{n_m} for all $n > 0$ and for all $n_m > n$. Taking the limit as m goes to infinity we have for $n > 0$, that

$$c_1 \|P_n(H^{11} - U^1 * Q^o * V^1)\|_1 + c_2 \|P_n(H^{22} - U^2 * Q^o * V^2)\|_2^2 \quad \leq$$
$$\lim_{m \to \infty} \nu_{n_m}.$$

It follows that

$$c_1\|H^{11} - U^1 * Q^o * V^1\|_1 + c_2\|H^{22} - U^2 * Q^o * V^2\|_2^2 \leq \lim_{m\to\infty} \nu_{nm}.$$

Thus we have shown that $\lim_{m\to\infty} \nu_{nm} = \nu$. Since ν_n is a monotonically increasing sequence, it follows that $\nu_n \nearrow \nu$. Note that $(\Phi^{11,nm}, \Phi^{22,nm})$ and $(\Phi^{11,o}, \Phi^{22,o})$ satisfy all the conditions stipulated in Lemma 10. Thus

$$c_1\|(\Phi^{11,nm})_p - (\Phi^{11,o})_p\|_1 + c_2\|\Phi^{22,nm} - \Phi^{22,o}\|_2^2 \to 0.$$

$\square$

Corollary 2. *Suppose in problem (3), $H := H^{11} = H^{22}$, $U := U^1 = U^2$ and $V := V^1 = V^2$, and $c_2 > 0$. Let Q^o be a solution of problem (3) and let $\Phi^{11,o} = \Phi^{22,o} := H - U * Q^o * V$. Then Φ^o is unique. If Q^n denotes the optimal solution to (4) then $\Phi^n := P_n(H - U * Q^n * V)$ is such that $\|\Phi^n - \Phi^o\|_2$ converges to zero.*

Proof: Let

$$A_{al} = \{\Phi : \Phi = H - U * Q * V \text{ with } \|Q\|_1 \leq \alpha\}.$$

Then A_{al} is a convex set. Thus

$$\nu = \inf_{\Phi \in A_{al}} c_1\|\Phi\|_1 + c_2\|\Phi\|_2^2. \tag{5}$$

Also, if $c_2 > 0$ then $c_1\|\Phi\|_1 + c_2\|\Phi\|_2^2$ is a strictly convex function of Φ. Thus we have that the minimizer to (5), given by Φ^o, is unique.

We now show that $\|\Phi^n - \Phi^0\|_2^2$ converges to zero. Suppose $\|\Phi^n - \Phi^0\|_2^2$ does not converge to zero. Then there exists a subsequence $\Phi^{nm} := P_{nm}(H - U * Q^{nm} * V)$ and $\epsilon > 0$, such that

$$\|\Phi^{nm} - \Phi^0\|_2^2 > \epsilon, \tag{6}$$

where $\|Q^{nm}\|_1 \leq \alpha$. Using arguments used in the proof of Theorem 3 we can conclude that there exists a subsequence $\{Q^{nm_k}\}$ of $\{Q^{nm}\}$ which converges pointwise to an optimal solution Q^1, of (3). Furthermore, if Φ^1 is defined to be $H - U * Q^1 * V$ then from arguments used in proof of Theorem 3 we have

$$c_1\|(\Phi^{nm_k})_p - (\Phi^1)_p\|_1 + \|\Phi^{nm_k} - \Phi^1\|_2^2 \to 0.$$

In particular, $\|\Phi^{nm_k} - \Phi^1\|_2^2$ converges to zero. However, Φ^1 is a solution to (5) and therefore Φ^1 is equal to Φ^o because the solution (5) is unique. Thus $\|\Phi^{nm_k} - \Phi^0\|_2^2$ converges to zero. which is a contradiction to (6). This proves that $\|\Phi^n - \Phi^0\|_2^2$ converges to zero. $\square$

Corollary 3. *There exists a solution Q^o to the following problem:*

$$\nu = \inf_{\|Q\|_1 \leq \alpha} \|H - U * Q * V\|_1.$$

Furthermore if Q^n denotes the optimal solution to

$$\nu_n = \inf_{\|Q\|_1 \leq \alpha} \|P_n(H - U * Q * V)\|_1,$$

then

$$\nu_n \nearrow \nu.$$

If p is such that $\|(H - U * Q^o * V)_p\|_1 = \|H - U * Q^o * V\|_1$ then there exists a subsequence of $\{H - U * Q^{n_m} * V\}$ of $\{H - U * Q^o * V\}$ such that

$$\|(P_{n_m}(H - U * Q^{n_m} * V))_p - (H - U * Q^o * V)_p\|_1 \to 0.$$

Proof: In the problem definitions of the main and the auxiliary problems given by (2) and (3) respectively, let $c_2 = 0$, $c_1 = 1$, $H^{11} = H$, $U^1 = U$, and $V^1 = V$. The results of the corollary follow directly from Theorem 3. □

The results of this corollary relate very well with the ones obtained in [3] where the following problem was addressed:

$$\inf_{Q \in \ell_1^{n_u \times n_y}} \max\{\|H - U * Q * V\|_1, \alpha\|Q\|_1\}.$$

Adding the $\mathcal{H}_2$ norm of the closed loop makes the lower bounds behave nicer because then the original suboptimal solution sequence converges in the $\mathcal{H}_2$ norm to the optimal (see Corollary 2).

2.4   Converging upper bounds

Let ν^n be defined by

$$\inf_Q c_1\|H^{11} - U^1 * Q * V^1\|_1 + c_2\|H^{22} - U^2 * Q * V^2\|_2^2$$

subject to   (7)

$$\|Q\|_1 \leq \alpha$$
$$Q(k) = 0 \text{ if } k > n.$$

It is clear that $\nu^n \geq \nu^{n+1}$ because any Q in $\ell_1^{n_u \times n_y}$ which satisfies the constraints in the problem definition of ν^n will satisfy the constraints in the problem definition of ν^{n+1}. For the same reason we also have $\nu^n \geq \nu$ for all relevant n.

Similar results as developed for the lower bounds can be proven for these upper bounds. The details are left to the reader.

3 $\ell_1/\mathcal{H}_2$ problem

In this section we formulate a problem which minimizes the ℓ_1 norm of a transfer function while restraining the $\mathcal{H}_2$ norm below a prespecified level. Similar to the development in the previous section we define an auxiliary problem which regularizes the main problem. Converging upper and lower bounds are established for the auxiliary problem.

3.1   Problem statement

Given a plant G and its Youla parameterization by the parameters H, U, and V let the matrices H, U, and V be partitioned according to equation (1). The $\ell_1/\mathcal{H}_2$ problem statement is: Given a plant G, positive real number γ solve the following problem;

$$\inf_{Q \in \ell_1^{n_u \times n_y}} \|H^{11} - U^1 * Q * V^1\|_1$$

subject to

$$\|H^{22} - U^2 * Q * V^2\|_2^2 \leq \gamma$$
$$Q \in \ell_1^{n_u \times n_y}.$$

$$(8)$$

The $\ell_1/\mathcal{H}_2$ *auxiliary* problem statement is: Given a plant G, positive real number γ solve the following problem;

$$\eta = \inf_{Q \in \ell_1^{n_u \times n_y}} \|H^{11} - U^1 * Q * V^1\|_1$$

subject to

$$\|H^{22} - U^2 * Q * V^2\|_2^2 \leq \gamma$$
$$\|Q\|_1 \leq \alpha.$$

$$(9)$$

The parameter Q in the optimization stated in (8) can be restricted to the set

$$\{Q \in \ell_1^{n_u \times n_y} \text{ such that } \|H^{11} - U^1 * Q * V^1\|_1 \leq \|H\|_1\}.$$

Thus for all relevant Q for the optimization in the main $\ell_1/\mathcal{H}_2$ problem (given by (8)) $\|U^1 * Q * V^1\|_1 \leq \|H^{11}\|_1$. Once we have the above bound available on $R := U^1 * Q * V^1$ the rest of the discussion in Subsection 2.2 is applicable to the main and the auxiliary $\ell_1/\mathcal{H}_2$ problems given by (8) and (9) respectively. This establishes the relevance of the auxiliary $\ell_1/\mathcal{H}_2$ problem. Now we establish a result which shows that problem (9) does not have any anamolous behavior with respect to α and γ.

Theorem 4. *Let S be the set:*

$$\{\gamma \in [0, \infty) : \text{ there exists } Q \in \ell_1^{n_u \times n_y} \text{ with } \|H^{22} - U^2 * Q * V^2\|_2^2 \leq \gamma\}.$$

Let $\eta : [0, \infty) \times S \to R$ be defined by

$$\eta(\alpha, \gamma) := \inf_{Q \in \ell_1^{n_u \times n_y}} \|H^{11} - U^1 * Q * V^1\|_1$$

subject to

$$\|H^{22} - U^2 * Q * V^2\|_2^2 \leq \gamma$$
$$\|Q\|_1 \leq \alpha.$$

where α is any non-negative real number and $\gamma \in S$. Then $\eta(\alpha, \gamma)$ is continuous with respect to α and γ on $(0, \infty) \times int(S)$.

Now we obtain converging upper and lower bounds to the auxiliary problem.

3.2   Converging lower bounds

Let η_n be defined by

$$\inf_{Q \in \ell_1^{n_u \times n_y}} \|P_n(H^{11} - U^2 * Q * V^2)\|_1$$

subject to   (10)

$$\|P_n(H^{22} - U^2 * Q * V^2)\|_2^2 \leq \gamma$$
$$\|Q\|_1 \leq \alpha.$$

It is clear that only the parameters of $Q(0), \ldots, Q(n)$ enter into the optimization problem and therefore (10) is a finite dimensional quadratic programming problem. Once optimal sequence $\{Q(0), \ldots, Q(n)\}$ is found, $Q = \{Q(0), \ldots, Q(n), 0, \ldots\}$ will be an FIR optimal solution to (10).

Theorem 5. *Suppose the constraint set in problem (9) is nonempty. Then problem (9) always has an optimal solution $Q^0 \in \ell_1^{n_u \times n_y}$. Furthermore,*

$$\eta_n \nearrow \eta.$$

Proof: We know that for any Q in $\ell_1^{n_u \times n_y}$, $\|P_n(H^{11} - U^2 * Q * V^2)\|_1 \leq \|P_{n+1}(H^{11} - U^2 * Q * V^2)\|_1$ and $\|P_n(H^{11} - U^2 * Q * V^2)\|_2^2 \leq \|P_{n+1}(H^{11} - U^2 * Q * V^2)\|_2^2$. Therefore $\eta_n \leq \eta_{n+1}$ for all $n = 1, 2, \ldots$. Thus $\{\eta_n\}$ forms an increasing sequence. Similarly it can be shown that for all n, $\eta_n \leq \eta$.

For $n = 1, 2, \ldots$, let $\{Q^n\} \in \ell_1^{n_u \times n_y}$ be FIR solutions of (10). As the sequence $\{Q^n\}$ is uniformly bounded by α in $\ell_1^{n_u \times n_y}$ it follows from Banach-Alaoglu theorem that there exists a subsequence $\{Q^{n_m}\}$ of $\{Q^n\}$ and $Q^o \in \ell_1^{n_u \times n_y}$ such that $Q_{ij}^{n_m} \to Q_{ij}^o$ in the $W(c_0^*, c_0)$ topology. Following arguments similar to the ones employed in the proof of Theorem 3 we can show that Q^o is an optimal solution to (9) and that $\eta_n \nearrow \eta$.   □

3.3   Converging upper bounds

Let η^n be defined by

$$\inf_{Q \in \ell_1^{n_u \times n_y}} \|H^{11} - U^1 * Q * V^1\|_1$$

subject to   (11)

$$\|H^{22} - U^2 * Q * V^2\|_2^2 \leq \gamma$$
$$\|Q\|_1 \leq \alpha$$
$$Q(k) = 0 \text{ if } k > n.$$

It is clear that $\eta^n \geq \eta^{n+1}$ because any Q in $\ell_1^{n_u \times n_y}$ which satisfies the constraints in the problem definition of η^n will satisfy the constraints in the problem definition of η^{n+1}. For the same reason we also have $\eta^n \geq \eta$ for all relevant n.

Similar results as developed for the lower bounds can be proven for these upper bounds. The details are left to the reader.

Analogous results can be established using the machinery developed here for the $\mathcal{H}_2/\ell_1$ problem. Some of these results are given in [7].

4   Conclusions

In this article we have studied two problems. In the first termed the combination problem a positive combination of the one norm of a transfer function and the square of the two norm of another transfer function is minimized. Converging upper and lower bounds were obtained. It was established that approximating solutions to the optimal can be obtained which converge to the optimal in the two norm. The relation to the pure ℓ_1 problem was also discussed.

In the second problem the ℓ_1 norm of a transfer function was minimized while restraining the $\mathcal{H}_2$ norm of another below a prespecified level. Results similar to the combination problem were obtained.

The machinery developed seems appropriate for other time domain norms like the ℓ_∞ norm of the response due to a given signal and possibly the $\mathcal{H}_\infty$ norm. A general framework in which the interplay of the $\mathcal{H}_2$, ℓ_1 $\mathcal{H}_\infty$ and other time domain measures is the topic of ongoing research.

5   Appendix

An exhaustive treatment of the mathematical results used in this article is given in [4].

Lemma 10. *Let* $\{(\Phi^{11,k}, \Phi^{22,k})\}$ *be a sequence such that*

$$(\Phi^{11,k}(t), \Phi^{22,k}(t)) \to (\Phi^{11,o}(t), \Phi^{22,o}(t)) \text{ for all } t$$

and furthermore

$$f(\Phi^{11,k}, \Phi^{22,k}) \le f(\Phi^{11,o}, \Phi^{22,o}) \text{ for all } k. \tag{12}$$

Let $\|\Phi^{11,o}\|_1 = \|(\Phi^{11,o})_p\|_1$ *where* $(\Phi^{11,o})_p$ *represents the* p^{th} *row of* $\Phi^{11,o}$. *Then*

$$c_1\|(\Phi^{11,k})_p - (\Phi^{11,o})_p\|_1 + c_2\|\Phi^{22,k} - \Phi^{22,o}\|_2^2 \to 0 \text{ as } k \to \infty.$$

The same conclusion holds if condition (12) is replaced with the following condition:

$$f(\Phi^{11,k}, \Phi^{22,k}) \to f(\Phi^{11,o}, \Phi^{22,o}). \tag{13}$$

Proof: The proof is left to the reader.   □

References

1. M. A. Dahleh and I. J. Diaz-Bobillo. *Control of Uncertain Systems: A Linear Programming Approach.* Prentice Hall, Englewood Cliffs, New Jersey, 1995.
2. N. Elia and M. A. Dahleh. Controller design with multiple objectives. *IEEE Trans. Automat. Control*, 42, no. 5:596–613, 1997.
3. M. Khammash. Solution of the ℓ_1 mimo control problem without zero interpolation. In *Proceedings of the IEEE Conference on Decision and Control.* pp: 4040-4045, Kobe, Japan, December 1996.
4. D. G. Luenberger. *Optimization by Vector Space Methods.* John Wiley and Sons, Inc., 1969.
5. M. V. Salapaka, M. Dahleh, and P. Voulgaris. Mixed objective control synthesis: Optimal $\ell_1/\mathcal{H}_2$ control. *SIAM Journal on Control and Optimization*, V35 N5:1672–1689, 1997.
6. M. V. Salapaka, M. Dahleh, and P. Voulgaris. Mimo optimal control design: the interplay of the $\mathcal{H}_2$ and the ℓ_1 norms. *IEEE Trans. Automat. Control*, 43, no. 10:1374–1388, 1998.
7. M. V. Salapaka, M. Khammash, and M. Dahleh. Solution of mimo $\mathcal{H}_2/\ell_1$ problem without zero interpolation. In *Proceedings of the IEEE Conference on Decision and Control.* Vol. 2, pp: 1546-1551, San Diego, CA, December 1997.
8. M. V. Salapaka, P. Voulgaris, and M. Dahleh. Controller design to optimize a composite performance measure. *Journal of Optimization Theory and its Applications*, 91 no. 1:91–113, 1996.
9. P. Voulgaris. Optimal $\mathcal{H}_2/\ell_1$ control via duality theory. *IEEE Trans. Automat. Control*, 4, no. 11:pp. 1881–1888, 1995.
10. D. C. Youla, H. A. Jabr, and J. J. Bongiorno. Modern wiener-hopf design of optimal controllers - part 2: The multivariable case. *IEEE Trans. Automat. Control*, 21, no. 3:pp. 319–338, 1976.

Robustness Synthesis in ℓ_1: A Globally Optimal Solution *

Mustafa Khammash[1], Murti V. Salapaka[1], and Tim Van Voorhis[2]

[1] Electrical & Computer Engineering Department
[2] Industrial and Manufacturing Systems Engineering
Iowa State University
Ames, IA 50011
USA

Abstract. This paper solves the problem of synthesis of controllers achieving *globally optimal* robust performance against structured time-varying and/or non-linear uncertainty. The performance measure considered is the infinity to infinity induced norm of a system's transfer function. The solution utilizes linear relaxation for finding the global optimal solution.

1 Introduction

System robustness has been the subject of extensive research. Various results concerning the stability and performance robustness analysis for several types of uncertainty models have been obtained in the literature. For the time-invariant norm bounded structured perturbations, the Structured Singular Value (SSV) theory [3,4] provides a nonconservative measure for stability. For time-varying norm bounded structured perturbations , [5–7] provide computable necessary and sufficient conditions for robust stability and performance when the induced-infinity norm is used to measure perturbation size, while [8,9] provide conditions for the induced 2-norm. The robustness synthesis problem, however, remains largely unsolved for any norm. Controllers which address the objectives of the SSV can be designed using the so called $D - K$ iterations (see e.g. [4]). When perturbations with induced ∞-norm bound are present, a $D - K$ iteration type procedure appears in [7]. However, neither method guarantees that a global minimum is achieved, and in general a local minimum is reached. The problem appears to be an inherently nonconvex one in either norm. In the Full Information or State-Feedback cases, however, the synthesis problem for H_∞ has been

* This research was supported by NSF grants ECS-9733802, ECS-9110764, ECS-9457485 and ISU grant SPRIGS-7041747

solved for any number of uncertainty blocks. See [12]. Recently, Yamada and Hara [14] provided an algorithm for approximately finding a global solution to the constantly scaled H_∞ control synthesis problem.

In this article, we address the output feedback synthesis problem when the signal norm is the infinity norm and the perturbations are structured time-varying and/or nonlinear systems with an induced ∞-norm bound. A *globally optimal* solution to the robustness synthesis problem is obtained. It is shown that the solution involves only solving certain linear programming problems.

2   Notation

ℓ_∞ is the space of bounded sequences of real numbers, ℓ_1 is the space of absolutely summable sequences of real numbers, $\ell_1^{p \times q}$ the $p \times q$ matrices of elements of ℓ_1. If $x \in \ell_1^{p \times q}$, $\|x\|_1 := \max_i \sum_j \|x_{ij}\|_1$. P_N is the truncation operator so that for any sequence x, $P_N x = y$ where $y(k) = x(k)$ whenever $k \le N$ and $y(k) = 0$ for $k > N$.

3   Problem Statement

Consider the system, described in Figure 1. where G, K, u, w, z, and y

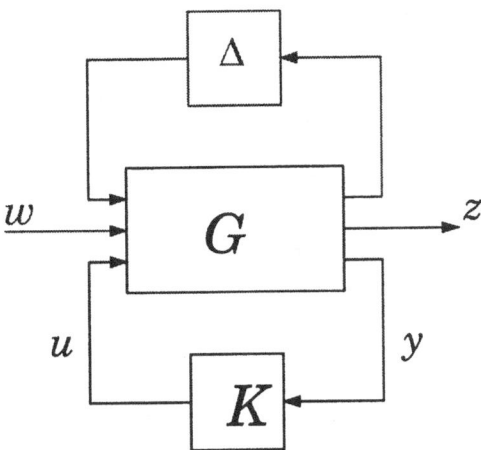

Fig. 1. Robust Performance Problem

are the generalized LTI plant, an LTI controller, the control input, the exogenous disturbance, the regulated output and the measured output respectively. Modeling uncertainty is described by the perturbation block Δ which is restricted to lie in the following set of admissible perturbations:

320   Mustafa Khammash et al.

$\underline{\Delta} := \{\Delta = diag(\Delta_1, \ldots, \Delta_n) \;\; : \;\; \Delta_i : \ell_\infty \Rightarrow \ell_\infty$ is causal, and $\|\Delta_i\| :=$
$\sup_{u \neq 0} \frac{\|\Delta_i u\|_\infty}{\|u\|_\infty} \leq 1\}$,

where the norm used is the ℓ_∞ norm. The perturbation may therefore
be nonlinear or time-varying. In this article we will restrict the study to
discrete-time systems. The system in Figure 1 is said to be robustly stable
if it is ℓ_∞-stable for all admissible perturbations, i.e. for all $\Delta \in \underline{\Delta}$. The
problem we shall address in this article is as follows:

Problem Statement: Find a linear finite dimensional controller K such
that:

1. The system achieves robust stability, and
2. The system achieves robust performance, i.e.

$$\|\mathcal{T}_{zw}\| < 1 \qquad \forall \Delta \in \underline{\Delta}$$

where $\mathcal{T}_{zw}$ is the map from w to z, and the norm used above is the
induced ℓ_∞ operator norm.

4   Conditions for Robustness

Before addressing the robustness synthesis problem stated in the previous
section we look at the relevant robustness analysis conditions. The robust
performance problem can be converted to a robust stability problem, by
adding a fictitious perturbation block Δ_p connecting the output z into the
input w (see[6]). Thus there is no loss of generality, if we address the robust
stability problem alone.

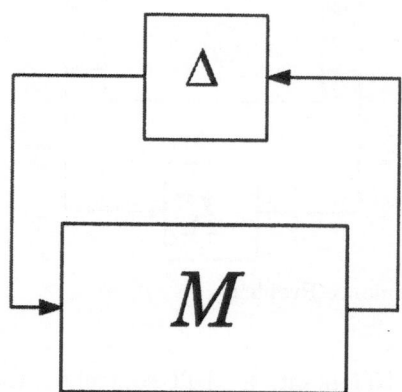

Fig. 2. Stability robustness problem

The robust stability problem is depicted in Figure 2 where G and any stabilizing controller K have been lumped into one system M. Let Φ be the impulse response matrix of M.

Since M is LTI, causal, and stable, Φ will belong to $\ell_1^{n \times n}$. Clearly, any stable weighting on the perturbation Δ can be absorbed into Φ. Each Φ_{ij} has an induced norm which can be computed arbitrarily accurately. In particular, $\|\Phi_{ij}\|_1 = |D_{ij}| + \sum_{k=0}^{\infty} |C_i A^k B_j|$, where A, B_i, C_j, D_{ij} are the constant matrices in the state-space description of M_{ij}. We can therefore define the following matrix:

$$\hat{\Phi} = \begin{bmatrix} \|\Phi_{11}\|_1 & \cdots & \|\Phi_{1n}\|_1 \\ \vdots & & \vdots \\ \|\Phi_{n1}\|_1 & \cdots & \|\Phi_{nn}\|_1 \end{bmatrix}.$$

The following robustness conditions will be the basis for the proposed synthesis method.

Theorem 1. *The system in Figure 3 achieves robust stability if and only if any one of the following conditions holds:*

1. $\rho(\hat{\Phi}) < 1$, *where $\rho(.)$ denotes the spectral radius.*
2. $\inf_{D \in \mathcal{D}} \|D^{-1} \hat{\Phi} D\|_1 < 1$ *where $\mathcal{D} := \{diag(d_1, \ldots, d_2) : d_i > 0\}$.*

5   Formulation as an Optimization Problem

The robustness *synthesis* problem can be stated as follows:

$$\inf_{D \in \mathcal{D}} \inf_{Q \in \ell_1} \left\| D^{-1} \Phi(Q) D \right\|_1 =: \gamma_* \qquad (1)$$

where $\Phi(Q) = H - U * Q * V$ is the standard parameterization of the closed-loop system using the Youla parameter Q (see [2]). For each fixed $D = diag(d_1, \ldots, d_n)$, problem (1) is a standard ℓ_1 norm-minimization problem. We say Φ is an achievable closed loop map if there exists a stable Q such that $\Phi = H - U * Q * V$. It can be shown that Φ is achievable if and only if $\mathcal{A}\Phi = b$ where $\mathcal{A} : \ell_1^{n \times n} \to \ell_1$ is a linear operator and $b \in \ell_1$. Both $\mathcal{A}$ and b can be determined based on H, U and V. Before recasting the main problem given by (1) we define for any $D \in \mathcal{D}$

$$\begin{aligned} \gamma(D) := \inf & \|D^{-1} \Phi D\|_1 \\ & \text{subject to} \\ & \mathcal{A}\Phi = b, \end{aligned} \qquad (2)$$

$$\begin{aligned} \overline{\gamma}^N(D) := \inf & \|D^{-1} \Phi D\|_1 \\ & \text{subject to} \\ & \mathcal{A}\Phi = b \\ & \Phi(k) = 0 \text{ for all } k \geq N \text{ and} \end{aligned} \qquad (3)$$

$$\underline{\gamma}^N(D) := \inf \|D^{-1}\Phi D\|_1$$
$$\text{subject to} \tag{4}$$
$$P_N A\Phi = P_N b.$$

It is to be noted that problems (3) and (4) can be shown to be finite dimensional linear programming problems. Using the above definitions we further define

$$\gamma_* := \inf_{D\in\mathcal{D}} \gamma(D), \tag{5}$$

$$\overline{\gamma}_*^N := \inf_{D\in\mathcal{D}} \overline{\gamma}^N(D), \tag{6}$$

$$\underline{\gamma}_*^N := \inf_{D\in\mathcal{D}} \underline{\gamma}^N(D) \tag{7}$$

$$\tag{8}$$

The following results can be established.

Lemma 11. $\overline{\gamma}_*^N \searrow \gamma_*$ and $\underline{\gamma}_*^N \nearrow \gamma_*$ as $N \to \infty$

The previous lemma suggests that if an effective solution procedure exists to solve for $\overline{\gamma}_*^N$ and $\underline{\gamma}_*^N$ then we can obtain converging upper and lower bounds to γ_*. The rest of the development in this section will be focussed on deriving a method to obtain $\overline{\gamma}_*^N$ for any given N. The methodology for finding $\underline{\gamma}_*^N$ can be identically developed. Note that in the problem statement of (3) only a finite number of variables are present. However, because of the special structure of the matrix A it can be shown that if only finite number of variables $\Phi(k)$ are involved then only a finite number of constraints posed by the equation $A\Phi = b$ are relevant to the optimization. Thus it can be shown that

$$\overline{\gamma}^N(D) = \inf \|D^{-1}\Phi D\|_1$$
$$\text{subject to}$$
$$P_{N'} A P_N (\Phi^+ - \Phi^-) = P_{N'} b \tag{9}$$
$$\Phi(k) = 0 \text{ for all } k \geq N,$$

where N' depends on N. The optimization associated with (9) can be cast into the finite dimensional linear program:

$$\overline{\gamma}^N(D) = \inf \alpha$$
$$\text{subject to}$$
$$\sum_j \sum_{k=1}^{N} d_j(\Phi_{ij}^+(k) + \Phi_{ij}^-(k)) \leq d_i\alpha \tag{10}$$
$$A^N(\Phi^+ - \Phi^-) = b^N$$
$$\Phi^+ \geq 0, \ \Phi^- \geq 0,$$

where $A^N := P_{N'} A P_N$ and $b^N := P_{N'} b$.

Thus the resulting finite dimensional optimization problem that needs to be solved has the following structure;

$$\overline{\gamma}_*^N = \inf \alpha$$

subject to

$$\sum_j d_j p_{ij} \le d_i \alpha$$
$$p_{ij} = \sum_{k=1}^N \Phi_{ij}^+(k) + \Phi_{ij}^-(k) \tag{11}$$
$$A^N(\Phi^+ - \Phi^-) = b^N$$
$$\Phi^+ \ge 0, \; \Phi^- \ge 0, \; p_{ij} \ge 0, \; \alpha \ge 0, d_j > 0.$$

For what follows, we will assume that *a priori* upper and lower bounds are available for the optimal values of the variables d_j, α and p_{ij} involved in the optimization. Thus $L_{ij} \le p_{ij} \le U_{ij}, 0 < L_{d_j} \le d_j \le U_{d_j}$, and $L_\alpha \le \alpha \le U_\alpha$. Such bounds are obtainable from the original problem data and will allow us to limit the optimization search to a bounded set in the parameter space. Accordingly the optimization problem of interest becomes

$$\mu := \inf \alpha$$

subject to

$$\sum_j d_j p_{ij} \le d_i \alpha$$
$$p_{ij} = \sum_{k=1}^N \Phi_{ij}^+(k) + \Phi_{ij}^-(k)) \tag{12}$$
$$A^N(\Phi^+ - \Phi^-) = b^N$$
$$\Phi^+ \ge 0, \; \Phi^- \ge 0$$
$$L_{ij} \le p_{ij} \le U_{ij}, \; L_{d_j} \le d_j \le U_{d_j}, L_\alpha \le \alpha \le U_\alpha.$$

6 Problem solution

In this section we provide the problem solution. Consider problem (12) posed on a smaller grid;

$$\mu(\ell_d, u_d) := \inf \alpha$$

subject to

$$\sum_j d_j p_{ij} \le d_i \alpha$$
$$p_{ij} = \sum_{k=1}^N \Phi_{ij}^+(k) + \Phi_{ij}^-(k) \tag{13}$$
$$A^N(\Phi^+ - \Phi^-) = b^N$$
$$\Phi^+ \ge 0, \; \Phi^- \ge 0$$
$$L_{ij} \le p_{ij} \le U_{ij}, \; \ell_{d_j} \le d_j \le u_{d_j}, L_\alpha \le \alpha \le U_\alpha.$$

where $\ell_d = (\ell_{d_1}, \ldots, \ell_{d_n})$, $u_d = (u_{d_1}, \ldots, u_{d_n})$. For our purposes the interval $[\ell_{d_j}, u_{d_j}]$ is a subset of the interval $[L_d, U_d]$. Note that since the variables in the statement of $\mu(\ell_d, u_d)$ are confined to smaller region than in μ, the above problem is being solved for a sub-problem on a grid.

In solving the subproblem, a relaxation scheme will be employed. For this purpose, the following lemma is needed:

Lemma 12. *If the variables $x_j \in R$ satisfy the conditions $\ell_j \leq x_j \leq u_j$ and $t_{ij} := x_i x_j$ then*

$$t_{ij} \geq u_j x_i + u_i x_j - u_i u_j, \tag{14}$$

$$t_{ij} \leq \ell_j x_i + u_i x_j - u_i \ell_j, \tag{15}$$

$$t_{ij} \leq u_j x_i + \ell_i x_j - \ell_i u_j, \tag{16}$$

$$t_{ij} \geq \ell_j x_i + \ell_i x_j - \ell_i \ell_j. \tag{17}$$

Furthermore, if variables $t_{ij} \in R$ satisfy (14), (15), (16), (17) and x_k satisfy $\ell_k \leq x_k \leq u_k$ then

$$|t_{ij} - x_i x_j| \leq \frac{1}{4}(u_i - \ell_i)(u_j - \ell_j). \tag{18}$$

Define $W_{ij} := \{(p_{ij}, d_j, w_{ij}) \in R^3 | \ (14), (15), (16), (17)$ are satisfied with $t_{ij}, x_i, x_j, u_i, \ell_i, u_j, \ell_j$ replaced by $w_{ij}, p_{ij}, d_j, U_{ij}, L_{ij}, u_{d_j}, \ell_{d_j}$ respectively$\}$. Furthermore, define $W_i := \{(\alpha, d_i, w_i) \in R^3 | \ (14), (15), (16), (17)$ are satisfied with $t_{ij}, x_i, x_j, u_i, \ell_i, u_j, \ell_j$ replaced by $w_i, \alpha, d_i, U_\alpha, L_\alpha, u_{d_j}, \ell_{d_j}$ respectively$\}$.

Thus it follows that

$$
\begin{aligned}
\mu(\ell_d, u_d) = \inf \ &\alpha \\
\text{subject to} \ & \\
&\sum_j w_{ij} \leq w_i \\
&(p_{ij}, d_j, w_{ij}) \in W_{ij} \\
&(\alpha, d_i, w_i) \in W_i \\
&w_{ij} = d_j p_{ij}, \ w_i = d_i \alpha \\
&p_{ij} = \sum_{k=1}^N \Phi_{ij}^+(k) + \Phi_{ij}^-(k)) \\
&A^N(\Phi^+ - \Phi^-) = b^N \\
&\Phi^+ \geq 0, \ \Phi^- \geq 0 \\
&L_{ij} \leq p_{ij} \leq U_{ij}, \ \ell_{d_j} \leq d_j \leq u_{d_j}, L_\alpha \leq \alpha \leq U_\alpha.
\end{aligned}
\tag{19}
$$

Note that in the above equation all the constraints are linear constraints except for the constraints $w_{ij} = p_{ij} d_j$ and $w_i = \alpha d_i$. Let $\mu_R(\ell_d, u_d)$ be the *relaxed* problem obtained by removing the nonlinear constraints from (19). Then

Lemma 13. *If the problem $\mu_R(\ell_d, u_d)$ is infeasible then so is the problem $\mu(\ell_d, u_d)$. If $\mu_R(\ell_d, u_d)$ is feasible then $\mu_R(\ell_d, u_d) \leq \mu(\ell_d, u_d)$.*

Now we will obtain an upper bound on μ from the solution to μ_R. Suppose $(p_{ij}^*, \alpha^*, d_j^*, w_{ij}^*, w_i^*, \Phi^{+*}, \Phi^{-*})$ are feasible variables for the relaxed problem with $\alpha^* = \mu_R$. We will construct feasible variables for (19) from these variables.

Let $w_{ij}^f := p_{ij}^* d_j^*$, and let $\alpha^f = \max_i \sum_j (w_{ij}^f / d_i^*)$ and let $w_i^f := d_i^* \alpha^f$. Further suppose that the index i_0 is such that $w_{i_0 j}^f / d_{i_0}^* = \alpha^f$. Then it follows that $(p_{ij}^*, \alpha^f, d_j^*, w_{ij}^f, w_i^f, \Phi^{+*}, \Phi^{-*})$ are feasible variables for (13). Thus $\alpha^f \geq \mu(\ell_d, u_d)$. Using the above construction of α^f one can establish the following result,

Lemma 14.

$$|\alpha^f - \alpha^*| \leq \frac{\frac{1}{4}\sum_j (u_{i_0 j} - \ell_{i_0 j})(u_d - \ell_d) + \frac{1}{4}(u_\alpha - \ell_\alpha)(u_d - \ell_d)}{d_{i_0}}.$$

Note that α^* is a lower bound to $\mu(\ell_d, u_d)$ and α^f is an upper bound to $\mu(\ell_d, u_d)$. Thus $\alpha^* \leq \mu(\ell_d, u_d) \leq \alpha^f$. The above lemma gives an estimate on how good an approximation is given by α^f and α^*. Note that for any optimization on a smaller grid the optimal value $\mu(\ell_d, u_d)$ is an upper bound on μ (because the optimization for μ is on the whole variable space whereas it is only on a portion of the variable space for $\mu(\ell_d, u_d)$). Thus the upper bound obtained on $\mu(\ell_d, u_d)$ is indeed an upper bound on μ.

Suppose, the gridding is performed to obtain a mesh such that for each grid $\|u_d - \ell_d\|_\infty = \epsilon$. Then the total number of problems to be solved equals $(\frac{U_d - L_d}{\epsilon})^n$ where n is the number of d variables. Let μ^* represent the minimum of the optimal values $\mu(\ell_d, u_d)$ on each grid. Then it is clear that $\mu = \mu^*$. Let μ_R^* denote the relaxation associated with μ^*. Let α^f denote the upper bound as obtained in Lemma 14. Then it follows that $\mu^* \leq \mu \leq \alpha^f$. Also, it follows that $\alpha^f - \mu \leq$

$$\frac{\frac{1}{4}\sum_j (U_{i_0 j} - L_{i_0 j})(u_d - \ell_d) + \frac{1}{4}(\overline{U}_\alpha - L_\alpha)(u_d - \ell_d)}{d_{i_0}} \leq \epsilon \left(\frac{\frac{1}{4}\sum_j (U_{i_0 j} - L_{i_0 j}) + \frac{1}{4}(U_\alpha - L_\alpha)}{L_d} \right).$$

Thus to obtain a value with an ϵ tolerance the number of problems to be solved is in the order of $\frac{1}{\epsilon^n}$. Also for any grid associated with a given ℓ_d, u_d, if $\mu_R(\ell_d, u_d)$ is greater than any upper bound on any other grid then that grid can be discarded. This can be used as the basis for a branch and bound algorithm whereby branches corresponding to regions in the d-parameter space can be fathomed as soon as the lower bound obtained from solving the relaxed problem for these regions becomes larger than the best available global upper bound for the problem. If this does not happen, further gridding on that region is performed, and hence new branches are formed. The process continues until all the branches which have not been fathomed have lower bounds equal to (or up to a given tolerance of) the best available upper bound, which also will be the global optimal value.

7   Robust Control Example

Consider the system in Figure 3 where

$$P = \frac{z + 2}{z^2(z - 0.5)(z + 0.5)}$$

and

$$W = \frac{0.02z^2 - 0.04z + 0.02}{z^2 + 1.56z + 0.64}.$$

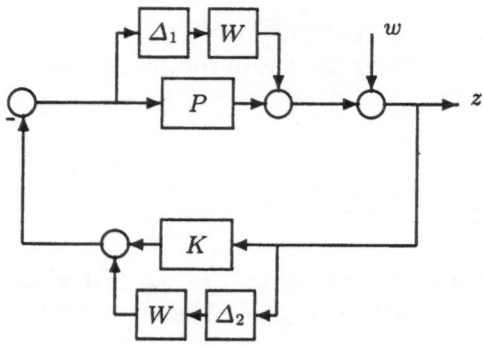

Fig. 3. Example system

The objective is to design (if possible) a controller which makes the worst case norm of the mapping from w to z less than one. This can posed as a robust stability problem by connecting z to w through a fictitious perturbation whose norm is allowed to be less than or equal to one, and then requiring the resulting system with 3 perturbation blocks to be robustly stable. The synthesis problem for that system is:

$$\inf_{D \in \mathcal{D}} \ \inf_{Q \in \ell_1} \ \left\| D^{-1} \Phi(Q) D \right\|_1 =: \gamma_* \tag{20}$$

where Φ is the impulse response matrix of the system "seen" by the three uncertainty blocks. When applying the RLT algorithm to this problem, the globally optimal value ($\pm$ 0.01) of the objective function was found to be 2.93. This was obtained after 71 iterations. After only 35 iterations the best available upper bound ($\pm$ 0.01) coincides with the global solution. However, the best available lower bound was 2.8, and additional iterations were needed only to verify that the best upper bound value cannot be improved upon. The figure below shows the branches that have been fathomed after 35 iterations (gray color). The lower bounds obtained for these branches guarantee that the global optimal cannot be achieved in that region. Further branching on the white region allows us to zoom in on the region where the global optimal is achieved.

References

1. F.A. Al-Khayyal, C. Larsen, and T. Van Voorhis, "A Relaxation Method for Quadratically Constrained Quadratic Programs," *Journal of Global Optimization*, **6**, pp. 215-230, 1995.
2. M. A. Dahleh and I. J. Diaz-Bobillo, *Control of Uncertain Systems: A Linear Programming Approach*, Prentice Hall, 1995.
3. J. C. Doyle, "Analysis of Feedback Systems with Structured Uncertainty," *IEE Proceedings*, Vol. 129, PtD, No. 6, pp. 242-250, November, 1982.

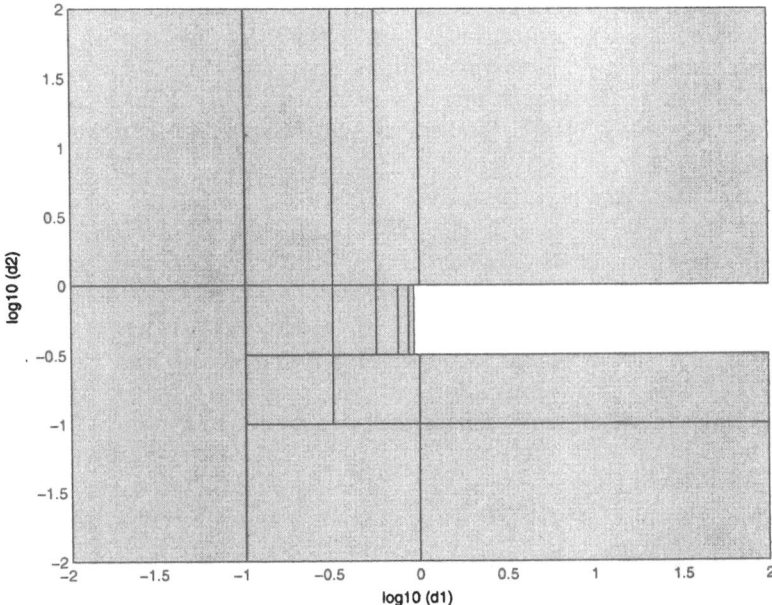

Fig. 4. Fathomed branches in the d parameter space (after 35 iterations)

4. J. C. Doyle, J. E. Wall, and G. Stein, "Performance and Robustness Analysis for Structured Uncertainty," *Proceedings of the 20th IEEE Conference on Decision and Control*, 1982, pp. 629-636.

5. M.A. Dahleh and Y. Ohta, "A Necessary and Sufficient Condition for Robust BIBO Stability," *Syst. and Contr. Lett.* **11**, 271-275, (1988).

6. M. Khammash and J. B. Pearson, "Performance Robustness of Discrete-Time Systems with Structured Uncertainty", *IEEE Transactions on Automatic Control*, vol. AC-36, no. 4, pp. 398-412, 1991.

7. M. Khammash and J. B. Pearson, "Analysis and Design for Robust Performance with Structured Uncertainty", *Systems and Control Letters*, **20** (1993) 179-187.

8. A. Megretski, "Necessary and Sufficient Conditions of Stability: A Multi-loop Generalization of the Circle Criterion," *IEEE Transactions on Automatic Control*, Vol. 38, pp. 753-756, May 1993.

9. J.S. Shamma, "Robust Stability with Time-Varying Structured Uncertainty," IEEE Transactions on Automatic Control, April 1994, pp. 714–724.

10. Mustafa H. Khammash, "Synthesis of Globally Optimal Controllers for Robust Performance to Unstructured Uncertainty," *IEEE Transactions on Automatic Control*, vol. 41, No. 4, pp. 189–198.

11. M. Khammash, "A New Approach to the Solution of the ℓ_1 Control Problem: The Scaled-Q Method," *IEEE Transactions on Automatic Control*, to appear.

12. A. Packard, K. Zhou, P. Pandey, J. Leonhardson, and G. Balas, " Optimal Constant I/O Similarity Scaling for Full-Information and State-Feedback Contro Problems," *Systems and Control Letters*, vol. 19, pp. 271-280, 1992.

13. H.D. Sherali and C.H. Tuncbilek, "A Global Optimization Algorithm for Polynomial Programming Problems Using a Reformulation-Linearization Technique," *Journal of Global Optimization*, **2**, pp. 101-112, 1992.
14. Y. Yamada, and S. Hara, "Global Optimization for H_∞ Control with Constant Diagonal Scaling," *IEEE Transactions on Automatic Control*, Vol 43, No. 2, pp. 191-203.

Optimal Control of Distributed Arrays with Spatial Invariance

Bassam Bamieh[1], Fernando Paganini[2], and Munther Dahleh[3]

[1] University of California, Santa Barbara, CA 93106, USA
[2] University of California, Los Angeles, CA 90095, USA
[3] Massachusetts Institute of Technology, Cambridge, MA 02139, USA

Abstract. We consider distributed parameter systems where the underlying dynamics are spatially invariant, and where the controls and measurements are spatially distributed. These systems arise in many applications such as the control of platoons, smart structures, or distributed flow control through the fluid boundary. For optimal control problems involving quadratic criteria such as LQR, $\mathcal{H}_2$ and $\mathcal{H}_\infty$, it is shown how to reduce the optimization to a family of problems over spatial frequency. We also show that optimal controllers have an inherent degree of decentralization, which leads to a practical distributed architecture. Under a more general class of performance criteria, a general result is given showing that optimal controllers inherit the spatial invariance structure of the plant.

1 Introduction

In most control system applications the spatially distributed aspect of the dynamics is treated as only *internal* to the system, since the control is implemented by a relatively small number of discrete actuators and sensors; for such systems a "lumped" finite state model usually suffices for control design. In contrast, this chapter concerns the situation where the spatially distributed aspect is unavoidable since it includes the input-output structure, not just the internal states; traditional examples of this include the control of vehicle platoons [24,7,29] and cross-directional control [12,22,4] in the paper processing industry. More recently, technological progress in the area of Micro Electro Mechanical Systems has raised the possibility of manufacturing large arrays of micro sensors and actuators with integrated control circuitry, as sketched in Fig. 1; such devices are already being used for fluid flow control [13,14]. This raises the possibility of control at essentially the distributed parameter level, which suddenly brings this largely mathematical theory [23,3,8] to the center stage of control design. For these systems the natural abstraction is a spatio-temporal system where all signals are indexed in both space and time [15]. Natural questions which arise are (i) how to design controllers for these systems with regard to global objectives; and (ii) how can these control algorithms be implemented in a distributed array.

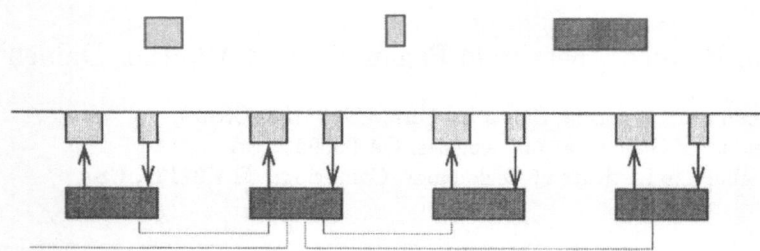

Fig. 1. Distributed control array

In this chapter we study these questions for problems with *spatial invariance*. This notion, a counterpart of time invariance, means that the dynamics are invariant with respect to translations of some spatial coordinates. One of our main results is that for spatially invariant plants, one can restrict attention to spatially invariant controllers without any loss in performance. Also, for optimal control problems involving quadratic criteria (LQR, $\mathcal{H}_2$ and $\mathcal{H}_\infty$), the infinite dimensional problem is reduced by spatial Fourier transform to a family of finite-dimensional problems over spatial frequency, which amount to solving a family of Riccati equations. We also show that the optimal infinite-dimensional controllers have an inherently semi-decentralized architecture (which we refer to as "localized"), consisting of a distributed infinite array of finite dimensional controllers with separation structure, and observer and state feedback operators which are spatial convolutions. We further show that the relevant convolution kernels have exponential rates of decay spatially, which allows us to argue that one can obtain a close to optimal control law for the array, where the communication requirements are restricted to a local neighborhood.

Our work is related to several earlier results in the literature, and we briefly mention here some connections. The earliest use of a spatial-invariance concept for control design is the work of [24] for infinite strings of vehicles, already bringing in spatial transforms (see also extensions in [7,21]). The work on discretized partial differential equations [5] can be viewed as similar to our results for the group $\mathbb{Z}_n$. This case has also appeared elsewhere in the literature in the context of block-circulant transfer function matrices [22], and is related to the "lifting" technique for N-periodic systems [19]. Other related work includes the literature on systems over rings [30,20,16,15], where stability and stabilizability were studied by means of parameterized families of finite dimensional tests, and the work of [10,11] on symmetries of linear dynamical systems, where it was shown that stabilization can always be achieved without "breaking symmetries" in the original plant.

The theory outlined in this chapter and developed in detail in [1] generalizes the above work by introducing a common framework for all such problems, extends the results based on stability to more general performance measures, and studies the localized structure of the control solutions.

2   Spatial Invariance and Diagonalization

In this section we define the class of problems under consideration and outline of the diagonalization method which can be applied to it. Our starting point is the following set of assumptions:

- The spatial variables (denoted by x) form a (locally compact, abelian) *group* $\mathbb{G}$: for example the real line $\mathbb{R}$, the circle $\mathbb{T}$, their discrete counterparts $\mathbb{Z}$ and $\mathbb{Z}_p$, and direct products of the above spaces, e.g. $\mathbb{R}^d$, $\mathbb{Z}^d$, or the cylinder $\mathbb{T} \times \mathbb{R}$. The group operation (denoted by $+$) introduces a natural notion of translation $x \mapsto x + x_o$.
- The dynamics are spatially invariant: i.e., the operators and equations describing the system commute with the translation operator $(T_{x_o} f)(x) = f(x - x_o)$, defined over functions on $\mathbb{G}$. We mostly consider $\mathcal{L}_2$ function spaces: $\mathcal{L}_2(\mathbb{G})$ is the space of square integrable functions over the group, with respect to the translation invariant (Haar) measure. Translation invariant operators are *diagonalized* by the Fourier transform $\mathcal{F}$, which associates a function $f(x)$ on $\mathbb{G}$ with a function $\hat{f}(\lambda)$ on the *dual group* $\hat{\mathbb{G}}$; for a general definition see [27], for our purposes it suffices to consider the table of standard transform pairs:

$\mathbb{G}$	$\mathbb{R}$	$\mathbb{T}$	$\mathbb{Z}$	$\mathbb{Z}_p$
$\hat{\mathbb{G}}$	$\mathbb{R}$	$\mathbb{Z}$	$\mathbb{T}$	$\mathbb{Z}_p$

The key property is that a translation invariant operator A on $\mathcal{L}_2(\mathbb{G})$ is isomorphic to *multiplication* operator $\hat{A} : \hat{f}(\lambda) \mapsto \hat{A}(\lambda)\hat{f}(\lambda)$ in the space $\mathcal{L}_2(\hat{\mathbb{G}})$.
- The actuators and sensors are fully distributed over the group; in particular for each location x and time t, we dispose of control inputs $u(x,t)$ and measurements $y(x,t)$.

A special instance of the above assumptions is the continuous time state-space description

$$\frac{\partial}{\partial t}\psi(x,t) = A\psi(x,t) + Bu(x,t), \tag{1}$$

$$y(x,t) = C\psi(x,t) + Du(x,t). \tag{2}$$

At each instant of time the signals u, y and ψ are in the vector-valued spaces $\mathcal{L}_2^p(\mathbb{G})$, $\mathcal{L}_2^m(\mathbb{G})$, and $\mathcal{L}_2^n(\mathbb{G})$ respectively. A, B, C, D are translation invariant operators on $\mathcal{L}_2(\mathbb{G})$ of appropriate dimensions, and are static. For example, if A, B, C, D are matrices whose elements are PDE operators (in x) with constant coefficients, then they are spatially invariant. In general, these operators will be unbounded, so the notion of a solution to (1) requires some care, and involves the theory of C_0 semigroups of operators; for a full discussion see [8]. The state variable $\psi(x,t)$ could be either:

- Finite dimensional for each x, t, meaning that the only distributed coordinates are those available for control. Examples of this include control of heat in a beam or surface by distributed heating/sensing, or the control of vehicle platoons.
- Infinite dimensional for each x, t, when there are other spatial coordinates to which the control has no access. An example would be the control of fluid flow by actuation on the boundary.

In this chapter we will focus on the first case, for which we will be able to provide the strongest results; generalizations are pointed out in the conclusions section.

We now point out the key observation that by taking a Fourier transform, a spatially invariant system (1-2) is *diagonalized* into the decoupled form

$$\frac{d}{dt}\hat{\psi}(\lambda, t) = \hat{A}(\lambda)\hat{\psi}(\lambda, t) + \hat{B}(\lambda)\hat{u}(\lambda, t) \tag{3}$$

$$\hat{y}(\lambda, t) = \hat{C}(\lambda)\hat{\psi}(\lambda, t) + \hat{D}(\lambda)\hat{u}(\lambda, t) \tag{4}$$

where $\hat{A}(\lambda)$, $\hat{B}(\lambda)$, $\hat{C}(\lambda)$, $\hat{D}(\lambda)$ are multiplication operators. Now, the transformed system (3-4) is in effect a decoupled family of standard finite dimensional LTI systems over the frequency parameter λ.

We now proceed to briefly discuss how some basic system-theoretic concepts can be treated with this diagonalization.

Stability. Consider first the stability of the autonomous equation

$$\frac{\partial}{\partial t}\psi = A\psi \tag{5}$$

with $\psi \in \mathcal{L}_2(\mathbb{G})$. Definitions of stability, asymptotic stability and exponential stability have been studied for such systems (see [3]), which extend, with some complications, the finite dimensional theory. In the translation invariant case, this question can be studied by means of the diagonalized system $\frac{d}{dt}\hat{\psi} = \hat{A}\hat{\psi}$. It turns out that checking exponential stability is *almost* equivalent to checking "pointwise" the stability of the decoupled systems.

Theorem 1. *If A is the generator of a strongly continuous semigroup, then the following two statements about the system (5) are equivalent:*

1. *The system is exponentially stable: there exist M, α such that*

$$\|\psi(t)\| \leq M\, e^{-\alpha t}\|\psi(0)\|, \quad t \geq 0,$$

2. *For each $\lambda \in \hat{G}$, $\hat{A}(\lambda)$ is stable, and the solution of the family of matrix Lyapunov equations*

$$\hat{A}^*(\lambda)P(\lambda) + P(\lambda)\hat{A}(\lambda) = -I$$

is bounded, i.e. $\sup_{\lambda \in \hat{G}} \|P(\lambda)\| < \infty$.

The above theorem then implies that checking exponential stability can be handled by finite dimensional tools plus a search over λ.

Stabilizability Similar statements can be made about the question of stabilizability of the system:

$$\frac{\partial}{\partial t}\psi = A\psi + Bu \tag{6}$$

The system (6) is *exponentially stabilizable* if there exists an operator $F :$ $\mathcal{D}(A) \longrightarrow \mathcal{D}(B)$ such that $A + BF$ is exponentially stable. It turns out that checking stabilizability can be done by a pointwise solution to a parameterized family of finite dimensional Riccati equations.

Theorem 2. *Let A be the generator of a C^o semigroup. Then, the system in (6) is exponentially stabilizable if and only if the following two conditions hold:*

1. *For all $\lambda \in \hat{G}$, the pair $\left(\hat{A}(\lambda), \hat{B}(\lambda)\right)$ is stabilizable.*
2. *The solution of the family of matrix Riccati equations*

$$\hat{A}^*(\lambda)P(\lambda) + P(\lambda)\hat{A}(\lambda) - P(\lambda)\hat{B}(\lambda)\hat{B}^*(\lambda)P(\lambda) + I = 0.$$

is bounded, i.e. $\sup_{\lambda \in \hat{G}} \|P(\lambda)\| < \infty$.

We remark here that the above two theorems imply that checking stability or stabilizability of spatially-invariant systems can be done by checking the same condition for the finite dimensional decoupled systems for every frequency $\lambda \in \hat{G}$. The extra condition of checking the boundedness of the solution to the Lyapunov and Riccati equation is not needed when the group $\hat{G}$ is compact (i.e. for the case of spatially discrete systems $G = \mathbb{Z}, \mathbb{Z}_n$).

3   Optimal Control with Quadratic Measures

We now turn our attention to optimal control problems for such systems. The main observation is that for quadratic-type cost criteria, the performance can be studied in the spatial Fourier domain, since by virtue of the Plancherel Theorem we have

$$\langle f, h \rangle = \int_G \langle f(x), h(x) \rangle dx = \int_{\hat{G}} \langle \hat{f}(\lambda), \hat{h}(\lambda) \rangle d\lambda = \langle \hat{f}, \hat{h} \rangle.$$

This transformation of norms leads to a diagonalization of the optimal control problem, as is now discussed.

3.1   The Distributed Linear Quadratic Regulator

We begin by studying the distributed LQR problem; there is of course an abundant literature (see [8,17]) on this problem, characterizing the optimum in terms of a solution to an operator Riccati equation, in an analogous fashion to the finite dimensional theory. Such equations are difficult to solve in general, however, and to our knowledge this has only been attempted for specific examples. What has not been emphasized in this literature is that, for the relatively large class of translation invariant problems, the problem diagonalizes exactly into a parameterized family of finite dimensional LQR problems. This is now explained; for simplicity only the infinite horizon problem is discussed.

Consider the problem of minimizing the functional

$$J = \int_0^\infty \langle Q\psi, \psi \rangle + \langle Ru, u \rangle dt \tag{7}$$

subject to the dynamics (1), and $\psi(x, 0) = \psi_0(x) \in \mathcal{L}_2(\mathbb{G})$. We assume that

- A, B, Q and R are translation invariant operators.
- Q and R are strictly positive definite operators.
- (A, B) is exponentially stabilizable.

By using spatial transforms, the problem can then be rewritten as the minimization of

$$J = \int_{\hat{\mathbb{G}}} \int_0^\infty \left(\hat{\psi}(\lambda, t)^* \hat{Q}(\lambda) \hat{\psi}(\lambda, t) + \hat{u}(\lambda, t)^* \hat{R}(\lambda) \hat{u}(\lambda, t) \right) dt \, d\lambda \tag{8}$$

subject to (3) and $\hat{\psi}(\lambda, 0) = \hat{\psi}_0(\lambda)$. Now it is clear from (8) and (3) that the problem decouples over λ, that is, it is "block-diagonal" with the blocks parameterized by λ. At a fixed λ it amounts to no more than a classical finite-dimensional LQR problem. Therefore the unique solution to this problem is achieved by the translation invariant state feedback $u = -R^{-1}B^*Px$, where $\hat{P}(\lambda)$ is the positive definite solution to the parameter-dependent algebraic Riccati equation

$$\hat{A}(\lambda)^* \hat{P}(\lambda) + \hat{P}(\lambda)\hat{A}(\lambda) + \hat{Q}(\lambda) - \hat{P}(\lambda)\hat{B}(\lambda)\hat{R}(\lambda)^{-1}\hat{B}(\lambda)^* \hat{P}(\lambda) = 0 \tag{9}$$

for $\lambda \in \hat{\mathbb{G}}$. The main observation here is that when A, B, Q, R are translation invariant operators, then the solution to the operator ARE in the LQR problem is also a translation invariant operator. The exact conditions under which this yields a stabilizing controller are:

Theorem 3. *Consider the LQR problem (7), (1), where A, B, Q, R are translation invariant operators, with $\hat{R}(\lambda) > 0$, $\hat{Q}(\lambda) \geq 0$. If (A, B), and $(A^*, Q^{\frac{1}{2}})$ are exponentially stabilizable, then*

1. *The solution to the family of matrix ARE's in (9) is uniformly bounded, i.e. $\sup_{\lambda \in \hat{\mathbb{G}}} P(\lambda) < \infty$.*

2. The translation invariant feedback operator $K = -R^{-1}B^*P$ is expo-
   nentially stabilizing.

Notice that the optimal controller is static, and is obtained in the spatial Fourier domain by $\hat{K}(\lambda) = -\hat{R}^{-1}(\lambda)\hat{B}^*(\lambda)\hat{P}(\lambda)$. While this is closely related to solving LQR problems by so-called modal decomposition, we should note that we do not advocate modal *truncation* as away of obtaining finite-dimensional approximations.

Instead, we will proceed further and analyze the properties of the resulting feedback operators in the original spatial variables. Inverting the transform, the controller becomes a static spatial convolution of the form

$$u(x,t) = K(x) *_x \psi(x,t).$$

Since truncations in $K(x)$ are directly linked to spatial localization (discussed below), this will yield a more natural approximation scheme than modal truncation.

3.2 $\mathcal{H}_2$ and $\mathcal{H}_\infty$ Control

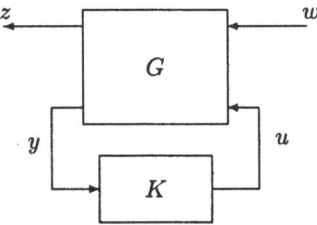

Fig. 2. Optimal disturbance rejection problem

The diagonalization method extends to the study of optimal disturbance rejection problems of the general form depicted in Fig. 2, as long as the performance measures are based on quadratic norms, as in the $\mathcal{H}_2$ or $\mathcal{H}_\infty$ ($\mathcal{L}_2$-induced) optimal control problems.

More specifically, let the plant G in the above figure be a linear, space/time invariant distributed system, which admits a state-space representation

$$\frac{\partial}{\partial t}\psi(x,t) = A\psi(x,t) + B_1w(x,t) + B_2u(x,t),$$
$$z(x,t) = C_1\psi(x,t) + \qquad\qquad D_{12}u(x,t),$$
$$y(x,t) = C_2\psi(x,t) + D_{21}w(x,t).$$

The feedback K, which is also distributed, must internally (exponentially) stabilize the system and minimize a certain norm of the closed loop map T_{zw}. As will be shown in more generality in Section 5, no performance loss occurs by restricting the design to controllers which are themselves space/time invariant. Under these circumstances, the closed loop $H := T_{zw}$ is a space/time invariant system, which can be represented by the two-variable transfer function

$$H(\lambda, s) = \hat{C}(\lambda)(sI - \hat{A}(\lambda))^{-1}\hat{B}(\lambda) + \hat{D}(\lambda)$$

where s denotes Laplace transform, and λ is spatial frequency.

Under these circumstances we can define the $\mathcal{H}_\infty$ norm as

$$\|H\|_\infty := \sup_{\lambda \in \hat{G}, \omega \in \mathbb{R}} \bar{\sigma}(\hat{H}(\lambda, j\omega)),$$

which also corresponds to the $\mathcal{L}_2$-induced operator norm, and the $\mathcal{H}_2$ norm

$$\|H\|_2^2 = \frac{1}{2\pi} \int_{\hat{G}} \int_{-\infty}^{\infty} \text{trace}(\hat{H}(\lambda, j\omega)^* \hat{H}(\lambda, j\omega)) \; d\lambda \; d\omega,$$

which can be interpreted as measure of the system response to white noise in multiple dimensions. We remark that in both cases the Hardy space (analytic) structure is with respect to the temporal-frequency variables alone.

The distributed parameter systems literature contains versions of the $\mathcal{H}_2$ and $\mathcal{H}_\infty$ problems, and solutions in terms of operator equations; once again, however, if we bring in the special structure of spatially invariant problems, we can exploit the frequency domain to obtain a substantial simplification in the solution.

In particular, for the $\mathcal{H}_2$ problem we have that under suitable technical assumptions [1], the optimal controller is described in the spatial frequency domain by

$$\frac{\partial}{\partial t}\hat{\psi}_\kappa(\lambda, t) = \left[\hat{A}(\lambda) + \hat{B}_2(\lambda)\hat{F}(\lambda) + \hat{L}(\lambda)\hat{C}_2(\lambda)\right] \hat{\psi}_\kappa(\lambda, t) - \hat{L}(\lambda)\hat{y}(\lambda, t)$$

$$\hat{u}(\lambda, t) = \hat{F}(\lambda)\hat{\psi}_\kappa(\lambda, t), \tag{10}$$

where $\hat{F}(\lambda) = -\hat{B}_2^*(\lambda)\hat{P}_1(\lambda)$ and $\hat{L}(\lambda) := -\hat{P}_2(\lambda)\hat{C}_2^*(\lambda)$, and $\hat{P}_1(\lambda)$, $\hat{P}_2(\lambda)$ are the solutions of suitable Riccati equations at each spatial frequency. In particular at each λ we recover the familiar structure of a state observer combined with state feedback. Once again, however, we will return to the original domain for issues of implementation, as discussed in the next section.

In a similar manner, the search for a controller achieving $\|T_{zw}\|_\infty < \gamma$ reduces to a parametrized family of Riccati equations, of the expected form (see [1]).

4   The Structure of Quadratically Optimal Controllers

The optimal controllers obtained in Section 3 have the following attractive features:

- They provide *global* performance guarantees. In particular, they will ensure overall stability.
- They can be effectively computed by a family of low dimensional problems across spatial frequency.

However we have not considered the issue of a implementation of the control algorithm. In this regard, rather than a highly complex *centralized* controller with information from all the distributed array, it would be desirable to have distributed intelligence as in Fig. 1, where each actuator runs a local algorithm with information from the neighboring sensors. In this section we analyze the optimal schemes from this perspective. Relevant questions are:

- (Section 4.1) Does the control law lend itself to a distributed architecture?
- (Section 4.2) To what degree is information from far away sensors required? Notice that this pertains to approximate diagonalization in the *original* spatial variables.

4.1   Local Controller Architecture

We now illustrate the surprisingly intuitive and appealing architecture of quadratically optimal controllers (by this we mean the LQR, $\mathcal{H}_2$ and the "central" $\mathcal{H}_\infty$ controller).

First note that the since the ARE solutions for all three problems are translation-invariant operators, then their controllers are spatially-invariant systems; in particular, the same algorithm must be run at each actuator location, the influence of each sensor depending on its position relative to the actuator.

To understand the structure of this algorithm, let us examine more closely the optimal $\mathcal{H}_2$ controller, obtained by inverse transform of the equations (10):

$$\frac{\partial}{\partial t}\psi_K(x,t) = A\psi_K(x,t) + B_2 u(x,t) + L(x) *_x [y(x,t) - C_2\psi_K(x,t)]$$

$$u(x,t) = F(x) *_x \psi_K(x,t).$$

The above implies the following structure of the optimal controller:

(a) A *distributed estimator* whose local state is $\psi_K(x,t)$. To propagate this state, one needs to know the outputs of neighboring estimators, and convolve the prediction errors with the kernel $L(x)$ (the size of this neighborhood is determined by the spread of L). We note that at a given $x_o \in \mathbb{G}$, the local controller state $\psi_K(x_o,t)$ has a physical interpretation; it is the estimate of the system's local state $\psi(x_o,t)$.

(b) The feedback at position x is given by $u(x,t)$ which is computed by convolving neighboring state estimates with the kernel $F(x)$ (the size of this neighborhood is determined by the spread of F).

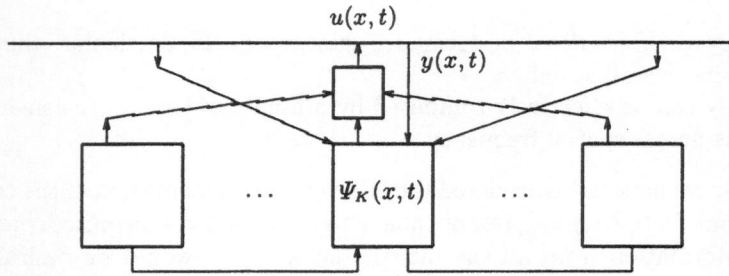

Fig. 3. Controller Architecture

Fig. 3 gives a pictorial illustration of this controller structure. Thus the optimal laws are directly amenable to a distributed implementation, with localized actuation and information passing. What determines the degree of localization, and thus the communication burden for the array, is the spread of the convolution operators L and F. Note that the system operators B_2 and C_2 are typically PDE operators, therefore localized. However in general the Riccati solutions P_1 and P_2 will not be differential operators ($\hat{P}_1(\lambda)$ and $\hat{P}_2(\lambda)$ are not rational in general, see the example in Section 4.2), and their convolution kernels will have a spread, reflecting the need of information passing within the array.

In the next subsection we will provide means of evaluating the spread of P_1 P_2; in particular we will see that these convolution kernels decay exponentially in space; thus the optimal control laws have an inherent degree of decentralization. From a practical perspective, the convolution kernels can be truncated to form "local" convolution kernels that have performance close to the optimal, and preserve the appealing architecture described above.

4.2 The Degree of Spatial Localization

We will study the localization issue for systems with unbounded spatial domains; for these we can ask the question of how the controller gains decay as we move away in space. For concreteness focus on the case $\mathbb{G} = \mathbb{R}$ (and $\hat{\mathbb{G}} = \mathbb{R}$), analogous ideas apply to the discrete case $\mathbb{G} = \mathbb{Z}$. We first consider an example of LQR optimization.

Example Consider the heat equation on an infinite bar with distributed heat injection

$$\frac{\partial}{\partial t}\psi(x,t) = c\frac{\partial^2}{\partial x^2}\psi(x,t) + u(x,t)$$

Here the group $\mathbb{G}$ is the real line. The standard Fourier transform yields the transformed system

$$\frac{d}{dt}\hat{\psi}(j\lambda,t) = -c\lambda^2\hat{\psi}(j\lambda,t) + \hat{u}(j\lambda,t)$$

Note that we are writing the Fourier transform as a function on the imaginary axis $j\mathbb{R}$; this system-theoretic notation will be useful when considering analytic continuation issues below.

Taking for example $Q = qI$ (multiple of the identity) and $R = I$ in the LQR cost (7), the corresponding (scalar) parameterized Riccati equation is

$$-2c\lambda^2 \hat{p}(j\lambda) - \hat{p}(j\lambda)^2 + q = 0$$

which has a positive solution

$$\hat{p}(j\lambda) = -c\lambda^2 + \sqrt{c^2\lambda^4 + q}.$$

An inverse Fourier transform would yield the convolution kernel $k(x) = -p(x)$ of the optimal state feedback $u = K\psi$. Note that even though the system and the cost are rational in λ, the optimal control is irrational. This in particular implies that it cannot be implemented by a "completely localized" PDE in x and t, but it must look at distant points to compute a spatial convolution,

$$u(x, t) = \int_{\mathbb{R}} k(x - \zeta)\psi(\eta) \, d\zeta.$$

Thus the spatial decentralization question is directly related to the decay rate of $p(x)$ as a function of x. It turns out that this can be studied by *analytic continuation* of the Fourier transform $\hat{p}(j\lambda)$ into the complex plane. In particular, $\hat{p}(j\lambda)$ can be extended to the function

$$\hat{p}(\sigma) = c\sigma^2 + \sqrt{c^2\sigma^4 + q}$$

of $\sigma \in \mathbb{C}$, which is analytic in a region of the complex plane which avoids the four branch cuts shown by the diagonal lines in Fig. 4. Thus, the Fourier transform of $\hat{p}$ can be analytically extended to the strip

$$\left\{ \sigma \in \mathbb{C} : |Re(\sigma)| < \frac{\sqrt{2}}{2} \left(\frac{q}{c^2} \right)^{\frac{1}{4}} \right\}.$$

Now, by Laplace transform theory this implies that $p(x)$ decays exponentially, i.e. that there exists $M > 0$ such that

$$|k(x)| \leq M e^{-\alpha|x|}, \quad \text{for any } \alpha < \frac{\sqrt{2}}{2} \left(\frac{q}{c^2} \right)^{\frac{1}{4}}.$$

Since $\{k(x)\}$ decays exponentially with $|x|$, it can be truncated to form a "localized" feedback convolution operator whose closed loop performance is close to the optimal. We note that in this particular problem, and interesting tradeoff seems to be in place: in the limit of "cheap" control (i.e. $q \longrightarrow \infty$) the analyticity region grows and the controller becomes more decentralized. It seems thus possible that there is an inherent tradeoff between actuator authority and controller decentralization.

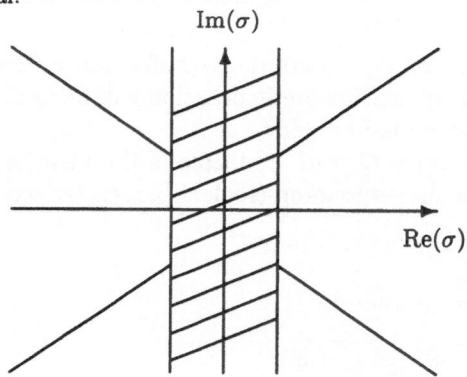

Fig. 4. Analytic continuation region for $\hat{p}(j\lambda) = -c\lambda^2 + \sqrt{c^2\lambda^4 + q}$

Given this example, we immediately inquire to what extent these conclusions generalize; this problem is studied in [1,26]. Given a Riccati equation over λ with solution $\hat{P}(\lambda)$, the main question is whether a suitable analytic continuation to a strip $|Re(\sigma)| \leq \alpha$ can be obtained. Conditions under which this holds are given in [1,26], as well as a method of bounding the parameter α (which measures the exponential decay).

5   A General Result on the Spatial Invariance of Optimal Controllers

In this section we divert from our $\mathcal{L}_2$ set up and consider problems for general $\mathcal{L}_p$ induced norms. We have seen that quadratically optimal controllers for spatially invariant systems are themselves spatially invariant. We will ask a similar question for more general $\mathcal{L}_p$ induced norms: given a spatially-invariant generalized plant, is the optimal controller necessarily spatially invariant. Fortunately, the answer is yes, and this implies a significant reduction in the complexity of the control design problem.

To state the result precisely, consider a set up in terms of the "standard problem" of robust control, as shown in Figure 2. All signals w, z, u, y are vector-valued signals indexed over the same group $\mathbb{G}$. The objective in this problem is to find a stabilizing controller that minimizes the $\mathcal{L}_p(\mathbb{G})$-induced norm from from w to z. For reference we note here that the $\mathcal{L}_p(\mathbb{G})$ norm is defined as

$$\|w\|_p := \left(\int_\mathbb{G} \int_\mathbb{R} |w_x(t)|^p dt\, dx \right)^{\frac{1}{p}} = \left(\int_\mathbb{G} \|w\|_p^p dx \right)^{\frac{1}{p}}.$$

In the usual notation, we will refer to the closed loop system in Fig. 2 by $T_{zw}(K)$. For any given p, the input-output sensitivity of the system is given by the $\mathcal{L}_p$ induced norm:

$$\|T_{zw}(K)\|_{p-i} := \sup_{w \in \mathcal{L}_p} \frac{\|z\|_p}{\|w\|_p}.$$

If the controller is stabilizing the above worst case gain will be finite.

Let LSI and LSV be the classes of Linear Spatially Invariant and Linear Spatially Varying (not necessarily stable) systems respectively. let us define the following two problems:

$$\gamma_{si} := \inf_{\substack{\text{stabilizing } K \\ K \in LSI}} \|T_{zw}(K)\|_{p-i};$$

$$\gamma_{sv} := \inf_{\substack{\text{stabilizing } K \\ K \in LSV}} \|T_{zw}(K)\|_{p-i},$$

which are the best achievable performances with LSI and LSV controllers respectively. We now can state the main result of this section:

Theorem 4. *If the generalized plant G is spatially invariant and has at least one spatially-invariant stabilizing controller, then the best achievable performance can be approached with a spatially-invariant controller. More precisely*

$$\gamma_{si} = \gamma_{sv}.$$

The above result is reminiscent of questions related to time-varying versus time-invariant compensation [18,28,6], where it was shown that for linear time invariant plants and induced norm performance objectives, time-varying controllers offer no advantage over time-invariant ones. In fact a similar technique of proof to that of [28,6] (which used averaging over time) can be applied here, since it only needs a group structure. For details see [1].

6 Conclusion

In this chapter we have outlined a framework for the study of control systems in which sensing and actuation are distributed over a spatial coordinate, with the respect to which the plant dynamics are invariant. We have focused the discussion on systems with *fully* distributed sensing/actuation, in the sense that all spatial variables are covered in the distribution and the remaining state is finite dimensional. The main conclusions extend, however, to the case where additional distributed variables exist, as in the case of boundary fluid control. For these systems, (i) Theorem 4 applies with no change, and (ii) the spatial Fourier method can still be used to diagonalize the optimal control problem in the case of quadratic criteria. The only difference is that in the latter case the problem at each spatial frequency remains infinite dimensional, and must be studied by some additional approximation. Nevertheless, diagonalization yields a significant reduction in complexity.

We conclude by mentioning some other related work and some open directions. In [9], a similar class of problems is tackled by using Linear Matrix Inequality tools for multi-dimensional and parameter dependent control synthesis; the resulting solution has an attractive localized architecture. In [25], a recursive information flow system is proposed to implement the truncated control laws of this paper, and the resulting problem of combined communications and control is analyzed. One important open question is the study of how to affect *by design* the level of localization. Another general research direction involves the consideration of finite geometries and boundary effects which often violate the strict definition of spatial invariance.

References

1. Bamieh B., Paganini F., Dahleh M.A., "Distributed Control of Spatially-Invariant Systems", to appear in *IEEE Trans. on Autom. Control*.
2. H.T. Banks, R.S. Smith and Y. Wang, *Smart Material Structures: Modeling, Estimation and Control*, Wiley, New York, 1996.
3. S. P. Banks, *State-Space and frequency-domain methods in the control of distributed parameter systems*, London, UK: Peter Peregrinus Ltd.
4. R.D. Braatz and J.G. VanAntwerp, "Robust Cross-Directional Control of Large Scale Paper Machines", in *Proc. 1996 IEEE Intl. Conf. Control Applications*, Dearborn, MI 1996.
5. R.W. Brockett and J.L. Willems, "Discretized Partial Differential Equations: Examples of Control Systems Defined on Modules", in *Automatica*, vol. 10, pp. 507-515, 1974.
6. H. Chapellat and M. Dahleh, "Analysis of time-varying control strategies for optimal disturbance rejection and robustness," *IEEE Transactions on Automatic Control*, vol. AC-37, November 1992.
7. K. C. Chu, "Decentralized control of high-speed vehicular strings," *Transportation Science*, pp. 361–384, November 1974.
8. R.F. Curtain and H. Zwart, *An introduction to infinite-dimensional linear systems theory*, Texts in applied mathematics, Vol 21, New York : Springer-Verlag, c1995.
9. D'Andrea R., "A Linear Matrix Inequality Approach to Decentralized Control of Distributed Parameter Systems", *Proceedings 1998 ACC*, Philadelphia, PA.
10. F. Fagnani and J.C. Willems, "Representations of Symmetric Linear Dynamical Systems", in *SIAM J. Control Optimization*, vol. 31, no. 5, pp. 1267-1293, September 1993.
11. F. Fagnani and J.C. Willems, "Interconnections and Symmetries of Linear Differential Systems".

12. E.M. Heaven, I.M. Jonsson, T.M. Kean, M.A. Manness and R.N. Vyse, "Recent Advances in Cross Machine Profile Control", in *IEEE Control Systems Magazine*, vol. 14, No. 5, October 1994.

13. C. Ho and Y. Tai, "REVIEW: MEMS and its applications for flow control," *ASME Journal of Fluid Engineering*, vol. 118, September 1996.

14. S. S. Joshi, J. L. Speyer, and J. Kim, "A system theory approach to the feedback stabilization of infinitesimal and finite-amplitude disturbances in plane poiseuille flow." To appear in *Journal of Fluid Mechanics*, 1997.

15. E. W. Kamen, "Stabilization of linear spatially-distributed continuous-time and discrete-time systems," in *Multidimensional Systems Theory* (N. K. Bose, ed.), Hingham, MA: Kluwer, 1985.

16. E. W. Kamen and P. P. Khargonekar, "On the control of linear systems whose coefficients are functions of parameters," *IEEE Transactions on Automatic Control*, vol. AC-29, no. 1, pp. 25–33, 1984.

17. B. van Keulen, $\mathcal{H}_\infty$-*Control for Distributed Parameter Systems*, Birkhauser, 1993.

18. P. Khargonekar and K. Poolla, "Uniformly optimal control of linear time-invariant plants: Nonlinear time-varying controllers", in *Systems and Control Letters*, vol. 6, pp. 303–308, Jan. 1986.

19. P.P. Khargonekar, K. Poolla and A. Tannenbaum, 'Robust control of linear time-invariant plants using periodic compensation', *IEEE Trans. Aut. Control*, Nov. 1985, v.30, No.11.

20. P. P. Khargonekar and E. Sontag, "On the relation between stable matrix fraction factorizations and regulable realizations of linear systems over rings," *IEEE Transactions on Automatic Control*, vol. AC-27, no. 3, pp. 627–638, 1982.

21. M. L. El-Sayed and P.S. Krishnaprasad (1981) "Homogeneous Interconnected Systems: An Example", *IEEE Trans. Aut. Control*, Vol AC-26, pp 894-901.

22. D. Laughlin, M. Morari and R.D. Braatz, "Robust Performance of Cross-Directional Basis-Weight Control in Paper Machines", in *Automatica*, vol. 29, pp. 1395-1410, 1993.

23. J.L. Lions, *Optimal Control of Systems Governed by Partial Differential Equations*, translated by S.K. Mitter, Springer-Verlag, New York, 1971.

24. S. M. Melzer and B. C. Kuo, "Optimal regulation of systems described by a countably infinite number of objects," *Automatica*, vol. 7, pp. 359–366, 1971.

25. Paganini F., "A Recursive Information Flow System for Distributed Control Arrays", to appear in Proc. 1999 American Control Conference.

26. Paganini F., Bamieh B., "Decentralization Properties of Optimal Distributed Controllers", submitted to the *37'th CDC*, 1998.

27. W. Rudin, *Fourier Analysis on Groups*. New York, NY: Interscience-Wiley, 1962.

28. J. Shamma and M. Dahleh, "Time-varying versus time-invariant compensation for rejection of persistent bounded disturbances and robust stabilization," *IEEE Transactions on Automatic Control*, vol. AC-36, pp. 838–748, July 1991.

29. B. Shu and B. Bamieh, "Robust H_2 control of vehicular strings." Submitted to *ASME J. Dynamics, Measurement and Control*, 1996.

30. E.D. Sontag, "Linear Systems over Commutative Rings: a Survey", *Richerche di Automatica*, Vol. 7, 1, July 1976.

Numerical Search of Stable or Unstable Element in Matrix or Polynomial Families: A Unified Approach to Robustness Analysis and Stabilization*

Boris T. Polyak and Pavel S. Shcherbakov

Institute of Control Science, Profsojuznaya 65, Moscow 117806, Russia

Abstract. In this paper, we develop a new approach to robustness analysis and design under parametric uncertainty. The heart of the approach is an iterative procedure of non-smooth optimization, which is aimed at moving the zeros of a parameter-dependent polynomial toward a prescribed domain. The method is based on the first-order and second-order formulae for zeros of perturbed polynomials and, hence, deals directly with perturbed zeros rather than coefficients. The salient feature of the method is that it applies to a broad class of problems involving any differentiable dependence of the coefficients of polynomials on the vector of parameters. A similar method based on the same ideas is developed for families of matrices. An immediate application to the control theory leads to solution of a number of fundamental problems: Robust stability of polynomial and matrix families; Maximal degree of stability; Stabilization via low-order controllers; Simultaneous stabilization; Robust stabilization. We illustrate the efficacy of the approach by several numerical examples.

1 Introduction

In this paper we deal with families of parameterized polynomials $p(s, q)$ or matrices $A(q)$, where $q \in \mathbb{R}^{\ell}$ is the vector of parameters, which will also be referred to as the vector of uncertainties. Mathematically, we pose the following two problems:

RM *Robustness Margins.* Let the "nominal" polynomial $p(s, 0)$ be stable and let $\|q\| \leq \gamma$ for some vector norm $\|\cdot\|$ and $\gamma > 0$; we write $q \in Q_{\gamma}$, where Q_{γ} is a ball of radius γ in the norm $\|\cdot\|$. Find the robustness margin

$$\gamma_{\max} \doteq \max\{\gamma : p(s, q) \text{ is stable } \forall q \in Q_{\gamma}\}.$$

---

* The work was supported by grants RFFI 96-01-00993 and INTAS IR-97-0782.

ST *Stabilization.* Suppose the "nominal" polynomial $p(s, 0)$ is unstable. Detect if there exists a $q \in \mathbb{R}^{\ell}$ such that $p(s, q)$ is stable; if the answer is positive, find a stabilizing value of q (possibly the smallest).

The same problems are posed for families of matrices.

Said another way, the RM problem is: Given a *stable* element in the family, find the closest unstable element and the distance between them. Similarly, the ST problem is: Given an *unstable* element in the family, find if there exist stable elements (possibly, a closest stable).

Many classical problems in control theory reduce to the two formulations above; the RM setup relates to the robustness analysis, while the ST setup relates to design. These include Static output feedback stabilization; Stabilization via low-order controllers; Simultaneous stabilization; Stability radius of matrices; Robust stability of polynomials, etc.

The approach developed in this paper is based on the perturbation theory for polynomials and matrices. With this approach, both RM and ST problems are treated on a uniform basis although we shall see that there is an intrinsic difference between them. The method deals directly with perturbations of zeros or eigenvalues rather than with those of coefficients, which is more adequate and provides the method with more flexibility. The approach applies to both statements for various types of dependence on the vector of parameters and various vector norms. The real case $q \in \mathbb{R}^{\ell}$ and the complex case $q \in \mathbb{C}^{\ell}$ are equally treatable as well as Hurwitz stability and Schur stability.

To the authors' knowledge, quite a few papers are attempting at similar approaches. In the pioneering works by E. Polak and coauthors (see, e.g. [1]), a method of nondifferentiable optimization was developed for control systems design with frequency-dependent singular value constraints and performance index. In contrast, we deal directly with frequency-independent eigenvalues and unconstrained problems. In [1], the authors did not distinguish between the ST and RM problems, and their method is inapplicable to the RM setup; apart from that, polynomial families are not considered as special case. Nevertheless, the works by E. Polak made a significant impact to the development of optimization-based numerical methods for design problems.

In [2], a linear programming (LP) approach was proposed for solving ST problems; the constraints in the LP problem were formulated in terms of the coefficients of polynomials, and the objective was to attain a target interval polynomial having a priori specified properties. The technique is not based on perturbation theory for zeros and eigenvalues.

2   Robust Stability of Polynomial Families

This section deals with the RM statement, and we propose an iterative method for finding an upper bound for the robust stability margin for families of polynomials.

2.1   Perturbations of zeros of polynomials

We first present a formula for the linear approximation of perturbed zeros of polynomials.

Proposition $\mathcal{P}$. *Let* $p(s, q)$ *be a polynomial in* s, *which depends on the vector* $q = (q_1, \ldots, q_\ell)^{\mathrm{T}} \in Q \subseteq \mathbb{R}^\ell$ *of parameters, and* $\deg p(s, q) = n = \mathrm{const}(q)$. *Suppose* $p(s, q)$ *is differentiable with respect to* q *at the point* $q = 0$ *and denote*

$$\pi_i(s) = \left. \frac{\partial p(s, q)}{\partial q_i} \right|_{q=0}, \qquad i = 1, \ldots, \ell. \tag{1}$$

Let $s_k \doteq s_k(0)$ *denote a simple zero of the polynomial* $p_0(s) \doteq p(s, 0)$. *Then for sufficiently small* q *there exists a zero* $s_k(q)$ *of the polynomial* $p(s, q)$ *such that*

$$s_k(q) = s_k - \frac{(\pi^k, q)}{r_k} + o(q), \tag{2}$$

where

$$\pi^k = (\pi_1^k, \ldots, \pi_\ell^k)^{\mathrm{T}}, \qquad \pi_i^k = \pi_i(s_k), \qquad r_k = p_0'(s)\big|_{s=s_k}. \tag{3}$$

These relationships are a particular case of Proposition $\mathcal{M}$ (see Sec. 3); they can be obtained by straightforward expanding in Taylor series. The polynomial $p_0(s) = p(s, 0)$ is called nominal, while $p(s, q)$ is referred to as a perturbed one.

Relationship (2) particularizes to some types of dependence on q, commonly encountered in control theory:

Affine. Let $p(s, q) = p_0(s) + \sum_{i=1}^\ell q_i p_i(s)$, then $\pi_i(s) = p_i(s)$, and (2) yields $s_k(q) \approx s_k - (p_i(s_k), q)/p_0'(s_k)$.

Multilinear. Let $p(s, q) = 1 + \prod_{j=1}^\ell (1 + (q_i - q_i^0)s)$, where $q_i^0 > 0$ are some nominal values of the parameters (assuming $p(s, 0)$ is stable). Then $\partial p(s, q)/\partial q_i|_{q_i=0} = s \prod_{j \neq i}(1 - q_j^0 s)$ so that $\pi_i^k = -s_k(1 - p_0(s_k))/(1 - q_i^0 s_k)$.

2.2   Stability conditions

Assume that $p(s, 0)$ is Hurwitz stable, i.e., $\max_k \mathrm{Re}\, s_k < 0$. Based on the linear approximation $s_k(q) \approx s_k - (\pi^k, q)/r_k$ given by (2), we formulate first-order conditions for a perturbed polynomial $p(s, q)$ to be stable. For a given q, we require $\mathrm{Re}\, s_k(q) < 0$ for all k:

$$\mathrm{Re}\, s_k - \mathrm{Re} \frac{(\pi^k, q)}{r_k} < 0 \quad \forall\, k,$$

or in a compact form:

$$\max_k \left(\mathrm{Re}\, s_k + (u^k, q) \right) < 0, \qquad u_i^k \doteq -\mathrm{Re}(\pi_i^k/r_k), \quad i = 1, \ldots, \ell. \tag{4}$$

If Schur stability is meant, that is, $\max_k |s_k| < 1$, then instead of (4) we have

$$\max_k |s_k + (u^k, q)| < 1, \qquad \text{where } u_i^k = -\pi_i^k/r_k, \quad i = 1, \dots, \ell.$$

2.3   Robust stability

Assume now that $p(s, 0)$ is stable and let $\| \cdot \|$ be some vector norm on $\mathbb{R}^\ell$; the goal is to find the maximal value of $\gamma > 0$ such that $p(s, q)$ is stable for all $\|q\| \leq \gamma$. We thus formulated the problem of finding the robustness margin $\gamma_{\max}$ for the polynomial family $p(s, q)$, $q \in \gamma Q$, where $Q \subset \mathbb{R}^\ell$ is a unit ball in $\| \cdot \|$-norm:

$$\gamma_{\max} = \max\{\gamma : \max_k \operatorname{Re} s_k(q) < 0 \text{ for all } q \in \gamma Q\}. \tag{5}$$

With (4) in hand, this leads to the following estimate for $\gamma_{\max}$ based on linear approximation (2):

$$\gamma_{\max} \approx \overline{\gamma} = \max\left\{\gamma : \max_k \max_{q \in Q}\left(\operatorname{Re} s_k + \gamma(u^k, q)\right) < 0\right\},$$

or, equivalently:

$$\overline{\gamma} = \max\left\{\gamma : \max_k\left(\operatorname{Re} s_k + \gamma \max_{q \in Q}(u^k, q)\right) < 0\right\}, \quad u_i^k \doteq -\operatorname{Re}(\pi_i^k/r_k). \tag{6}$$

The innermost max is computable for commonly used norms, and we have closed-form solutions to problem (6):

l_∞-**norm (interval uncertainties).** Let $|q_i| \leq 1$, then $\max_{q \in Q}(u^k, q) = \sum_{i=1}^\ell |u_i^k|$ and $\overline{q}_i = \operatorname{sign}(u_i^k)$, where $\overline{q} = (\overline{q}_1, \dots, \overline{q}_\ell)$ is the maximizer. From (6) we obtain $\overline{\gamma} = \min_k\left(-\operatorname{Re} s_k/\sum_{i=1}^\ell |u_i^k|\right)$, $\overline{q}_i = \overline{\gamma}\operatorname{sign}(u_i^m)$, where $m = \arg\min_k\left(-\operatorname{Re} s_k/\sum_{i=1}^\ell |u_i^k|\right)$.

Euclidean norm (spherical uncertainties). Let $\|q\| = (\sum_{i=1}^\ell q_i^2)^{1/2} \leq 1$, then $\max_{q \in Q}(u^k, q) = \|u^k\|$, $\overline{q} = u^k/\|u^k\|$, and from (6) we obtain $\overline{\gamma} = \min_k(-\operatorname{Re} s_k/\|u^k\|)$ and $\overline{q} = \overline{\gamma} u^m/\|u^m\|$, where $m = \arg\min_k\left(-\operatorname{Re} s_k/\|u^k\|\right)$.

Similar expressions can be obtained for other l_p norms and for the case of Schur stability.

2.4   Upper bound for the robustness margin

Problem (5) stated in Sec. 2.3 can be reformulated in the following way. We introduce the function $\eta(q) = \max_k \operatorname{Re} s_k(q)$ and note that (5) is equivalent to finding a minimum norm solution to the problem $\eta(q) \geq 0$:

$$\text{Find } \min \|q\| \text{ subject to } \eta(q) \geq 0.$$

We propose an iterative method for solving this problem by using linearized conditions of stability (4). Namely, instead of solving problem (5) for $\eta(q)$, we solve problem (6) for the linearized function $\overline{\eta}(q) = \max_k(\mathrm{Re}\, s_k + (u^k, q))$ (4) and take its solution $\overline{q}$ as a first-step approximation to the solution of (5). The next step is performed with $p(s, 0)$ being changed for $p(s, \overline{q})$ and so forth. We formulate the algorithm for the case of Hurwitz stability; it is assumed that $\deg p(s, q) = n$, $q \in \mathbb{R}^\ell$.

2.5   Algorithm I

Step 1. $\widehat{q} = 0$.

Step 2. Compute zeros s_k of $p(s, 0)$ and the vectors u^k according to (1), (3), (4).

Step 3. Compute $h^k \doteq \max\limits_{\|q\| \le 1}(u^k, q)$ and the maximizer $q^k \doteq \arg \max\limits_{\|q\| \le 1}(u^k, q)$.

Compute $\gamma \doteq \min\limits_k(-\mathrm{Re}\, s_k / h^k)$ and $m = \arg\min_k(-\mathrm{Re}\, s_k / h^k)$; take q^m as the direction and γ as the step-size: $dq \doteq \gamma \cdot q^m$.

Step 4. If $p(s, dq)$ is stable, change variables $q := q - dq$ (i.e., adopt $p(s, dq)$ as a "new" $p(s, 0)$) and put $\widehat{q} := \widehat{q} + dq$. Go to Step 2.

Otherwise find $\alpha_{\min} = \min\{\alpha > 0 : p(s, \alpha \cdot dq) \text{ is unstable}\}$ and put $\widehat{q} := \widehat{q} + \alpha_{\min} \cdot dq$.

Step 5. Adopt $\overline{\gamma} \doteq \|\widehat{q}\|$ as an upper bound for the robustness margin.

At each step, the algorithm seeks the minimum norm increment dq of the vector of parameters that vanishes the maximal real part of linear approximation (2) to zeros of $p(s, q)$. That is, the zeros of $p(s, q)$ are shifted toward the right-half plane iteratively in q.

Geometric interpretation: At the j-th step, we solve the linearized problem (6) at the point $q^{(j-1)}$ ($\widehat{q}$ at Steps 2-3 of the Algorithm) and obtain $\gamma^{(j)}$ ($\overline{\gamma}$ at Step 3). We then build a ball of radius $\gamma^{(j)}$ in the norm $\|\cdot\|$ centered at $q^{(j-1)}$ and check if it is entirely inside the stability domain $\mathcal{D}$ for the actual problem (Step 4). If the answer is positive, take a critical point q^{cr} on the surface of this ball, which is closest to the boundary of $\mathcal{D}$ and set $q^{(j)} \doteq q^{cr}$ (renewed $\widehat{q} + dq$ at Step 4). If the answer is negative, shrink the ball until it fits $\mathcal{D}$; the critical point q^{cr} obtained at the last step is the solution: it is on the boundary of $\mathcal{D}$, and $\|q^{cr}\|$ is an upper estimate for $\gamma_{\max}$.

Practically, the algorithm exposes a very fast convergence to a destabilizing point $\widehat{q}$ on the boundary of the stability domain of the family, and the distance between $p(s, 0)$ and $p(s, \widehat{q})$ is quite precise an estimate for $\gamma_{\max}$. Clearly, the method is locally optimal and, generally speaking, it does not converge to a global solution, i.e., $\widehat{q}$ is not the closest destabilizing point, therefore, $\overline{\gamma} \ge \gamma_{\max}$. However, it was tested over a great variety of problems and showed a very good performance. The following way to "globalize" the algorithm was implemented in the computer code. We repeatedly run the algorithm starting from randomly generated initial points q^i in a small neighborhood of $q = 0$ and obtain destabilizing points $\widehat{q}^i$ and the corresponding randomized solutions $\overline{\gamma}^i$ for $\overline{\gamma}$; then pick the smallest solution obtained, call it a *refined estimate* $\widetilde{\gamma}$.

2.6   Numerical examples

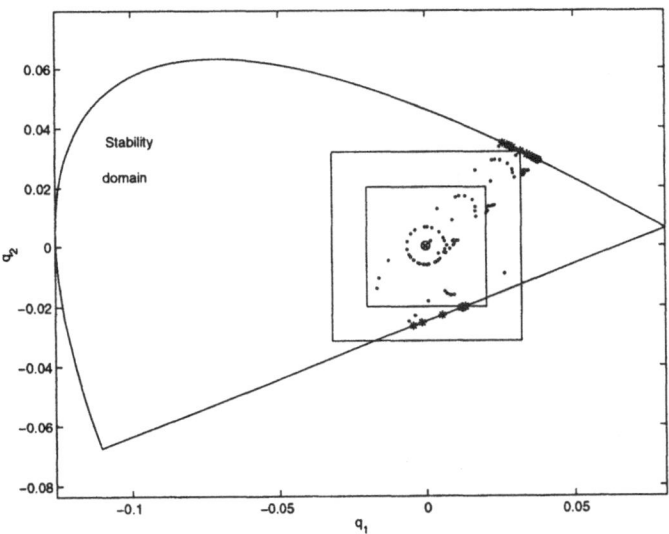

Fig. 1. Robustness margins in Example 1

Example 1. Affine uncertainty. We consider the two-dimensional case $q \in \mathbb{R}^2$ and the interval affine family $p(s,q) = p_0(s) + q_1 p_1(s) + q_2 p_2(s)$, $|q_i| \leq \gamma$, with stable nominal $p_0(s)$. The exact solution is provided by the frequency domain method in [3]. For the case of two parameters, the results are well visualized by using the D-decomposition technique [4]. Figure 1 depicts the stability domain $\mathcal{D}$ in the space of q, obtained with D-decomposition; the larger square corresponds to $\overline{\gamma} = 0.0319$ obtained after the first ($q^0 = [0; \ 0]$) run of the algorithm (converged after 6 steps); the smaller square corresponds to the refined estimate $\widetilde{\gamma} = 0.0201$. The stars on the boundary of $\mathcal{D}$ relate to destabilizing points $\widehat{q}^i$ corresponding to randomly generated points q^i (a dotted circle around zero). Finally, other dots correspond to the trajectories of repeated runs of the algorithm. In this example, polynomials $p_i(s)$, $i = 0, 1, 2$, having degrees 5, 4, and 3, respectively, were generated randomly:

$$p_0 = [1.0000 \quad 1.9282 \quad 1.3622 \quad 0.4406 \quad 0.0637 \quad 0.0030];$$

$$p_1 = [\quad 0 \quad -0.5626 \quad 0.1922 \quad 0.9098 \quad -0.7358 \quad -0.0464];$$
$$p_2 = [\quad 0 \quad\quad 0 \quad 0.7214 \quad -0.0810 \quad -0.7314 \quad 0.1206].$$

The exact value of the robustness radius is $\gamma_{max} = 0.0181$.

Example 2. Multilinear dependence. This example is borrowed from [5]; it reduces to the interval polynomial family $p(s, q) = p_0(s) + q_1 p_1(s) + q_2 p_2(s) + q_1 q_2 p_{12}(s) + q_3 p_3(s)$ with multilinear dependence on q, where $|q_1| \leq 0.2\gamma$, $|q_2| \leq 0.3\gamma$, $|q_3| \leq 0.1\gamma$, and

$$p_0(s) = s^4 + 20s^3 + 124s^2 + 1040s + 1600, \quad p_1(s) = s^3 + 16s^2 + 60s,$$

$$p_2(s) = s^3 + 14s^2 + 40s, \quad p_{12}(s) = s^2 + 10s, \quad p_3(s) = 800s + 1600,$$

with stable nominal polynomial $p_0(s)$. Application of Algorithm I with $q^0 = [0;\ 0;\ 0]$ gives $\overline{\gamma} = 3.4173$ after 3 iterations, while the solution provided in the original paper [5] is $\gamma_{max} = 3.44$.

3   Robust Stability of Matrices

The method elaborated in the previous section for polynomial families admits a straightforward extension to the matrix case. In this section we consider the following problem: Given a parameterized matrix family $A(q)$, $\|q\| \leq \gamma$, such that $A(0)$ is stable, find the robustness radius

$$\gamma_{max} = \max\{\gamma :\ A(q) \text{ is stable } \forall \|q\| \leq \gamma\}. \tag{7}$$

3.1   Perturbation formula for eigenvalues

The exposition to follow leans on the following result on perturbations of eigenvalues.

Proposition M. *Let $A(q)$ be a real $n \times n$ matrix that depends on the vector of real parameters $q = (q_1, \ldots, q_\ell)$. Assume that $A(q)$ is differentiable at $q = 0$ and denote*

$$D_i \doteq \left. \frac{\partial A(q)}{\partial q_i} \right|_{q=0}, \quad i = 1, \ldots, \ell.$$

Let $\lambda_k \doteq \lambda_k(0)$ be an algebraically simple eigenvalue of $A(0)$ and let x_k and y_k denote the right and the left eigenvectors of $A(0)$ associated with λ_k. Then, for sufficiently small q, there exists an eigenvalue $\lambda_k(q)$ of the matrix $A(q)$ such that

$$\lambda_k(q) = \lambda_k + \sum_{i=1}^{\ell} \frac{y_k^* D_i x_k}{y_k^* x_k} q_i + o(q).$$

This result is a slight modification of Theorem 6.3.12 [6].

Remark 1. We say that x and y are right and left eigenvectors of A associated with its eigenvalue λ if $Ax = \lambda x$ and $y^*A = \lambda y^*$. The notion $*$ stands for the complex conjugation for vectors and matrices: $A^* \doteq \overline{A}^T$.

Remark 2. Since any eigenvector is defined to the accuracy of an arbitrary scalar multiplier, we normalize every right eigenvector x_k so that $x_k^* x_k = 1$; similarly, left eigenvectors can be normalized to $y_k^* x_k = 1$. (It can be shown that $y_k^* x_m = 0$ for $k \neq m$, while $y_k^* x_k \neq 0$ for an algebraically simple eigenvalue λ_k.)

With the scaling of Remark 2, we rewrite the relationship above:

$$\lambda_k(q) \approx \lambda_k + (v^k, q), \qquad v^k = (v_1^k, \ldots, v_\ell^k), \qquad v_i^k = y_k^* D_i x_k. \qquad (8)$$

We thus obtain an approximation to perturbed eigenvalues that is linear in parameters. This general form particularizes to various simple situations. For instance, if we deal with an affine family $A(q) = A_0 + \sum_{i=1}^{\ell} q_i A_i$, then $A(0) = A_0$, $D_i = A_i$.

3.2 Upper bound for the robust stability margin

Further considerations mimic the logic used in the previous section for polynomial families. We consider Hurwitz stability of matrices $\max_k \operatorname{Re} \lambda_k(A) < 0$ and introduce vectors w^k, $k = 1, \ldots, n$, such that $w_i^k = \operatorname{Re}(y_k^* D_i x_k)$, where x_k, y_k, and D_i are defined in Proposition $\mathcal{M}$, and x_k and y_k are scaled as in Remark 2. Next, assuming that the unperturbed matrix $A(0)$ is stable and using the perturbation formula (8), we obtain linearized conditions of stability for a perturbed matrix $A(q)$ analogous to (4) (with u^k being changed for w^k). Finally, on the basis of these conditions, we immediately obtain an estimate for the robustness radius $\gamma_{\max}$ by introducing some vector norm $\| \cdot \|$ on $\mathbb{R}^\ell$ and letting the perturbation vector q vary in the ball of radius γ in this norm:

$$\overline{\gamma} = \max\left\{ \gamma : \max_k (\operatorname{Re} \lambda_k + \gamma \max_{\|q\| \leq 1} (w^k, q)) < 0 \right\}, \quad w_i^k = \operatorname{Re}(y_k^* D_i x_k). \quad (9)$$

Note that this formulation admits for the following statement of the problem: Find the maximal value of γ retaining the stability of $A + \gamma \Delta$, where A is stable and the perturbation matrix Δ is bounded in some norm. For instance, with Frobenius matrix norm $\|A\|_F = \left(\sum_{i,j} a_{ij}^2 \right)^{1/2}$ we obtain

$$\overline{\gamma} = \min_k \frac{-\operatorname{Re} \lambda_k}{\|W^k\|_F}, \qquad W_{ij}^k = \operatorname{Re}(\overline{y}_i^k x_j^k), \quad i, j = 1, \ldots, n.$$

Here, x_i^k is the i-th component of the eigenvector x^k, and $\overline{y}_j^k$ denotes the complex conjugate number. The destabilizing uncertainty is $\widehat{\Delta} = \frac{\overline{\gamma} W^m}{\|W^m\|_F}$, where $m = \arg\min_k (-\operatorname{Re} \lambda_k / \|W^k\|_F)$.

We now turn back to (9) and note that using linearization (8)–(9) of the original problem (7) yields the same algorithm for computing an upper bound for the robustness margin of a matrix family, as in the polynomial case. At each step of this procedure, eigenvalues of $A(q)$ are shifted toward the instability domain $\operatorname{Re}\lambda(q) \geq 0$ in the same way as were zeros of the polynomial $p(s, q)$; namely, by vanishing the maximal real part of the linear approximation of perturbed eigenvalues. The only difference is in the coefficient vector of the underlying linear form (w^k in (9) vs. u^k in (6)).

3.3   Examples

Example 1. Interval uncertainty. $A = A_0 + \Delta$, $|\Delta_{ij}| \leq \gamma$. In [7], a Metzlerian matrix $A_0 \in \mathbb{R}^{40 \times 40}$ was considered with $a_{ii} = -60$, $a_{ij} = 2$ for $i < j$, $a_{ij} = 1$ for $i > j$. The exact value of the interval stability radius was found to be $\gamma_{\max} = 0.0913$. We obtain $\overline{\gamma} = 0.0913$ after 2 steps.

Example 2. Spherical uncertainty. $A = A_0 + \Delta$, $\|\Delta\|_F \leq \gamma$. For A_0 of general form, only lower estimates for robustness radius are available, e.g., γ_L [8], [9] and γ_B [10]. We take the nominal A_0 to be a 10×10 matrix -grcar(10) from the TestMatrix Toolbox in Matlab [11] and obtain the following results. Lower estimates: $\gamma_L = 0.0125$; $\gamma_B = 0.0297$. After 3 steps our method gives $\overline{\gamma} = 0.1163$. In general, to test the quality of $\overline{\gamma}$ in situations where no exact solution is available, we suggest Monte Carlo sampling. For this example, 5,000 random perturbations $\|\Delta\|_F \leq \overline{\gamma}$ detected no unstable matrices $A + \Delta$. This estimate seems to be quite accurate: for the quadrupled radius $\gamma = 4\overline{\gamma}$, less than 1% of unstable samples were detected!

Examples presented in Sec. 2 and 3 are typical and constitute small part of the massive body of numerical experiments conducted with Algorithm I in the polynomial and matrix cases. The method shows excellent performance both in terms of accuracy and rate of convergence.

4   Stabilization

In this section we consider a family of polynomials $p(s, q)$ parameterized by $q \in \mathbb{R}^\ell$ such that $\deg p(s, q) = n = \text{const}\,(q)$ and $p(s, 0)$ is unstable. The problem is to find if there exists a $\widehat{q}$ such that $p(s, \widehat{q})$ is stable.

4.1   Main idea

We basically use the same idea of iteratively shifting the zeros of $p(s, q)$ so as to satisfy the conditions $\operatorname{Re} s_k(q) < 0$, $k = 1, \ldots, n$. This iterative procedure is based on the perturbation formula for zeros of $p(s, q)$, derived in Sec. 2.1 and its extensions to the case of multiple roots. The essential difference between the RM and the ST problems is that in order to satisfy the target condition $\operatorname{Re} s_k(q) \geq 0 \, \forall k$ in the RM problem, it is sufficient to move only one zero to the right (the one with the maximal real part), whereas to satisfy the target condition $\operatorname{Re} s_k(q) < 0 \, \forall k$ in the ST problem, *all* zeros must be moved to the left-half plane. As a result, in the ST problem, the method may have to shift *several zeros at once* at some step, which is, however, not a serious obstacle and can be performed using Proposition $\mathcal{P}$. What is more important is that there might be multiple zeros among them, and this is quite often the case. This necessitates a more subtle account for perturbations. Indeed, the linear approximation to perturbed zeros, obtained in Proposition $\mathcal{P}$, is only valid under the assumption that the zero is simple or, equivalently, the first derivative $p_0'(s_k)$ is nonvanishing. In Sec. 4.4, we derive second-order conditions, which are also linear in parameters. These conditions provide a simple way to account for multiple zeros and to perform a step of the method by solving a well-defined problem of quadratic programming.

4.2   A quadratic programming problem

Let us consider a step of the method. At the current point $q = 0$, we built linear approximations $\operatorname{Re} s_k(q) \approx (a^k, q) - b_k$ and solve the problem

Find $\min \|dq\|$ subject to $(a^k, dq) - b_k \leq 0, \quad k = 1, \ldots, n,$

whose solution defines the step of the method. That is, at each step we solve a quadratic programming problem (QP)

$$\min \|dq\| \qquad \text{subject to} \ \ A \cdot dq \leq b, \tag{10}$$

where the matrix A of constraints is composed of a^k, and the vector b is composed of b_k.

Let $\eta \doteq \max_i \operatorname{Re} s_i$; then the zero s_k is called *rightmost* if $\operatorname{Re} s_k = \eta$, i.e., it has maximal real part. Clearly, only rightmost zeros should form the constraints in the QP problem (10), hence, only such zeros are considered below.

4.3   Conditions of stability

Suppose $p(s, 0)$ is unstable, i.e., there exist unstable zeros $\operatorname{Re} s_k(0) \geq 0$. We introduce accuracy $\delta > 0$, and a polynomial is considered stable if $\max_k \operatorname{Re} s_k < -\delta$. Let Δ_k be the perturbation of a rightmost zero s_k, caused by dq. To make this zero stable, we require $\operatorname{Re}(s_k + \Delta_k) \leq -\delta$ or, equivalently,

$$\operatorname{Re} \Delta_k \leq -\mu_k, \qquad \mu_k \doteq -(\operatorname{Re} s_k + \delta). \tag{11}$$

The key issue is a proper representation of perturbations Δ_k as linear functions of dq.

4.4 Linearized conditions

Below, we distinguish between simple and multiple rightmost zeros and derive different perturbation formulae, both linear in parameters.

First-order approximation. Assume that s_k is a simple zero, i.e., $p(s_k, 0) = 0$; $p'(s, 0)|_{s=s_k} \neq 0$. By Proposition $\mathcal{P}$, the following linearization of condition (11) is readily available:

$$(w^k, dq) \leq -\mu_k, \qquad w_i^k = -\mathrm{Re}\frac{\pi_i(s_k)}{p_0'(s_k)}, \qquad \pi_i(s) = \left.\frac{\partial p(s, q)}{\partial q_i}\right|_{q=0},$$

$$p_0'(s_k) = p'(s, 0)|_{s=s_k}. \tag{12}$$

Each of such inequalities constitute a constraint in the QP problem (10), i.e., we put $a^k = w^k$ and $b_k = -\mu_k$. If s_k is complex, then s_k and its complex conjugate s_k^* contribute the same constraints to the QP (10) so that one of them can be omitted. Hence, the number of constraints is equal to the number of different real parts of rightmost zeros.

Second-order approximation. Let a zero s_k have multiplicity 2, that is, $p(s_k, 0) = 0$; $p'(s, 0)|_{s=s_k} = 0$; $p''(s, 0)|_{s=s_k} \neq 0$. Approximating $p(s, q)$ at the point $(s_k, 0)$ to the accuracy of second-order terms gives

$$\frac{1}{2}\Delta_k^2 p_0''(s_k) + \Delta_k \sum_{i=1}^{\ell} dq_i \cdot \pi_i'(s_k) + \sum_{i=1}^{\ell} dq_i \cdot \pi_i(s_k) = 0,$$

$$p_0''(s_k) \doteq p''(s, 0)|_{s=s_k}, \qquad \pi_i'(s_k) = \pi_i'(s)|_{s=s_k}.$$

Denoting

$$u_i^k = \frac{\pi_i'(s_k)}{p_0''(s_k)}, \qquad v_i^k = \frac{2\pi_i(s_k)}{p_0''(s_k)} \tag{13}$$

and

$$\alpha \doteq (u^k, dq), \qquad \beta \doteq (v^k, dq),$$

(we omit subscript k in α and β) we obtain the quadratic equation in Δ_k

$$\Delta_k^2 + 2\alpha\Delta_k + \beta = 0$$

having solutions

$$\Delta_k^{(1,2)} = -\alpha \pm \sqrt{\alpha^2 - \beta}.$$

Now assume that s_k is pure real and impose linear conditions on α and β ensuring the fulfillment of (11). We have two options to guarantee (11) by choosing $\Delta_k^{(1,2)}$ either real or complex.

Option 1 is to split the s_k into a pair of complex conjugate zeros $s_k^+ = \hat{s}_k + jt_k$ and $s_k^- = \hat{s}_k - jt_k$, where $\hat{s}_k \leq -\delta$ and $t_k \neq 0$. In that case it is sufficient to set $\alpha^2 - \beta < 0$ and $\alpha = \mu_k$. For instance, the conditions $\alpha = \mu_k$ and $\beta \geq \mu_k^2$ guarantee the fulfillment of (11), and we arrive at the following linear conditions on dq:

$$\begin{cases} (u^k, dq) = \mu_k, \\ (-v^k, dq) \leq -\mu_k^2. \end{cases} \tag{14}$$

Option 2 is to split s_k into 2 pure real zeros $s_k^+ \leq -\delta$ and $s_k^- \leq -\delta$, for which purpose we set $\alpha \geq \mu_k$ and $\beta = \mu_k^2$ ensuring the relaxed conditions $s_k^+ \leq -\delta/2$ and $s_k^- \leq -\delta$. With this choice, condition (11) is not satisfied, although we do reduce Re s_k. We thus have

$$\begin{cases} (-u^k, dq) \leq -\mu_k, \\ (v^k, dq) = \mu_k^2. \end{cases} \tag{15}$$

With both options, we obtain linear constraints (14) or (15) on dq leading to the reduction of the real part of a multiple zero s_k. A procedure is developed to decide which of the two options to choose (respectively, a zero is called multiple zero of type 1 or type 2).

We summarize. At each step of the method we distinguish between the three categories of rightmost zeros: (i) simple zeros, either complex or real; (ii) multiple zeros of type 1; (iii) multiple zeros of type 2, and compose inequalities (12) or (14), (13), or (15), (13), respectively. These inequalities altogether constitute the constraints in the quadratic programming problem (10). A solution to this QP problem provides a step of the method such that zeros are shifted towards the left-half plane. Note that every multiple zero leads to the pair of constraints (14) or (15) so that the total number of constraints is equal to the number of different real parts of rightmost zeros (accounting for their multiplicity). We formulate the following algorithm.

4.5   Algorithm II

Step 1. Specify numerical constants and set $\widehat{q} = 0$.

Step 2. Calculate $\eta = \max_k$ Re s_k, where s_k are the zeros of $p(s, 0)$.
   If $\eta \leq -\delta$, stop; $\widehat{q}$ is a stabilizing value of parameters.
   Otherwise, identify rightmost zeros, calculate the vectors w^k, u^k, v^k according to (12), (13), compose inequalities (12), (14), (15).

Step 3. Solve QP (10) with constraints (12), (14), (15) to obtain a solution dq.

Step 4. Find $\gamma_{\min} = \arg \min_{0 \leq \gamma \leq 1} \eta(\gamma)$, where $\eta(\gamma) = \max_k$ Re s_k and s_k are zeros of $p(s, \gamma \cdot dq)$. Change variables $q := q - \gamma_{\min} \cdot dq$ (i.e., adopt $p(s, \gamma_{\min} \cdot dq)$ as a "new" $p(s, 0)$) and put $\widehat{q} := \widehat{q} + \gamma_{\min} \cdot dq$. Go to Step 2.

4.6   Discussion

Algorithm II is derived under the assumptions that there are no complex multiple zeros and that pure real zeros have multiplicity not higher than 2.

So, generally speaking, there is no guarantee for the algorithm to converge to a stable point. However, our rich computational experience to date shows that the violation of these assumptions is rather a pathology, while the occurrence of a single multiple real root is an intrinsic feature of ST problems. Nevertheless, to escape a higher multiplicity trap at some point q, a randomization $q := q + \xi$ can be used, where ξ is a small noise.

Algorithm II is simple, practically efficient, and it applies to diverse problems in control theory, which is illustrated by the examples below.

4.7   Numerical examples

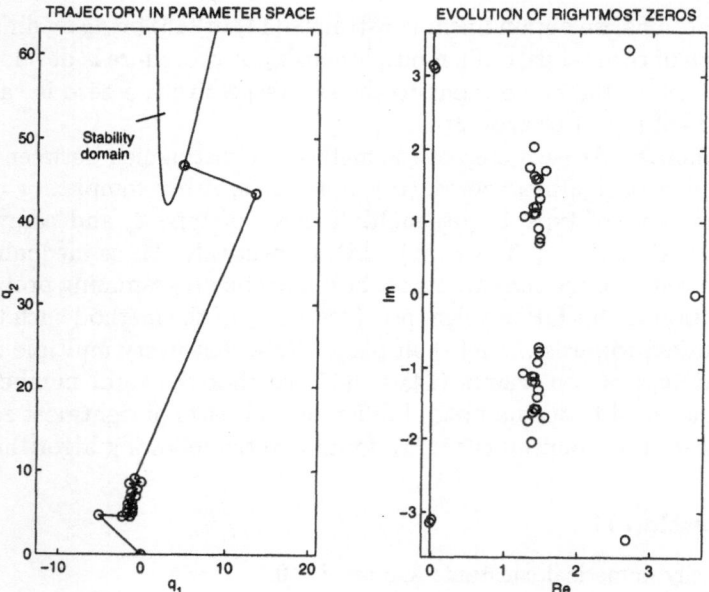

Fig. 2. The trajectory of Algorithm II and evolution of rightmost zeros for Example 1

Example 1. Static Output Feedback Stabilization. Given a linear system $\dot{x} = Fx + Gu$, $y = Hx$, find if there exists a control law $y = Ku$ stabilizing the system. An example in [12] involves a 1×2 gain matrix $K = [q_1 \ q_2]$ so that the problem reduces to finding a stable polynomial in the family $p(s,q) = p_0(s) + q_1 p_1(s) + q_2 p_2(s)$, where $p_0(s) = s^3 - 13s$; $p_1(s) = s^2 - 5s$; $p_2(s) = s + 1$. A very lengthy derivation in [12] based on the decidability algorithms of Tarski and Seidenberg yields a stability point $\widehat{q} = [2; 50]$. We first note that the problem contains two uncertain parameters so that the classical D-decomposition technique can be used to describe the whole stability domain of $p(s,q)$ (Figure 2, left). However, to illustrate our method, we start from $q^0 = [0;\ 0]$ and obtain a stabilizing point $\hat{q} = [4.7631;\ 46.7713]$ after 20 steps. Note that in general, static output feedback stabilization problems are NP-hard [13].

Example 2. Stabilization via PID Controller. Given a SISO plant specified by transfer function $P(s) = 1/(s+2)(s+1)^2(s-1)$, find a stabilizing PID controller of the form $C(s,q) = q_1 + q_2 s + q_3/s$. The problem reduces to the ST formulation by considering the closed-loop characteristic polynomial $p(s,q) = s(s+2)(s+1)^2(s-1) + q_1 s + q_2 s^2 + q_3$. Application of Algorithm II with initial point $q^0 = [0;\ 0;\ 0]$ yields a stabilizing point $\hat{q} = [2.1793;\ 5.5783;\ 0.3472]$ after 24 steps.

Example 3. Matrix Stabilization. Algorithm II admits a straightforward modification to the matrix case. We exemplify its behavior by finding a stable element in the affine matrix family $A(q) = A_0 + q_1 A_1 + q_2 A_2 + q_3 A_3$ with unstable nominal A_0 and perturbation matrices A_1, A_2, A_3, given by

$$A_0 = \begin{pmatrix} 2 & 2 & 0 \\ -1 & 0 & 0 \\ -1 & 0 & 2 \end{pmatrix} ; \quad A_1 = \begin{pmatrix} -0.4582 & 0.4027 & 0.9691 \\ 0.7370 & -0.4511 & -0.3452 \\ -0.7406 & 0.8160 & 0.7331 \end{pmatrix} ;$$

$$A_2 = \begin{pmatrix} -0.5890 & 0.5471 & -0.6725 \\ 0.1761 & 0.5744 & 0.4972 \\ -0.7262 & -0.4928 & -0.8219 \end{pmatrix} ; \quad A_3 = \begin{pmatrix} -0.9571 & -0.3868 & -0.1505 \\ 0.4578 & -0.6560 & 0.3161 \\ -0.1786 & 0.4769 & 0.5364 \end{pmatrix} .$$

Eigenvalues of A_0 are $\lambda_1 = 2$; $\lambda_{2,3} = 1 \pm j$; the entries of A_i, $i = 1,2,3$, were generated randomly uniformly on $[-1, 1]$. We start with $q^0 = [0;\ 0;\ 0]$ and after 11 steps achieve a stabilizing point $\hat{q} = [-4.5757;\ 2.3821;\ 7.5503]$ such that the eigenvalues of $A(\hat{q})$ are $\lambda_1 = -5.2960$; $\lambda_{2,3} = -0.01 \pm j1.8385$.

5   Conclusion

The approach developed is expected to be a powerful tool for solving fundamental problems in control theory which admit no solution by existing techniques. It has a simple underlying idea and applies to a wide range of problems involving differentiable dependence on the parameters. Although it does not possess a guaranteed convergence to the global solution, the results of numerical experiments look very promising: The algorithms are numerically stable and expose a very good performance for a large number of test problems, both in terms of accuracy and rate of convergence.

The method can be further refined by using slightly different second-order conditions (for Algorithm II) and by optimizing the algorithms numerically (finer step-size correction, choice of constants, etc).

References

1. Polak, E., Wardi, Y. (1982) Nondifferentiable optimization algorithm for designing control systems having singular value inequalities. *IEEE TAC*, **18**, 3, 267–283.
2. Keel, L. H., Bhattacharyya, S. P. (1997) A linear programming approach to controller design. In: *Proc. 36th CDC*, San Diego, CA, 2139–2148.
3. Tsypkin, Ya. Z., Polyak, B.T. (1991) Frequency domain approach to robust stability of continuous systems. In: *Systems and Control: Topics in Theory and Applications* (F. Kozin and T. Ono, eds), MITA Press, Osaka, 389–399.
4. Siljiak, D. D. (1969) *Nonlinear Systems*. Wiley, New York.
5. de Gaston, R. R. E., Safonov, M. G. (1988) Exact calculation of the multiloop stability margin. *IEEE TAC*, **33**, 2, 156–171.
6. Horn, R. A., Johnson, C. R. (1986) *Matrix Analysis*. Cambridge University Press, Cambridge.
7. Polyak, B. T., Panchenko, O. B. (1997) A probabilistic approach to robust stability of interval matrices. *Sov. Phys. Dokl., Ser. Control Theory*, **353**, 4, 456–458.
8. Bhattacharyya, S. P., Chapellat, H., Keel, L. H. (1995) *Robust Control: The Parametric Approach*, Prentice Hall PTR, Upper Saddle River, NJ.
9. Shcherbakov, P. S. (1998) A sufficient condition of robust stability of uncertain matrices. *Avtomat. Telemekh.*, 8, 71–79.
10. Emel'yanov, S. V., Korovin, S. K., Bobylev, N. A. (1997) On a particular problem in robustness theory. *Sov. Phys. Dokl.*, **357**, 4, 747–749.
11. Higham, N. J. (1995) *The Test Matrix Toolbox for* MATLAB *(version 3.0)*, Numerical Analysis Report No. 276, The Univ. of Manchester, Centre for Comput. Math.
12. Anderson, B. D. O., Bose, N. K., Jury, E. I. (1975) Output feedback stabilization and related problems—solution via decision methods. *IEEE TAC*, **20**, 1, 53–66.
13. Blondel, V., Tsitsiklis, J. N. (1997) NP-hardness of some linear control design problems. *SIAM J. Contr. Optim.*, **35**, 6, 2118–2127.

A Convex Approach to a Class of Minimum Norm Problems

Graziano Chesi[1], Alberto Tesi[2], Antonio Vicino[1], and Roberto Genesio[2]

[1] Dipartimento di Ingegneria dell'Informazione, Università di Siena
   Via Roma 56, 53100 Siena, Italy
   E-mail: vicino@ing.unisi.it
[2] Dipartimento di Sistemi e Informatica, Università di Firenze
   Via di S. Marta 3, 50139 Firenze, Italy
   E-mail: atesi@dsi.unifi.it

Abstract. This paper considers the problem of determining the minimum euclidean distance of a point from a polynomial surface in R^n. It is well known that this problem is in general non-convex. The main purpose of the paper is to investigate to what extent Linear Matrix Inequality (LMI) techniques can be exploited for solving this problem. The first result of the paper shows that a lower bound to the global minimum can be achieved via the solution of a one-parameter family of LMIs. Each LMI problem consists in the minimization of the maximum eigenvalue of a symmetric matrix. It is also pointed out that for some classes of problems the solution of a single LMI provides the lower bound. The second result concerns the tightness of the bound. It is shown that optimality of the lower bound can be readily checked via the solution of a system of linear equations. In addition, it is pointed out that lower bound tightness is strictly related to some properties concerning real homogeneous forms. Finally, an application example is developed throughout the paper to show the features of the approach.

1 Introduction

It is a well known fact that a large number of system analysis and control problems amount to computing the minimum distance of a point from a surface in a finite dimensional space. Several issues in robustness analysis of control systems, such as the μ and the related stability margin computation [1]-[3], and the problem of estimating the domain of attraction of equilibria of nonlinear systems [4],[5] fall in this class of problems. Unfortunately, it is also known that the large majority of such problems can be formulated as non-convex optimization programs, whose solution is in general extremely difficult, if not impossible, to obtain with reasonable computational effort.

Recently, powerful convex optimization techniques have been devised for problems in the form of Linear Matrix Inequalities (LMIs) [6]. Such techniques have been succesfully employed in connection with suitable changes of variables for convexifying some classes of optimization problems.

The main purpose of this paper is to show how LMI techniques can be exploited for solving a class of minimum norm problems. More specifically, the problem of determining the minimum euclidean distance of a point from a polynomial surface in R^n is considered. It is first shown that a lower bound to the global minimum can be achieved via the solution of a one-parameter family of LMIs, once a suitable change of variables has been performed. In particular, each LMI problem requires the minimization of the maximum eigenvalue of a symmetric matrix. In some cases the solution of a single LMI provides the lower bound. Tightness of this lower bound is investigated, providing a computationally simple optimality test and showing a strict relation between lower bound optimality and some properties of real homogeneous forms. A simple numerical example is developed throughout the paper to illustrate the features of the approach.

The paper is organized as follows. In Section 2 the minimum norm problem is formulated. Section 3 introduces a canonical form of the considered problem. The lower bound based on LMI techniques is given in Section 4. Section 5 concerns the tightness of the lower bound. Some concluding comments end the paper in Section 6.

Notation.
R^n: real n-space;
R_0^n: $R^n \setminus \emptyset$;
$x = (x_1, \ldots, x_n)'$: vector of R^n;
$R^{n,n}$: real $n \times n$-space;
$A = [a_{ij}]$: matrix of $R^{n,n}$;
A': transpose of A;
A^{-1}: inverse of A;
Ker$[A]$: null space of A;
E_n: $n \times n$ identity matrix;
$A > 0$ $(A \geq 0)$: positive definite (semidefinite) matrix;
$\lambda_M\{A\}$: maximum real eigenvalue of A;
$\| \cdot \|_{2,W}$: weighted euclidean norm.

2   Problem formulation and preliminaries

In order to state the problem, we need to introduce the following definitions.

Definition 1.
A map $f^{(d)} : R^n \longrightarrow R$ is said a (real n-variate homogeneous) form of degree d if

$$f^{(d)}(x) = f^{(d)}(x_1, \ldots, x_n) = \sum_{\substack{i_1 \geq 0, \ldots, i_n \geq 0}}^{i_1 + \ldots + i_n = d} f_{i_1, \ldots, i_n} x_1^{i_1} \cdots x_n^{i_n}$$

where $f_{i_1,\ldots,i_n} \in R$ are the coefficients and d is a non-negative integer.

With some abuse of notation, we will sometimes denote a form simply by f, dropping the explicit dependence on the degree.

Definition 2.
A (real n-variate homogeneous) form f is said positive definite (semidefinite) in a certain region if $f(x) > 0$ ($f(x) \geq 0$) for all x in that region.
A (real n-variate homogeneous) form f is said positive definite (semidefinite) if $f(x) > 0$ ($f(x) \geq 0$) for all $x \in R_0^n$.

Observe that a positive definite form is necessarily of even degree. Moreover, a (real n-variate) polynomial $p(x)$ can always be written as the sum of forms of suitable degree, i.e.,

$$p(x) = p(x_1,\ldots,x_n) = \sum_{k=0}^{N} f^{(k)}(x_1,\ldots,x_n)$$

where $f^{(k)}$ are (real n-variate homogeneous) forms of degree k.

Now, we introduce a definition on polynomials which will be used to avoid the trivial case when the solution of the considered optimization problem is zero.

Definition 3.
A polynomial $p(x) = \sum_{k=0}^{N} f^{(k)}(x)$ is said locally positive definite if its form of minimum degree is positive definite.

We can now formulate our problem.

Problem I.
Let $p(x)$ be a locally positive definite polynomial and W a symmetric positive definite matrix. Compute the minimum (weighted) euclidean distance of the origin from the surface defined by $p(x) = 0$, i.e., solve the constrained optimization problem

$$\rho^* = \min_{x \in R_0^n} \|x\|_{2,W}^2$$

subject to   (1)

$$p(x) = 0.$$

Remark 1.
The above optimization problem can incorporate multiple polynomial constraints in an obvious way. Suppose that the constraints are $p_1(x) = 0$

and $p_2(x) = 0$. Thus, an equivalent scalar polynomial constraint is $p_1^2(x) + p_2^2(x) = 0$.

Several analysis and control problems can be cast in the form of Problem I. To name but a few, consider the computation of the l_2 parametric stability margin of a control system affected by parametric uncertainty [3], the estimation of the domain of attraction of equilibria of nonlinear systems via quadratic Lyapunov functions [4], the D-stability of real matrices [7] that plays a key role in the analysis of singularly perturbed systems [8], the computation of the region of validity of optimal linear $\mathcal{H}_\infty$ controllers for nonlinear systems [9]. In the sequel we consider the following illustrative example concerning the estimation of the domain of attraction of an equilibrium point of a nonlinear system.

Example.
Consider the second order nonlinear system

$$\begin{cases} \dot{x}_1 = -0.5x_1 - 0.95x_1x_2^2 - 0.2x_2^3 \\ \dot{x}_2 = -0.5x_2 + 2x_1^3 + 0.7x_2^3 \end{cases}$$

where $x = (x_1, x_2)'$ is the state vector.
It is easily verified that the origin in an asymptotically stable equilibrium point. Then, it is well kwown that the largest ellipsoidal estimate of the domain of attraction of the origin, that can be computed via the quadratic Lyapunov function $V(x) = x_1^2 + x_2^2$, is given by $V(x) = \rho^*$, where ρ^* is the solution of the optimization problem

$$\rho^* = \min_{x \in R_0^n} V(x)$$

subject to   (2)

$$\frac{\partial V(x)}{\partial x} \dot{x} = -x_1^2 - x_2^2 + 4x_1^3x_2 - 1.9x_1^2x_2^2 - 0.4x_1x_2^3 + 1.4x_2^4 = 0.$$

It is readily checked that the above problem is in the form of Problem I, once

$$W = E_2$$

and

$$p(x) = x_1^2 + x_2^2 - 4x_1^3x_2 + 1.9x_1^2x_2^2 + 0.4x_1x_2^3 - 1.4x_2^4.$$

Note that the optimization problem (2) has a local minimum in addition to the global one, as depicted in Fig. 1.

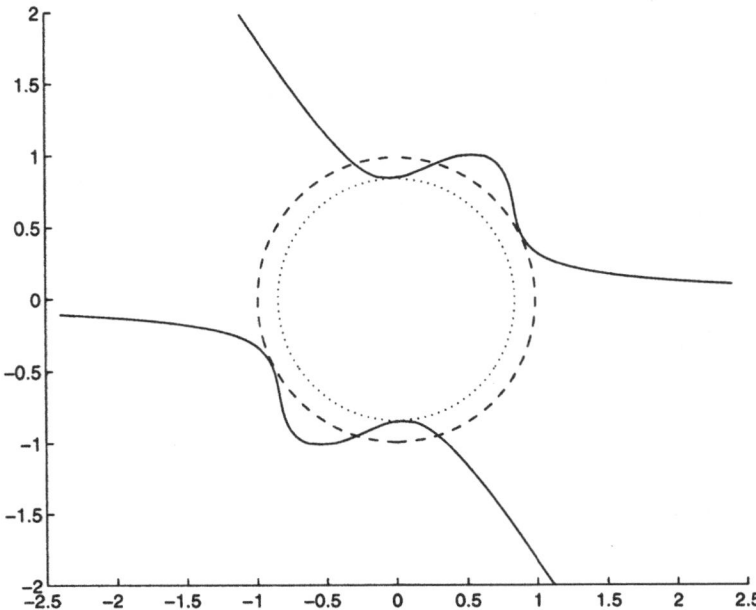

Fig. 1. Curve defined by $p(x) = 0$ (solid), global minimum (dotted) and local minimum (dashed).

3   A canonical optimization problem

Our aim is to investigate how convex optimization techniques such as LMIs can be exploited for approximating the solution of Problem I. To this purpose, we introduce a canonical optimization problem where the constraint is given by a single polynomial consisting of even forms only.

Problem II.
Let $u(x)$ be a positive definite quadratic form and $w(x)$ be a locally positive definite polynomial of the form

$$w(x_1, \ldots, x_n) = \sum_{k=0}^{m} v^{(2k)}(x_1, \ldots, x_n),$$

where $v^{(2k)}$ are given forms of even degree. Solve the following constrained optimization problem

$$c^* = \min_{x \in R_0^n} u(x)$$

subject to   (3)

$$w(x) = 0.$$

At a first sight, Problem II appears a simplified version of Problem I. However, it is straightforward to show that any problem of the form (1) can be equivalently written in the form of (3).

Proposition 1.
Let $p(x)$ and W be given. Then, $u(x)$ and $w(x)$ can be constructed such that $\rho^* = c^*$.

Proof. It is easily checked that if

$$u(x) = x'Wx$$

and

$$w(x) = \begin{cases} p(x) & \text{if } p(x) = p(-x) \quad \forall\, x \\ p(x)p(-x) & \text{if } p(x) \neq p(-x) \quad \text{for some } x, \end{cases}$$

then $\rho^* = c^*$. $\square$

Example continued.
In this case it turns out that if

$$u(x) = x_1^2 + x_2^2$$

and

$$w(x) = p(x),$$

then $\rho^* = c^*$. Therefore, Problem II becomes

$$c^* = \min_{x \in R_0^n} x_1^2 + x_2^2$$

subject to

$$x_1^2 + x_2^2 - 4x_1^3 x_2 + 1.9x_1^2 x_2^2 + 0.4x_1 x_2^3 - 1.4x_2^4 = 0.$$

4 An LMI-based lower bound

In this section, our primary goal is to provide a lower bound to the optimal solution c^* that can be computed via LMI techniques. Due to space limitation, some of the proofs are omitted (see [10] for details).

Figure 2 shows a geometrical view of the tangency problem (3). Since $w(x)$ is locally positive definite, it turns out that c^* can be computed as the smallest positive value of c such that $w(x)$ loses positivity for some x belonging to the level set

$$\mathcal{U}_c = \{x \in R^n : u(x) = c\}.$$

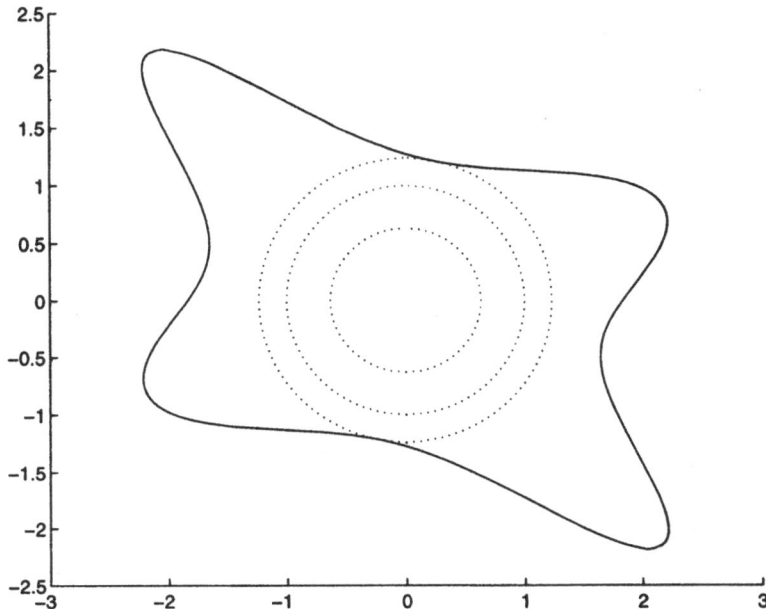

Fig. 2. Geometrical view of Problem II.

This fact suggests that the computation of c^* could be performed by a sequence of positivity tests of $w(x)$ on the ellipsoids $u(x) = c$. To this purpose, let us consider the real form of degree $2m$

$$\hat{w}(x; c) = \sum_{k=0}^{m} v^{(2k)}(x) \frac{u^{m-k}(x)}{c^{m-k}} \tag{4}$$

where c is any positive real. We have the following result.

Lemma 1.
For a given positive c, the following two statements are equivalent:

i) $w(x) > 0$ $\forall\, x \in \mathcal{U}_c$;
ii) $\hat{w}(x; c) > 0$ $\forall\, x \in R_0^n$. $\square$

The lemma above suggests that a constrained polynomial positivity test is equivalent to a positivity test of a form.

Now, we rewrite $\hat{w}(x; c)$ in a form more convenient for our purposes. To proceed, let $x \in R^n$ and consider the sub-vector $x_{[k]} \in R^{n-k+1}$ composed of the last $n - k + 1$ coordinates of x, i.e.,

$$x_{[k]} = (x_k, x_{k+1}, \cdots, x_n)'. \tag{5}$$

Obviously, $x_{[1]} = x$. Furthermore, let us introduce the vector defined recursively as follows

$$x_{[k]}^{\{m\}} := \begin{cases} (x_k x_{[k]}^{\{m-1\}'}, x_{k+1} x_{[k+1]}^{\{m-1\}'}, \cdots, x_n x_{[n]}^{\{m-1\}'})' & m > 0 \\ 1 & m = 0. \end{cases} \qquad (6)$$

For ease of notation, we denote the vector $x_{[1]}^{\{m\}}$, composed of all the distinct monomials of degree m built with the n coordinates of x, by $x^{\{m\}}$. Note that $x^{\{m\}} \in R^{\sigma(n,m)}$ where

$$\sigma(n,m) := \binom{n+m-1}{n-1}.$$

Since $\hat{w}(x;c)$ is a real form of degree $2m$, it is easy to show that

$$\hat{w}(x;c) = x^{\{m\}'} \Omega_{2m}(c) x^{\{m\}} \qquad (7)$$

where $\Omega_{2m}(c) \in R^{\sigma(n,m),\sigma(n,m)}$ is a suitable symmetric matrix.

It turns out that the matrix $\Omega_{2m}(c)$ is not uniquely defined. Indeed, consider the set of matrices

$$\mathcal{L} := \left\{ L \in R^{\sigma(n,m),\sigma(n,m)} : \quad L = L' \text{ and } x^{\{m\}'} L x^{\{m\}} = 0 \ \forall x \in R^n \right\}.$$

Such a set has the following property.

Lemma 2.
$\mathcal{L}$ is a linear space and its dimension $d_{\mathcal{L}}$ is

$$d_{\mathcal{L}} = \frac{1}{2}\sigma(n,m)\left[\sigma(n,m)+1\right] - \sigma(n,2m). \qquad \square$$

Therefore, the set $\mathcal{L}$ admits a linear parameterization (see [10] for details) that can be exploited to describe all the matrices satisfying condition (7).

Lemma 3.
Let $\hat{\Omega}_{2m}(c)$ be any symmetric matrix such that

$$\hat{w}(x;c) = x^{\{m\}'} \hat{\Omega}_{2m}(c) x^{\{m\}}.$$

Let $L(\alpha)$ be a linear parameterization of $\mathcal{L}$ and define

$$\Omega_{2m}(c;\alpha) := \hat{\Omega}_{2m}(c) + L(\alpha). \qquad (8)$$

Then,

$$\hat{w}(x;c) = x^{\{m\}'} \Omega_{2m}(c;\alpha) x^{\{m\}} \qquad \forall \alpha \in R^{d_{\mathcal{L}}}, \ \forall x \in R^n. \qquad \square \quad (9)$$

Note that, for a fixed c, the symmetric matrix $\Omega_{2m}(c;\alpha)$ depends affine linearly on α.

We can now give the main result of this section.

Theorem 1.
Let

$$\hat{c} = \sup c$$

such that (10)

$$\min_{\alpha} \lambda_{max} \left[-\Omega_{2m}(c; \alpha) \right] \leq 0.$$

Then, $\hat{c}$ is a lower bound of c^*, i.e.,

$$\hat{c} \leq c^*.$$

Proof. From Lemma 1, it is straightforward to verify that problem (3) can be rewritten as

$$c^* = \sup c$$

such that

$$\hat{w}(x; c) \geq 0.$$

Now, let $\hat{\alpha}$ be such that

$$\lambda_{max} \left[-\Omega_{2m}(\hat{c}; \hat{\alpha}) \right] = 0.$$

From Lemma 2 and Lemma 3 it follows that $\hat{w}(x; \hat{c}) \geq 0$ and thus the proof is complete. □

The above theorem states that a lower bound for Problem II can be computed via the solution of a one-parameter family of convex optimization problems. Specifically, for a fixed value of the parameter c, each problem requires the minimization of the maximum eigenvalue of a symmetric matrix with respect to the vector α (see (8) and (10)), a problem that can be solved via a standard LMI [11].

Example continued.
Since $n = 2$ and $m = 2$, $\hat{w}(x; c)$ has the form (see (4))

$$\hat{w}(x; c) = (x_1^2 + x_2^2)(x_1^2 + x_2^2)c^{-1} - 4x_1^3 x_2 + 1.9x_1^2 x_2^2 + 0.4x_1 x_2^3 - 1.4x_2^4$$

and $x^{\{m\}}$ reduces to (see (5) and (6))

$$x^{\{2\}} = (x_1^2, \ x_1 x_2, \ x_2^2)'.$$

From Lemma 2 it turns out that $\mathcal{L}$ is a one-dimensional linear space which admits the following parameterization

$$L(\alpha) = \begin{bmatrix} 0 & 0 & -\alpha \\ 0 & 2\alpha & 0 \\ -\alpha & 0 & 0 \end{bmatrix} \tag{11}$$

where $\alpha \in R$. It is easily checked that expression (9) of Lemma 3 holds for

$$\Omega_4(c; \alpha) = \hat{\Omega}_4(c) + L(\alpha)$$

where

$$\hat{\Omega}_4(c) = \begin{bmatrix} c^{-1} & -2 & 0 \\ -2 & 2c^{-1} + 1.9 & 0.2 \\ 0 & 0.2 & c^{-1} - 1.4 \end{bmatrix}$$

and $L(\alpha)$ is given by (11). Therefore, according to Theorem 1, a lower bound to c^* is provided by the following convex optimization problem

$$\hat{c} = \sup c$$

such that

$$\min_{\alpha \in R} \lambda_{max} \begin{bmatrix} -c^{-1} & 2 & \alpha \\ 2 & -2c^{-1} - 1.9 - 2\alpha & -0.2 \\ \alpha & -0.2 & -c^{-1} + 1.4 \end{bmatrix} \leq 0.$$

Note that for a fixed c the above problem amounts to minimize the maximum eigenvalue of a symmetric matrix with respect to the scalar α.

To conclude the section, we consider a special class of problems (3) for which only a single LMI problem is needed for computing $\hat{c}$. Such a class $\mathcal{C}$ includes polynomials $w(x)$ with two terms, as precisely defined below.

Definition 4. A polynomial w is said to belong to the class $\mathcal{C}$ if

$$w(x) = v^{(2l)}(x) + v^{(2m)}(x)$$

where $l \in \{0, 1\}$.

The class $\mathcal{C}$ has the following property (see [10] for a proof).

Lemma 4.
Let $w \in \mathcal{C}$ be locally positive definite. Then, there exist a positive definite matrix M_0 and a symmetric matrix M_1 such that

$$\Omega_{2m}(c, \alpha) = c^{l-m} M_0 + M_1 + L(\alpha). \qquad \Box \tag{12}$$

We are now ready to state the next theorem.

Theorem 2.
Assume that $w \in C$ and let

$$\sigma = \min_{\alpha} \lambda_{max} \left[(\hat{M}_0')^{-1} (-M_1 - L(\alpha))(\hat{M}_0)^{-1} \right] \tag{13}$$

where $\hat{M}_0$ is such that

$$M_0 = \hat{M}_0' \hat{M}_0.$$

Then,

$$\hat{c} = \sigma^{\frac{1}{l-m}}. \tag{14}$$

Proof. It immediately follows from the optimizatiom problem (10) and equation (12). $\square$

Therefore, if $w \in C$, the lower bound is computed via the minimization of the maximum eigenvalue of a symmetric matrix with respect to the vector α.

Example continued.
It is easily verified that Lemma 4 applies with $l = 1$, $m = 2$, and

$$M_0 = \begin{bmatrix} 1 & 0 & 0 \\ 0 & 2 & 0 \\ 0 & 0 & 1 \end{bmatrix}$$

$$M_1 = \begin{bmatrix} 0 & -2 & 0 \\ -2 & 1.9 & 0.2 \\ 0 & 0.2 & -1.4 \end{bmatrix}.$$

Therefore, solution of (13) yields

$$\sigma = 1.4082$$

and the corresponding optimal value of the scalar α

$$\hat{\alpha} = 0.0829.$$

Thus, from (14) we have

$$\hat{c} = 0.7101.$$

5   Optimality properties of the LMI-based lower bound

In the previous section an LMI-based lower bound $\hat{c}$ has been computed. The key step in the development of such a bound is the positivity condition of the matrix $\Omega_{2m}(c;\alpha)$. Obviously, such a condition is more stringent than the actual condition of positivity of the corresponding form $\hat{w}(x;c)$. This fact is responsible for making the lower bound $\hat{c}$ in general suboptimal, i.e., strictly lower than the optimal solution c^*.

In this section, we give a result for checking if the lower bound is achieved, i.e., $\hat{c} = c^*$.

Theorem 3.
Let $\hat{\alpha}$ be such that

$$\lambda_{max}\left[-\Omega_{2m}(\hat{c};\hat{\alpha})\right] = 0.$$

Then, the following statements are equivalent

i) $\hat{c} = c^*$

ii) $\exists \hat{x} \in R^n : \begin{cases} \hat{x}^{\{m\}} \in \text{Ker}\left[\Omega_{2m}(\hat{c};\hat{\alpha})\right] \\ u(\hat{x}) = \hat{c} \end{cases}$.

The theorem above suggests the following simple test for checking optimality of $\hat{c}$. Compute a vector of the null space of $\Omega_{2m}(\hat{c};\hat{\alpha})$ and verify if it is equal to $x^{\{m\}}$ for some $x \in R^n$. It can be shown that such a verification amounts to solving a system of n linear equations [10].

Example continued.
Simple numerical computations yield

$$\text{Ker}\left[\Omega_4(\hat{c};\hat{\alpha})\right] = \{\gamma(0.0016, -0.0403, 0.9992)', \ \gamma \in R\}.$$

Since $n = 2$ and $m = 2$, condition ii) of Theorem 3 holds if there exist $x = (x_1, x_2)' \in R^2$ and $\gamma \in R$ such that

$$\begin{cases} x_1^2 & = 0.0016\gamma \\ x_1 x_2 & = -0.0403\gamma \\ x_2^2 & = 0.9992\gamma \\ x_1^2 + x_2^2 & = 0.7101. \end{cases} \tag{15}$$

It is straighforward to verify that the vectors

$$\hat{x} = \pm(0.0339, -0.8420)'$$

solves (15) for $\gamma = 0.7096$. Therefore, the lower bound is optimal, i.e.,

$$c^* = 0.7101$$

and $\hat{x}$ are indeed the tangency points of the dotted circle and the solid curve $(p(x) = 0)$ depicted in Fig. 1.

Finally, note that the solution of (15) is obtained in two successive steps: i) the first and third equations, with γ determined exploiting the fourth equation, are solved for x_1 and x_2; ii) the second equation is verified for those x_1 and x_2.

Theorem 3 provides a simple computational test for checking optimality of the lower bound. It is however important to investigate structural properties of $w(x)$ under which the optimality condition is verified. We have the following general result (see [10] for a proof).

Theorem 4.
Consider Problem II and let
$$w^*(x; c^*) := \sum_{k=0}^{m} v^{(2k)}(x) \frac{u^{m-k}(x)}{c^{*m-k}}.$$
Then, $\hat{c} = c^*$ if and only if $w^*(x; c^*)$ can be written as a sum of squares of real forms of degree m.

The above theorem can be exploited togheter with some properties of real forms [12] to make it possible to derive a-priori conditions ensuring optimality of the lower bound [10].

6   Conclusion

In this paper the problem of determining the minimum euclidean distance of a point from a polynomial surface in R^n is considered. As a first result it is shown that a lower bound to the global minimum is achieved via the solution of a one-parameter family of LMIs, once a suitable change of variables is performed. Each LMI consists in the minimization of the maximum eigenvalue of a symmetric matrix. It is also pointed out that in some cases the solution of a single LMI provides the lower bound. Tightness of this lower bound has been investigated. It is shown that optimality of the lower bound can be readily checked via the solution of a system of linear equations. In addition, it is shown out that lower bound tightness is strictly related to some properties concerning real forms. A simple example is developed throughout the paper to illustrate the main features of the approach.

References

1. J.C. Doyle, J.E. Wall, G. Stein, " Performance and robustness analysis for structured uncertainty," *Proc. 21st IEEE Conf. on Decision and Control*, Orlando, Florida, pp. 629-636, 1982.

2. J.C. Doyle, " Analysis of feedback systems with structured uncertainties," *IEE Proceedings-D Control Theory and Applications*, vol. 129, pp. 242-250, 1982.

3. S.P. Bhattacharyya, H. Chapellat and L. H. Keel, *Robust Control: The Parametric Approach*, Prentice Hall PTR, NJ, 1995.

4. R. Genesio, M. Tartaglia, and A. Vicino, "On the estimation of asymptotic stability region. State of the art and new proposals", *IEEE Transactions on Automatic Control*, vol. 30, pp. 747-755, 1985.

5. H. D. Chiang and J. S. Thorp, "Stability regions of nonlinear dynamical systems: a constructive methodology", *IEEE Transactions on Automatic Control*, vol. 34, pp. 1229-1241, 1989.

6. S. Boyd, L. El Ghaoui, E. Feron and V. Balakrishnan, *Linear Matrix Inequalities in System and Control Theory*, Siam, Philadelphia, 1994.

7. D. Hershkowitz, "Recent directions in matrix stability," *Linear Algebra and its Applications*, vol. 171, pp. 161-186, 1992.

8. H. K. Khalil, *Nonlinear Systems*, McMillan Publishing Company, New York, 1992.

9. A.J. van der Schaft, "$\mathcal{L}_2$-gain analysis of nonlinear state feedback $\mathcal{H}_\infty$-control," *IEEE Transactions on Automatic Control*, vol. 37, pp. 770-784, 1992.

10. G. Chesi, A. Tesi, A. Vicino, R. Genesio, "On convexification of some minimum distance problems," *Research Report DSI-19/98*, Dipartimento di Sistemi e Informatica, Firenze, 1998.

11. Yu. Nesterov and A. Nemirovsky, *Interior Point Polynomial Methods in Convex Programming: Theory and Applications*, Siam, Philadelphia, 1993.

12. G. Hardy, J.E. Littlewood, G. Pólya, *Inequalities: Second edition*, Cambridge University Press, Cambridge, 1988.

Quantified Inequalities and Robust Control

Chaouki T. Abdallah[1], Marco Ariola[2], Peter Dorato[1]; and
Vladimir Koltchinskii[3]

[1] Department of EECE
   University of New Mexico
   Albuquerque, NM 87131, USA.
[2] Dipartimento di Informatica e Sistemistica
   Universitá degli Studi di Napoli Federico II
   Napoli, Italy.
[3] Department of Mathematics and Statistics
   University of New Mexico
   Albuquerque, NM 87131, USA.

Abstract. This paper studies the relationship between quantified multivariable polynomial inequalities and robust control problems. We show that there is a hierarchy to the difficulty of control problems expressed as polynomial inequalities and a similar hierarchy to the methods used to solve them. At one end, we have quantifier elimination methods which are exact, but doubly exponential in their computational complexity and thus may only be used to solve small size problems. The Branch-and-Bound methods sacrifice the exactness of quantifier elimination to approximately solve a larger class of problems, while Monte Carlo and statistical learning methods solve very large problems, but only probabilistically. We also present novel sequential learning methods to illustrate the power of the statistical methods.

1 Introduction

Many control approaches can be reduced to *decidability problems* or to *optimization questions*, both of which can then be reduced to the question of finding a real vector satisfying a set of inequalities. This paper deals mainly with robust control design problems where these inequalities are quantified multivariate polynomial functions of the unknown variables. Given a reasonable amount of resources, Such questions may not be answerable exactly for large-size problems, but recent research has shown that we can "approximately" answer these questions "most of the time", and have "high confidence" in the correctness of the answers.

Many theories come together to explain the results in this paper: Decision theory over the reals, computational complexity, and statistical learning theory. Decision theory was initiated by Tarski [28] who showed that the first-order theory of the real numbers is decidable. Tarski's idea was to eliminate the quantifiers in order to arrive at a semi-algebraic expression which is equivalent to the original quantified expression over the reals. This will be used in this paper to solve "small size" control problems [16]. Computational complexity relies on recent work (1980s) [17] dealing with the efficiency of algorithms and has led to the classification of problems into solvable and hard problems. Work in [9,10,14,24,25] has called attention to the difficulty of solving some control problems from a computational point of view. An approach which is tailored to "mid-size" difficult control problems consists of branch-and-bound algorithms such as the Bernstein Branch and Bound (BBB) [22,23,38]. Finally, statistical learning theory originated during the 1960s and 1970s, and merged with the idea of empirical risk minimization advanced under the theory of empirical processes [31] and with Monte Carlo ideas. The results have been used in [21,26,4,29,5–7,11–13,19], to solve "large-size" *robust analysis* problems while Vidyasagar used learning theory to solve large-size *robust design* problems [32,35]. The approaches discussed in this paper, the *quantifier elimination, branch and bound*, and *statistical learning theory* approaches, may be used in situations where analytical necessary or sufficient conditions are lacking.

The paper is organized as follows: In section 2 we present an overview of the robust control problem dealt with as a decision question. Section 3 reviews decidable but computationally difficult control problems. Section 4 presents a short overview of Branch and Bound methods as one remedy to the computational difficulty of control problems, while section 5 presents stochastic methods as another remedy. The remainder of the paper concentrates on recent results by the authors in improving the stochastic methods to deal with robust control problems. Applications to control design are presented in section 6 while our conclusions are given in section 7.

2   Robust Control Problems

It may be argued that many (if not most) robust control problems may be rephrased as either decision problems or existence questions which admit a yes or no answer, or optimization of performance questions. Note that these two problems are hard to solve "exactly". We will focus first on the control problem design of designing a fixed-order controller for an uncertain single input single output (SISO), linear time invariant (LTI) system shown in Figure 1 which may be stated as a decision problem is as follows,

Problem 1: *Given a real rational function $G(s, X)$, where $X = [x_1 \ x_2 \ \cdots \ x_k]$ is a k-dimensional real vector, does there exist an l-dimensional real vector $Y = [y_1 \ y_2 \ \cdots \ y_l]$, $[\underline{y_i} \leq y_i \leq \overline{y_i}]$, $1 \leq i \leq l$, in the real rational $C(s, Y)$ such that for all (x_i) $[\underline{x_i} \leq x_i \leq \overline{x_i}]$, $1 \leq i \leq k$, the closed-loop* *system $T(s, X, Y)$ satisfies some performance objectives placed on a scalar performance index $\Psi(X, Y)$, including closed-loop stability?*

Note that if either X or Y are known, the problem simplifies to an analysis

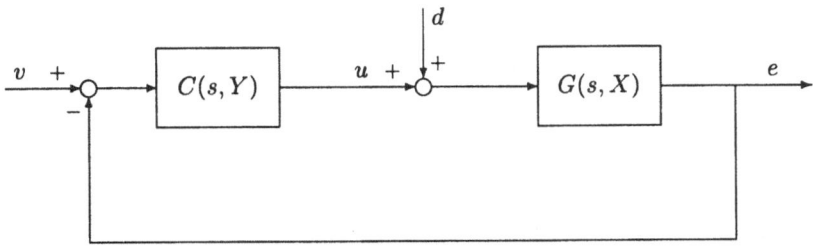

Fig. 1. Feedback Structure for Problem 1

problem. The general decision control problem is very hard because it leads to a nonlinear, partial differential Hamilton-Jacobi-Bellman (HJB) equation (for nonquadratic performance objectives). Note that some control problems (such as the simultaneous stabilization problem of more than 2 plants) are actually undecidable. The control problems we study are converted to a decision problem relating to the satisfiability of multivariate polynomial inequalities which are then decidable using Tarski's decision theory. Many practical robust control design problems can be reduced to the study of *quantified Boolean* formulae of the type

$$\forall (X \in \mathcal{X})[p_1(X, Y) > 0 \wedge p_2(X, Y) > 0 \cdots \wedge p_s(X, Y) > 0] \tag{1}$$

where $\forall$ denotes the logic "for all" operator, and $\wedge$ denotes the logic "and" operator. The functions $p_i(X, Y)$ are multivariate polynomial functions, in the components of the vectors X and Y. The unquantified variable Y in the formula (1) typically represents controller design parameters, while the quantified variable X represent uncertain plant parameters, or frequency variables (for linear problems). A sample of the application of decision theory to linear control problems may be found in [3,16,18]

3   Decision Theory in Control

Until recently, it was felt that decidable problems are practically solved and thus not very interesting. In particular, this would mean that fixed-order controller design are effectively solved. The introduction of Computational complexity theory has since changed this misconception. Computational complexity theory is often used to establish the tractability or intractability of computational problems, and is concerned with the determination of the intrinsic computational difficulty of these problems [17].

One important concept in this theory is that of a *polynomial-time algorithm*. In practice, such an algorithm can be feasibly implemented on a real computer. This is in contrast to an *exponential-time algorithm*, which is only feasible if the problem being solved is extremely small. Unfortunately, it turns out that QE is doubly exponential [8] and thus the control problems introduced in the previous section are not practically solved. The complexity class $\mathcal{P}$ consists of all decision problems that can be decided in polynomial-time, using a Turing machine model of computation. The complexity class $\mathcal{NP}$ consists of all decision problems that can be decided algorithmically in *nondeterministic* polynomial-time. An algorithm is nondeterministic if it is able to choose or guess a sequence of choices that will lead to a solution, without having to systematically explore all possibilities. This model of computation is not realizable, but it is of theoretical importance since it is strongly believed that $\mathcal{P} \neq \mathcal{NP}$. In other words, these two complexity classes form an important boundary between the tractable (or easy) and intractable (or difficult) problems. A problem is said to be $\mathcal{NP}$-hard if it is as hard as any problem in $\mathcal{NP}$. Thus, if $\mathcal{P} \neq \mathcal{NP}$, the $\mathcal{NP}$-hard problems can only admit deterministic solutions that take an unreasonable (i.e., exponential) amount of time, and they require (unattainable) nondeterminism in order to achieve reasonable (i.e., polynomial) running times.

Recently, many decidable control problems have been shown to be $\mathcal{NP}$-complete (or $\mathcal{NP}$-hard) [25,24,10,14,30]. For given $N > 2$ linear systems, restricting the stabilizing compensator to be static (or dynamic but of a given order) makes their simultaneous stabilization problem decidable (although inefficiently) using the Tarski-Seidenberg approach as discussed before. The question then becomes how to deal with decidable but difficult control problems? We actually have 2 possibilities: 1) Limit the class of systems (linear, minimum-phase, passivity, etc.), or 2) Soften the goal for the class of systems we are interested in. The traditional approach of control engineers has been to limit the class of systems under study, but a recent movement to soften the goal as illustrated by the Bernstein-Branch-and-Bound (BBB) and stochastic methods is beginning to take hold.

4   Bernstein-Branch-and-Bound Methods

It is known that multivariate Bernstein polynomials can be used to estimate the range (Maximum and minimum value) of a polynomial function over a given "box" (simple interval ranges for the polynomial variables). In the BBB approach to the quantified MPI problem, the bounds on any given polynomial function $p_i(X, Y)$, i.e. $\underline{p}_i < p_i(X, Y) < \bar{p}_i$ are used to keep boxes where all the inequalities are satisfied and prune boxes that where at least one inequality is not satisfied. For example, a box is kept if for all i, $\underline{p}_i > 0$, and a box is pruned if $\bar{p}_i < 0$ for some i. If neither of these two cases hold the box is subdivided (branching), and the sub-boxes are explored via the polynomial bounds (bounding) as before. The final solution set Y is obtained by putting together boxes where for given values of Y, all the inequalities $p_i(X, Y) > 0$ are satisfied for all $X \in \mathcal{X}$. It can be shown that as the "size" of the boxes decreases, the difference $\bar{p}_i - \underline{p}_i$ approaches zero. Thus one can compute the region Y as accurately as desired by exploring arbitrarily small boxes [23].

Example 1 (Learning stability regions using BBB) *This problem has been presented by different authors and in different forms. Our treatment follows that of [2,23]. Given is a polynomial $p(s)$*

$$s^3 + (a_1 + a_2 + 1)s^2 + (a_1 + a_2 + 3)s + (6a_1 + 6a_2 + 2a_1 a_2 + 1 + r^2)$$

with the appropriate value of r we are to find the stability region in the coefficients space. This is obtained from Maxwell's criterion as the region defined by the inequalities,

$$0 < (a_1 + a_2 + 1)(a_1 + a_2 + a_3) - (6a_1 + 6a_2 + 2a_1 a_2 + 1 + r^2)$$
$$0 < a_1 + a_2 + 1$$
$$0 < a_1 + a_2 + a_3$$
$$0 < (a_1 + a_2 + a_3) - (6a_1 + 6a_2 + 2a_1 a_2 + 1 + r^2)$$

If we use a BBB approach, we obtain Figure 2, which is approximating a circular instability region of radius $r = 0.25$, and given by, $(a_1 - 1/2)^2 + (a_2 - 1/2)^2 < (1/2)^2$ Thus, the dark region in Figure 2 contains points where the inequalities are guaranteed to hold.

5   Stochastic Methods

Randomized algorithms is yet another approach which forces us to soften the goal but allows us to consider a larger class of control problems [33]. We can thus answer more questions albeit probabilistically which sometimes may be good enough. In order to fix ideas, we formulate an LTI robust control problem as follows [33],

Problem 2: *Given a closed-loop system $T(s, X, Y)$ with a performance measure $\Psi(X, Y)$, where X, Y are random real-valued vectors, find a vector Y_0, if one exists, of controller parameters which has a high probability of minimizing the expected value with respect to X of an appropriate function $f(X, Y)$ of $\Psi(X, Y)$.*

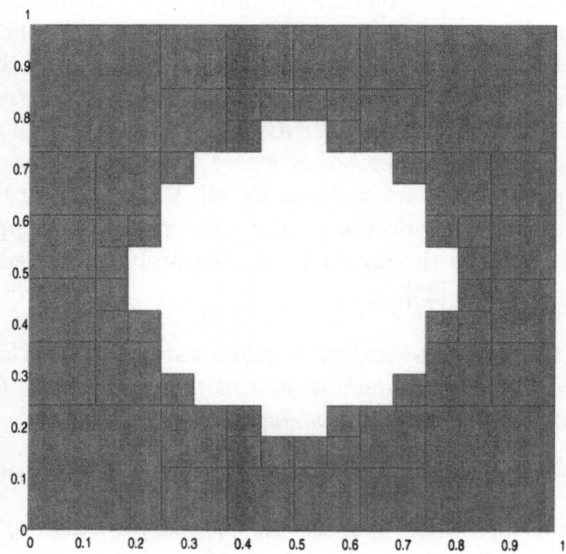

Fig. 2. Stability region of Example 1 using BBB

The related decision problem is to ascertain the existence of a vector Y_0 such that a certain level γ is achieved by $\mathbb{E}f(X, Y)$. Note that our problem has been changed from a deterministic decision problem to a probabilistic optimization problem. Also note that the randomness of X and Y is used to open the door for Monte-Carlo and statistical learning methods. Finally, we have converted a worst-case scenario (guaranteed-cost) into an average-case problem. In the context of stabilization, let $\Psi(X, Y) = 0$ if $T(s, X, Y)$ is stable and $\Psi(X, Y) = 1$ otherwise. By minimizing $\Psi(X, Y)$ we are actually maximizing the volume (or number in case of finite number of plants) which may be stabilized with $K(s, Y_0)$. In fact, let

$$g_Y(X) = \begin{cases} 1 & \Psi(X, Y) = 1 \\ 0 & \Psi(X, Y) = 0 \end{cases}$$

and $\mathcal{G} = \{g_Y(.); Y \in \mathcal{Y}\}$. The purpose of control is thus to choose Y_0 and thus the corresponding controller to stabilize the maximum number of plants. That may be achieved by minimizing the expected value $f(Y) = \mathbb{E}[g_Y(X)]$. Another interpretation is that we can then ascertain with confidence $1 - f(Y_0)$ that the controller $K(Y_0)$ stabilizes a random $G(X)$.

In [33], Vidyasagar introduced the following types of minima, in order to use statistical learning theory to design fixed-order robust controllers, which minimize the performance index in Problem 2.

Definition 1 *Let* $R : \mathcal{Y} \longrightarrow \mathbb{R}$ *and* $\varepsilon > 0$ *be given. A number* $R_0 \in \mathbb{R}$ *is said to be an approximate near minimum of* R *to accuracy* ε, *if* $|R_0 - \inf_{Y \in \mathcal{Y}} R(Y)| \leq \varepsilon$

Definition 2 *Suppose* $R : \mathcal{Y} \longrightarrow \mathbb{R}$, Q *is a given probability measure on* $\mathcal{Y}$, *and* $\alpha > 0$ *be given. A number* $R_0 \in \mathbb{R}$ *is a probable near minimum of* R *to level* α *if there exists a measurable set* $\mathcal{S} \subseteq \mathcal{Y}$ *with* $Q(\mathcal{S}) \leq \alpha$ *such that* $\inf_{Y \in \mathcal{Y}} R(Y) \leq R_0 \leq \inf_{Y \in \mathcal{Y} \setminus \mathcal{S}} R(Y)$ *where* $\mathcal{Y} \setminus \mathcal{S}$ *is the complement of the set* $\mathcal{S}$ *in* $\mathcal{Y}$.

This leads to our first stochastic algorithm.

Algorithm 1 *Given: A probability measure* Q *on* $\mathcal{Y}$, *a measurable function* $f : \mathcal{Y} \longrightarrow \mathbb{R}$, *a level parameter* $\alpha \in (0,1)$, *and a confidence parameter* $\delta \in (0,1)$. *Choose*

$$n \geq \frac{\log(1/\delta)}{\log[1/(1-\alpha)]}$$

and generate i.i.d. samples $Y_1, Y_2, \cdots, Y_n \in \mathcal{Y}$ *according to* Q. *Let* $\underline{f} = \min_{1 \leq i \leq n} f(Y_i)$ *Then, with confidence at least* $1 - \delta$, $\underline{f}$ *is a probable near minimum of* f *to level* α. *In fact* Y_0 *where* $\underline{f} = f(Y_0)$ *is a minimizer.*

This algorithm applies very generally, and is a basic Monte Carlo algorithm.

Example 2 (Learning stability regions) *This this same problem as that of example 1 which was solved using a BBB algorithm. The result is shown in Figure 3*

5.1   Theory of Empirical Processes

Let $(S, \mathcal{A})$ be a measurable space and let $\{X_n\}_{n \geq 1}$ be a sequence of independent identically distributed (i.i.d) observations in this space with common distribution P. We assume that this sequence is defined on a probability space $(\Omega, \Sigma, \mathbb{P})$. Denote by $\mathcal{P}(S) := \mathcal{P}(S, \mathcal{A})$ the set of all probability measures on $(S, \mathcal{A})$. Suppose $\mathcal{P} \subset \mathcal{P}(S)$ is a class of probability distributions such that $P \in \mathcal{P}$. In particular, if one has no prior knowledge about P, then $\mathcal{P} = \mathcal{P}(S)$. In this case, we are in the setting of *distribution free learning*. One application of statistical learning theory is to the problem of *risk minimization*. This basic problem plays an important role in randomized (Monte Carlo) algorithms for robust control problems, as has been shown by Vidyasagar [35]. Given a class $\mathcal{F}$ of $\mathcal{A}$-measurable functions f from S into $[0,1]$, the risk functional is defined as

$$R_P(f) := P(f) := \int_S f dP := \mathbb{E}f(X), \ f \in \mathcal{F}.$$

The goal is to find a function f_P that minimizes R_P on $\mathcal{F}$. Typically, the distribution P is unknown and the solution of the risk minimization problem

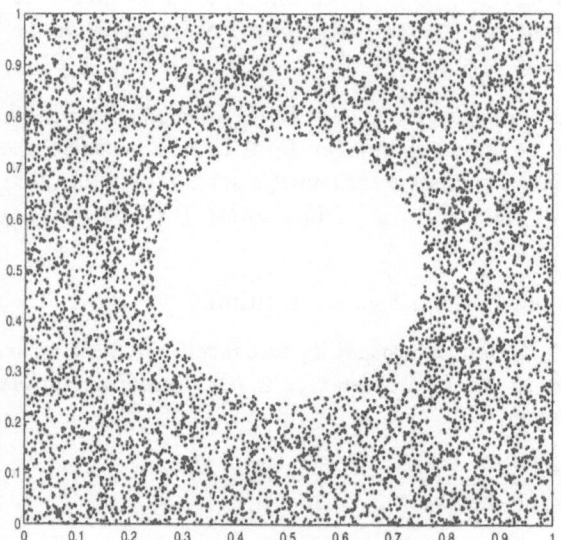

Fig. 3. Monte Carlo simulation for a hole in the stability region

is to be based on a sample $(X_1, \ldots, X_n)$ of certain size n of independent observations from P. In this case, the goal of statistical learning is more modest: given $\varepsilon > 0, \delta \in (0, 1)$, to find an estimate $\hat{f}_n \in \mathcal{F}$ of f_P, based on the data $(X_1, \ldots, X_n)$, such that

$$\sup_{P \in \mathcal{P}} \mathbb{P}\{R_P(\hat{f}_n) \geq \inf_{f \in \mathcal{F}} R_P(f) + \varepsilon\} \leq \delta. \tag{2}$$

In other words, one can write that with probability $1 - \delta$, $R_P(\hat{f}_n)$ is within ε of $\inf_{f \in \mathcal{F}} R_P(f)$. The method of *empirical risk minimization* is widely used in learning theory. Namely, the unknown distribution P is replaced by *the empirical measure P_n*, defined as

$$P_n(A) := \frac{1}{n} \sum_{k=1}^{n} I_A(X_k), \ A \in \mathcal{A}$$

where $I_A(x) = 1$ for $x \in A$ and $I_A(x) = 0$ for $x \notin A$. The risk functional R_P is replaced by the empirical risk R_{P_n}, defined by

$$R_{P_n}(f) := P_n(f) := \int_S f dP_n := \frac{1}{n} \sum_{k=1}^{n} f(X_k), \ f \in \mathcal{F}.$$

The problem is now to minimize the empirical risk R_{P_n} on $\mathcal{F}$. Let $f_{P_n} \in \mathcal{F}$ be a function that minimizes R_{P_n} on $\mathcal{F}$. In what follows f_{P_n} is used as our learning algorithm, i.e. $\hat{f}_n := f_{P_n}$. Determining the sample complexity of the empirical risk minimization method is definitely one of the central and most challenging problems of statistical learning theory (see, e.g., [15], or Vidyasagar [33] for the relevant discussion in the context of robust control problems). A reasonable upper bound for the sample complexity can be obtained by finding the minimal value of n for which the expected value $\mathbb{E}f(X)$ is approximated uniformly over the class $\mathcal{F}$ by the empirical means with given accuracy ε and confidence level $1 - \delta$. More precisely, denote

$$N(\varepsilon, \delta) := N_{\mathcal{F},\mathcal{P}}^L(\varepsilon, \delta) := \min\left\{ n \geq 1 : \sup_{P \in \mathcal{P}} \mathbb{P}\{\|P_n - P\|_{\mathcal{F}} \geq \varepsilon\} \leq \delta \right\},$$

where $\| \cdot \|_{\mathcal{F}}$ is the sup-norm in the space $\ell^\infty(\mathcal{F})$ of all uniformly bounded functions on $\mathcal{F}$. Let us call the quantity $N(\varepsilon; \delta)$ *the (lower) sample complexity of empirical approximation* on the class $\mathcal{F}$. Unfortunately, the quantity $N_{\mathcal{F},\mathcal{P}}^L(\varepsilon, \delta)$ is not known for most of the nontrivial examples of function classes and only rather conservative upper bounds for this quantity are available. These bounds are expressed in terms of various entropy characteristics and combinatorial quantities, such as VC-dimensions, which themselves are not always known precisely and replaced by the upper bounds [33].

Going back to our control motivation, we note that our problem involves also the finding of the minimum of a certain performance objective or more precisely, finding the controller parameters which will correspond to such minimum. What these randomized algorithms provide is a probably approximate near minimum of a stochastic process R (say, $R := R_{P_n}$, see the definition above) with confidence $1 - \delta$, level α and accuracy ε (see [33]).

Definition 3 *Suppose $R : \mathcal{Y} \to \mathbb{R}$, that P is a given probability measure on $\mathcal{Y}$, and that $\alpha \in (0,1)$, $\delta \in (0,1)$ and $\varepsilon > 0$ are given. A number R_0 is a probably approximate near minimum of R with confidence $1 - \delta$, level α and accuracy ε, if*

$$Prob\left\{ \inf_{Y \in \mathcal{Y}} R(Y) - \varepsilon \leq R_0 \leq \inf_{Y \in \mathcal{Y}\setminus S} R(Y) + \varepsilon \right\} \geq 1 - \delta$$

with some measurable set $S \subseteq \mathcal{Y}$ such that $Q(S) \leq \alpha$.

Sequential Learning Algorithms Next, we review our recent work on sequential algorithms for a general problem of empirical risk minimization. The full results appear elsewhere [20]. This approach does not depend on the explicit calculation of the VC-dimension [31,34], although its finiteness remain critical to the termination of the design algorithm. The sequential algorithms chosen are based on *Rademacher bootstrap. An important feature of our approach is the randomness of the sample size for which a given accuracy of learning is achieved with a guaranteed probability. Thus, the sample complexity of our method of learning is rather a random variable. Its value is* *not known in advance and is to be determined in the process of learning. The lower bound for this random variable is the value of the sample size which the sequential learning algorithm starts working with. The upper bounds for the random sample complexity are of the same order of magnitude as the standard conservative upper bounds for the sample complexity of empirical risk minimization algorithms. The sequential method of learning, described below, is designed to overcome some of the difficulties encountered with the standard learning methods. We start with some basic definitions.*

Definition 4 *Let $\{\Sigma_n\}_{n \geq 1}$ be a filtration of σ-algebras (i.e. for all $n \geq 1$ $\Sigma_n \subset \Sigma_{n+1}$) such that $\bar{\Sigma}_n \subset \Sigma, n \geq 1$ and X_n is Σ_n-measurable. Less formally, Σ_n consists of the events that occur by time n (in particular, the value of random variable X_n is known by time n).*

Definition 5 *A random variable τ, taking positive integer values, will be called a stopping time iff, for all $n \geq 1$, we have $\{\tau = n\} \in \Sigma_n$. In other words, the decision whether $\tau \leq n$, or not, depends only on the information available by time n.*

Given $\varepsilon > 0$ and $\delta \in (0,1)$, let $\bar{n}(\varepsilon, \delta)$ denote the initial sample size of our learning algorithms. We assume that $\bar{n}$ is a non-increasing function in both ε and δ. Denote by $\mathcal{T}(\varepsilon, \delta) := \mathcal{T}_{\mathcal{F}, \mathcal{P}}(\varepsilon, \delta)$ the set of all stopping times τ such that $\tau \geq \bar{n}(\varepsilon; \delta)$ and

$$\sup_{P \in \mathcal{P}} \mathbb{P}\{\|P_\tau - P\|_{\mathcal{F}} \geq \varepsilon\} \leq \delta.$$

If now $\tau \in \mathcal{T}(\varepsilon, \delta)$ and $\hat{f} := f_{P_\tau}$ is a function that minimizes the empirical risk based on the sample $(X_1, \ldots, X_\tau)$ it is possible to prove that

$$\sup_{P \in \mathcal{P}} \mathbb{P}\left\{R_P(f_{P_\tau}) \geq \inf_{f \in \mathcal{F}} R_P(f) + 2\varepsilon\right\} \leq \delta.$$

The questions, though, are how to construct a stopping time from the set $\mathcal{T}(\varepsilon, \delta)$, based only on the available data (without using the knowledge of P) and which of the stopping times from this set is best used in the learning algorithms. The following definition could be useful in this connection.

Definition 6 *A parametric family of stopping times $\{\nu(\varepsilon, \delta) : \varepsilon > 0, \delta \in (0,1)\}$ is called strongly (statistically) efficient for the class $\mathcal{F}$ with respect to $\mathcal{P}$ if and only if (iff) there exist constants $K_1 \geq 1, K_2 \geq 1$ and $K_3 \geq 1$ such that for all $\varepsilon > 0$ and $\delta \in (0,1)$, $\nu(\varepsilon, \delta) \in \mathcal{T}(K_1\varepsilon, \delta)$ and for all $\tau \in \mathcal{T}(\varepsilon, \delta)$, $\sup_{P \in \mathcal{P}} \mathbb{P}\{\nu(K_2\varepsilon, \delta) > \tau\} \leq K_3\delta.$*

Thus, using strongly efficient stopping time $\nu(\varepsilon; \delta)$ allows one to solve the problem of empirical approximation with confidence $1 - \delta$ and accuracy $K_1\varepsilon$. With probability at least $1 - K_3\delta$, the time required by this algorithm is less than the time needed for any sequential algorithm of empirical approximation with accuracy $\varepsilon/(K_1 K_2)$ and confidence $1 - \delta$.

Definition 7 *We call a family of stopping times* $\{\nu(\varepsilon, \delta) : \varepsilon > 0, \delta \in (0,1)\}$ *weakly (statistically) efficient for the class* $\mathcal{F}$ with respect to $\mathcal{P}$ iff there exist constants $K_1 \geq 1, K_2 \geq 1$ and $K_3 \geq 1$ such that for all $\varepsilon > 0$ and $\delta \in (0,1)$ $\nu(\varepsilon, \delta) \in \mathcal{T}(K_1\varepsilon, \delta)$ and $\sup_{P \in \mathcal{P}} \mathbb{P}\{\nu(K_2\varepsilon, \delta) > N(\varepsilon; \delta)\}\} \leq K_3\delta$.

Using weakly efficient stopping time $\nu(\varepsilon; \delta)$ also allows one to solve the problem of empirical approximation with accuracy $K_1\varepsilon$ and confidence $1-\delta$. With probability at least $1 - K_3\delta$, the time required by this algorithm, is less than the sample complexity of empirical approximation with accuracy $\varepsilon/(K_1 K_2)$ and confidence $1 - \delta$. Note that, under the assumption $N(\varepsilon; \delta) \geq \bar{n}(\varepsilon; \delta)$, we have $N(\varepsilon, \delta) \in \mathcal{T}(\varepsilon, \delta)$. Hence, any strongly efficient family of stopping times is also weakly efficient. In fact, it turns out that there exists a weakly efficient family of stopping times that is not strongly efficient [20]. We show below how to construct efficient stopping times for empirical risk minimization problems. The construction is based on a version of bootstrap. Let $\{r_n\}_{n\geq 1}$ be a *Rademacher sequence* (i.e. a sequence of i.i.d. random variables taking values $+1$ and -1 with probability $1/2$ each). We assume, in addition, that this sequence is independent of the observations $\{X_n\}_{n\geq 1}$. Let $\lfloor \cdot \rfloor$ denote the floor of the argument, and define

$$\nu(\varepsilon, \delta) := \nu_{\mathcal{F}}(\varepsilon, \delta)$$

$$:= \min\{n : \|n^{-1} \sum_{j=1}^{n} r_j \delta_{X_j}\|_{\mathcal{F}} \leq \varepsilon, \ n = 2^k \bar{n}(\varepsilon, \delta), \ k = 0, 1, \ldots\}$$

Theorem 1 *Suppose that* $\bar{n}(\varepsilon, \delta) \geq \lfloor \frac{4}{\varepsilon^2} \log(\frac{4}{\delta}) \rfloor + 1$. *Then, for all* $\varepsilon > 0, \delta \in (0,1)$, $\nu(\varepsilon; \delta) \in \mathcal{T}(K_1\varepsilon; \delta)$ *with* $K_1 = 5$. *Moreover, suppose that* $N(\varepsilon, \delta) \geq \bar{n}(\varepsilon, \delta) \geq \lfloor \frac{4}{\varepsilon^2} \log(\frac{4}{\delta}) \rfloor + 1$. *Then* $\{\nu_{\mathcal{F}}(\varepsilon, \delta) : \varepsilon > 0, \delta \in (0, 1/2)\}$ *is a weakly efficient family of stopping times for any class* $\mathcal{F}$ *of measurable functions from S into $[0, 1]$ with respect to the set $\mathcal{P}(S)$ of all probability distributions on S.*

The following algorithm summarizes this approach and was proven in [20].

Algorithm 2 *Given:*

- *Sets $\mathcal{X}$ and $\mathcal{Y}$,*
- *Probability measures P on $\mathcal{X}$ and Q on $\mathcal{Y}$,*
- *A measurable function $f : \mathcal{X} \times \mathcal{Y} \longrightarrow [0, 1]$, and*
- *An accuracy parameter $\varepsilon \in (0, 1)$, a level parameter $\alpha \in (0, 1)$, and a confidence parameter $\delta \in (0, 1)$.*

Let $R_P(\cdot) = \mathbb{E}_P[f(X, \cdot)]$ and

$$R_{P_n}(\cdot) = \frac{1}{n} \sum_{j=1}^{n} f(X_j, \cdot)$$

Then,

1. *Choose m independent controllers parameters according to Q where*

$$m \geq \frac{\log(2/\delta)}{\log[1/(1-\alpha)]}$$

2. *Choose n independent plants parameters according to P, where* $K_1 = 5$, *and*

$$n = \left\lfloor \frac{4K_1^2}{\varepsilon^2} \log\left(\frac{8}{\delta}\right) \right\rfloor + 1$$

3. *Evaluate the stopping variable*

$$\gamma = \max_{1 \leq i \leq m} \left| \frac{1}{n} \sum_{j=1}^{n} r_j f(X_j, Y_i) \right|$$

where r_j are Rademacher random variables, (also independent of the plant sample). If $\gamma > \frac{\varepsilon}{K_1}$, add n more independent plants with parameters having distribution P to the plant samples, set $n := 2n$ and repeat step 3

4. *Choose the controller which minimizes the cost function R_{P_n}. Then with confidence at least $1 - \delta$, this controller minimizes R_P to a level α and accuracy ε.*

6   Applications To Robust Control Design

The benchmark problem was originally proposed in [37]. The plant consists of a two-mass/spring system with non-collocated sensor and actuator. The system can be represented in dimensionless state-space form as

$$\begin{bmatrix} \dot{x}_1 \\ \dot{x}_2 \\ \dot{x}_3 \\ \dot{x}_4 \\ \dot{u} \end{bmatrix} = \begin{bmatrix} 0 & 0 & 1 & 0 & 0 \\ 0 & 0 & 0 & 1 & 0 \\ \frac{-k}{m_1} & \frac{k}{m_1} & \frac{-c}{m_1} & \frac{c}{m_1} & \frac{f}{m_1} \\ \frac{k}{m_2} & \frac{-k}{m_2} & \frac{c}{m_2} & \frac{-c}{m_2} & 0 \\ 0 & 0 & 0 & 0 & \frac{-1}{\tau} \end{bmatrix} \begin{bmatrix} x_1 \\ x_2 \\ x_3 \\ x_4 \\ u \end{bmatrix} + \begin{bmatrix} 0 \\ 0 \\ 0 \\ 0 \\ \frac{1}{\tau} \end{bmatrix} u_c + \begin{bmatrix} 0 \\ 0 \\ 0 \\ \frac{1}{m_2} \\ 0 \end{bmatrix} w \qquad (3)$$

$$y = x_2 + v \qquad (4)$$

$$z = x_2 \qquad (5)$$

where x_1 and x_2 are the positions of the masses, u is the control input force, w is the plant disturbance, y is the sensor measurement corrupted by the noise v, and z is the output to be controlled. The six uncertain parameters are assumed to be uniform independent random variables in the following intervals: $0.5 < k < 2$, $0.5 < m_1 < 1.5$, $0.5 < m_2 < 1.5$, $0 < c < 0.1$, $0.9 < f < 1.1$ and $0.001 < \tau < 0.4$. We shall denote by $X \in \mathcal{X} \subseteq \mathbb{R}^6$ the vector of these uncertain parameters $X = [k \ m_1 \ m_2 \ c \ f \ \tau]^T$. Three design problems were proposed in [37].

Pb 1: The first problem requires:

1. Closed-loop stability for $m_1 = m_2 = 1$ and $0.5 < k < 2$;
2. A 15 s settling time for unit disturbance impulse for the nominal plant $m_1 = m_2 = k = 1$;
3. The minimization of the control effort and of the controller complexity.

Pb 2: In the second design problem, the unit disturbance impulse is replaced by a sinusoidal disturbance of frequency 0.5 rad/s and constant but unknown amplitude and phase. Asymptotic rejection of the disturbance is required within 20 s for $m_1 = m_2 = 1$ and $0.5 < k < 2$.

Pb 3: The third design problem is the same as Pb 1 except that the parameters m_1, m_2 and k are uncertain with nominal value 1.

Many controllers were proposed for this problem. They are collected and analyzed in [27], where the authors, after evaluating the nominal performance, carry out a Stochastic Robustness Analysis in order to analyze the behavior when the plant parameters change. Based on the specifications, our target is to design a fixed-structure controller such that 1) The nominal plant is stabilized, 2) The 15 seconds settling time specification is satisfied for the nominal plant, 3) The control effort does not exceed a one unit saturation limit in response to a unit w disturbance, for the nominal plant, and 4) A certain cost function is minimized. This cost function accounts for the closed-loop stability and the performance in the presence of parameter variations. Note that due to teh size and complexity of this problem, both decision theory methods and BBB methods are certain to fail. Using the randomized Algorithm 2, we shall design two fixed-order controllers for the plant (3)–(5). In this section, we consider a second-order then a third-order controller and show that higher-order controllers, achieve better performance than lower-order ones. Denoting by $Y \in \mathcal{Y} \subseteq \mathbb{R}^l$ the vector of controller coefficients, the two chosen controllers have the following structures

$$K_1(s, Y) = \frac{a_0 s^2 + a_1 s + a_2}{s^2 + b_1 s + b_2}, \quad Y = \begin{bmatrix} a_0 & a_1 & a_2 & b_1 & b_2 \end{bmatrix}^T \tag{6}$$

$$K_2(s, Y) = \frac{a_1 s^2 + a_2 s + a_3}{s^3 + b_1 s^2 + b_2 s + b_3}, \quad Y = \begin{bmatrix} a_1 & a_2 & a_3 & b_1 & b_2 & b_3 \end{bmatrix}^T \tag{7}$$

The coefficients of the controllers are chosen to have uniform distributions. For the controller (6) these coefficients take values in the intervals

$$a_0 \in [0.5, 10], \ a_1 \in [-2, -0.5], \ a_2 \in [-0.3, -0.1], \ b_1 \in [1, 5], \ b_2 \in [1, 6]$$

whereas for the controller (7) they take values in

$$a_1 \in [-50, 50], \ a_2 \in [-120, -40], \ a_3 \in [-40, -10],$$
$$b_1 \in [70, 170], \ b_2 \in [80, 160], \ b_3 \in [100, 140]$$

In order to use the randomized algorithm methodology, this problem has been reformulated in the following way (see also [35], [20]). Let us define a cost function

$$\Psi(Y) = max\{\psi_1(Y), \psi_2(Y)\} \tag{8}$$

where

$$\psi_1(Y) = \begin{cases} 1 \text{ if the nominal plant is not stabilized or if the settling} \\ \quad \text{time and the control limit specifications are} \\ \quad \text{not both satisfied for the nominal plant,} \\ 0 \text{ otherwise} \end{cases} \tag{9}$$

and

$$\psi_2(Y) = E\left(\zeta(X,Y)\right) \tag{10}$$

where E indicates the *expected value* with respect to X, and

$$\zeta(X,Y) = \begin{cases} 1 \text{ if the randomly generated plant is not stabilized} \\ \frac{2}{3} \text{ if both the control limit and the settling time} \\ \quad \text{specifications are not satisfied} \\ \frac{1}{3} \text{ if either the control limit or the settling time} \\ \quad \text{specification is not satisfied} \\ 0 \text{ otherwise} \end{cases}$$

Our aim is to minimize the cost function (8) over $\mathcal{Y}$. The optimal controller is then characterized by the vector of parameters Y^* for which

$$\Psi^* := \Psi(Y^*) = \inf_{Y \in \mathcal{Y}} \Psi(Y) \tag{11}$$

Finding the vector Y^* which minimizes (11) would imply the evaluation of the expected value in (10) and then the minimization of (8) over the set $\mathcal{Y}$. What we shall find is a probably approximate near minimum of $\Psi(Y)$ with confidence $1 - \delta$, level α and accuracy ϵ. In both cases of our controllers, the procedure needed just one iteration to converge, i.e. $k = 1$. Therefore, for $\delta = 0.05$, $\alpha = 0.005$ and $\epsilon = 0.1$, m evaluated to 736 controllers and n evaluated to $50,753$ plants. In Figure 4, the stopping variables in the two cases are shown. The *suboptimal* controllers are

$$K_1(s) = \frac{1.1110s^2 - 1.7393s - 0.2615}{s^2 + 3.6814s + 2.9353} \tag{12}$$

$$K_2(s) = \frac{31.9432s^2 - 76.6527s - 12.7876}{s^3 + 92.1586s^2 + 123.3358s + 131.8229} \tag{13}$$

and the corresponding values of the cost function are $\hat{\Psi}_m(Y_{opt}) = 0.2683$ for (12) and $\hat{\Psi}_m(Y_{opt}) = 0.2062$ for (13). As expected, with a more complex controller we get a better result.

We shall compare the performance of the two controllers (12)–(13) with those of 4 other controllers analyzed in [27] with the same structure as the ones proposed here.

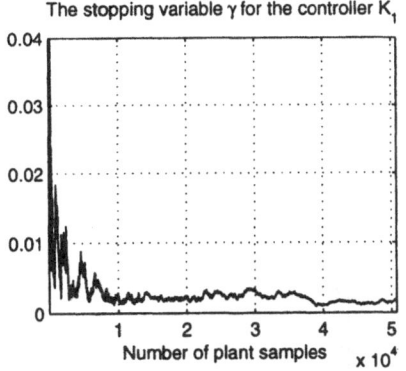

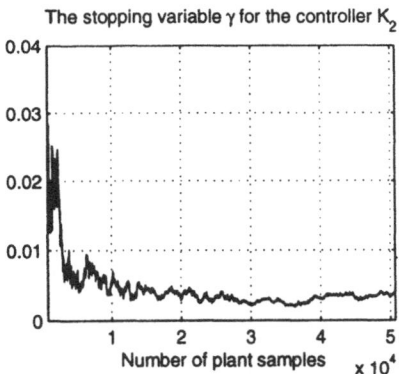

Fig. 4. The stopping variables

First an analysis of the nominal performance is carried out. This analysis is summarized in Table 1, which shows the settling time T_S, the magnitude of the maximum control (u_{max}) used when rejecting the impulse disturbance w, the gain margin (GM), the phase margin (PM), the magnitude of the steady state response (SR) to the 0.5 rad/s sinusoidal disturbance and the covariance of control response (U_{cov}) to a measurement noise v with unit standard deviation. The two controllers designed with the statistical approach compare favorably with the others: as expected, the specifications on the settling time T_S and on the maximum control u_{max} are met. Moreover as a result of the chosen cost function (8), the two controllers K_1 and K_2 show good stability margins.

Controller	Order	T_S (s)	u_{max}	GM (dB)	PM (deg)	SR (dB)	U_{cov}
K_1	2	12.3	0.408	5.75	31.9	11.6	∞
C	2	19.7	0.468	3.27	26.5	13.3	∞
E	2	18.2	0.884	2.39	22.0	17.1	∞
K_2	3	13.2	0.450	3.64	37.7	10.1	6.33
A	3	21.0	0.514	2.56	26.7	10.1	6.30
B	3	19.5	0.469	3.27	26.8	13.2	13.02

Table 1. Nominal Performance

The performance in the presence of the parameter variations are quantified using a Monte Carlo evaluation. According to the distributions of the parameters, 20,000 plants (see [27] for the choice of this number) are randomly generated and estimates of the following three metrics are calculated

1. P_I: Probability of instability.
2. P_{T_S}: Probability of exceeding the settling time.

3. P_u: Probability of exceeding the control limit.

As shown in Table 2, the two controllers designed with the statistical approach exhibit a better behavior in all the three cases. The data for the controllers A, B, C and E are taken from [27]

Controller	P_I	P_{T_S}	P_u
K_1	0.002	0.803	0.002
C	0.041	0.874	0.041
E	0.125	0.999	0.409
K_2	0.033	0.547	0.033
A	0.165	0.793	0.165
B	0.039	0.963	0.047

Table 2. Robust Performance

7 Conclusions

This paper presented an overview of the interplay between quantified multinomial inequalities and robust control problems. We started out by pointing out that both existence and performance problems in robust control design for a large class of systems may be mathematically framed as a question of solving quantified multinomial inequalities. We then studied the decidability and computational difficulty of these problems and concluded that depending on the size of the control problem, different methods exist to solve them. At one end of the hierarchy, quantifier elimination methods can solve small size control problems exactly but are highly inefficient as the size of the problem grows. Branch-and-Bound methods may be used for medium size problems, and stochastic methods may be used for large-size problems. We concentrated on the stochastic methods since they are less known in the control literature and introduced a new sequential learning algorithm based on Bootstrap sampling. It should be noted that the sequential learning methodology presented in this paper can be used in many other application areas: one only needs to have an efficient analysis tool in order to convert it to an efficient design methodology. Our future research is concentrating at the theoretical level in obtaining better optimization algorithms and at the application level in designing software modules for linear and nonlinear control design.

References

1. C. Abdallah, P. Dorato, W. Yang, R. Liska, and S. Steinberg. Application of quantifier elimination theory to control system design. In *Proceedings of the*

4th IEEE Mediterranean Symposium on Control & Automation, pages 41–44, Chania, Crete, Greece, 1996.

2. J. Ackermann, H. Hu, and D. Kaesbauer. Robustness analysis: A case study. In *Proc. IEEE Conf. on Dec. and Control*, pages 86–91, Austin, TX, 1988.

3. B. Anderson, N. Bose, and E. Jury. Output Feedback and related problems-Solution via Decision Methods. *IEEE Trans. on Automatic Control*, AC-20:53–65, 1975.

4. E. Bai, R. Tempo, and M. Fu. Worst case properties of the uniform distribution and randomized algorithms for robustness analysis. In *Proc. IEEE American Control Conf.*, pages 861–865, Albuquerque, NM, 1997.

5. B. Barmish and C. Lagoa. The uniform distribution: A rigorous justification for its use in robustness analysis. *Mathematics of Control, Signals, and Systems*, 10:203–222, 1997.

6. B. Barmish, C. Lagoa, and R. Tempo. Radially truncated uniform distributions for probabilistic robustness of control systems. In *Proc. IEEE American Control Conf.*, pages 853–857, Albuquerque, NM, 1997.

7. B. Barmish and R. Tempo. Probabilistic robustness: A new line of research. Tutorial Workshop, CDC San Diego, CA, 1997.

8. S. Basu, R. Pollack, and M. Roy. On the combinatorial and algebraic complexity of quantifier elimination. In *Proceedings of the 35th Symposium on Foundations of Computer Science*, pages 632–641, Santa Fe, NM, 1994.

9. V. Blondel. *Simultaneous Stabilization of Linear Systems*. Springer-Verlag, London, 1st edition, 1994.

10. V. Blondel and J. Tsitsiklis. NP-hardness of some linear control design problems. *SIAM Journal of Control and Optimization*, 35:2118–2127, 1997.

11. X. Chen and K. Zhou. On the probabilistic characterization of model uncertainty and robustness. In *Proc. IEEE Conf. on Dec. and Control*, pages 3816–3821, San Diego, CA, 1997.

12. X. Chen and K. Zhou. A probabilistic approach for robust control. In *Proc. IEEE Conf. on Dec. and Control*, pages 4894–4895, San Diego, CA, 1997.

13. X. Chen and K. Zhou. Constrained optimal synthesis and robustness analysis by randomized algorithms. In *Proc. IEEE American Control Conf.*, pages 1429–1433, Philadelphia, PA, 1998. Vol. 33.

14. G. Coxson. *Computational Complexity of Robust Stability and Regularity in Families of Linear Systems*. PhD thesis, The University of Wisconsin-Madison, 1993.

15. L. Devroy, L. Györfi, and G. Lugosi. *A probabilistic theory of pattern recognition*. Springer-Verlag, Berlin, 1996.

16. P. Dorato, W. Yang, and C. Abdallah. Robust Multi-Objective Feedback Design by Quantifier Elimination. *J. Symbolic Computation*, 24:153–159, 1997.

17. M. Garey and D. Johnson. *Computers and Intractability: A Guide to the Theory of NP-Completeness*. W.H. Freeman and Co., New York, N.Y., 1979.

18. M. Jirstrand. *Algebraic methods for modeling and design in control*. PhD thesis, Linköping University - Sweden, 1996. Thesis Nr. 540.

19. P. Khargonekar and A. Tikku. Randomized algorithms for robust control analysis and synthesis have polynomial complexity. In *Proc. IEEE Conf. on Dec. and Control*, pages 3470–3475, Kobe, Japan, 1996.

20. V. Koltchinskii, C. T. Abdallah, M. Ariola, P. Dorato, and D. Panchenko. Statistical Learning Control of Uncertain Systems: It is better than it seems. Technical Report EECE 99-001, EECE Department, The University of New Mexico, 1999. Submitted to Trans. Auto. Control, Feb. 1999.

21. L.H. Lee and K. Poolla. Statistical validation for uncertainty models. In B. Francis and A.R. Tannenbaum, editors, *Feedback Control, Nonlinear Systems, and Complexity*, pages 131–149, Springer Verlag, London, 1995.

22. S. Malan, M. Milanese, and M. Taragna. Robust anaysis and design of control systems using interval arithmetic. *Automatica*, 33:1364–1372, 1997.

23. S. Malan, M. Milanese, M. Taragna, and J. Garloff. b^3 algorithm for robust performance analysis in presence of mixed parametric and dynamic perturbations. In *Proc. IEEE Conf. on Dec. and Control*, pages 128–133, Tucson, AZ, 1992.

24. A. Nemirovskii. Several NP-hard problems arising in robust stability analysis. *Mathematics of Control, Signals, and Systems*, 6:99–105, 1993.

25. S. Poljak and J. Rohn. Checking robust nonsigularity is NP-hard. *Mathematics of Control, Signals, and Systems*, 6:1–9, 1993.

26. L. Ray and R. Stengel. A Monte Carlo approach to the analysis of control system robustness. *Automatica*, 29:229–236, 1993.

27. R. Stengel and C. I. Marrison. Robustness of Solutions to a Benchmark Control Problem. *Journal of Guidance, Control, and Dynamics*, 15(5):1060–1067, 1992.

28. A. Tarski. *A Decision Method for Elementary Algebra and Geometry*. Univ. of California Press, Berkeley, 1951. 2nd Ed.

29. R. Tempo, E. Bai, and F. Dabbene. Probabilistic robustness analysis: Explicit bounds for the minimum number of samples. *Systems and Control Letters*, 30:237–242, 1997.

30. O. Toker and H. Özbay. "On the NP-hardness of solving bilinear matrix inequalities and simultaneous stabilization with static output feedback". *Proc. IEEE American Control Conf.*, pages 2056–2064, June 1995.

31. V. Vapnik and A. Chervonenkis. Weak convergence of empirical processes. *Theory of Probability and its Applications*, 16:264–280, 1971.

32. M. Vidyasagar. *A Theory of Learning and Generalization with Applications to Neural Networks and Control Systems*. Springer-Verlag, Berlin, 1996.

33. M. Vidyasagar. Statistical learning theory and its applications to randomized algorithms for robust controller synthesis. In *European Control Conference (ECC97)*, volume Plenary Lectures and Mini-Courses, pages 162–190, Brussels, Belgium, 1997.

34. M. Vidyasagar. *A Theory of Learning and Generalization*. Springer-Verlag, London, 1997.

35. M. Vidyasagar. Statistical learning theory and randomized algorithms for control. *IEEE Control Systems Magazine*, 18(6):69–85, 1998.

36. E. Walter and L. Jaulin. Guaranteed characterization of stability domains via set inversion. *IEEE Trans. Aut. Control*, 39(4):886–889, 1994.

37. B. Wie and D. S. Bernstein. A Benchmark Problem for Robust Control Design. In *Proceedings of the 1990 American Control Conference*, pages 961–962, San Diego, CA, May 1990.

38. M. Zettler, and J. Garloff Robustness analysis of polynomial parameter dependency using Bernstein expansion. *IEEE Trans. Aut. Control*, 43:425–431, 1998.

Dynamic Programming for Robust Control: Old Ideas and Recent Developments

Franco Blanchini[1] and Stefano Miani[2]

[1] Dipartimento di Matematica ed Informatica,
   Università degli Studi di Udine, via delle Scienze 208, 33100 Udine - ITALY
[2] Dipartimento di Matematica ed Informatica,
   Università degli Studi di Udine, via delle Scienze 208, 33100 Udine - ITALY

Abstract. This paper presents a brief summary of some old ideas concerning the control of uncertain systems based on dynamic programming. We show some recent developments and mention some open problems in the area.

1 Introduction

Dynamic programming is a well established approach to deal with dynamic systems control problems. Its range of application is so wide that it is rather difficult to find unifying textbooks which cover all the different areas of application. This fact is so crucial that the term "dynamic programming" is often used with quite different meanings depending on the specific application field. An excellent unifying textbook is [13] to which the reader is referred for a complete overview of the past history of the topic.

In particular several control problems under lack of information have been often faced via dynamic programming. A typical framework is that of dynamic games: a player (the controller), whose goal is that of assuring some dynamic system performance specifications, acts simultaneously with another player (the nature) whose goal is the opposite one. Again the literature on this problem is quite extensive (see [45], [26] [3]) and a comprehensive survey is beyond the scope of this paper.

In this work we consider a specific class of problems which can be viewed as dynamic games of systems with unknown–but–bounded disturbances. These problems concern in general systems having two inputs, the control and the disturbance, which both take their values in assigned sets. The goal of the control is to keep the state bounded or to drive it to a target set despite the action of the disturbance. This problem is known as min–max reachability of target sets or target tubes and it was first considered in the early 70s [10] [35]. It was almost completely abandoned since the solution techniques require hard computational effort, apparently not compatible with the computer technology of the time. More recently these techniques have been reconsidered and, in view of the improved computer performances, they are still subject of investigation.

In this paper we describe these techniques and we show recent developments which are receiving a renewed interest in the control community.

2   A basic problem

Consider the following linear system

$$x(k+1) = F(x(k), u(k), d(k)) \tag{1}$$

where $x(k) \in R^n$ is the system state, $u(k) \in R^m$ is the control input and $d(k) \in R^q$ is a disturbance (exogenous) input. The two inputs simultaneously act on the system with opposite goals.

Essentially there are two different basic assumptions which can be made on the nature of the disturbance d. The input $d(k)$ can indeed be regarded as a stochastic signal of which proper characterizing parameters are specified (form of distribution, variance, average, etc.) or it can be assumed unknown but bounded, i.e. it is assumed to take values in an assigned constraint set. The former specification leads to the stochastic dynamic programming for which the reader is referred to [13] for details. The second approach leads to the so called worst–case disturbance rejection problem which is the main concern of this paper. Let us then consider the following constraint

$$d(k) \in D \subset R^q \tag{2}$$

which is the only information we have on d. Normally constraints are present also for the state and the control input:

$$u(k) \in U \subseteq R^m, \tag{3}$$

$$x(k) \in X \subseteq R^n. \tag{4}$$

The constraints (3) and (4) are of a similar form but their nature is essentially different from those expressed by (2). In fact while the former represent the disturbance characterization, the latter usually represent system limitations that must be included in the design constraints. In the following we assume that X, U, D are compact sets and F is continuous.

The basic problem we focus on in this context is the so called min–max reachability problem reported next in its finite and infinite version.

Problem 1. **Finite–time min–max reachability problem** Given $\bar{k} \geq 0$, find a control strategy $u = \Phi(x)$ and a feasible initial condition set $X_{ini} \subset X$ such that if $x(0) \in X_{ini}$ then $u(k) \in U$, and $x(k) \in X$ for all $0 \leq k \leq \bar{k}$.

Problem 2. **Infinite–time min–max reachability problem** Find a control strategy $u = \Phi(x)$ and a feasible initial condition set $X_{ini} \subset X$ such that if $x(0) \in X_{ini}$ then $u(k) \in U$, and $x(k) \in X$ for all $k \geq 0$.

The basic principles for the solution to the above problems are due to the seminal works of the 70s' [35] [10] [11] [12] and they are based on the following time–reversed construction.

Construction of the infinite–time reachability set

1. Set $k = 0$, $S_0 = X$.
2. Compute the one–step controllability set to S_{-k}

$$\tilde{S}_{-k-1} \doteq \{x : \exists u \in U : \quad F(x, u, d) \in S_{-k}, \quad \forall d \in D\}$$

3. Compute $S_{-k-1} \doteq \tilde{S}_{-k-1} \bigcap X$
4. Set $k = k + 1$ and, if $k < \bar{k}$, go to Step 2, else Stop.

This procedure produces a sequence of nested sets $S_{-k-1} \subset S_{-k}$, $k = 1, 2, \ldots$ (if $\bar{k} = \infty$ the procedure might be endless and we will discuss some stopping criteria later). It is easy to show that all these sets are closed, since X, U and D are compact and F is continuous. The sets S_{-k} have the following property: if $x(0) \in S_{-\bar{k}}$ then there exists a feedback strategy $u = u(x)$ such that $x(k) \in X$, $k = 0, 1, \ldots, \bar{k}$.

In particular for $\bar{k} = \infty$,

$$S_{-\infty} = \bigcap_{k=0}^{-\infty} S_{-k}$$

is the infinite–time min–max reachability set. The non–emptiness of this set guarantees the existence of a solution to the infinite–time reachability problem according to the next proposition.

Proposition 9. *The Infinite–time reachability problem has a solution if and only if*

$$S_{-\infty} \neq \emptyset$$

and any set $X_{ini} \subseteq S_{-\infty}$ is a feasible initial condition set.

The control strategy $\Phi(x)$ is achieved taking for each $x \in S_{-\infty}$ a value u inside this set

$$\Omega(x) \doteq \{u : \quad F(x, u, d) \in S_{-\infty}, \quad \forall d \in D\} \tag{5}$$

The selection $u \in \Omega(x)$ is necessary and sufficient to keep the state in $S_{-\infty}$, which is in turn necessary and sufficient to keep it inside X over an infinite horizon, for all possible disturbances $d \in D$.

The construction leads in principle to the problem solution. The difficulties connected with its application are anyway apparent: there is an infinite number of sets which have to be computed together with their intersection. Furthermore the selection $u(k) \in \Omega(x(k))$, which has to be performed on–line, may require hard computation.

This procedure can be implemented using linear programming methods if linear systems of the following form

$$x(k+1) = Ax(k) + Bu(k) + Ed(k), \tag{6}$$

are considered and X, D and U are polyhedral sets. Indeed in this case the sets S_{-k} turn out to be polyhedral. Unfortunately, $S_{-\infty}$, being the intersection of infinitely many polyhedral sets, is convex but not necessarily polyhedral. However, this set can be arbitrarily closely approximated by a polyhedron, for instance by one of the S_{-k}, for sufficiently large k. Moreover, since a stopping criterion for the procedure is $S_{-k-1} = S_{-k}$, there are cases in which $S_{-\infty}$ is itself polyhedral and equals $S_{-k} = S_{-k-1}$.

The set $S_{-\infty}$ has the property of being *controlled invariant*: this means that if $x(0) \in S_{-\infty}$ then $x(k) \in S_{-\infty}$ for all $d(k) \in D$, if a suitable control action is applied. In fact $S_{-\infty}$ is the largest controlled invariant set included in X, in the sense that any other controlled–invariant set in X is included in $S_{-\infty}$. It is easy to see that when $S_{-\infty}$ is a polyhedral set the determination of the on–line control strategy $u(k) \in \Omega(x(k))$ is a linear programming feasibility problem. Indeed assume that $S_{-\infty}$ admits a representation of the form

$$S_{-\infty} = \{x : f_i x \leq g_i, \quad i = 1, 2, \ldots, r\},$$

then

$$\Omega(x) = \{u : f_i(Ax + Bu) \leq g_i - \delta_i, \quad i = 1, 2, \ldots, r\}, \tag{7}$$

where the numbers δ_i are defined as

$$\delta_i \doteq \max_{d \in D} f_i Ed$$

Example 2. Consider the following linear system of the form (6) with

$$A = \begin{bmatrix} 1 & 1 \\ -1 & 1 \end{bmatrix} \quad B = \begin{bmatrix} 0 \\ 1 \end{bmatrix} \quad E = \begin{bmatrix} 1 \\ 1 \end{bmatrix}$$

with the sets

$$X = \{x : |x_1| \leq 1, \quad |x_2| \leq 1\},$$
$$U = \{u : |u| \leq 1\}$$

and

$$D = \{d : |d| \leq 0.2\}$$

The procedure produces the following sequence:

$$S_{-1} = \{x : |x_1| \leq 1, \quad |x_2| \leq 1, \quad |x_1 - x_2| \leq 1.8, \quad |x_1 + x_2| \leq 0.8\}$$
$$S_{-2} = \{x : |x_1| \leq 1, \quad |x_2| \leq 0.7, \quad |x_1 + x_2| \leq 0.8\}$$
$$S_{-3} = \{x : |x_1| \leq 1, \quad |x_2| \leq 0.7, \quad |x_1 + x_2| \leq 0.8, \quad |x_1 - x_2| \leq 1.5\}$$
$$S_{-4} = S_{-3}$$

which is depicted in figure 1.

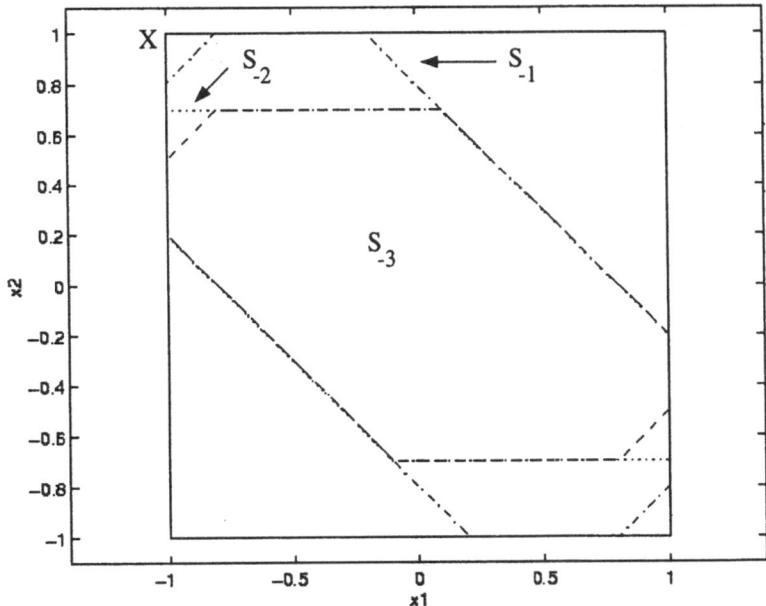

Fig. 1. The sequence of sets for example 2

Since $S_{-3} = S_{-4}$ the largest controlled invariant set is $S_{-\infty} = S_{-3}$. The set $\Omega(x)$ is given by

$$\Omega(x) = \{u : \quad |-x_1 + x_2 + u| \leq .5, \quad |2x_2 + u| \leq 0.4, \quad |2x_1 - u| \leq 1.5\}$$

This means that given $(x_1, x_2) \in S_{-\infty}$ there exists u which satisfies the inequalities above and any of such u is a feasible control in the sense that it keeps x in $S_{-\infty}$ which is necessary and sufficient for $x \in X$.

The technique above is not necessarily confined to time invariant systems. However for time–varying systems the problem is much more complicated since in general one needs to compute the reduced target tube, a generalization of the infinite–time reachability set [35] [10]. A particular case, in which the system is time–varying but periodic, has been discussed in [25].

The problem of the min–max reachability problem has been formulated above under state feedback (complete information). It can also be formulated as an output feedback problem (partial information) by assuming that only the following output measurements are available

$$y(k) = Cx(k) + w(k), \tag{8}$$

where $w(k) \in W$ is an output noise. In this case the state feedback compensator can be used in connection with a set membership state estimator [35] [10] whose recursive definition is reported next.

First note that the state $x(k)$ of (6) can be partitioned as

$$x(k) = x_u(k) + x_d(k)$$

where $x_u(k)$ is the solution with zero initial condition and input $u(k)$ (thus $x_u(k)$ is available to the controller) and $x_d(k)$ is the solution with input $d(k)$ and the assigned initial condition. Then $x_d(k)$ is unknown and must be estimated. Thus, without restriction, it is possible to consider the system that generates $x_d(k)$

$$x_d(k + 1) = Ax_d(k) + Ed(k)$$
$$y(k) = Cx_d(k) + w(k)$$

If it is assumed that a confinement set for the initial state $x(0)$ is given, say $x_d(0) \in X_d(0|0) \doteq X_0$, then it is possible to compute forward in time the k-step confinement set $X_d(k|k)$ according to the following algorithm.

Construction of the set–membership state estimator

1. Set $X_d(0|0) \doteq X_0$ and $k = 1$.
2. Compute the one step controllability set

$$X_d(k|k - 1) \doteq \{x = Ax' + Ed, \quad \forall x' \in X_d(k - 1|k - 1), \text{ and } \forall d \in D\}$$

3. Compute the subset of $X_d(k|k - 1)$ which is compatible with the measurement $y(k)$

$$X_d(k|k) \doteq X_d(k|k - 1) \bigcap \{x : -Cx + y(k) \in W\}$$

The double index of $X_d(k|h)$ is referred to the fact that it includes all the states at time k compatible with the measurements up to the time h. The study and construction of worst–case estimation techniques has recently been the subject of many papers, see for instance [68] [52] [51] [30]. Assuming that $x(0) \in X_0$, at the instant k the state of the system will be any vector of the form (if we reconsider the effect of the control)

$$\hat{x}(k) = x_u + \hat{x}_d, \quad \hat{x}_d \in \hat{X}_d(k|k) \tag{9}$$

(where the notation $\hat{x}_d$ now refers to the fact that we are estimating x_d), namely a state that *can have been generated* by the input sequences $u(k)$

and $d(k)$ starting from X_0 and compatible with the output measurements. Now, for control purposes, one needs to choose on–line a control value $u(k)$ which assures that the following inclusion

$$Ax(k) + Bu(k) + Ed(k) \in S_\infty,$$

is satisfied for all $d \in D$ and for all

$$x(k) = x_u(k) + \hat{x}_d(k), \quad \text{with} \quad \hat{x}_d(k) \in X_d(k|k)$$

because any of such vector is possible. Thus the vector $u(k)$ must be taken in such a way that

$$u(k) \in \Gamma(k|k) \doteq \bigcap_{x \in X_d(k|k)} \Omega(x + x_u)$$

where Ω is the set defined in (5). Assume now that one of the vectors $\hat{x}(k)$ in (9) is taken as actual state (note that this choice is arbitrary). If the estimation error can be bounded as

$$\delta(k) \doteq \hat{x}(k) - x(k), \quad \delta(k) \in \Delta(k)$$

then this leads to

$$A\hat{x}(k) + Bu(k) + Ed(k) - A\delta(k) \in S_\infty, \quad \forall \quad d(k) \in D, \quad \text{and } \delta(k) \in \Delta(k),$$

which means that, virtually, the estimation error can be considered as a new disturbance entering the systems through the matrix A. Here we find the main trouble of the approach as far as it concerns the output feedback problem: the set $X(k|k)$ and then the set $\Delta(k)$ must be updated on–line. This leads to a very hard computational burden. This is why (see [35] [10]) bounding sets for $\Delta(k)$ are normally computed so that $\Delta(k) \subset \hat{\Delta}(k)$, where the sets $\hat{\Delta}(k)$ are of appropriate shape (for instance ellipsoids or parallelotopes [30]). This over bounding clearly leads to a conservative solution.

Although all these concepts are well established in the past literature, there is a renewed attention for this kind of techniques. In particular, several related problems such as the construction of invariant (controlled–invariant) sets require techniques which are very similar in spirit, although they consider different problems with different goals. In the next sections we present some of these techniques and we enlighten possible new research directions.

3   The disturbance rejection problem

The mentioned technique can be as well used as an alternative approach to solve l_1 disturbance rejection problems. The l_1 norm of an input–output map $z = Wd$ is defined as

$$\sup_{\|d\|_{l_\infty} \leq 1} \|z\|_{l_\infty}$$

where the l_∞-norm of a vector sequence $z(k)$ is defined as

$$\|z\|_{l_\infty} \doteq \sup_{k \geq 0} \max_i |z_i(k)|.$$

The minimization of the l_1-norm of a closed–loop plant over the set of all linear stabilizing compensator can be solved via linear programming. Dynamic programming offers a different approach that leads to nonlinear compensators.

Consider the dynamic system

$$x(k+1) = Ax(k) + Bu(k) + Ed(k)$$
$$z(k) = Fx(k) + Gd(k)$$

and suppose one seeks a control $u(x)$ which, for an assigned $\mu > 0$, achieves the goal

$$\|z(k)\|_\infty \leq \mu$$

for all $\|d(k)\|_\infty \leq 1$. This means that it is required that

$$\|Fx(k) + Gd(k)\|_\infty \leq \mu$$

a condition which is equivalent to the fact that the state X is included in the following set

$$X = \left\{ x: \quad |F_i x(k)| \leq \mu - \max_{\|d\| \leq 1} |G_i d|, \quad i = 1, 2, \ldots, p. \right\},$$

where F_i and G_i are the ith rows of F and G respectively. The set X, included in the state space, is a constraint set for the state and then the disturbance rejection problem reduces to that of keeping the state $x(k)$ inside the set X. Under state feedback this reduces to the computation of the infinite–time reachability set as explained above. Additional constraints on the control of the form $u \in U$ can be obviously included. The links between this technique and the more recent l_1 theory have been established in [62] [63] [22] [23] for the state feedback case. The output feedback case has been considered in [65].

The modern l_1 synthesis, formulated in [32], leads to optimal *linear compensators*. However, it is known that nonlinear compensators may outperform linear ones as far as it concerns the minimization of the l_1-norm [66]. Thus, at least theoretically, the set-based mentioned methods which involve dynamic programming techniques, have considerable interest. Unfortunately, at the current status, their implementation is far from being practically realistic. A practically reasonable compromise can be achieved by the ellipsoidal approximation of the estimation-invariant set already proposed by [10] and [35] [69]. The interest for ellipsoidal bounding sets is receiving a renewed interest thanks to the efficient LMI tools recently developed for the problem of persistent disturbance rejection [27] [1].

4   Lyapunov functions and invariant sets for robust control

Lyapunov's second method, along with its applications to the analysis and synthesis of control strategies for uncertain systems, is a milestone in the history of robust control. The Lyapunov approach for robust analysis and synthesis started twenty years ago with the works [49] [38] [46] [5]. The robust quadratic stability analysis and stabilization problems were the ones on which researchers attention first concentrated [61] [70] [9] [4] [56] [57] [60] [37] (for a tutorial exposition the reader is referred to [31]). The quadratic function approach is strongly connected to the $\mathcal{H}_\infty$ control as shown in [42].

Thanks to the Riccati equation [57] and LMI-based techniques [27] that render them easily tractable, quadratic functions have become a very popular tool for robust analysis and synthesis. Unfortunately the robust stability/stabilizability conditions based on quadratic functions are conservative. This was observed often in the past [71] [28] [14] [53] In [24] it is shown that the quadratic stabilizability margin can be infinitely smaller than the true stabilizability margin for Linear Parameter Varying (LPV) systems.

Since for LPV systems it is known (see [49] [47]) that stability (stabilizability) implies the existence of a Lyapunov (Control–Lyapunov) function, a more general class of Lyapunov functions is necessary to obtain non–conservative conditions. One of this class is that of the polyhedral Lyapunov functions [28] [14] [53] [55] [50] [17] [19]. The existence of one of such Lyapunov function is a necessary condition for stability of linear LPV systems [28] [53] and for robust stabilizability of LPV systems [17].

The construction of polyhedral Lyapunov functions is in general harder than the construction of quadratic functions. The basic motivation is the absence of powerful tools such as LMI or Riccati–type equations (see [18] for details). However, polyhedral Lyapunov functions can be constructed by means of iterative procedures involving computational geometry. Examples of these procedures are in [28] [53] [50] [55] for the robust analysis problem and in [16] [17] for the robust synthesis. Some of these procedures are similar in spirit to the dynamic programming procedure presented in the previous section.

For the sake of illustration we show now how to compute the unit ball of a polyhedral Lyapunov function for an LPV uncertain polytopic system of the form

$$x(k+1) = A(w(t))x(k) + B(w(t))u(k),$$

with

$$A(w) = \sum A_i w_i,$$

$$B(w) = \sum B_i w_i,$$

$$\sum_{i=1}^{r} w_i = 1, \quad w_i \geq 0$$

where A_i and B_i are assigned matrices. To this aim we need the following definition. Let the real number $0 < \lambda \leq 1$ be given.

Definition 1. A convex and compact set P containing the origin in its interior, is λ–contractive, if for all $x \in P$ there exists $u(x)$ such that

$$A(w)x + B(w)u(x) \in \lambda P$$

A contractive set can be always seen as the unit ball of a Lyapunov function. We define polyhedral Lyapunov function a Lyapunov function whose unit ball is polyhedral. One of such functions, for a 0 symmetric polytope P, can be expressed as follows

$$\Psi(x) = \|Fx\|_\infty$$

for some full column rank matrix F.

The following procedure allows for the construction of a polyhedral Lyapunov function $\Psi(x)$, if such function exists, for a given uncertain system [16].

Construction of a polyhedral Lyapunov function

1. Set $k = 0$, $R_0 = X$ where X is any *arbitrary* 0–symmetric polytope including the origin in its interior. Fix $0 \leq \lambda < 1$ and a tolerance $\epsilon > 0$ such that $\lambda + \epsilon < 1$.
2. Compute the one–step controllability set to λR_{-k}

$$\tilde{R}_{-k-1} \doteq \{x : \exists u : A_i x + B_i u \in \lambda R_{-k}, \quad \forall\, i = 1, 2, \ldots r\}$$

3. Compute $R_{-k-1} \doteq \tilde{R}_{-k-1} \bigcap X$.
4. If R_{-k-1} is $\lambda + \epsilon$–contractive, then stop otherwise go to Step 2.

The sets R_{-k} are polyhedral and they are nested in that $R_{-k-1} \subset R_{-k}$ If the system admits a Lyapunov function assuring a speed of convergence λ, then the procedure stops in a finite number of steps giving a polytope S_{-k} which is $\lambda + \epsilon$–contractive. Thus, relaxing the speed of convergence from λ to $\lambda + \epsilon$ assures the finite stopping criterion. Conversely if such function does not exist with the prescribed λ and ϵ the sequence of sets S_{-k} collapse to a set with empty interior. Note that the convergence of the procedure to a set with non–empty interior, if it occurs at all, is assured for any initial set R_0 as above. Obviously the final set does depend on R_0. See [20] for more details on the procedure and the stopping criterion.

Assume that the Lyapunov function $\Psi(x)$ is derived. A suitable feedback control can be derived by considering the piecewise linear control introduced in [39]. For continuous–time systems

$$\dot{x}(t) = A(w)x(t) + B(w)u(t) \tag{10}$$

the procedure above can be used by means of the Euler approximating System (EAS)

$$x(k + 1) = [I + \tau A(w)]x(k) + \tau B(w)u(k) \tag{11}$$

Indeed if a λ–contractive polyhedral set is found for the EAS (11) then the same associated Lyapunov function assures that [17]

$$\dot{\Psi}(x(t)) \le -\beta \Psi(x(t))$$

for the continuous–time system (10) with

$$\beta = \frac{1-\lambda}{\tau}$$

In this case, the Gutman and Cwikel control [39], which turns out to be Lipschitz continuous, can be still applied [17]. A feedback control can be also found by "smoothing" the polyhedral Lyapunov function [21].

The procedure is similar in spirit to the dynamic programming one presented in Section 2. However, it is conceptually different since now we do not require that the initial state is inside the unit ball of the Lyapunov function. Convergence to zero is indeed assured, if the procedure stops successfully for any arbitrary initial state.

Example 3. Consider the following example of an open–loop unstable linear continuous–time system of the considered form with

$$A_1 = \begin{bmatrix} 0 & .5 \\ 1 & -.5 \end{bmatrix}, \ A_2 = \begin{bmatrix} -1 & .4 \\ 0 & .8 \end{bmatrix}, \ B_1 = \begin{bmatrix} 0 \\ 2 \end{bmatrix}, \ B_2 = \begin{bmatrix} 2 \\ 1 \end{bmatrix}.$$

We searched for a .98-contractive set for the EAS of this system when $\tau = .1$. After more than 100 iterations the procedure stopped successfully, producing a region with 10 vertices (two of which are very close each other and cannot be distinguished in figure 2), thus assuring the existence of a Lyapunov function for the continuous time system with decreasing factor $\beta = .1$. The level surface of the derived function is depicted in figure 2. A piecewise-linear stabilizing control that can be associated to this system is

$$u = K^{(i)}x$$

where i is the index of the "sector" including x

$$i = \arg\max |F^{(k)}x|$$

where $F^{(k)}$ is the kth row of F, the 5×2 matrix characterizing the Lyapunov function $\Psi(x) = \|Fx\|_\infty$, and where $K^{(i)}$ is a linear gain associated to the i-th sector (see [20] for details). The matrices F and K (the latter is achieved by stacking the $K^{(i)}$s) are

$$F = \begin{bmatrix} 1.020 & 0.051 \\ 0.000 & 1.000 \\ 1.000 & 0.000 \\ 0.806 & -0.389 \\ 0.459 & -1.082 \end{bmatrix} \quad K = \begin{bmatrix} -2.499 & -4.850 \\ -1.693 & -5.599 \\ -4.455 & -9.740 \\ -23.06 & -47.10 \\ -2.250 & -5.499 \end{bmatrix}$$

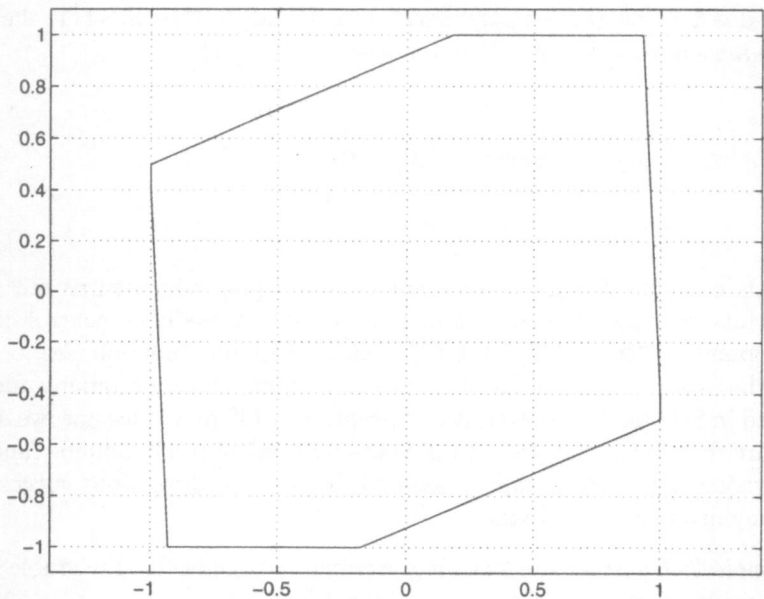

Fig. 2. Lyapunov function level surface

5 Related problems

Besides the robust stabilization of uncertain systems there are several related problems which can be solved by means of techniques related to those discussed here. One of these is the control under state and control constraints. Consider for instance a linear system

$$x(k + 1) = Ax(k) + Bu(k)$$

with $u \in U$ and $x \in X$. Assume that one wishes to find a subset of states of X that can be driven to the origin without violating the constraints. In this case the dynamic programming approach still works. For instance the methods presented in the paper [40] [44] [54] resort to dynamic programming techniques to solve the problem.

Other popular methods to deal with the problem are those based on the so called receding horizon technique introduced many years ago [58] [33] [29] [43]. The basic idea of the method is the following: given the current state $X(k) \in X$, a sequence of inputs $u(h) \in U$, $h = k, k + 1, \ldots, k + T - 1$ is computed for a given T, in such a way that $X(h) \in X$ for $h = k, k+1, \ldots, k+ T$. Then the first member of the sequence input is applied $u(k)$ and a new sequence is computed on the basis of the new state $x(k + 1)$. Clearly, this scheme does not assure that constraint violations will not occur in the future since a finite horizon computation is performed. The computed sequence, however, can be optimized in such a way that good system performances are assured.

Very recently these techniques have been used along with those based on invariant sets [67], [48] [25] [15] [7] [8] to assure both stability of the scheme and constraints satisfaction.

The techniques based on dynamic programming, such as [40] [44] [54] are based on the computation of the largest controlled–invariant set inside X, thus they provide necessary and sufficient conditions for the constraint fulfilment (i.e. the initial state must be included in such set). However such maximal set may have an extremely complex representation which in turn reflects on the control complexity.

These receding horizon schemes have the advantage that they do not require the computation of the largest invariant set, thus they result in a lower complexity. Clearly these schemes are conservative, and they may fail to provide feasible solutions even if such solutions exists.

6   Final discussions

Encouraged by the fast development of the computer technology, several dynamic programming methods for the solution of control and estimation problems in the presence of unknown–but–bounded noise introduced in the past have been very recently reconsidered.

The basic ideas for the constructions of robust controllers and robust estimator are still valid in several fields such as the persistent disturbance rejection problems, the robust state observers and the construction of non–quadratic Lyapunov functions.

There are still several important problems still unsolved that can be approached by means of the presented methods. As we have seen the connection state estimator–controller leads to strong complexity problems. Thus the trade–off between optimality and complexity is still a crucial problem. Recently some gain–scheduling problems have been approached by recursive procedures that have strong analogies with those discussed here [64]. This is an open application area.

The stabilization of uncertain LPV systems can be achieved by means of polyhedral Lyapunov functions in a non–conservative way. The problem here is that normally the complexity of such function is very high. A reduction of the complexity is expected by considering an extended class of function such as the piecewise–quadratic. Such kind of function have been applied in the stability analysis of hybrid systems [59]. Another drawback of the non–quadratic Lyapunov functions are the difficulties which are encountered in considering output instead of full state feedback feedback stabilization problems. How non–quadratic Lyapunov functions can be constructed for the measurement feedback case is an open question.

References

1. J.Abedor, K.Nagpal and K.Poolla, "A linear matrix inequality approach to peak-to-peak gain minimization. Linear matrix inequalities in control theory and applications", Internat. J. Robust Nonlinear Control, 6, no. 9-10, pp. 899–927, 1996.

2. N.E.Barabanov, "Lyapunov indicator of discrete inclusion" , Autom. and Rem. Contr. parts I,II,III, Vol. 49, no. 2, pp. 152-157; no. 4, pp. 283-287; no. 5, pp. 558-566, 1988.

3. T.Basar, "Dynamic noncooperative games", Acad. Press, New York, 1992.

4. B.R.Barmish, I.Petersen and A.Feuer, "Linear Ultimate Boundedness Control of Uncertain Systems", Automatica, Vol. 19, No. 5, pp. 523-532, 1983.

5. B.R.Barmish, M.Corless and G.Leitmann, "A new class of stabilizing controllers for uncertain dynamicsly systems", SIAM J. on Contr. and Optim. Vol. 21, No.2, March 1983.

6. B.R.Barmish and J.Sankaran, "The propagation of parametric uncertainty via polytopes", IEEE Trans. Automat. Control, Vol. 24, no. 2, pp. 346–349, 1979.

7. A.Bemporad, A.Casavola and E.Mosca, "Nonlinear Control of constrained linear systems with predictive reference management", IEEE Trans. on Autom. Contr., Vol. 42, no. 3 March, pp. 340-349, 1997.

8. A.Bemporad and E.Mosca, "Fulfilling hard constraints in uncertain linear systems by reference managing", Automatica, Vol. 34, no. 4, 451–461, 1998.

9. J.Bernussou, P.L.D.Peres and J.Geromel, "On a convex parameter space method for linear control design of uncertain systems", SIAM J. on Contr. and Opt., 29, pp. 381-402, 1991.

10. D.P.Bertsekas and I.B Rhodes "On the minmax reachability of target set and target tubes", Automatica, Vol. 7, pp. 233–247, 1971.

11. D.P.Bertsekas and I.B.Rhodes, "Recursive state estimation for a set-membership description of uncertainty", IEEE Trans. Automatic Control, Vol. 16, pp. 117–128, 1971.

12. D.P.Bertsekas, "Infinite-time reachability of state-space regions by using feedback control". IEEE Trans. Automatic Control, Vol. 17, no. 5, pp. 604–613, 1972.

13. D.P.Bertsekas, "Dynamic programming and optimal control", Athena Scientific, Belmont, Ma, 1995.

14. A.Bhaya and F.C.Mota "Equivalence of stability concepts for discrete time-varying systems" International J. on Robust and Nonlinear Control, Vol. 4, no. 6, p. 725-740, 1994.

15. F.Blanchini, "Control Synthesis of Discrete Time Systems with Control and State Bounds in the Presence of Disturbance" *Journal of Optimization Theory and Applications*, vol. 65, no. 1, p. 29-40, April 1990.

16. F.Blanchini "Ultimate boundedness control for discrete-time uncertain system via set-induced Lyapunov functions", IEEE Trans. on Autom. Contr., Vol. 39, no. 2, 428-433, 1994.

17. F.Blanchini "Nonquadratic Lyapunov function for robust control", Automatica, Vol. 31, no 3, p. 451-461, 1995.

18. F.Blanchini, " Set invariance in control – a survey, Automatica, to appear, 1999.

19. F.Blanchini and S.Miani, (1995a) "On the transient estimate for linear systems with time-varying uncertain parameters", IEEE Trans. on Circ. and Syst., Part I, to appear.

20. F.Blanchini and S.Miani, " Piecewise-linear functions in robust control", in Robust control via variable structure and Lyapunov methods, Edited by F. Garofalo and L. Glielmo, Springer Verlag, pp. 213-240, 1996.

21. F.Blanchini and S.Miani, "A universal class of smooth functions for robust control", IEEE Transactions on Automatic Control, March 1999, to appear.

22. F.Blanchini and M.Sznaier, "Persistent disturbance rejection via static state feedback", IEEE Trans. on Autom. Contr., Vol. 40, no. 6, pp. 1127-1131, 1995.

23. F.Blanchini, S.Miani and M.Sznaier, "Worst case l_∞ to l_∞ gain minimization: dynamic versus static state feedback", Proceedings of *Proceedings of the 35th Conference on Decision and Control,* 1996, Special session on Set Theoretic Methods in Control and Estimations, pp. 2395-2400

24. F.Blanchini and A.Megretski, "Robust state feedback control of lti systems: nonlinear is better than linear", *IEEE Transactions on Automatic Control,* 1999, accepted for publication.

25. F.Blanchini and W.Ukovich, "A Linear Programming .Approach to the Control of Discrete-Time Periodic System with State and Control Bounds in the Presence of Disturbance", Journal of Optimization Theory and Applications, Vol. 73, no. 3, p. 523-539, September 1993.

26. F.Blaquiere, F.Gerard and G.Leitmann, "Quantitative and qualitative games", Academic Press, 1969.

27. S.Boyd, L.El Ghaoui, E.Feron and V.Balakrishnan, "Linear matrix inequality in system and control theory", SIAM Studies in Applied Mathematics, Philadelphia, 1994.

28. R.K.Brayton C.H.Tong, "Constructive stability and asymptotic stability of dynamical systems", IEEE Trans. on Circ. and Syst., Vol. CAS-27, no. 11, pp. 1121-1130, 1980.

29. T.S.Chang and D.E.Seborg, "A Linear Programming Approach for Multivariable Feedback Control with Inequality Constraints", International Journal of Control, Vol. 37, No. 3, pp. 583-597, 1983.

30. L.Chisci, A.Garulli and G.Zappa, "Recursive state bounding by parallelotopes", Automatica, vol. 32, pp. 1049-1056, 1996.

31. M.Corless, "Robust analysis and controller design via quadratic Lyapunov functions". In A. Zinober, editor, Variable structure and Lyapunov Control, Vol 193, Lecture notes in in control and information science, Chapter 9, pp. 181–203, Springer Verlag, 1994.

32. M.A.Dahleh and J.B.Pearson, "l^1–Optimal feedback controllers for MIMO discrete–time systems," IEEE Trans. Automat. Contr., Vol. AC-32, no.4, pp. 314–322, 1987.

33. J.H.De Vleger, H.B Verbuggen and P.M Bruijn, "A time–optimal control algorithm for digital computer control", Automatica, Vol. 18, no. 2, pp. 219–244, 1982.

34. P.Dorato, R.Tempo and G.Muscato, "Bibliography on robust control", Automatica, Vol. 29, no.1, p. 201-214, 1993.

35. J.D.Glover and F.C.Schweppe, "Control of linear dynamic systems with set constrained disturbances", IEEE Trans. on Autom. Contr., Vol AC-16, no. pp. 411–423, 1971.

36. B.Grumbaum, "Convex polytopes", John Wiley & Sons, New York, 1969.
37. K.Gu, Y.H.Chen, M.A. Zohdy and N.K.Loh, "Quadratic stabilizability of uncertain systems: a two level optimization setup", Automatica, Vol. 27, no. 1, pp. 161–165, 1991.
38. S.Gutman, "Uncertain dynamic systems – a Lyapunov min–max approach", IEEE Trans. on Autom. Contr., Vol. 24, no. 3, pp. 437–443, 1979.
39. P.O.Gutman and M.Cwikel, 'Admissible sets and feedback control for discrete-time linear systems with bounded control and states", IEEE Trans. on Autom. Contr., Vol. AC-31, No. 4, pp. 373-376, 1986.
40. P.O.Gutman and M.Cwikel, "An algorithm to find maximal state constraint sets for discrete–time linear dynamical systems with bounded control and states", IEEE Trans. on Autom. Contr., Vol. 32, No. 3, pp. 251-254, 1987.
41. S.Gutman and G.Leitmann, "On a class of linear differential games", Int. Jour. of Opt. Th. and Appl., Vol. 17, no. 5/6, 1975.
42. P.Kargonekar, I.R.Petersen and K.Zhou, "Robust stabilization of uncertain systems and H_∞ optimal control", IEEE Trans. on Autom. Contr., Vol. AC-35, pp. 356-361, 1990.
43. S.S.Keerthi and E.G.Gilbert, "Optimal infinite-horizon feedback laws for a general class of constrained discrete-time systems: stability and moving-horizon approximations", J. Optim. Theory Appl., Vol. 57, no. 2, pp. 265–293, 1988.
44. S.S.Keerthi and E.G.Gilbert, "Computation of minimum-time feedback control laws for discrete-time systems with state-control constraints", IEEE Trans. on Autom. Contr., Vol. 32, no. 5, pp. 432-435, 1987.
45. Krasovskii and Krasovskii "Control under lack of information", Birkauser, 1995.
46. G.Leitmann, "Guaranteed asymptotic stability for some linear systems with bounded uncertainties", Trans. of The A.S.M.E. Vol. 101, p. 212-216, 1979.
47. Y.Lin, E.D.Sontag and Y.Wang, "A smooth converse Lyapunov theorem for robust stability", SIAM J. on Contr. and Opt., Vol. 34, no. 1, pp. 124-160, 1996.
48. D.Q.Mayne and H.Michalska, "Receding horizon control of nonlinear systems". IEEE Trans. Automat. Control, 35, no. 7, pp. 814–824, 1990.
49. A.M.Meilakhs, "Design of stable systems subject to parametric perturbations", Automation and Remote Contr., Vol. 39, no. 10, pp. 1409–1418, 1979.
50. A.N. Michael, B.H.Nam and V.Vittal, "Computer generated Lyapunov Functions for Interconnected Systems: Improved Results with applications to power systems", IEEE Trans. on Circ. and Syst. Vol. 31, no. 2, 1984.
51. M.Milanese and G.Belforte, "Estimation Theory and Uncertain Intervals evaluation in the presence of Unknown bub Bounded Errors: Linear Families of Models end Estimators", IEEE Trans. on Autom. Contr., Vol. AC-27, pp. 408–414, 1982.
52. M.Milanese and A.Vicino "Optimal Estimation Theory for Dynamic Systems with Set Membership Uncertainty: an Overview", Automatica, Vol. 27, pp. 997–1009, 1991.
53. A.P.Molchanov and E.S.Pyatnitskii, "Lyapunov functions specifying necessary and sufficient conditions of absolute stability of nonlinear nonstationary control system", Autom. and Rem. Contr.,parts I,II,III, Vol. 47, no. 3, pp. 344-354; no. 4, pp.443-451; no. 5, pp. 620-630, 1986.

54. R.Morris and R.F.Brown, Extension of Validity of the GRG Method in Optimal Control Calculation, IEEE Transactions on Automatic Control, Vol. 21, pp. 420–422, 1976.

55. Y.Ohta, H.Imanishi, L.Gong and H.Haneda, "Computer generated Lyapunov functions for a class of nonlinear systems", IEEE Trans. on Circ. and Syst. Vol. 40, no. 5, 1993.

56. I.R.Petersen, "Quadratic stabilizability of uncertain systems: existence of a nonlinear stabilizing control does not imply the existence of a linear stabilizing control", IEEE Trans. on Autom. Contr. Vol. 30, no. 3, pp. 292-293, 1995.

57. I.R.Petersen and C.Hollot, "A Riccati equation approach to the stabilization of uncertain systems", Automatica, Vol. 22, p. 397-411, 1986.

58. A.I.Propoi, "Use of linear programming methods for synthesizing sampled data automatic systems", Autom. and Remote Control, vol. 24, pp. 837–844, 1963.

59. A.Rantzer and M.Johansson, "Computation of piecewise quadratic functions for hybrid systems", IEEE Trans. on Autom. Contr. Vol. 43, no. 4, pp. 555–559, 1998.

60. M.A.Rotea and P.P.Kargonekar, "Stabilization of uncertain systems with norm-bounded uncertainty: a Lyapunov function approach", SIAM J. on Contr. and Opt., Vol. 27, p.1462-1476, 1989.

61. D.D.Siljak "Parameter space methods for robust control design: a guided tour", IEEE Trans. on Autom. Contr., Vol 34, no. 7, p.674-688, 1989.

62. J.S.Shamma "Nonlinear state feedback for l_1 optimal control", Syst. and Contr. Lett., Vol. 21, pp. 40-41, 1994.

63. J.S.Shamma, "Optimization of the l^∞-induced norm under full state feedback", IEEE Trans. Automat. Control 41, no. 4, pp. 533–544, 1996.

64. J.S.Shamma and D.Xiong, "Control of Rate Constrained Linear Parameter Varying Systems", Proc. of the 34th CDC, New Orleans, La, Dec 1995, pp. 2515–2520.

65. J.S.Shamma and K.Y.Tu, "Set–valued observer and optimal disturbance rejection", IEEE Trans, to appear, 1999.

66. A.A.Stoorvogel, "Nonlinear L_1 optimal controllers for linear systems", IEEE Trans. Automat. Control, Vol. 40, no. 4, pp. 698-690, 1996.

67. M.Sznaier and M.Damborg, "Control of Constrained Discrete Time Linear Systems Using Quantized Controls", Automatica, Vol. 25, no. 4, pp. 623-628, 1989.

68. R.Tempo, "Robust Estination and Filtering in the Presence of Bounded Noise", IEEE Trans. On Autom. Contr., Vol. 21, pp. 420–422, 1988.

69. P.B.Usoro, F.C.Schweppe, L.A.Gould and Wormley D.N., "A Lagrange Approach to Set Theoretic Control Synthesis", IEEE Trans. on Autom. Control, Vol. 27, No. 2, pp. 393-399, 1982.

70. R.K.Yedavalli and Z.Liang, "Reduced conservatism is stability robustness bounds by state transformations", IEEE Trans. on Autom. Contr. Vol. 31, pp. 863-866, 1986.

71. A.L.Zelentsowsky, "Nonquadratic Lyapunov functions for robust stability analysis of linear uncertain systems", IEEE Trans. on Autom. Control, vol. 39, no. 1, pp. 135-138, 1994.

Robustness of Receding Horizon Control for Nonlinear Discrete-time Systems

Giuseppe De Nicolao, Lalo Magni, and Riccardo Scattolini

Dipartimento di Informatica e Sistemistica, Universty of Pavia, Italy, via Ferrata 1, 27100 Pavia, Italy

Abstract. In this paper robustness analysis and synthesis of state-feedback nonlinear Receding-Horizon (*RH*) control schemes is considered. In particular, robustness properties based on the monotonicity of the cost function and on inverse optimality are discussed. A particular attention is devoted to a new *RH* synthesis approach based on the solution of a finite-horizon dynamic game. This control law guarantees that the L_2 gain of the closed-loop is less than or equal to a given number γ in a prescribed neighbourhood of the equilibrium state.

1 Introduction

The widespread industrial success of Receding-Horizon (*RH*) control is due to the simplicity of its rationale as well as to its ability to deal with nonlinearities and constraints on the control and state variables. In *RH* control of discrete-time systems, at any time instant t the current control $u(t)$ is obtained by solving a finite-horizon optimization problem over the interval $[t, t+N]$ where N is the optimization horizon. At the next time instant $t+1$, the new control $u(t + 1)$ is found by translating the optimization horizon and solving a new problem over $[t + 1, t + N + 1]$.

Although many predictive control alghorithms, see e.g. DMC [10] or GPC [8], are currently used in industrial applications, it is well known that closed-loop stability is not guaranteed for a generic finite-horizon cost function, as shown in [2]. In the linear case, stability can be obtained by complementing the cost function with a terminal zero-state constraint, see e.g. [17] for a state-space formulation or [9] and [26] for input-output versions. For the stable modes, the need of equality constraints can be removed by introducing a terminal penalty equal to the infinite-horizon cost due to zero control [29]. It is interesting to note that a fairly general stability theory for linear *RH* control can be developed by referring to the monotonicity properties of a suitable difference Riccati equation initialized with the terminal penalty matrix [2].

In the linear case, the importance of RH control is diminished by the fact that, in absence of nonlinearities and constraints, it is hardly better than infinite-horizon (IH) linear quadratic (LQ) control. For constrained systems, the link between RH and LQ control has been clarified in [34], where it is shown that the IH optimal control law can be found as the solution of a finite-horizon problem where the terminal penalty is equal to the IH cost of the unconstrained LQ problem. Other contributions along this line are reported in [31] and [7].

The RH control of nonlinear systems is of practical interest since the IH optimisation problem is computationally intractable even when state and input constraints are absent. In this context, the first stability results were reported in [3], [16], and [24] where it is shown that the zero-state terminal constraint guarantees closed-loop stability. However, the presence of the terminal equality constraint places an heavy requirement on the on-line optimising controller. This motivated the development in [25] of the dual-mode RH controller, which replaces the equality constraint with an inequality one, namely that $x(t + N)$ belongs to a suitable neighbourhood of the origin where the nonlinear system is stabilised by a linear control law. Another stabilisation method, see [37], is based on a terminal contractive constraint requiring that the norm of the terminal state $x(t + N)$ is sufficiently smaller than the norm of $x(t)$.

More recently, schemes have been proposed that combine a terminal penalty with a terminal inequality constraint. In particular, in [27] and [4] it has been shown that stabilisation can be enforced by a suitable quadratic terminal penalty. On the other hand, in [14] and [12] closed-loop stability of the equilibrium was established using a (nonquadratic) terminal penalty equal to the cost-to-go associated with a locally stabilising linear control law.

In view of all these approaches, the stability issue of RH control can be now considered as solved also for non linear systems. On the contrary, the robustness properties of nonlinear RH control algorithms have still to be examined in depth and, even more important, the synthesis of RH techniques with guaranteed robustness is still a largely unsolved problem.

The main purpose of the present contribution is to provide a concise review of the existing results on the robustness of state-feedback nonlinear RH control schemes and present a new RH synthesis approach based on nonlinear H_∞ control concepts. Specifically, the paper is organised as follows. In Section 2 the RH nonlinear control problem is formally stated and the closed-loop stability issue is addressed. Section 3 reviews some approaches to the robustness analysis of RH control, while in Section 4 a synthesis method is presented. This approach provides a (suboptimal) RH solution to the classical H_∞ nonlinear control problem, consisting of the design of a control law such that a prescribed disturbance attenuation level is achieved.

2   Nonlinear RH control

Consider the nonlinear discrete-time system

$$x(k+1) = f(x(k), u(k), w(k)), \quad k \geq 0 \tag{1}$$
$$z(k) = h(x(k), u(k), w(k)) \tag{2}$$

where $x \in \mathbb{R}^n$ is the state, $u \in \mathbb{R}^m$ is the control variable, $w \in \mathbb{R}^p$ is the disturbance, $w \in l_2([0, T])$, for every positive integer T, $z \in R^s$ is the output, f and h are C^2 functions with $f(0, 0, 0) = 0$, $h(0, 0, 0) = 0$, and $x(0) = x_0$. In this and in the following section we do not consider the disturbance (i.e. $w(k) = 0$, $k > 0$). Moreover, for simplicity, throughout the paper control and state constraints will not be dealt with.

The standard Receding Horizon (RH) control law is based on a Finite-Horizon (FH) optimisation problem. In particular, letting $u_{t,t+N-1} := [u(t), u(t+1), \ldots u(t+N-1)]$, $Q > 0$ and $R > 0$, we will consider cost functions of the type

$$J_{FH}(x, u_{t,t+N-1}, N) = \sum_{k=t}^{t+N-1} \{x(k)'Qx(k) + u(k)'Ru(k)\}$$
$$+ V_f(x(t+N)) \tag{3}$$

to be minimized at any time instant t with respect to $u_{t,t+N-1}$, subject to (1), and the terminal constraint

$$x(t+N) \in X_f \subset R^n \tag{4}$$

Different stabilising algorithms (see Paragraph 2.1 below) are characterized by the choices of the terminal penalty function $V_f(x)$ and the terminal region X_f.

Associated with (3), (4) it is possible to define an NRH (Nonlinear Receding Horizon) control strategy in the following way: at every time instant t, define $\bar{x} = x(t)$ and compute the optimal solution $u^o_{t,t+N-1}$ for the FH problem (3) subject to (1) and (4); then apply the control $u(t) = u^o(\bar{x})$ where $u^o(\bar{x})$ is the first column of $u^o_{t,t+N-1}$. The NRH control law is implicitly defined by the function $u = \kappa^{RH}(\bar{x})$ where $\kappa^{RH}(\bar{x}) = u^o(\bar{x})$. Although the FH minimization of (3) has to be performed at each time instant, this is much more viable than solving an IH problem.

Note that a fundamental problem in nonlinear RH control concerns the feasibility issue, that is the possibility, given the initial state $\bar{x}$, to find a (finite) control sequence such that the state and control trajectories fulfil prespecified constraints and the state at the end of the optimisation horizon belongs to X_f. The set of states $\bar{x}$ such that the problem is feasible is called the output admissible set. The size of the output admissible set, as well as the complexity of the optimisation problem, grow with N so that there is a trade-off between complexity, feasibility and performance in the choice of the optimization horizon N. For a deeper discussion on these points, the interested reader is referred to [13].

2.1   Closed-loop stability

In agreement with the results on RH control of linear systems [1], [2], it is well known that, if $V_f(x) \equiv 0$ and $X_f = R^n$, for a given N it may well happen that the RH controller yields an unstable closed-loop system. Nevertheless, by a proper design of the "tuning knobs" V_f and X_f it is possible to ensure closed-loop stability with a finite horizon N.

In the LQ case a fairly complete stability theory is available which is based on the so-called Fake Riccati analysis [28]. The main point is to choose $V_f(x)$ and X_f so as to force the monotonicity of the solution of a relevant difference Riccati equation. Once monotonicity is established, it follows that

$$J_{FH}^o(x, N) = \min_{u_{t,t+N-1}} J_{FH}(x, u_{t,t+N-1}, N)$$

is a Lyapunov function for the closed-loop and stability is guaranteed.

An analogous rationale can be extended to the nonlinear case. In fact, the closed-loop stability of the origin in most RH schemes is proven by showing that $J_{FH}^o(x, N)$ is a Lyapunov function. For this purpose, the main point is to demonstrate that

$$J_{FH}^o(f(x, u^o(x), 0), N) < J_{FH}^o(x, N) \tag{5}$$

which can be done in two steps. First, by optimality it always holds that

$$
\begin{aligned}
&J_{FH}^o(f(x, u^o(x), 0), N - 1) \\
&= J_{FH}^o(x, N) - x'Qx - u^o(x)'Ru^o(x) < J_{FH}^o(x, N)
\end{aligned}
\tag{6}
$$

The second step is to show that

$$J_{FH}^o(\xi, N) \leq J_{FH}^o(\xi, N - 1), \quad \forall \xi \in \Xi \tag{7}$$

where Ξ is a neighbourhood of the origin in R^n. From (7), letting $\xi = f(x, u^o(x), 0)$ and recalling (6), one sees that (5) follows. Now, the inequality (7), which will be hereafter termed as *monotonicity property*, is fulfilled only for a suitable choice of $V_f(x)$ and X_f. Different choices of $V_f(x)$ and X_f that enforce monotonicity are proposed in [3], [16], [24], [27], [4], [14], [12].

3   Robustness analysis

Robustness analysis for nonlinear systems is considerably more complex than in the linear case. In this section, two approaches will be considered: monotonicity-based robustness and the use of inverse optimality to establish robustness margins.

By the way, it is worth mentioning that, whenever the equilibrium is exponentially stable, it remains an asymptotically stable fixed point also in the presence of disturbances asymptotically decaying to zero [32]. This result is of particular importance for establishing closed-loop stability in the output feedback case [20].

3.1   Monotonicity-based robustness

The methods whose stability proof is based on monotonicity enjoy some sort of embedded robustness. In fact, the keystone of the stability argument is the decrease of the optimal value of the FH cost, i.e. $J^o_{FH}(f(\bar{x}, \kappa^{RH}(\bar{x}), 0), N) < J^o_{FH}(\bar{x}, N)$. Now, assume that, due to modelling errors, the real system is

$$x(k+1) = f_r(x(k), u(k), w(k)), \quad x(t) = \bar{x}, \quad k \geq t \tag{8}$$

with $f_r(0,0,0) = 0$ and $f_r(\cdot, \cdot, \cdot) \neq f(\cdot, \cdot, \cdot)$. In other words, the nominal model (1) used to compute $\kappa^{RH}(x)$ through the solution of the FH problem (3), (4) is only an approximate representation of the real system. In view of this discrepancy, there is no guarantee that

$$J^o_{FH}(f_r(\bar{x}, \kappa^{RH}(\bar{x}), 0), N) < J^o_{FH}(\bar{x}, N) \tag{9}$$

and closed-loop stability may be lost. However, recalling (6), (7) if the RH controller at hand enjoys the monotonicity property, it holds that

$$J^o_{FH}(f(\bar{x}, \kappa^{RH}(\bar{x}), 0), N) - J^o_{FH}(\bar{x}, N) \leq -\bar{x}'Q\bar{x} - \kappa^{RH}(\bar{x})'R\kappa^{RH}(\bar{x})$$

Hence, there is some margin and, provided that $\|f(x, u, 0) - f_r(x, u, 0)\|$ is not "too large", compared to $\|(x, u)\|$, (9) will hold so that stability is preserved in a suitable region also in the perturbed case. Note, however, that a thorough robustness analysis should take into account the possibility that the model error causes the state to drift outside the output admissible set, that is into regions where the optimal control problem is infeasible.

3.2   Robustness based on inverse optimality

A whole class of stability margins can be derived by showing that the RH controller coincides with the optimal control law associated with a suitably defined IH nonlinear optimization problem. To this purpose, consider an RH control law $u = \kappa^{RH}(x)$ that enjoys the monotonicity property and let

$$l(x) := x'Qx + J^o_{FH}(x, N-1) - J^o_{FH}(x, N)$$

In view of monotonicity, $l(x) > 0$ for all $x \neq 0$ belonging to the output admissible set. Then, consider the "fake" IH cost function

$$J^F_{IH}(\bar{x}, u(\cdot)) = \sum_{k=t}^{\infty} \{l(x(k)) + u(k)'Ru(k)\} \tag{10}$$

Under regularity assumptions, it is possible to show that $\kappa^{RH}(x)$ coincides with the optimal IH control law solving (10), see e.g. [23] where the continuous-time case is considered. As such, $\kappa^{RH}(x)$, enjoys all the stability margins of IH regulators. For example, in the case of discrete time systems, the robustness properties with respect to gain and additive perturbations reported in [11] can easily be proven. This approach is even more interesting for the continuous-time formulations of nonlinear RH control. In fact, if the plant is affine in u (i.e. $\dot{x} = a(x) + b(x)u$) one can guarantee the celebrated "infinite gain margin-50% gain reduction tolerance" as well tolerances to suitable classes of dynamic input uncertainties [15], [33].

4   Robustness synthesis

This section describes a new approach that directly considers a disturbance attenuation specification in the synthesis of the RH control law [21]. This approach can be seen also as a practical way to solve the nonlinear H_∞ problem.

It is well known that the H_∞ theory provides an excellent theoretical framework for dealing with nonlinear stability and robustness issues. On the other hand the derivation of the optimal control law calls for the solution of a Hamilton-Jacobi-Isaacs equation [19], a difficult computational task which impairs the application to real systems. The Receding Horizon (RH) state-feedback approach presented in this section aims at overcoming this problem, at least partially, by replacing it with the solution of a finite-horizon two-players differential game. The state-feedback control law here derived yields a closed-loop system with an L_2 gain less than or equal to a given number γ in a suitable region $\Omega^{RH}(N)$ of the state space which depends on the length of the optimization horizon N. For any generic value $N > 1$ one has to solve a variational min-max problem (which can still imply a hard computational burden), while for $N = 1$ one is faced with an easily tractable min-max problem.

The derivation proceeds as follows. First, for a generic discrete-time nonlinear system, by assuming a suitable monotonicity property of the game cost, it is shown that the L_2-gain of the closed-loop is less than or equal to γ in $\Omega^{RH}(N)$. Then, under the assumption that the system under control is affine, it is proven that the monotonicity property is guaranteed by a proper choice of the terminal weight and the terminal inequality constraint in the min-max problem. These results extend to the discrete-time case some previous work for continuous-time nonlinear systems [5], [22]. The main advantage of the present formulation with respect to [5] is the use of feedback policies instead of open-loop policies for the minimizing player. In this way the use of a feedback precompensator is not needed any more. The H_∞ RH control of linear unconstrained systems has been previously studied in [35] and [18]. For the linear H_∞ RH problem with constraints see also [6] and [30].

4.1   Receding horizon formulation of nonlinear H_∞ control law

Consider a two-player game between the controller and the nature, where the controller chooses the input $u(k)$ as a function of the current state $x(k)$

so as to ensure that the effect of the disturbance $w(\cdot)$ on the output $z(\cdot)$ is sufficiently small for any choice of $w(\cdot)$ made by nature. This H_∞ problem is of significant practical interest, because of the possibility to formulate and solve the robust stability problem in this setup (see Sect.4.3 for more details).

The derivation of the RH control law is based on the solution of a finite-horizon differential game, where u is the input of the minimizing player (the controller) and w is the input of the maximizing player (the nature).

In the following, according to the RH paradigm, the finite time interval $[t, t+N-1]$ will be considered. In this interval the controller has to choose a sequence of feedback control strategies $\kappa_{t,t+N-1} := [\kappa_0(x(t)), \ldots, \kappa_{N-1}(x(t+N-1))]$ where $\kappa_i(\cdot) : \mathbb{R}^n \to \mathbb{R}^m$, is called *policy*. On the other hand, the sequence of disturbances chosen by nature is denoted by $w_{t,t+N-1} := [w(t), \ldots, w(t+N-1)]$. Differently from the standard RH approach, in this case it is in general not convenient to consider open-loop control strategies since open-loop control would not account for changes in the state due to unpredictable inputs played by the nature (see also [30]). Hence, at each time t, the minimizing player optimizes his sequence $\kappa_{t,t+N-1}$ of policies, i.e. minimization is carried out in an infinite-dimensional space. Conversely, in open-loop it would be sufficient to minimize with respect to the sequence $[u(t), u(t+1), \ldots, u(t+N-1)]$ of future control actions, a sequence which belongs to a finite-dimensional space.

Let define

$$S(z, w) = \frac{1}{2}\left\{\gamma^2 \|w\|^2 - \|z\|^2\right\}$$

The *Finite-Horizon Optimal Dynamic Game* (*FHODG*) consists of the minimization with respect to $\kappa_{t,t+N-1}$, and the maximization with respect to $w_{t,t+N-1}$, of the cost function

$$J(\bar{x}, \kappa_{t,t+N-1}, w_{t,t+N-1}, N) = -\sum_{i=t}^{t+N-1} S(z(i), w(i)) + V_f(x(t+N))$$

subject to the terminal constraint

$$x(t+N) \in X_f \subset R^n$$

and subject to (1)-(2) with $x(t) = \bar{x}$, and $u(i) = \kappa_{i-t}(x(i))$; $V_f(x)$ is a nonnegative function with $V_f(0) = 0$ and the terminal region X_f is a set containing the origin as interior point. Here, γ is a constant, which can be interpreted as the disturbance attenuation level.

Then, for a given initial condition $\bar{x} \in R^n$, denote by

$$\kappa_{t,t+N-1}^o = \arg \min_{\kappa_{t,t+N-1}} \max_{w_{t,t+N-1}} J(\bar{x}, \kappa_{t,t+N-1}, w_{t,t+N-1}, N)$$

and

$$w_{t,t+N-1}^o = \max_{w_{t,t+N-1}} J(\bar{x}, \kappa_{t,t+N-1}^o, w_{t,t+N-1}, N)$$

the saddle point solution, if exists, of the $FHODG$.

According to the RH paradigm, the value of the feedback control as a function of $\bar{x}$ is obtained by solving the $FHODG$ and setting

$$\kappa^{RH}(\bar{x}) = \kappa_0^o(\bar{x})$$

where $\kappa_0^o(\bar{x})$ is the first column of $\kappa_{t,t+N-1}^o := [\kappa_0^o(x(t)), \ldots, \kappa_{N-1}^o(x(t+N-1))]$.

Definition 2. : Let $\mathcal{K}(\bar{x}, N)$ be the set of all policies $\kappa_{t,t+N-1}$ such that starting from $\bar{x}$, it results $x(t+N) \in X_f$ for every admissible disturbance sequences $w_{t,t+N-1}$.

Definition 3 (Playable set). [36] Let $\Omega^{RH}(N)$ be the set of initial states $\bar{x}$ such that $\mathcal{K}(\bar{x}, N)$ is nonempty.

4.2 Sufficient conditions for stability and L_2-gain attenuation

This paragraph presents some standing assumptions and sufficient conditions for the scheme presented in this section to be stabilizing and achieve the desired L_2-gain attenuation. A particular setup under which these conditions can be proven to be satisfied will be described in paragraph 4.3.

In the following, the optimal value of the $FHODG$ will be denoted by $V(\bar{x}, N)$, i.e. $V(\bar{x}, N) := J(\bar{x}, \kappa_{t,t+N-1}^o, w_{t,t+N-1}^o, N)$.

Our aim is to ensure that the closed-loop system

$$\Sigma^{RH} : \begin{cases} x(k+1) = f(x(k), \kappa^{RH}(x(k)), w(k)) \\ z(k) = h(x(k), \kappa^{RH}(x(k)), w(k)) \end{cases}$$

has L_2-gain less than or equal to γ in the sense specified below. In what follows $\bar{\Omega}$ is a subset of $\Omega^{RH}(N)$ which contains the origin as interior point.
Assumption A1: *The system (1)-(2) is zero-state detectable in $\bar{\Omega}$ i.e.,* $\forall x(t) \in \bar{\Omega}$,

$$z(k)|_{w=0} = h(x(k), u(k), 0) = 0, \quad \forall k \geq t \implies \lim_{k \to \infty} x(k) = 0$$

Condition C1: $V(x, N)$ *is nonnegative,* $\forall x \in \bar{\Omega}$.

Note that due to the indefiniteness of $S(z, w)$ this is not trivially satisfied contrary to what happens for most other conventional (i.e. non-H_∞) RH schemes.
Condition C2: $V(x, N)$ *is continuous in the origin.*
Condition C3 (Monotonicity property): $V(x, N+1) \leq V(x, N), \forall x \in \bar{\Omega}$.

In many important cases the space of controls for the maximizing player is not the whole of l_2. For example the space may be limited by assumed limitations on the magnitude of w, or by other restrictions implied for example by condition C1 (see Paragraph 4.3 for this special case). The space of admissible disturbances is denoted by $\mathcal{W}$.

Condition C4: For every $w \in \mathcal{W}$, $\bar{\Omega}$ is a positively invariant nonempty set for the system Σ^{RH}.

Definition 4. Suppose that γ is a given positive real number. A system

$$\Sigma : \begin{cases} x(k+1) = f_c(x(k), w(k)), & x(0) = x_0, \quad k \geq 0 \\ z(k) = h_c(x(k), w(k)) \end{cases} \tag{11}$$

is said to have L_2-gain less than or equal to γ in $\Omega \subseteq \mathbb{R}^n$ if, for every $w \in \mathcal{W}$

i) Ω is a positively invariant set for the system Σ;
ii) there exists a finite quantity $\beta(x_0)$ such that $\forall x_0 \in \Omega$, $\forall T \geq 0$,

$$\sum_{i=0}^{T} \|z(i)\|^2 \leq \gamma^2 \sum_{i=0}^{T} \|w(i)\|^2 + \beta(x_0)$$

Theorem 1. *Under Assumption A1 and Conditions C1–C4 the system Σ^{RH} is internally stable and has L_2-gain less than or equal to γ, in $\bar{\Omega}$.*

This proof, as well as the proofs of the subsequent results, can be found in [21].

4.3 A robustly stable receding horizon control scheme

For affine nonlinear systems affected by a particular class of uncertainties, the monotonicity property is now ensured following the guidelines proposed in [5] for continuous-time systems. Specifically, a quadratic terminal penalty function and a terminal region defined as a suitable level set of the penalty function are used. In [5], an open-loop optimization problem is first solved and a precompensation feedback control law is then used to achieve robustness. The main drawback of this approach is that the playable set can be very small and it is not even possible to guarantee that it is larger than the terminal region X_f. Conversely, a closed-loop min-max problem is here considered. In this way, it is possible to guarantee that the playable set contains X_f. Moreover no precompensation is needed in order to design the controller. On the other hand, the use of an infinite dimensional decision space (due to the minimization with respect to policies) makes the min-max problem much more complex.

Consider the system

$$x(k+1) = f_1(x(k)) + F_2(x(k))u(k) + F_3(x(k))w(k) \tag{12}$$

$$z(k) = \begin{bmatrix} h_1(x(k)) \\ u \end{bmatrix}$$

where f_1, F_2, F_3 and h_1 are C^2 functions with $f_1(0) = 0$ and $h_1(0) = 0$. For convenience, the corresponding discrete-time linearized system is represented as

$$x(k+1) = F_1 x(k) + F_2 u(k) + F_3 w(k)$$
$$z(k) = \begin{bmatrix} H_1 x(k) \\ u \end{bmatrix}$$

where $F_1 = \left.\frac{\partial f_1}{\partial x}\right|_{x=0}$, $F_2 = F_2(0)$, $F_3 = F_3(0)$, $H_1 = \left.\frac{\partial h_1}{\partial x}\right|_{x=0}$. Given a square $n \times n$ matrix P, define also the symmetric matrix

$$R = R(P) = \begin{bmatrix} R_{11} & R_{12} \\ R_{21} & R_{22} \end{bmatrix}$$

where

$$R_{11} = F_2'PF_2 + I$$
$$R_{12} = R_{21}' = F_2'PF_3$$
$$R_{22} = F_3'PF_3 - \gamma^2 I$$

Assumption A2: It is assumed that the disturbance w represents an uncertainty satisfying

$$w = \Delta z \tag{13}$$

where Δ is an arbitrary nonlinear dynamic system, with state x^Δ, having a finite $L_2 - gain < \frac{1}{\gamma}$ with $\beta(x^\Delta(0)) = 0$.

Given the initial state $\bar{x}$ and a policy vector $\kappa_{t,t+N-1} \in \mathcal{K}(\bar{x}, N)$, the disturbance vector $w_{t,t+N-1}$ will be said to be admissible if, along the trajectory of (12), with $u(i) = \kappa_{i-t}(x(i))$ one has

$$\sum_{i=0}^{T} \|w(t+i)\|^2 \leq \frac{1}{\gamma^2} \sum_{i=0}^{T} \|z(t+i)\|^2, \quad \forall T < N$$

The corresponding set of admissible disturbances $w_{t,t+N-1}$ is denoted by $\mathcal{W}(\bar{x}, \kappa_{t,t+N-1})$.

Proposition 10. : *Suppose there exists a positive definite matrix P such that:*

(i) $R_{11} > 0$, $R_{22} < 0$

(ii) $-P + F_1'PF_1 + H_1'H_1 - \begin{pmatrix} F_2'PF_1 \\ F_3'PF_1 \end{pmatrix}' R^{-1} \begin{pmatrix} F_2'PF_1 \\ F_3'PF_1 \end{pmatrix} < 0$

Then, there exists a nonempty neighbourhood Ω of the origin that, for all w satisfying Assumption A2, is a positively invariant set for system (12) with control law $u = \kappa^*(x)$ where

$$\begin{bmatrix} \kappa^*(x) \\ \xi^*(x) \end{bmatrix} = -R(x)^{-1} \begin{bmatrix} F_2(x)'Pf_1(x) \\ F_3(x)'Pf_1(x) \end{bmatrix}$$

with

$$R(x) = \begin{bmatrix} F_2(x)'PF_2(x) + I & F_2(x)'PF_3(x) \\ F_3(x)'PF_2(x) & F_3(x)'PF_3(x) - \gamma^2 I \end{bmatrix} = \begin{bmatrix} r_{11} & r_{12} \\ r_{21} & r_{22} \end{bmatrix}$$

Observe that P can be computed by solving a discrete-time H_∞ algebraic Riccati equation.

Proposition 11. : *Under Assumption A2, if $V_f(x) = \frac{1}{2}x'Px$ and $X_f = \Omega$, where P and Ω are defined in Proposition 10 then Conditions C1, C3, C4 hold with $\bar{\Omega} = \Omega^{RH}(N)$.*

Theorem 2. : *Let A1-A2 and C2 hold and suppose there exists a positive definite matrix P such that (i) and (ii) of Proposition 10 hold. Then, letting $V_f(x) = \frac{1}{2}x'Px$, there exists a nonempty neighbourhood Ω of the origin such that with $X_f = \Omega$,*

$$\Sigma_a^{RH} : \begin{cases} x(k+1) = f_1(x(k)) + F_2(x(k))\kappa^{RH}(x(k)) + F_3(x(k))w(k) \\ z(k) = \begin{bmatrix} h_1(x(k)) \\ \kappa^{RH}(x(k)) \end{bmatrix} \end{cases}$$

is internally stable and has L_2-gain less than or equal to γ in $\Omega^{RH}(N)$.

The most important parameter to be tuned is the optimization horizon N, which can be chosen to find a compromise between computational complexity (which grows with N) and the extent of the playable region (i.e. $\Omega^{RH}(N+1) \supseteq \Omega^{RH}(N) \supseteq ... \supseteq \Omega^{RH}(1) \supseteq \Omega$). Note however that for $N = 1$ no closed-loop optimization on strategies is required (the finite-horizon dynamic game is just solved in a finite-dimensional space).

5   Conclusions

In the recent years several stabilising nonlinear RH control algorithms have been proposed and by now the issue of closed-loop stability is well established. The study of robustness, conversely, is still an open field of research. Two robustness analysis methods based on monotonicity and inverse optimality have been reviewed. It is, however, the issue of robust synthesis that is at the same time more appealing and more involved. The H_∞ approach described in the paper is a promising one but still the need to perform optimisation in an infinite-dimensional policy space limits substantially its practical applicability. A possible remedy could be to restrict the optimisation within a suitable finite-dimensional subset of the policy space.

Acknowledgement: The authors acknowledge the partial financial support by MURST Project "Algorithms and architectures for the identification and control of industrial systems"

References

1. R. Bitmead, M. Gevers, I. Petersen, and R.J.Kaye. Monotonicity and stabilizability properties of solutions of the Riccati difference equation: Propositions, lemmas, theorems, fallacious conjectures and counterexamples. *System & Control Letters*, 5:309–315, 1985.
2. R. Bitmead, M. Gevers, and V. Wertz. *Adaptive Optimal Control: The Thinking Man's GPC*. Prentice Hall, 1990.
3. C. Chen and L. Shaw. On receding horizon feedback control. *Automatica*, 18(3):349–352, 1982.
4. H. Chen and F. Algöwer. A quasi-infinite horizon nonlinear model predictive control scheme with guaranteed stability. *Automatica*, 34:1205–1217, 1998.
5. H. Chen, C. Scherer, and F. Allgöwer. A game theoretical approach to nonlinear robust receding horizon control of constrained systems. In *American Control Conference '97*, 1997.
6. H. Chen, C. Scherer, and F. Allgöwer. A robust model predictive control scheme for constrained linear systems. In *Proc. DYCOPS 5, Corfu, Greece*, 1998.
7. D. Chmielewski and V. Manousiouthakis. On constrained infinite-time linear quadratic optimal control. *System & Control Letters*, 29:121–129, 1996.
8. D. Clarke, C. Mothadi, and P. Tuffs. Generalized predictive control- part I and II. *Automatica*, 23:137–160, 1987.
9. D. Clarke and R. Scattolini. Constrained receding horizon predictive control. *Proc. IEE Part D*, 138:347–354, 1991.
10. C. Cuttler and B. Ramaker. Dynamic matrix control-a computer control algorithm. In *Automatic Control Conference*, 1980.
11. G. De Nicolao, L. Magni, and R. Scattolini. On the robustness of receding-horizon control with terminal constraints. *IEEE Trans. Automatic Control*, 41:451–453, 1996.
12. G. De Nicolao, L. Magni, and R. Scattolini. Stabilizing predictive control of nonlinear *ARX* models. *Automatica*, 33:1691–1697, 1997.
13. G. De Nicolao, L. Magni, and R. Scattolini. Stability and robustness of nonlinear receding-horizon control. In *NMPC Workshop - Assessment and Future Directions*, pages 77–90, Ascona, Switzerland, 1998.
14. G. De Nicolao, L. Magni, and R. Scattolini. Stabilizing receding-horizon control of nonlinear time-varying systems. *IEEE Trans. on Automatic Control*, AC-43:1030–1036, 1998.
15. S. Glad. Robustness of nonlinear state feedback- a survey. *Automatica*, 23:425–435, 1987.

16. S. Keerthi and E. Gilbert. Optimal, infinite-horizon feedback laws for a general class of constrained discrete-time systems. *J. Optimiz. Th. Appl.*, 57:265–293, 1988.

17. W. Kwon and A. Pearson. On feedback stabilization of time-varying discrete-linear systems. *IEEE Trans. on Automatic Control*, 23:479–481, 1978.

18. S. Lall and K. Glover. A game theoretic approach to moving horizon control. In D. Clarke, editor, *Advances in Model-Based Predictive Control*, pages 131–144. Oxford University Press, 1994.

19. W. Lin and C. Byrnes. H_∞-control of discrete-time nonlinear systems. *IEEE Trans. on Automatic Control*, pages 494–510, 1996.

20. L. Magni, G. De Nicolao, and R. Scattolini. Output feedback receding-horizon control of discrete-time nonlinear systems. In *IFAC Nonlinear Control Systems Design Symposium*. Enschede, The Netherlands, 1998.

21. L. Magni, G. De Nicolao, R. Scattolini, and F. Allgöwer. H_∞ receding horizon control for nonlinear discrete-time systems. *Submitted to Systems & Control Letters*.

22. L. Magni, H. Nijmeijer, and A. van der Schaft. A receding-horizon approach to the nonlinear H_∞ control problem. *Submitted to Automatica*.

23. L. Magni and R. Sepulchre. Stability margins of nonlinear receding horizon control via inverse optimality. *System & Control Letters*, 32:241–245, 1997.

24. D. Mayne and H. Michalska. Receding horizon control of nonlinear systems. *IEEE Trans. on automatic Control*, 35:814–824, 1990.

25. H. Michalska and D. Mayne. Robust receding horizon control of constrained nonlinear systems. *IEEE Trans on Automatic Control*, 38:1512–1516, 1993.

26. E. Mosca and J. Zhang. Stable redesign of predictive control. *Automatica*, 28:1229–1233, 1992.

27. T. Parisini and R. Zoppoli. A receding-horizon regulator for nonlinear systems and a neural approximation. *Automatica*, 31:1443–1451, 1995.

28. M. Poubelle, R. Bitmead, and M. Gevers. Fake algebraic riccati technique and stability. *IEEE Trans. on Automatic Control*, (33):379–381, 1988.

29. J. Rawlings and K. Muske. The stability of constrained receding horizon control. *IEEE Trans on Automatic Control*, 38:1512–1516, 1993.

30. P. Scokaert and D. Mayne. Min-max feedback model predictive control for constrained linear systems. *IEEE Trans. on Automatic Control*, 43:1136–1142, 1998.

31. P. Scokaert and J. Rawlings. Constrained linear quadratic regulation. *IEEE Transactions on Automatic Control*, 43:1163–1169, 1998.

32. P. Scokaert, J. Rawlings, and E. Meadows. Discrete-time stability with perturbations: Application to model predictive control. *Automatica*, 33:463–470, 1997.

33. R. Sepulchre, M. Jankovic, and P. Kokotovic. *Constructive Nonlinear Control*. Springer-Verlag, 1996.

34. M. Sznaier and M. Damborg. Control of constrained discrete time linear systems using quantized controls. *Automatica*, 25:623–628, 1989.

35. G. Tadmor. Receding horizon revisited: An easy way to robustly stabilize an LTV system. *System & Control Letters*, 18:285–294, 1992.

36. T. Vincent and W. Grantham. *Nonlinear and Optimal Control Systems*. John Wiley & Sons, 1997.

37. T. Yang and E. Polak. Moving horizon control of nonlinear systems with input saturation, disturbances and plant uncertainty. *Int. J. Control*, 58:875–903, 1993.

Nonlinear Representations and Passivity Conditions in Discrete Time

Salvatore Monaco[1] and Dorothée Normand-Cyrot[2]

[1] Dipartimento di Informatica e Sistemistica
   Università di Roma "La Sapienza"
   Via Eudossiana 18, 00184 Rome, Italy.
[2] Laboratoire des Signaux et Systèmes, CNRS-ESE
   Plateau de Moulon, 91190 Gif-sur-Yvette, France.

Abstract. Based on the idea of describing discrete-time dynamics as coupled difference and differential equations, we here study conditions under which a given nonlinear discrete-time system is passive and lossless.

1 Introduction

A new representation of discrete-time dynamics as coupled difference and differential equations has been recently proposed by the authors themselves in [8]. The difference equation models, through jumps, the free evolution or a nominal one associated with some constant control value while the differential equation models the influence of the control. The differential structure is at the basis of a geometric study which enables us to stress some intriguing analogies with continuous-time dynamics, in particular regarding the analycity of the control dependency compared to the time dependency. This analogy is here strenghtened referring to the concept of dissipativity and passivity ([7]).

The basic studies in [9,10,4], where the concepts of dissipativity and passivity have been investigated for continuous-time state-space representations of nonlinear systems, provide a physical interpretation of stability and stabilizability problems. On these bases a method for global asymptotic stabilization of nonlinear systems has recently been proposed in [2,3].

One of the major difficulties to develop the theory in discrete time is the absence of conditions to check whether a nonlinear system is passive or not. A basic result in discrete time has recently been stated in [1] where a discrete-time version of the Kalman-Yakubovitch-Popov Lemma was proposed for the class of lossless systems which admit state affine representations, thus providing the basis for the study of global stabilization in discrete time [1].

The present paper deals with passivity and losslessness for nonlinear discrete-time systems whose dynamics admit the beforementioned representation ([8]). In this context we give sufficient conditions for passivity and necessary and sufficient conditions for losslessness. The latter restores the ones given in [1] for state affine systems. The conditions stated are discussed in some particular situations and compared to the well-known linear ones.

2 Discrete-time dynamics as coupled difference and differential equations

Definition 5. (Analytically Parametrized Discrete - Time Dynamics) Let M be an analytic manifold, U_0 a neighbourhood of zero, and assume the maps analytic in their arguments; given a function $F_0 : M \to M$ and a parametrized vector field G^0 on M, an **APDTD** is defined by the two equations

$$x^+ = F_0(x) \tag{1}$$

$$\frac{d}{du}(x^+(u)) = G^0(x^+(u), u); \quad x^+(0) = x^+. \tag{2}$$

As a matter of fact a nonlinear difference equation of the usual form

$$x_{k+1} = F(x_k, u_k) \tag{3}$$

can be recovered by integrating (2) between 0 and u_k for the initial condition $x_k^+(0)$ computed from (1) setting $x_k^+ = x_k^+(0) := F_0(x_k)$. One obtains,

$$x_{k+1} := x_k^+(u_k) = x^+(0) + \int_0^{u_k} G^0(x^+(v), v)dv$$

where $F : M \times U \to M$ is analytic and such that $F(., 0) = F_0(.)$.

According to Definition 1 the state at time $(k+1)$ is computed by mixing up a difference equation generating a jump "after" time (k), together with a differential equation which models the control effect.

It must be pointed out that F_0 and G^0 could be assumed parametrized by $\bar{u}$, say $F_{\bar{u}}(x)$ and $G^0(\bar{u})(x, u)$, to describe a dynamics around an assigned constant control value $\bar{u}$.

Remark 1. The following example, from [11], shows that our approach can be usefully applied to model real phenomena. In [11] the authors make reference to a classical Leslie model describing the natural growth of age-structured populations adapted to the case of controlled populations. The starting point of the modelling procedure is the uncontrolled evolution described by the linear drift

$$F_0(x(k)) = Ax(k)$$

in which the i-th component $x_i(k)$ of the n-dimensional vector $x(k)$ is the number of individuals of age i at the beginning of year k, and the $(n \times n)$ *Leslie matrix* A has the well known structure with entries the strictly positive *survival coefficients*, $s_i < 1, i = 0, \ldots, n - 1$, and the nonnegative *fertility coefficients*, $f_i, i = 0, \ldots, n$. The growth is controlled by removing individuals (forest exploitation, harvesting, hunting,..). Thus the removing process must be modelled together whith the growing process in order to obtain a complete description of the process. It is quite intuitive to understand that assuming that the removing season starts when the reproduction is over and that the *harvesting effort*, u, is constant, the rate of change of the population due to harvesting is proportional to the number of individuals, which is equivalent, in our context, to assume

$$G^0(x, u) = -Qx$$

where Q is the matrix of the *catchability coefficients* and can be assumed diagonal, $Q = diag[d_i]_1^n$.

With this in mind and following our approach the following representation is obtained

$$x^+ = Ax \tag{4}$$

$$\frac{d}{du}(x^+(u)) = -Qx^+(u); \quad x^+(0) = x^+. \tag{5}$$

We will show later on that the model (4-5) restitutes the one computed in [11].

The discrete-time evolution

As far as the relations between $F(x, u)$ in (3) and $G^0(x, u)$ in (2) are concerned, consider their expansions in powers of u around $u = 0$

$$G^0(x, u) = G_1^0(x) + \sum_{i \geq 1} \frac{u^i}{i!} G_{i+1}^0(x) \tag{6}$$

$$F(x, u) := F_0(x) + \sum_{i \geq 1} \frac{u^i}{i!} F_i(x) \tag{7}$$

Let us define the operator $\mathcal{D}_u := \frac{d}{du} + L_{G^0(x,u)}$, where L_G stands for the Lie derivative associated with G (i.e. $L_G(F) := J_x[F] \times G$), and $J_x[.]$ denotes the Jacobian with respect to x of the function into the brackets. It has been shown in [8] that (4) admits the power expansion (7) with

$$F_i(x) = \mathcal{D}_u^{i-1}(G^0(x, u))\big|_{u=0}\big|_{x=F_0(x)} \tag{8}$$

Specifying (8) one finds for the first terms

$$F_2(x) = [\frac{dG^0(.,v)}{dv} + L_{G^0(.,v)}(G^0(x,v))]|_{v=0}|_{x=F_0(x)}$$

$$= G_2^0(x)|_{x=F_0(x)} + L_{G_1^0(.)}(G_1^0(x))|_{x=F_0(x)}$$

$$F_3(x) = G_3^0(x)|_{x=F_0(x)} + 2L_{G_1^0(.)}(G_2^0(x))|_{x=F_0(x)}$$

$$+L_{G_2^0(.)}(G_1^0(x))|_{x=F_0(x)} + L_{G_1^0(.)} \circ L_{G_1^0(.)}(G_1^0(x))|_{x=F_0(x)}$$

$$\ldots$$

The evolution over one step described by (3), takes an exponential form, namely the exponential representation of the flow, $\phi_u^{G^0}(x)$, of the differential equation (2). As shovn in [8], the solution of the differential equation (2) initialized at (1) takes the form

$$x^+(u) = \phi_u^{G^0}(x) = e^{uG^0(.,u)}(Id)|_{x^+(0)} \tag{9}$$

where $uG^0(x,u) := M \to T_x M$ is a Lie element in the $(G_i^0(.) : i \geq 1)$'s whose expansion is given by

$$uG^0(.,u) = \sum_{i \geq 1} \frac{u^i}{i!} B_i(G_1^0, \ldots, G_i^0) \tag{10}$$

where B_i stands for a homogeneous Lie polynomial of degree i in its arguments when $G_i^0(.)$ is said, by convention, of degree i.

The proof of this result, detailed in [5], works out by performing the logarithmic expansion of

$$F(x,u) = e^{uD_v}(Id)|_{v=0}|_{x=F_0(x)} =$$

$$= [x + uG_1^0(x) + \sum_{i \geq 2} \frac{u^i}{i!}(D_v^{i-1}(G^0(x,v)))|_{v=0}]|_{x=F_0(x)}$$

The expression of the (B_i)'s as homogeneous polynomials in the (G_i^0)'s are given in [5] from which one deduces their Lie polynomials expressions. For the first terms, one finds

$$uG^0(.,u) = uB_1(G_1^0) + \frac{u^2}{2!}B_2(G_1^0, G_2^0) + \frac{u^3}{3!}B_3(G_1^0, G_2^0, G_3^0) + O(u^4)$$

with

$$B_1 = G_1^0, B_2 = G_2^0,$$
$$B_3 = G_3^0 + 1/2[G_1^0, G_2^0], B_4 = G_4^0 + [G_1^0, G_3^0],$$

where $[.,.]$ indicates the Lie bracket of two vector fields.

It is now clear that starting from Definition 1., adiscrete-time dynamics of the form (3) can be computed by integrating the differential equation (2) with suitable initialization (1) deduced from the free evolution. Reciprocally, starting from an analytic discrete-time dynamics $x_{k+1} = F(x_k, u_k)$, assuming the existence of a function G^0, analytic on M, parametrized by u and satisfying condition (2), x_{k+1} can be interpreted as the result of the integration between 0 and u_k of (2) with initialization in $F_0(x_k)$. Let us finally note that the invertibility of $F_0(.)$ (or more generally of $F_{\bar{u}}(.) = F(., \bar{u})$ when given a nominal control value) is a necessary and sufficient condition for getting two identical formulations since in such a case the existence and unicity of $G^0(x, u)$ are proven.

Remark 2. We can now go back to our example and compute the exponential representation associated with the differential representation (4-5). One has

$$x(k + 1) = F(x(k), u(k)) = e^{-u(k)Qx} Id|_{x=Ax(k)}$$

By computing, the state equation become

$$x(k + 1) = Ax(k) - u(k)QAx(k) + \frac{u(k)^2}{2} Q^2 Ax(k) + \cdots = e^{-u(k)Q} Ax(k)$$

which is the model proposed in [11] making use of quite complex and less intuitive arguments.

3   Passivity and losslessness

Starting from the representations introduced in the previous section, it is possible, following [7], to characterize dissipativity in terms of a set of sufficient conditions thus generalizing the Kalman-Yakubovitch-Popov conditions given in the linear case. In the lossless case, these conditions also become necessary without requiring any quadraticity assumption as in [1]. In the sequel it will be assumed that the dynamics (1) be drift invertible.

Let us consider the single input nonlinear dynamics (1) with output $H : M \times \mathbb{R} \to M$ analytic in both its arguments

$$y_k = H(x_k, u_k) \tag{11}$$

Assume the existence of at least one equilibrium pair and without loss of generality, $H(0,0) = 0$. F and H being analytic with respect to the control variable u, one may introduce their expansions

$$F(., u) := F_0(.) + \sum_{i \geq 1} \frac{u^i}{i!} F_i(.) \tag{12}$$

and

$$H(.,u) := H_0(.) + \sum_{i \geq 1} \frac{u^i}{i!} H_i(.) \tag{13}$$

where $F_i(.)$ and $H_i(.)$, defined from M to $\mathbb{R}$ for all $i \geq 0$, are analytic.

A real valued function, $W(u,y)$, defined on $U \times Y$ such that $W(0,y) = 0$ and such that for any $N \geq 0$, any u_k in $\mathcal{U}_0$ for $k \geq 0$ and x_0 in X

$$\sum_{k=0}^{N} |W(u_k, y_k)| < \infty$$

will be called the supply rate. The following definition is recalled

Definition 6. A system is said to be dissipative with a supply rate W if there exists a C^0 nonnegative function $V : X \to R$, called the storage function, such that for all u_k in $\mathcal{U}_0$, all x_k in X,

$$V(x_{k+1}) - V(x_k) \leq W(u_k, y_k) \tag{14}$$

(14) is equivalent to

$$V(x_{N+1}) - V(x_0) \leq \sum_{i=0}^{N} W(u_i, y_i) \tag{15}$$

for all $N \geq 0$ and u_i in $\mathcal{U}_0$ for $i = (1, \cdots, N)$, i.e. the stored energy along any trajectory from x_0 to x_N does not exceed the supply.

A - Nonlinear passivity in discrete time

Definition 7. A system is said to be passive if it is dissipative with supply rate $u_k y_k$, and storage function such that $V(0) = 0$, i.e., if it verifies for all u_k in $\mathcal{U}_0$, x_k in M, the inequality

$$V(x_{k+1}) - V(x_k) \leq u_k y_k \tag{16}$$

Remark 3. From (16), a passive system with a positive definite storage function is Lyapunov stable. Reciprocally, V is not increasing along trajectories such that $y_k = 0$. Since this constraint defines the zero dynamics, one deduces that a passive system with a positive storage function V has a Lyapunov stable zero dynamics.

A set of sufficient conditions for characterizing passivity can easily be obtained by rewriting the product $u_k y_k$ as an integral of $x^+(u)$. More precisely, one looks for a function $Y(x^+(u), u)$, such that the following equality holds

$$u_k y_k = \int_0^{u_k} Y(x^+(v), v) dv \tag{17}$$

or under differentiation with respect to u such that

$$Y(x^+(u), u) := H(x, u) + u\frac{\partial H}{\partial u}(x, u) =$$
$$(H_0 + \sum_{i \geq 1} \frac{(i+1)u^i}{i!} H_i)(x). \tag{18}$$

It results that Y is analytic in both arguments and admits with respect to u the expansion

$$Y(., u) = (Y_0 + \sum_{i \geq 1} \frac{u^i}{i!} Y_i)(.)$$

To express the Y_i in terms of the output map $H(x, u)$, one rewrites (18) as

$$e^{u\mathcal{G}^0(.,u)}(Y(., u)) := (H_0 + \sum_{i \geq 1} \frac{(i+1)u^i}{i!} H_i)(.)\big|_{F_0^{-1}(.)}.$$

Equating the terms of the same power in δ, one gets

$$H_0 \circ F_0^{-1} := Y_O$$
$$2H_1 \circ F_0^{-1} := Y_1 + L_{G_1^0}(Y_O) = Y_O^* + L_{G_1^0}(Y_O)$$
$$3H_2 \circ F_0^{-1} := Y_2 + L_{G_2^0}(Y_O) + L_{G_1^0}^2(Y_O) + 2L_{G_1^0}(Y_1)$$
$$:= (Y_1 + L_{G_1}(Y_O))^* + L_{G_1}(Y_1 + L_{G_1}(Y_O))$$
$$\cdots$$
$$iH_{i-1} \circ F_0^{-1} := ((i-1)H_{i-2} \circ F_0^{-1})^* + L_{G_1^0}(iH_{i-1} \circ F_0^{-1})$$
$$\cdots$$

where $*$ stands for a formal rule which acts on any polynomial in the $(L_{G_i^0})'s$ and the $(Y_i)'s$ as follows

$$(L_{G_{i_1}} \circ ... \circ L_{G_{i_p}}(Y_i))^* = \tag{19}$$
$$\sum_{i_j} L_{G_{i_1}} \circ ... \circ L_{G_{i_j+1}} ... \circ L_{G_{i_p}}(Y_i)$$
$$+L_{G_{i_1}} \circ ... \circ L_{G_{i_p}}(Y_{i+1}). \tag{20}$$

According to these notations, for all $k \geq 0$, all $u_k \in \mathcal{U}_0$, all x_k, the passivity inequality (14) takes the form

$$V(F_0(x_k)) - V(x_k) + \int_0^{u_k} L_{G^0(.,v)}(V)(x^+(v))dv$$
$$\leq \int_0^{u_k} Y(x^+(v), v)dv \tag{21}$$

It is now possible to state the proposed "KYP property"

Definition 8. (KYP) The system (1, 11, 2) is said to satisfy the KYP property if there exists a C^1 nonnegative function V, with $V(0) = 0$, such that for all x,

(i) $V(F_0(x)) - V(x) \leq 0$
(ii) for all $u \in \mathcal{U}_0$, all x in M, $uL_{G^0(\cdot,u)}(V)(x) \leq uY(x,u)$

The following result is thus immediate ([7]):

Proposition 12. *Any discrete-time system given by (1, 11, 2) which satisfies the KYP property is passive with a storage function V.*

B - Lossless systems

Definition 9. A passive system is said to be lossless if for all $u_k \in \mathcal{U}_0$, all x_k in X

$$V(x_{k+1}) - V(x_k) = u_k y_k \tag{22}$$

Such an equality holds for any k in N, $u_k \in \mathcal{U}_0$, x_k in X if, and only if, for all $N \geq 0$

$$V(x_{N+1}) - V(x_0) = \sum_{i=0}^{N} u_i y_i$$

In the framework here set, one easily obtains

Theorem 1. *A system is lossless if and only if there exists a C^1 nonnegative function V, with $V(0) = 0$, such that for all x*

(i) $V(F_0(x)) = V(x)$
(ii) $L_{G_i^0}(V)(x) = Y_{i-1}(x)$ for all $i \geq 1$

Conditions (ii) alone chracterize the losslessness in the case od driftless dynamics.

Condition (ii) can be expressed in terms of the output by (17) and substituting Y_i for $L_{G_{i+1}^0}(V)$ because of the lossless condition, thus finding

$$H_0 \circ F_0^{-1} = L_{G_1^0}(V)$$
$$2H_1 \circ F_0^{-1} = L_{G_2^0}(V) + (L_{G_1^0})^2(V)$$
$$3H_2 \circ F_0^{-1} = L_{G_3^0}(V) + (L_{G_1^0})^3(V)$$
$$+L_{G_2} \circ L_{G_1}(V) + 2L_{G_1} \circ L_{G_2}(V)$$
$$(i+1)H_i \circ F_0^{-1} = (iH_{i-1} \circ F_0^{-1})^* + iL_{G_1^0}(H_{i-1} \circ F_0^{-1})$$

where any term of the form $L_{G_{i_1}} \circ ... \circ L_{G_{i_p}}(V)^*$ appearing in $(iH_{i-1} \circ F_0^{-1})^*$ is computed according to

$$L_{G_{i_1}} \circ ... \circ L_{G_{i_p}}(V)^* = \sum_{i_j} L_{G_{i_1}} \circ ... \circ L_{G_{i_{j+1}}} \circ ... \circ L_{G_{i_p}}(V).$$

Corollary 2 *System (1, 11, 2) is lossless if, and only if there exists a C^1 nonnegative function V, with $V(0) = 0$, such that for all x*

(i) $V(F_0(x)) = V(x)$
(ii)' *for all $i \geq 1$ with $H_0 \circ F_0^{-1} = L_{G_1^0}(V)$*

$$(i+1)H_i \circ F_0^{-1} = (iH_{i-1} \circ F_0^{-1})^* + iL_{G_1^0}(H_{i-1} \circ F_0^{-1})$$

Remark 4. The situation studied in [1], corresponds to assuming linearity with respect to u in (1, 11) (i.e. $F_i = 0$ and $H_i = 0$ for $i \geq 2$) and the quadraticity in u of $V(F(x,u))$ which presently corresponds to $L_{G_{i_1}^0} \circ ... \circ L_{G_{i_p}^0}(V) = 0$ for $p \geq 1$ and $\sum_{i_1}^{i_p}(i_j) \geq 3$. In such a case the conditions in [1] are recovered

(i) $V(F_0(x)) - V(x) = 0$
(ii) $L_{G_1^0}(V)(x) = H_0 \circ F_0^{-1}(x)$
(iii) $L_{G_2^0}(V)(x) + (L_{G_1^0})^2(V)(x) = 2H_1 \circ F_0^{-1}(x)$,

which, in addition, imply the quadraticity in u of $V(F(x,u))$ since $H_2 = 0$.

A - A particular nonlinear case

Let us now examine the situation in which (2) holds with $G^0(x,u)$ not depending on u, i.e., $G_i^0 = 0$ for $i \geq 2$. One gets

$$F(x,u) = e^{uG_1^0}(Id)\big|_{F_0(x)} \tag{23}$$

The equalities (23) can be simplified as

$$H_0 \circ F_0^{-1} := Y_0$$

$$2H_1 \circ F_0^{-1} := Y_1 + L_{G_1^0}(Y_0) = Y_0^* + L_{G_1^0}(Y_0)$$

$$3H_2 \circ F_0^{-1} := Y_2 + L_{G_1^0}^2(Y_0) + 2L_{G_1^0}(Y_1)$$

$$:= (Y_1 + L_{G_1}(Y_0))^* + L_{G_1}(Y_1 + L_{G_1}(Y_0))$$

$$\cdots$$

$$iH_{i-1} \circ F_0^{-1} := ((i-1)H_{i-2} \circ F_0^{-1})^* + L_{G_1^0}(iH_{i-2} \circ F_0^{-1})$$

$$\cdots$$

where the rule (*) reduces to

$$(L_{G_1^0}^j(Y_i))^* = L_{G_1^0}^j(Y_{i+1}).$$

According to these notations, inequality (16) becomes,

$$V(F_0(x_k)) - V(x_k) + \int_0^{u_k} L_{G_1^0(\cdot)}(V)(x^+(v))dv$$

$$\leq \int_0^{u_k} Y(x^+(v), v)dv \tag{24}$$

The set of sufficient conditions for passivity are now for all $u \in \mathcal{U}_0$, all x in M,

(i) $V(F_0(x)) - V(x) \leq 0$
(ii) $uL_{G_1^0}(V)(x) \leq uY(x, u)$

and (ii) becomes $L_{G_1^0}(V)(x) \leq uY_0(x)$, if $Y(x, u) = Y_0(0)$.
In the lossless case, one immediately obtains

Proposition 13. *A discrete-time system of the form (1,12, 23) is lossless if, and only if there exists a C^1 nonnegative function V, with $V(0) = 0$, such that for all x*

(i) $V(F_0(x)) = V(x)$
(ii) $L_{G_1^0}(V)(x) = Y_0(x)$
(iii) $Y_i(x) = 0$ for all $i \geq 1$

or equivalently such that for all x (i) holds true, and

(ii)' $L_{G_1^0}(V) = H_0 \circ F_0^{-1}$
(iii)' $iL_{G_1^0}(H_{i-1} \circ F_0^{-1})\big|_{F_0} = (i+1)H_i \circ F_0^{-1}$ for all $i \geq 1$.

Remark 5. If $H_i(x) = 0$ for $i \geq 1$ then

$$Y(x, u) = e^{-uG_1^0}(Y_0)$$

so that conditions $(ii)'$ and $(iii)'$ become

(ii)' $L_{G_1^0}(V) = H_0 \circ F_0^{-1}$
(iii)' $L_{G_1^0}^i(V) = 0$ for all $i \geq 2$.

and a discrete-time system without direct input-ouput link, may be lossless ([1]).

B - Linear Systems

Given a linear system defined on $\mathbb{R}^n$

$$x_{k+1} = Ax_k + u_k B := F(x_k, u_k) \tag{25}$$
$$y_k = Cx_k + u_k D \tag{26}$$

where A, B, C, D are matrices of suitable dimensions; by considering derivatives w.r.to u, one easily computes $G_1^0(x, u) = B = $ Cst so that $G_i(x) = 0$ for $i \geq 2$. It follows that in the present differential geometric framework, (25) can be rewritten as

$$\frac{d(x^+(u))}{du} = B \tag{27}$$
$$x^+(0) = Ax$$

Starting from x_k, one first computes $x_k^+(0) = Ax_k$ and then integrates (27) from 0 to u_k thus getting

$$x_k^+(u_k) := x_{k+1} = Ax_k + \int_0^{u_k} Bdv = Ax_k + u_k B.$$

Definition 10. A discrete-time linear system with invertible drift term is said to be passive if and only if there exists a nonnegative matrix P such that the storage function $V(x) := \frac{1}{2}x^T P x$ satisfies for all x and u

$$\frac{1}{2}(x^T A^T + u^T B^T)P(Ax + Bu) - \frac{1}{2}x^T P x$$

$$\leq u^T C x + \frac{1}{2}u^T(D^T + D)u \tag{28}$$

Specifying the conditions for $G_1^0(x) = B$, one recovers:

Proposition 14. *A discrete-time linear system is passive with storage function* $V(x) := \frac{1}{2}x^T P x$ *if there exists a nonnegative matrix* P *such that*

(i) $A^T P A - P \leq 0$
(ii) $B^T P A = C$
(iii) $B^T P B - (D^T + D) \leq 0$

Corollary 3 *A linear discrete-time system is lossless if and only if there exists a nonnegative matrix* P *such that (i), (ii) and (iii) of proposition 3 hold true with the equqlity sign.*

Remark 6. As well known, these definitions require that D is nonzero otherwise condition (iii) cannot be satisfied with P positive. This implies that a linear continuous time system which is passive cannot be said passive as a discrete-time system issued from usual sampling. This pathology suggests the introduction of slightly different concepts of passivity in discrete-time that the ones discussed in the present paper.

References

1. C. I. Byrnes and W. Lin, Losslessness, Feedback Equivalence and Global Stabilization of Discrete-Time Nonlinear Systems, *IEEE Trans. on Aut. Cont.*, **AC-39**, 1994, 83–98.
2. C.I.Byrnes and A.Isidori, Asymptotic stabilization of minimum-phase nonlinear systems, *IEEE Trans. Aut. Cont.*, **AC-36**, 1991, 1122–1137.
3. C. I. Byrnes, A. Isidori and J. C. Willems, Passivity, Feedback Equivalence and the Global Stabilization of Minimum Phase Nonlinear Systems, *IEEE Trans. on Aut. Cont.*, **AC-36**, 1991, 1228–1240.
4. D. Hill and P. Moylan, The Stability of Nonlinear Dissipative Systems, *IEEE Trans. Aut. Cont.*, **AC-21**, 1976, 708–711.
5. S. Monaco and D. Normand-Cyrot, A unifying representation for nonlinear discrete-time and sampled dynamics, *Journal of Math. Systems, Estimation and Control*, 101–103 (summary) 1995.
6. S. Monaco and D. Normand-Cyrot, Geometric properties of a class of nonlinear dynamics, Proc. ECC-97, Bruxelles, 1997.
7. S. Monaco and D. Normand-Cyrot, On the conditions of passivity and losslessness in discrete-time, Proc. ECC-97, Bruxelles, 1997.

8. S. Monaco and D. Normand-Cyrot, Discrete-time state representations: a new paradigm, *Perspectives in Control*, (D. Normand-Cyrot Ed.), Birkhauser, 1998, 191-204.
9. J. C. Willems, Dissipative Dynamical Systems Part I: General Theory, *Arch. Ration. Mechanical Anal.*, **45**, 1972, 325–351.
10. J. C. Willems, Dissipative Dynamical Systems Part II: Linear Systems with Quadratic Supply Rates, *Arch. Ration. Mechanical Anal.*, **4**, 1972, 352–393.
11. S. Muratori and S. Rinaldi, *Structural properties of controlled population models*, Sys. and Cont. Lett., n.10, pp. 147-153,1996.

Robust Ripple-Free Output Regulation and Tracking under Structured or Unstructured Parameter Uncertainties

Osvaldo Maria Grasselli, Laura Menini, and Paolo Valigi

Dipartimento di Informatica, Sistemi e Produzione - Università di Roma "Tor Vergata" via di Tor Vergata, 00133 Roma, Italy - Fax. +39 06 72597430 Tel. +39 06 72597430 e-mail: grasselli@disp.uniroma2.it

Abstract. The problem of the robust continuous-time asymptotic tracking and disturbance rejection for a multivariable sampled-data system is studied, for the case when the available description of the continuous-time plant has a known dependence on some "physical" parameters and possibly only some of the scalar outputs must track corresponding reference signals. The necessary and sufficient conditions for the existence of a hybrid control system for which exponential stability with a prescribed rate of convergence and a continuous-time convergence to zero of the tracking error are guaranteed at least in a neighbourhood of the nominal parameters, and, possibly, the convergence is dead-beat at the nominal ones, are reported, together with the necessity of a wholly continuous-time internal model of the exogenous signals.

1 Introduction

The control of sampled-data systems has been extensively studied (see, e.g., [1–3,7,9]). In particular, it is known that, if decoupling, or robust output regulation and tracking, is to be obtained for a continuous-time linear time-invariant plant through a digital control system, a ripple between sampling instants may arise [6,12,24,25,27–29,18,20,22]. This ripple is undesirable (e.g., decoupling with some additional ripple is not really decoupling). Although it can be reduced by reducing the sampling interval, this is not always allowed. On the other hand, the ripple is modulated by the exogenous signals, so that its amplitude is unbounded if they are unbounded. Therefore, for the problem of asymptotic tracking and output regulation, it was recognized in different settings that a purely continuous-time internal model of the exogenous signals (with the only exception of the constant ones, if zero order holders are used) is needed in order to avoid the ripple [12,28,29,18,22] and that such a model must be effective for all values of the parameters of the plant for which output regulation and/or asymptotic tracking have to be achieved, otherwise the purely discrete-time compensator will guarantee asymptotic tracking and output regulation at the sampling instants only.

Robust solutions were given in [6,12] to such a problem under the implicit assumption of independent perturbations of the entries of the matrices appearing in the description of the plant to be controlled, and in [19] under uncertainties of some "physical" parameters. In [18,22] the problem was completely solved for classes of exponential-sinusoidal disturbance functions and reference signals for the latter case, i.e. when the actual dependence of the plant description on some uncertain "physical" parameters is known, and only a subset of the scalar outputs must track corresponding exogenous reference signals: for this setting, the conditions were given for the existence of a hybrid control system (including samplers and zero-order holders) that guarantees a ripple-free asymptotic tracking and output regulation at the nominal parameters and at least in some neighbourhood of them.

Specifically, in [18] the dead-beat convergence of both the free state response and the tracking error was required, at the nominal parameters and, under some circumstances, also in a neighbourhood of the nominal parameters. Here, for the same setting, the solution of the same kind of problem will be presented, under a more natural weakened form of the nominal dead-beat convergence requirement, to be satisfied in some neighbourhood of the nominal parameters. Also the results in [22], concerning an exponential convergence requirement with a prescribed rate of convergence, will be taken into account here in a comprehensive presentation, involving, for both kinds of convergence, the necessity of a continuous-time internal model of the exogenous signals in the forward path of the feedback control system, the necessary and sufficient conditions for the existence of a hybrid control system satisfying the requirements, and a sketch of the design procedure of such a control system. In addition, also the case of independent perturbations of the entries of matrices appearing in the state-space representation of the plant will be explicitly considered, for completeness.

2 Preliminaries

Consider the linear time-invariant plant P described by

$$\dot{x}(t){=}A(\beta)x(t) + B(\beta)u(t) + \sum_{i=1}^{\mu} M_i(\beta)d_i(t), \tag{1}$$

$$y(t){=}C(\beta)x(t) + \sum_{i=1}^{\mu} N_i(\beta)d_i(t), \tag{2}$$

where $t \in \mathbb{R}$ is time, $x(t) \in \mathbb{R}^n$, $u(t) \in \mathbb{R}^p$ is the control input, $d_i(t) \in \mathbb{R}^{m_i}$, $i = 1, 2, ..., \mu$, are the unmeasurable and unknown disturbance inputs,

$y(t) \in \mathbb{R}^q$ is the output to be controlled — which is assumed to be measurable — and $A(\beta)$, $B(\beta)$, $C(\beta)$, $M_i(\beta)$, $N_i(\beta)$, $i = 1, 2, ...\mu$, are matrices with real entries depending on a vector $\beta \in \Omega \subseteq \mathbb{R}^h$ of some (possibly "physical") parameters, which are subject to variations and/or uncertain. The nominal value β_0 of β is assumed to be an interior point of the bounded set Ω. It is assumed that each of the first $\overline{q}$ components of $y(t)$ must track the corresponding component of the reference vector $r(t) \in \mathbb{R}^{\overline{q}}$, $\overline{q} \le q$. Hence, the error signal $e(t) \in \mathbb{R}^{\overline{q}}$ for P is defined by

$$e(t) := Vr(t) - y(t), \quad V := [I \ 0]^T, \tag{3}$$

where I is the identity matrix. Denote by $\mathbb{Z}^+$ the set of nonnegative integers. The classes $\mathcal{R}$ of reference signals $r(\cdot)$ to be asymptotically tracked and $\mathcal{D}_i$ of disturbance functions $d_i(\cdot)$, $i = 1, 2, ..., \mu$, to be asymptotically rejected are of sinusoidal-exponential type and defined as follows:

$$\mathcal{R} := \mathcal{R}_1 \oplus \mathcal{R}_2 \oplus \cdots \oplus \mathcal{R}_{\overline{\mu}}, \tag{4}$$

$$\mathcal{R}_i := \left\{ r(\cdot) : r(t) = \delta e^{\alpha_i t} + \delta^* e^{\alpha_i^* t}, \forall t \ge 0, \delta \in \mathbb{C}^{\overline{q}} \right\}, \quad i = 1, 2, ..., \overline{\mu}, \tag{5}$$

$$\mathcal{D}_i := \left\{ d_i(\cdot) : d_i(t) = \delta e^{\alpha_i t} + \delta^* e^{\alpha_i^* t}, \forall t \ge 0, \delta \in \mathbb{C}^{m_i} \right\}, i = 1, 2, ..., \mu, \tag{6}$$

for some positive integer $\overline{\mu} \le \mu$ and some $\alpha_i \in \mathbb{C}$, $i = 1, 2, ..., \mu$, where $*$ means complex conjugate. The pairs $(M_i(\cdot), N_i(\cdot))$ and $(M_j(\cdot), N_j(\cdot))$ may coincide for some $i, j, i \ne j$, so that classes of disturbance inputs of the same kind as the class $\mathcal{R}$ of the reference signals, i.e., containing several modes, can be taken into account. The α_i's, $i = 1, 2, ..., \mu$, are assumed to be all distinct, and to have all non-negative real and imaginary parts, while the pair $(M_i(\cdot), N_i(\cdot))$ is assumed to be non-zero for each $i \in \{\overline{\mu} + 1, ..., \mu\}$. For convenience, the above defined exogenous signals $r(\cdot) \in \mathcal{R}$, $d_i(\cdot) \in \mathcal{D}_i$, $i = 1, 2, ..., \mu$, can be equivalently considered as output free motions of a properly defined exosystem G, whose state will be denoted by $z(t) \in \mathbb{R}^l$,

$$z(t) = e^{Ft} z(0), \tag{7}$$

according to

$$r(t) = L_0 z(t), \quad d_i(t) = L_i z(t), \quad i = 1, 2, ..., \mu, \tag{8}$$

so that, viceversa, $z(0) \in \mathbb{R}^l$ yields $r(\cdot) \in \mathcal{R}$, $d_i(\cdot) \in \mathcal{D}_i$, $i = 1, 2, ..., \mu$.

2.1 Statement of the Problem

When a digital control system is used for plant P, it is reasonable to require for the error $e(t)$, not only the asymptotic convergence to zero at the sampling times (as clarified in the Introduction), but the stronger ripple-free (i.e., continuous-time) asymptotic convergence to zero, that is

$$\lim_{t \to +\infty} e(t) = 0; \tag{9}$$

and to require, in addition, that such a property is guaranteed at least for all the values of the vector β belonging to some neighbourhood of β_0. If, in particular, the dead-beat convergence of the error is needed at the nominal parameters, this should be required for the whole continuous-time error response, and not only at the sampling times.

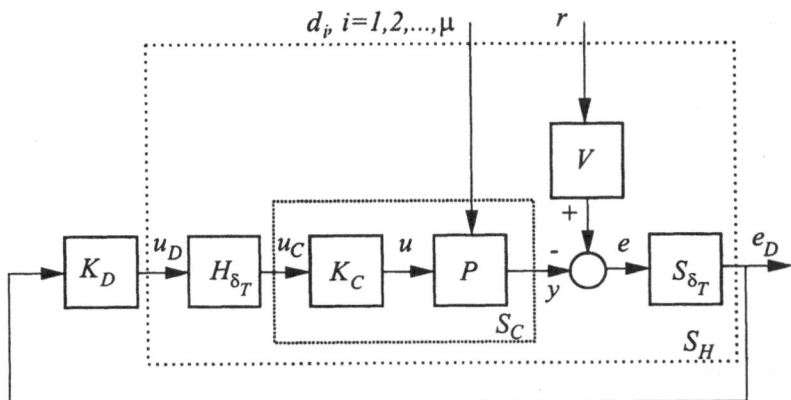

Fig. 1. The hybrid control system Σ

For such a kind of problem, in [12,28,29,18,22] the necessity of a continuous-time internal model of the exogenous signals in the forward path of the feedback control system was stated in different settings, as it was mentioned in the Introduction. If such an internal model is not entirely contained in the plant P, a continuous-time subcompensator K_C to be connected in series with P should provide the missing part (except, possibly, for the constant exogenous signals, if any). This justifies the use of the control scheme depicted in Fig. 1, where K_D is a discrete-time strictly proper dynamic subcompensator, K_C is a continuous-time dynamic subcompensator, H_{δ_T} and S_{δ_T} are a multivariable zero-order holder and a multivariable sampler, respectively, which are assumed to be synchronised, and H_{δ_T} is assumed to have the holding period equal to the sampling interval δ_T; namely, the sampler S_{δ_T} and the holder H_{δ_T} are described, respectively, by

$$e_D(k) = e(k\delta_T), \quad k \in \mathbb{Z},$$
$$u_C(t) = u_D(k), t \in (k\delta_T, \ (k+1)\delta_T], \quad k \in \mathbb{Z}, \quad u_C(t), u_D(k) \in \mathbb{R}^m;$$

S_{δ_T} can actually be substituted by two synchronised samplers, of $Vr(t)$ and $y(t)$, respectively.

Obviously, the asymptotic convergence of the free state response of the overall hybrid control system Σ represented in Fig. 1 is needed, in addition to requirement (9) on the tracking error. If a prescribed rate of convergence is required for the exponential decay, as it is natural, the corresponding control problem can be formally stated as follows.

Problem 1 *Find, if any, a linear time-invariant discrete-time strictly proper compensator K_D and a linear time-invariant continuous-time dynamic compensator K_C, so that the hybrid control system Σ depicted in Fig. 1 satisfies the following two requirements for $\beta = \beta_0$:*

(a_1) α-convergence for a prescribed $\alpha \in \mathbb{R}$, $\alpha \geq 0$, namely the existence of $\overline{\alpha}, c \in \mathbb{R}$, $\overline{\alpha} > \alpha$, $c > 0$, such that, for each initial state $x_\Sigma(0)$ of Σ, the free state response $x_\Sigma(t)$ of Σ satisfies $\|x_\Sigma(t)\| < c\,e^{-\overline{\alpha}t}$, for all $t \geq 0$;

(b_1) for each disturbance function $d_i(\cdot) \in \mathcal{D}_i$, $i = 1, 2, \ldots, \mu$, for each reference signal $r(\cdot) \in \mathcal{R}$, and for each initial state $x_\Sigma(0)$ of Σ, $\lim_{t \to +\infty} e(t) = 0$;

and, moreover,

(c_1) there exists a neighbourhood $\Psi \subseteq \Omega$ of β_0 such that conditions (a_1) and (b_1) hold for all $\beta \in \Psi$.

Notice that (a_1), for $\alpha = 0$, coincides with the mere asymptotic stability.

A different control problem arises when the dead-beat convergence of the continuous-time error response is needed at the nominal parameters. In this case it is natural to require also, for $\beta = \beta_0$, the dead-beat convergence of the free state response of the overall control system Σ, instead of (a_1). Obviously, these two requirements have to be suitably weakened for $\beta \neq \beta_0$; with this respect, the form of requirement (c_2) in the control problem stated below seems to be more natural than requirement (3) in [18].

Problem 2 *Find, if any, a linear discrete-time time-invariant strictly proper compensator K_D, and a linear time-invariant continuous-time dynamic compensator K_C, so that the hybrid control system Σ, depicted in Fig. 1, satisfies the following two requirements for $\beta = \beta_0$:*

(a_2) for each initial state $x_\Sigma(0)$ of Σ, the free state response of Σ is zero for all $t \geq \hat{t}$, for some $\hat{t} \in \mathbb{R}$, $\hat{t} > 0$;

(b_2) for each disturbance function $d_i(\cdot) \in \mathcal{D}_i$, $i = 1, 2, \ldots, \mu$, for each reference signal $r(\cdot) \in \mathcal{R}$, and for each initial state $x_\Sigma(0)$ of Σ, $e(t) = 0$ for all $t \geq \overline{t}$, for some $\overline{t} \in \mathbb{R}$, $\overline{t} > 0$;

and, moreover,

(c_2) for each $\alpha \in \mathbb{R}$, $\alpha \geq 0$, there exists a neighbourhood $\Psi_\alpha \subseteq \Omega$ of β_0 such that conditions (a_1) and (b_1) of Problem 1 hold for all $\beta \in \Psi_\alpha$.

Both Problems 1 and 2 will be studied under two technical assumptions.

Assumption 1 *There exists a closed neighbourhood $\Psi_a \subseteq \Omega$ of β_0 such that all the entries of $A(\beta)$, $B(\beta)$, $C(\beta)$ are continuous functions of β in Ψ_a and*

$$\text{rank} \begin{bmatrix} A(\beta) - \alpha_i I & B(\beta) \\ C(\beta) & 0 \end{bmatrix} = \text{rank} \begin{bmatrix} A(\beta_0) - \alpha_i I & B(\beta_0) \\ C(\beta_0) & 0 \end{bmatrix},$$
$$\forall \beta \in \Psi_a, i = 1, 2, ..., \mu. \tag{10}$$

Denote by $\sigma(A(\beta))$ the set of the eigenvalues of $A(\beta)$ and define

$$\Gamma(\beta) := \sigma(A(\beta)) \cup \{\alpha_1, \alpha_1^*, \alpha_2, \alpha_2^*, ..., \alpha_\mu, \alpha_\mu^*\}. \tag{11}$$

Assumption 2 *For each element γ of $\Gamma(\beta_0)$, none of the values $\gamma + \jmath 2\pi i/\delta_T$, $i \neq 0$, $i \in \mathbb{Z}$, is an element of $\Gamma(\beta_0)$, where $\jmath$ is the imaginary unit.*

Remark 1 Assumption 2 has a standard form (see, e.g., [9]). Its role will be clarified by the subsequent Remark 2. It implies that none of the values $\jmath 2\pi i/\delta_T$, $i \neq 0$, $i \in \mathbb{Z}$, is an element of $\Gamma(\beta_0)$, since the real matrix $A(\beta_0)$ has nonreal eigenvalues in complex conjugate pairs. □

Proposition 1 *[16] If there exists a neighbourhood $\overline{\Psi}_a \subseteq \Omega$ of some $\overline{\beta}_0 \in \Omega$ such that all the entries of $A(\beta)$, $B(\beta)$, $C(\beta)$ are continuous functions of β in $\overline{\Psi}_a$, then there exists an interior point β_0 of $\overline{\Psi}_a$ such that Assumption 1 holds.*

Then, not only a proper choice of δ_T allows Assumption 2 to be satisfied, but, by Proposition 1, Assumption 1 too can be satisfied by a proper choice of some β_0 near the initial choice $\overline{\beta}_0$ of it, provided that the matrices involved are continuous in some neighbourhood $\overline{\Psi}_a$ of $\overline{\beta}_0$.

2.2 Necessity of a continuous-time internal model of the exogenous signals

Call S_C the series connection of K_C and P (see Fig. 1), denote its state by $x_C(t) \in \mathbb{R}^{n_c}$, and denote by $\eta_{S_C}(t, \beta, x_C(0), u_C(\cdot))$, $z(0))$ the $y(t)$ response, at time t, of the continuous-time system S_C, from its initial state $x_C(0)$, to the control function $u_C(\cdot)$ and to the initial state $z(0)$ of the exosystem G, for the actual value of vector β. In a similar way, denote by $w_D(k)$ the state of the discrete-time compensator K_D, $k \in \mathbb{Z}$, and by $\theta_\Sigma(t, \beta, x_C(0), w_D(0), z(0))$ the $e(t)$ response, at time t, of the hybrid system Σ, from the initial states $x_C(0)$ of S_C and $w_D(0)$ of K_D, to the initial state $z(0)$ of the exosystem G, for the actual value of vector β.

In addition, call S_H the hybrid system constituted by the connection of the holder H_{δ_T}, the subcompensator K_C, the plant P, the comparison device and the sampler S_{δ_T} (see Fig. 1). Further, call S_D the discrete-time state-space model of S_H for $r(\cdot) = 0$ and $d_i(\cdot) = 0$, $i = 1, \ldots, \mu$, i.e., the system described by a pair of equations of the following type:

$$x_D(k + 1) = A_D(\beta)x_D(k) + B_D(\beta)u_D(k), \quad x_D(0) = x_C(0) \tag{12}$$

$$e_D(k) = -C_D(\beta)x_D(k), \tag{13}$$

yielding $x_D(k) = x_C(k\delta_T)$, where all the entries of matrices $A_D(\beta)$, $B_D(\beta)$, $C_D(\beta)$, are continuous functions of β in Ψ_a, under Assumption 1 [18].

A more complete discrete-time state-space model $\bar{S}_D$ of S_H allows to take into account nonzero disturbance functions $d_i(\cdot) \in \mathcal{D}_i$, $i = 1, 2, \ldots, \mu$, and reference signals $r(\cdot) \in \mathcal{R}$ [18], through a discrete-time model G_D of the continuous-time exosystem G, namely

$$z_D(k+1) = e^{F\delta_T} z_D(k), \quad z_D(0) = z(0), \tag{14}$$

yielding $z_D(k) = z(k\delta_T)$. By considering such a discrete-time model $\bar{S}_D$ of S_H [18], which is affected by $z_D(k)$ as an additional input, call Σ_D the discrete-time system constituted by the feedback connection of K_D and $\bar{S}_D$.

Remark 2 Now it can be stressed that the role of Assumption 2 is to guarantee that for $\beta = \beta_0$ there is no loss of observability in the series connection of G_D and $\bar{S}_D$ with respect to the connection of G, S_C and the comparison device, and, for $z(0) = 0$, no loss of reachability and observability in S_D with respect to S_C if the spectrum of $A_C(\beta_0)$ is wholly contained in $\Gamma(\beta_0)$ (this can be seen by properly extending the Corollaries 3.1 and 3.2 of [26]). This allows to put an internal model of the continuous-time exogenous signals into K_C (see Lemma A2 of the Appendix for a simple form of the internal model principle) while preserving in S_C and in S_D the structural properties of the plant P for $\beta = \beta_0$ (see the sufficiency proof of the subsequent Theorem 1) and guaranteeing the effectiveness of such an internal model as a continuous-time one, and not only as a discrete-time one.

Indeed, although the internal model principle can be of great help whenever the control system under consideration is wholly continuous-time or discrete-time [13,16], and even periodic [14,17], for the hybrid control system Σ considered here the sufficiency and, especially, the precise derivation of the necessity of a continuous-time internal model of the exogenous signals in S_C for a pair (K_C, K_D) to be a solution of Problem 1 or Problem 2 needs some special care to be obtained, just because of its hybrid nature.   □

However, the proof of the following lemma (concerning the necessity of such a continuous-time internal model) is omitted for the sake of brevity, since for item (i) it coincides with the proof of Lemma 2 of [22] (although Lemma 2 of [22] was stated only for requirement (c_1) of Problem 1), while the proof of item (ii) can be obtained through the additional comment that, if Assumption 2 holds, then G_D (or $z_D(k)$) has the constant mode if and only if G (or $z(t)$) has.

Lemma 15. *If the hybrid control system Σ represented in Fig. 1 satisfies either requirement (c_1), or requirement (c_2), then for each $\beta \in \Psi$, or for each $\beta \in \Psi_\alpha$, and for each initial state $\bar{z} \in \mathbb{R}^l$ of the exosystem G:*

(i) *there exists an initial state $[\bar{x}_C^T \ \bar{w}_D^T]^T$ of Σ such that the corresponding response $\bar{u}_C(t)$ of Σ to $\bar{z}$ in the $u_C(t)$ variable is constant for all $t > 0$, and the following relation is satisfied:*

$$V L_0 e^{Ft} \bar{z} - \eta_{S_C}(t, \beta, \bar{x}_C, \bar{u}_C(\cdot), \bar{z}) = 0, \quad \forall t \geq 0, t \in \mathbb{R}; \tag{15}$$

(ii) *under Assumption 2, if $\alpha_i \neq 0$ for all $i = 1, 2, \ldots, \mu$, or, more generally, if $\bar{z}$ is such that $e^{Ft}\bar{z}$ does not contain the constant mode, then such a constant control function $\bar{u}_C(t)$ is zero for all $t > 0$.*

By Lemma 15 – which is an extension of similar results in [12,18,28,29] to the setting here considered, – if Σ satisfies requirement (c_1), or requirement (c_2), then for each $\beta \in \Psi$, or for each $\beta \in \Psi_\alpha$, system S_C must contain a continuous-time internal model of all disturbance functions and reference signals (except, possibly, for constant signals, whose internal model could be provided by K_D), in the sense of (A.10) of Lemma A2 of the Appendix, and therefore a continuous-time subcompensator K_C – to be connected in series with P as in Fig. 1 – could be needed, as previously mentioned.

An algebraic characterization of the existence of a complete continuous-time internal model of all disturbance functions and reference signals in the continuous-time part S_C of the hybrid control system Σ, for a given parameter vector $\beta \in \Omega$, is given by Lemma 3(a) of [22], whereas its part (b) gives a similar algebraic characterization of the possibility of satisfying relation (15) with a constant control function $\overline{u}_C(\cdot)$, for the case of constant disturbance functions and reference signals, if any. In addition, Lemma 4 of [22] gives conditions that must be satisfied by the given plant P in order that a complete internal model of the exogenous signals can be included in the forward path of the control system Σ depicted in Fig. 1, for a given parameter vector $\beta \in \Omega$, by a proper choice of the continuous-time subcompensator K_C, and, possibly (for constant signals only), of K_D.

3   Main Result

The necessary and sufficient conditions for the existence of a solution of Problem 1 were given by Theorem 1 of [22], and are stated by the following theorem for completeness. It provides the necessary and sufficient conditions for the existence of a solution of Problem 2; they coincide with the conditions of Theorem 1 of [18], although the requirement 3) there considered is different from the previously defined requirement (c_2).

Theorem 1 (a) *There exist two compensators K_D and K_C that constitute a solution of Problem 2, under Assumptions 1 and 2, if and only if:*

(i) the triplet $(A(\beta_0), B(\beta_0), C(\beta_0))$ is reachable and observable;

(ii) there exists a neighbourhood $\Psi_b \subseteq \Omega$ of β_0 such that

$$\text{Im} \begin{bmatrix} M_i(\beta) \\ N_i(\beta) \end{bmatrix} \subseteq \text{Im} \begin{bmatrix} A(\beta) - \alpha_i I & B(\beta) \\ C(\beta) & 0 \end{bmatrix}, \ \forall \beta \in \Psi_b, i = 1, 2, \ldots, \mu, \quad (16a)$$

$$\text{Im} \begin{bmatrix} 0 \\ V \end{bmatrix} \subseteq \text{Im} \begin{bmatrix} A(\beta) - \alpha_i I & B(\beta) \\ C(\beta) & 0 \end{bmatrix}, \ \forall \beta \in \Psi_b, i = 1, 2, \ldots, \overline{\mu}. \quad (16b)$$

(b) *There exist two compensators K_D and K_C that constitute a solution of Problem 1, under Assumptions 1 and 2, if and only if condition (ii) holds, and, in addition:*

(iii) the triplet $(A(\beta_0), B(\beta_0), C(\beta_0))$ *satisfies the following conditions:*

$$\text{rank}\left[\, A(\beta_0) - \lambda I \;\; B(\beta_0) \,\right] = n, \quad \forall \lambda \in \mathbb{C}, \; \text{re}(\lambda) \geq -\alpha, \tag{17a}$$

$$\text{rank}\begin{bmatrix} A(\beta_0) - \lambda I \\ C(\beta_0) \end{bmatrix} = n, \quad \forall \lambda \in \mathbb{C}, \; \text{re}(\lambda) \geq -\alpha. \tag{17b}$$

The proof of Theorem 1(b) is omitted (see [22]); however, it seems worth to recall that condition (17a) is equivalent to the existence of a matrix $K \in \mathbb{R}^{p \times n}$ such that the real parts of all the eigenvalues of $A(\beta_0) + B(\beta_0)K$ are smaller than $-\alpha$, and that condition (17b) has a dual meaning. These properties wil be called α-stabilizability and α-detectability, respectively.

Proof of Theorem 1(a). (Necessity) Since requirement (a_2) is satisfied, S_D is controllable and reconstructable for $\beta = \beta_0$; this implies reachability and observability of S_C for $\beta = \beta_0$, and, hence, condition (i).

Since requirement (c_2) is satisfied, for any $\alpha \in \mathbb{R}, \alpha \geq 0$, and for any $\beta \in \Psi_\alpha$, consider an arbitrary initial state $\bar{z}$ of G such that $e^{Ft}\bar{z}$ contains only one nonconstant [or constant] mode characterized by $\alpha_i \neq 0$ [or $\alpha_i = 0$], for some $i \in \{1, 2, \ldots, \mu\}$, thus implying $d_j(\cdot) = 0$ for all $j \neq i$. By Lemma 15, and by a suitable application of Lemma A1 of the Appendix, with $\overline{\Sigma}, \overline{x}_0$ and $\overline{u}(\cdot)$ replaced by S_C, $\overline{x}_C$ and the disturbance function $d_i(\cdot)$ generated by $z(0) = \bar{z}$, there exists an initial state $\tilde{x}_C$ of S_C such that

$$V L_0 e^{Ft}\bar{z} - \eta_{S_C}(t, \beta, \tilde{x}_C, 0, \bar{z}) = 0, \quad \forall t \geq 0, \quad t \in \mathbb{R}, \tag{18}$$

[or such that

$$V L_0 e^{Ft}\bar{z} - \eta_{S_C}(t, \beta, \tilde{x}_C, \overline{u}_C(\cdot), \bar{z}) = 0, \quad \forall t \geq 0, \quad t \in \mathbb{R}, \tag{19}$$

for some constant control function $\overline{u}_C(\cdot)$] and such that the output response $\eta_{S_C}(t, \beta, \tilde{x}_C, 0, \bar{z})$ [or $\eta_{S_C}(t, \beta, \tilde{x}_C, \overline{u}_C(\cdot), \bar{z})$] of S_C and the corresponding state response contain only the mode characterized by α_i. Then, equation (18) [(19)], the state equation of S_C (which is similar to equation (1) of plant P), and the hypothesis $\text{re}[\alpha_i] \geq 0$, together wih the assumed α-convergence of Σ for the chosen α (see requirement (a_1)), imply that the state response of S_C, from $x_C(0) = \tilde{x}_C$, to the null control function $u_C(\cdot)$ [to the constant control function $u_C(\cdot) = \overline{u}_C(\cdot)$] and to the initial state $z(0) = \bar{z}$ of the exosystem G, has the form $\tilde{x}_C e^{\alpha_i t}$, if $\alpha_i \in \mathbb{R}$, or $\hat{x}_C e^{\alpha_i t} + \hat{x}_C^* e^{\alpha_i^* t}$, for some $\hat{x}_C$ such that $\hat{x}_C + \hat{x}_C^* = \tilde{x}_C$, if $\alpha_i \notin \mathbb{R}$. Thus, the freedom in the choice of $\bar{z}$, and the arbitrariness of index $i \in \{1, 2, \ldots, \mu\}$ and of $\beta \in \Psi_\alpha$, together with Lemmas 3 and 4 of [22], imply (16) with $\Psi_b = \Psi_\alpha$.

(Sufficiency) If Assumptions 1 and 2 and conditions (i) and (ii) hold, choose a dynamic continuous-time subcompensator K_C, having the complex numbers $\alpha_1, \alpha_1^*, \alpha_2, \alpha_2^*, \ldots, \alpha_\mu, \alpha_\mu^*$ as the only eigenvalues and such that the resulting system S_C is reachable and observable at $\beta = \beta_0$ and satisfies the following conditions

$$\text{Im}\begin{bmatrix} M_{C,i}(\beta) \\ N_{C,i}(\beta) \end{bmatrix} \subseteq \text{Im}\begin{bmatrix} A_C(\beta) - \alpha_i I \\ C_C(\beta) \end{bmatrix}, \quad \forall \beta \in \Psi_c, i = 1, 2, ..., \mu, \quad (20a)$$

$$\text{Im}\begin{bmatrix} 0 \\ V \end{bmatrix} \subseteq \text{Im}\begin{bmatrix} A_C(\beta) - \alpha_i I \\ C_C(\beta) \end{bmatrix}, \quad \forall \beta \in \Psi_c, i = 1, 2, ..., \overline{\mu}. \quad (20b)$$

where $A_C(\beta)$, $C_C(\beta)$, $M_{C,i}(\beta)$ and $N_{C,i}(\beta)$, $i = 1, 2, \ldots, \mu$, have for S_C the same meanining that $A(\beta)$, $C(\beta)$, $M_i(\beta)$ and $N_i(\beta)$, $i = 1, 2, \ldots, \mu$ have for plant P, and $\Psi_C \subseteq \Omega$ is some neighbourhood of β_0 (for the existence of such a compensator K_C and a design procedure of it, see [13,16]). Therefore, the discrete-time system S_D is controllable and reconstructible at $\beta = \beta_0$ (see, e.g., Corollaries 3.1 and 3.2 of [26]); hence, choose K_D so that all the state free motions of Σ_D are dead-beat convergent for $\beta = \beta_0$ (i.e., so that all the eigenvalues of system Σ_D are equal to zero for $\beta = \beta_0$). This trivially implies that requirement (a_2) is satisfied by Σ for $\beta = \beta_0$.

In addition, (20) and the application of Lemma 3 of [18] for each $i = 1, 2, \ldots, \mu$, imply that, for each $\beta \in \Psi_c$ and for each $\bar{z} \in \mathbb{R}^l$, there exists $\bar{x}_C \in \mathbb{R}^{n_c}$ such that (15) holds with $\bar{u}_C(\cdot) = 0$. Therefore:

$$\forall \beta \in \Psi_C, \forall \bar{z} \in \mathbb{R}^l, \exists \bar{x}_C \in \mathbb{R}^{n_c}:$$
$$\theta_\Sigma(t, \beta, \bar{x}_C, 0, \bar{z}) = 0, \quad \forall t \geq 0, \quad t \in \mathbb{R}. \quad (21)$$

Hence, by linearity, for each $\beta \in \Psi_C$, for each $\bar{z} \in \mathbb{R}^l$, for the state $\bar{x}_C \in \mathbb{R}^{n_c}$ corresponding to β and to $\bar{z}$, and for each initial state $[x_C^T(0) \ w_D^T(0)]^T$ of Σ, the following relation holds:

$$\theta_\Sigma(t, \beta, x_C(0), w_D(0), \bar{z}) = \theta_\Sigma(t, \beta, x_C(0) - \bar{x}_C, w_D(0), 0), \forall t \geq 0, t \in \mathbb{R}. \quad (22)$$

By virtue of the choice of K_D, (22) proves that requirement (b_2) is satisfied for $\beta = \beta_0$.

Then, notice that, for each $\beta \in \Omega$, $\Gamma(\beta)$ contains the spectrum of $A_C(\beta)$ and that, by the continuity of $A(\beta)$ in Ψ_a, there exists a neighbourhood $\Psi_d \subseteq \Omega$ of β_0 such that Assumption 2 rewritten with β instead of β_0 holds for each $\beta \in \Psi_d$.

Now, consider any $\alpha \in \mathbb{R}$, $\alpha \geq 0$, and define $\rho := e^{-\alpha \delta_T}$. Since, by Assumption 1 and Lemma 1 of [18], all the entries of $A_D(\beta)$, $B_D(\beta)$ and $C_D(\beta)$ are continuous functions of β in Ψ_a, the choice of K_D guarantees also the existence of a neighbourhood $\Psi_{e,\alpha} \subseteq \Omega$ of β_0 such that, for all $\beta \in \Psi_{e,\alpha}$, all the eigenvalues of system Σ_D are in modulus smaller than ρ. Therefore, by an extension of Theorem 4 of [9], this implies that for all $\beta \in \Psi_d \cap \Psi_{e,\alpha}$ the overall hybrid control system Σ satisfies requirement (a_1). Moreover, since for each $\beta \in \Psi_C$ and for each $\bar{z} \in \mathbb{R}^l$ there exists $\bar{x}_C \in \mathbb{R}^{n_c}$ such that (22) holds for all initial states $(x_C^T(0) \ w_D^T(0))^T$ of Σ, this proves that Σ satisfies requirement (b_1) for all $\beta \in \Psi_C \cap \Psi_d \cap \Psi_{e,\alpha}$. Hence, (a_1) and (b_1) are satisfied for all $\beta \in \Psi_C \cap \Psi_d \cap \Psi_{e,\alpha}$. $\square$

Remark 3 The design procedure of K_C and K_D sketched in the suffiency proof of Theorem 1(a) (giving solution to Problem 2) coincides with the design procedure specified in Remark 5 of [18]. If the choice of K_D is modified by designing K_D so that all the eigenvalues of Σ_D are smaller than $\rho := e^{-\alpha\delta_T}$ for $\beta = \beta_0$, this will provide a solution to Problem 1. It is stressed that K_C is designed as the series connection of μ subcompensators, i.e., one for each mode of the exosystem, in order to spare scalar holders, whereas the parallel connection of the internal models corresponding to each mode could be preferable for periodic discrete-time systems [17].   □

Remark 4 Under the hypothesis of continuity of the matrices $A(\beta)$, $B(\beta)$, $C(\beta)$, $M_i(\beta)$, $N_i(\beta)$, $i = 1, 2, \ldots, \mu$, Remark 2,3 in [16] and Remark 5 in [18] deal with the computation of good estimates of the largest subset Ψ_a, or Ψ_b, or Ψ_c of Ω such that conditions (10), or (16), or (20) (after K_C has been designed), respectively, hold, and with these largest subsets.   □

3.1 The case of unstructured uncertainties

It is worth mentioning that the above reported theory can be applied and/or adapted to a different type of parameter uncertainty, which is often considered in the literature for the problem of asymptotic tracking and output regulation (see, e.g., [4,5,8,10,11] and the references therein); namely, independent peturbations of all the entries of (some of) the matrices characterizing the plant representation, which in this case is assumed to be known at the nominal parameters only (i.e., no knowledge is assumed to exist about the dependence of the plant description on some "physical" parameters). Therefore, in order to explicitly give account of this case, let the only available model of plant P be its nominal representation; that is, let $A(\beta)$, $B(\beta)$, $C(\beta)$, $M_i(\beta)$, $N_i(\beta)$, $i = 1, 2, \ldots, \mu$, be substituted by A, B, C, M_i, N_i, $i = 1, 2, \ldots, \mu$, respectively, in (1), (2). Obviously, with these new notations, this new type of uncertainty can be taken into account in the previous setting by considering all the entries of the matrices that appear in (1) and (2) and are uncertain, as the components of vector β.

Then, call Problems 1' and 2' the above defined Problems 1 and 2 when referred to this new type of uncertainty characterization of the description of plant P. For such a modified setting, Lemma 15, together with the previously mentioned Lemmas 3 and 4 of [22], still hold. Then, the substitution of Assumption 1 with the following one gives rise to the subsequent theorem.

Assumption 3 *All the entries of the nominal matrices A and B (and, possibly, other matrices) appearing in equations (1) and (2) are subject to independent perturbations, and, in addition, rank $C = q$ (i.e., full row-rank).*

Theorem 2 *There exist two compensators K_D and K_C that constitute a solution of Problem 1', or, respectively, 2', under Assumptions 2 and 3, if and only if:*

(iv) the triplet (A, B, C) is α-stabilizable and α-detectable,
or, respectively,
(v) the triplet (A, B, C) is reachable and observable,
and, in addition,

$$\text{rank} \begin{bmatrix} A - \alpha_i I & B \\ C & 0 \end{bmatrix} = n + q, \; i = 1, 2, \ldots, \mu. \tag{23}$$

Proof The necessity follows from the restatement of Theorem 1 with the new notations (since it still holds in the new setting), by noting that the conditions of Theorem 1 are necessary irrespective of Assumption 1, and by deducing condition (23) from Assumption 3 and condition (ii) of Theorem 1 rewritten with the new notations. The sufficiency is easily obtained by noting that condition (23) implies both condition (ii) of Theorem 1 and Assumption 1, rewritten in the new setting.   □

See Remark 2.4 in [16] and Remark 4 in [18] for a comparison between conditions (ii) of Theorem 1 and (23), the genericity of the latter and the significance of the former.

4   Conclusions

Since the conditions of Theorems 1 and 2 are the same as (or similar to) the solvability conditions of similar problems for wholly continuous-time control systems [4,5,8,10,11,13,16] (although they must be derived through wholly different arguments, because of the technically awkward hybrid framework), and even the design procedure of K_C is just the same as the design procedure of an internal model of the exosignals - to be connected in series with a continuous-time plant - that is available for these control systems, the theory here reported, as well as its ongoing extensions to the case of multirate sampled-data systems [23] and to larger classes of exogenous signals [15], has merely one new special feature with respect to the corresponding previous results in the continuous-time setting: Lemma 15 (and its ongoing extensions to broader settings), i.e., the necessity of a continuous-time internal model of the exogenous signals, in order to achieve a ripple-free error response.

Acknowledgment

This work was supported by Ministero dell'Università e della Ricerca Scientifica e Tecnologica.

Appendix

Consider the linear time-invariant system $\overline{\Sigma}$ described by

$$\Delta \overline{x}(t) = \overline{A} \overline{x}(t) + \overline{B} \overline{u}(t), \tag{A.1}$$

$$\overline{y}(t) = \overline{C} \overline{x}(t) + \overline{D} \overline{u}(t), \tag{A.2}$$

where $t \in T$ is time, Δ denotes either the differentiation operator (if $T = \mathbb{R}$) or the one-step forward shift operator (if $T = \mathbb{Z}$), $\overline{x}(t) \in \mathbb{R}^{\overline{n}}$, $\overline{u}(t) \in \mathbb{R}^{\overline{p}}$, $\overline{y}(t) \in \mathbb{R}^{\overline{q}}$ and $\overline{A}, \overline{B}, \overline{C}, \overline{D}$ are matrices with real entries. Denote by $\overline{\varphi}(t, \overline{x}_0, \overline{u}(\cdot))$ and $\overline{\eta}(t, \overline{x}_0, \overline{u}(\cdot))$ the state and output responses, respectively, at time t of system $\overline{\Sigma}$ from the initial state $\overline{x}(0) = \overline{x}_0$ to the input function $\overline{u}(\cdot)$; denote by $\overline{U}$ the class of input functions $\overline{u}(\cdot)$ consisting of the free responses of a linear exosystem $\overline{G}$ described by $\Delta \overline{z}(t) = \overline{F}\overline{z}(t)$, $\overline{u}(t) = \overline{L}\overline{z}(t)$, $\overline{F} \in \mathbb{R}^{\overline{\nu} \times \overline{\nu}}$, where $\overline{F}$ has all the eigenvalues in the closed right half-plane (if $T = \mathbb{R}$), or outside the open disk of unit radius (if $T = \mathbb{Z}$). Lastly, assuming that $\overline{u}(\cdot) \in \overline{U}$, consider the unique decomposition of $\overline{\varphi}(t, \overline{x}_0, \overline{u}(\cdot))$ $(\overline{\eta}(t, \overline{x}_0, \overline{u}(\cdot)))$ into the sum of $\overline{\varphi}^{ss}(t, \overline{x}_0, \overline{u}(\cdot))$ $(\overline{\eta}^{ss}(t, \overline{x}_0, \overline{u}(\cdot)))$ and $\overline{\varphi}^{t}(t, \overline{x}_0, \overline{u}(\cdot))$ $(\overline{\eta}^{t}(t, \overline{x}_0, \overline{u}(\cdot)))$, consisting, respectively, of the modes of $\overline{u}(\cdot)$ and of the modes of system $\overline{\Sigma}$ that are not modes of $\overline{u}(\cdot)$. If system $\overline{\Sigma}$ is asymptotically stable, then the former is independent of $\overline{x}_0$ and is the steady-state response of $\overline{\Sigma}$, and the latter is the transient response of $\overline{\Sigma}$.

Lemma A1 *[22] For each $\overline{u}(\cdot) \in \overline{U}$, and for each $\overline{x}_0 \in \mathbb{R}^{\overline{n}}$:*
a) there exists a unique $\overline{x}_1 \in \mathbb{R}^{\overline{n}}$ such that

$$\overline{\varphi}(t, \overline{x}_1, \overline{u}(\cdot)) = \overline{\varphi}^{ss}(t, \overline{x}_0, \overline{u}(\cdot)), \quad \forall t \geq 0, \tag{A.3}$$

$$\overline{\varphi}(t, \overline{x}_0 - \overline{x}_1, 0) = \overline{\varphi}^{t}(t, \overline{x}_0, \overline{u}(\cdot)), \quad \forall t \geq 0; \tag{A.4}$$

b) in addition, the following relations hold:

$$\overline{\eta}(t, \overline{x}_1, \overline{u}(\cdot)) = \overline{\eta}^{ss}(t, \overline{x}_0, \overline{u}(\cdot)), \quad \forall t \geq 0, \tag{A.5}$$

$$\overline{\eta}(t, \overline{x}_0 - \overline{x}_1, 0) = \overline{\eta}^{t}(t, \overline{x}_0, \overline{u}(\cdot)), \quad \forall t \geq 0. \tag{A.6}$$

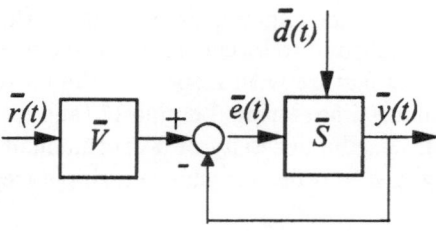

Fig. 2. The feedback system $\overline{\Sigma}$

Now, assume that $\overline{u}(t) = [\overline{r}^T(t) \ \overline{d}^T(t)]^T$, with $\overline{r}(t) \in \mathbb{R}^{\overline{\ell}}$, $\overline{d}(t) \in \mathbb{R}^{\overline{m}}$ and $\overline{\ell} + \overline{m} = \overline{p}$, and that system $\overline{\Sigma}$ has the feedback structure depicted in Figure 2, where the block denoted by $\overline{V}$ is a linear static link represented by the constant matrix $\overline{V}$, $\overline{S}$ is a linear time-invariant system described by:

$$\Delta \overline{x}(t) = \overline{A}_{\overline{S}}\overline{x}(t) + \overline{B}_{\overline{S}}\overline{e}(t) + \overline{M}_{\overline{S}}\overline{d}(t) \tag{A.7}$$

$$\overline{y}(t) = \overline{C}_{\overline{S}}\overline{x}(t) + \overline{D}_{\overline{S}}\overline{e}(t) + \overline{N}_{\overline{S}}\overline{d}(t) \tag{A.8}$$

with $(I + \overline{D}_{\overline{S}})$ nonsingular, and $\overline{r}(t)$, $\overline{d}(t)$ and $\overline{e}(t)$ can have the meaning of the reference, disturbance and error vector, respectively. In addition, denote by $\overline{\mathcal{R}}$ and $\overline{\mathcal{D}}$ the classes of functions $\overline{r}(\cdot)$ and $\overline{d}(\cdot)$, respectively, induced from $\overline{\mathcal{U}}$ by the above introduced partition of $\overline{u}(t)$. Lastly, denote by $\overline{\theta}_{\overline{\Sigma}}(t, \overline{x}_0, \overline{r}(\cdot), \overline{d}(\cdot))$ the $\overline{e}(t)$ variable response at time t of system $\overline{\Sigma}$ from the initial state $\overline{x}(0) = \overline{x}_0$, to the input functions $\overline{r}(\cdot)$ and $\overline{d}(\cdot)$, and denote by $\overline{\eta}_{\overline{S}}(t, \overline{x}_0, \overline{e}(\cdot), \overline{d}(\cdot))$ the output response $\overline{y}(t)$ at time t of system $\overline{S}$ from the initial state $\overline{x}(0) = \overline{x}_0$, to the input functions $\overline{e}(\cdot)$ and $\overline{d}(\cdot)$.

The following lemma [21,19] can be easily deduced from Lemma A1, and states a simple form of the *internal model principle* .

Lemma A2 *For each pair of functions $\overline{d}(\cdot) \in \overline{\mathcal{D}}$ and $\overline{r}(\cdot) \in \overline{\mathcal{R}}$, if $det(I + \overline{D}_{\overline{S}}) \neq 0$ and if the system $\overline{\Sigma}$ represented in Figure 2 is asymptotically stable, then*

$$\lim_{t \to \infty} \overline{\theta}_{\overline{\Sigma}}(\ t\ , \overline{x}_0, \overline{r}(\cdot), \overline{d}(\cdot)) = 0, \tag{A.9}$$

$$\Updownarrow$$

$$\exists \overline{x}_1 \in \mathbb{R}^{\overline{n}} : \overline{V}\overline{r}(t) - \overline{\eta}_{\overline{S}}(t, \overline{x}_1, 0, \overline{d}(\cdot)) = 0, \quad \forall t \geq 0. \tag{A.10}$$

References

1. J. Ackerman. *Sampled-Data Control Systems*. Springer-Verlag, 1985.
2. T. Chen and B. A. Francis. $\mathcal{H}_2$-optimal sampled-data control. *IEEE Trans. Automatic Control*, 36(4):387–397, 1991.
3. T. Chen and B. A. Francis. *Optimal Sampled-Data Control Systems*. Springer-Verlag, 1995.
4. E. J. Davison. The robust control of a servomechanism problem for linear time-invariant multivariable systems. *IEEE Trans. Aut. Control*, AC-21:25–34, 1976.
5. C. A. Desoer and Y.T. Wang. On the minimum order of a robust servocompensator. *IEEE Trans. Aut. Control*, 23:70–73, 1978.
6. R. Doraiswami. Robust control strategy for a linear time-invariant multivariable sampled-data servomechanism problem. *IEE Proceedings, Pt. D*, 129:283–292, 1982.
7. C.E. Dullerud. *Control of Uncertain Sampled-Data Systems*. Birkhäuser, 1996.
8. B. A. Francis. The linear multivariable regulator problem. *SIAM J. Control*, 15:486–505, 1977.
9. B. A. Francis and T. T. Georgiou. Stability theory for linear time-invariant plants with periodic digital controllers. *IEEE Trans. Aut. Control*, AC-33:820–832, 1988.
10. B. A. Francis and W. M. Wonham. The internal model principle for linear multivariable regulators. *Appl. Math. Opt.*, 2:170–194, 1975.
11. B. A. Francis and W. M. Wonham. The internal model principle of control theory. *Automatica*, 12:457–465, 1976.

12. G. F. Franklin and A. Emami-Naeini. Design of ripple-free multivariable robust servomechanism. *IEEE Trans. Automatic Control*, AC-31(7):661–664, 1986.
13. O. M. Grasselli and S. Longhi. Robust output regulation under uncertainties of physical parameters. *Systems and Control Letters*, 16:33–40, 1991.
14. O. M. Grasselli and S. Longhi. Robust tracking and regulation of linear periodic discrete-time systems. *Int. J. of Control*, 54(3):613–633, 1991.
15. O. M. Grasselli, S. Longhi, and A. Tornambè. Robust ripple-free tracking and regulation under uncertainties of physical parameters for ramp-exponential signals. In *32th IEEE Int. Conf. Decision and Control*, volume 1, pages 783–784, San Antonio, Texas, U.S.A., December 1993.
16. O. M. Grasselli, S. Longhi, and A. Tornambè. Robust tracking and performance for multivariable systems under physical parameter uncertainties. *Automatica*, 29(1):169–179, 1993.
17. O. M. Grasselli, S. Longhi, A. Tornambè, and P. Valigi. Robust output regulation and tracking for linear periodic systems under structured uncertainties. *Automatica*, 32(7):1015–1019, 1996.
18. O. M. Grasselli, S. Longhi, A. Tornambè, and P. Valigi. Robust ripple-free regulation and tracking for parameter dependent sampled-data systems. *IEEE Trans. Automatic Control*, 41(7):1031–1037, 1996.
19. O. M. Grasselli, S. Longhi, and A. Tornambé. Robust ripple-free tracking and regulation for sampled-data systems under uncertainties of physical parameters. In *Proc. of the European Control Conference, 1993*, vol. 2, pages 432–437, Groningen, The Netherlands, June 1993.
20. O. M. Grasselli and L. Menini. Continuous-time input-output decoupling for sampled-data systems. In *Proc. of the 6th IEEE Mediterranean Conf. on Control and Systems.* pages 829–834, Alghero, Italy, June 1998.
21. O. M. Grasselli and F. Nicolò. Modal synthesis of astatic multivariable regulation systems. In *Proc. of the 2nd IFAC Symp. on Multiv. Tech. Control Systems, paper 1.1.4*, Düsseldorf, Germany, October 1971.
22. O.M. Grasselli, L. Menini, and P. Valigi. Robust continuous-time asymptotic tracking and regulation for sampled-data systems with structured parametric uncertainties. In *Preprints of the 2nd IFAC Symposium on Robust Control Design*, pages 359—364, Budapest, Hungary, June, 25–27 1997.
23. O.M. Grasselli, L. Menini, and P. Valigi. Robust continuous-time tracking for multi-rate digital control systems. To appear in *Preprints of the 1999 IFAC World Congress*, Beijing, China, July 5 – 9 1999.
24. S. Hara and H. K. Sung. Ripple-free conditions in sampled-data control systems. In *Proc. of the 30th Conf. on Decision and Control*, pages 2670–2671, Brighton, (UK), December 1991.
25. K. Ichikawa. Finite-settling-time control of continuous-time plants. *Systems and Control Letters*, 9:341–343, 1987.
26. S. Longhi. Structural properties of multirate sampled-data systems. *IEEE Trans. Automatic Control*, 39(3):692–696, 1994.
27. O. A. Sebakhy. State-space design of ripple-free dead-beat multivariable systems. *Int. J. of Systems Science*, 20:2673–2694, 1989.
28. S. Urikura and A. Nagata. Ripple-free deadbeat control for sampled-data systems. *IEEE Trans. Automatic Control*, AC-32:474–482, 1987.
29. Y. Yamamoto. A function space approach to sampled data control systems and tracking problems. *IEEE Trans. Automatic Control*, 39(4):703–713, 1994.

An Experimental Study of Performance and Fault–Tolerance of a Hybrid Free–Flight Control Scheme

Antonio Bicchi

Interdept. Research Centre "E. Piaggio", Università di Pisa, via Diotisalvi, 2, 56125 Pisa, Italia

Abstract. In this paper we first describe an optimal coordinated conflict management scheme for a simplified model of air traffic, tending to minimize fuel consumption while guaranteeing safety against collisions. A decentralized implementation of such a scheme is then introduced, and its features as a hybrid control scheme are described. Finally, we discuss the tradeoff between performance and fault tolerance that goes with decentralization, and assess it by extensive simulation trials.

1 Introduction

In this paper, we investigate the advantages and disadvantages of a proposed decentralized Air Traffic Management (ATM) system, in which aircraft are allowed a larger degree of autonomy in deciding their route ("free flight") instead of following prespecified "sky freeways". A decentralized management scheme is described, which incorporates a hybrid control system switching among different optimal control solutions according to changes in the information structure between agents. The performance and robustness of this decentralization scheme is assessed by means of simulation trials, showing that, whereas optimality of plans is strictly nonincreasing, robustness to system failures is likely to improve. This work (based on [3]) is intended as a prepaparatory study towards the design and implementation of a *Flight Management System* (FMS) as part of an *Air Traffic Management* (ATM) system architecture allowing for multi-aircraft coordination maneuvers, which are guaranteed to be safe.

A number of issues should be considered when deciding on the appropriate level of centralization. An obvious one is the *optimality* of the resulting design. Even though optimality criteria may be difficult to define for the air traffic problem it seems that, in principle, the higher the level of centralization the closer one can get to the globally optimal solution. However, the complexity of the problem also increases in the process; to implement a centralized design one has to solve a small number of complex problems as opposed to large number of simple ones. As a consequence the implementation of a centralized solution requires a greater effort on the part of the designer to produce control algorithms and greater computational power to execute them. Another issue that needs to be considered is *fault tolerance* . The greater the responsibility assigned to a central controller the more dramatic are likely to be the consequences if this controller fails. In this respect there seems to be a clear advantage in implementing a decentralized design: if a single aircraft's computer system fails, most of the ATM system is still intact and the affected aircraft may be guided by voice to the nearest airport. Similarly, a distributed system is better suited to handling increasing numbers of aircraft, since each new aircraft can easily be added to the system, its own computer contributing to the overall computational power. Finally, the issue of *flexibility* should also be taken into account. A decentralized system will be more flexible from the point of view of the agents, in this case the pilots and airlines. This may be advantageous for example in avoiding turbulence or taking advantage of favorable winds, as the aircraft will not have to wait for clearance from ATC to change course in response to such transients or local phenomena.

2   Decentralized ATMS

In order to quantitatively evaluate the effects of decentralization on performance and robustness, we introduce a much simplified, yet significant model of the air traffic control problem. Consider the problem of steering N agents (aircraft) among $2N$ given via–points (see Figure 1). Aircraft are assumed to move on a planar trajectory (in fact, all fly within a given altitude layer), and their cruise speed is constant. The kinematic motion of the i–th aircraft is modeled as

$$\begin{bmatrix} \dot{x}_i \\ \dot{y}_i \\ \dot{\theta}_i \end{bmatrix} = \begin{bmatrix} \cos\theta_i \ u_i \\ \sin\theta_i \ u_i \\ \omega_i \end{bmatrix} \tag{1}$$

where $\xi_i = [x_i, y_i, \theta_i]^T$ is the state vector (comprised of x, y position and heading angle θ), u_i = constant is the cruise speed. Start and goal via–points are $\xi_{i,s}$ and $\xi_{i,g}$, respectively. We assume for simplicity that motions are synchronized, i.e. all aircraft are at the start via–point at the same time, and denote by T_i the time at which the the i–th aircraft reaches its goal. The cost to be minimized in choosing the control inputs (yaw rates) ω_i is the total time–to–goal, or equivalently the total path length (proportional to fuel consumption),

$$J = \sum_{i=1}^{N} \int_0^{T_i} dt = \sum_{i=1}^{N} T_i \tag{2}$$

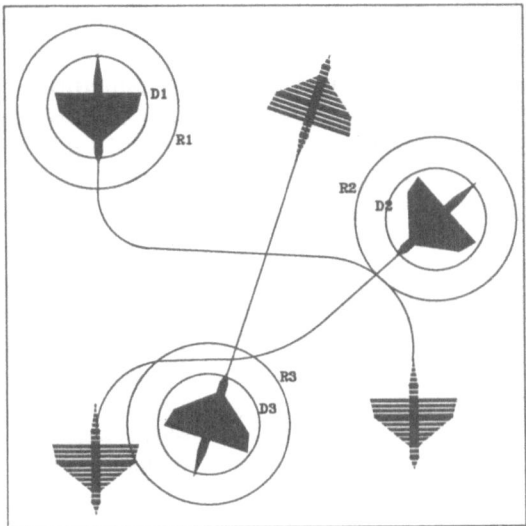

Fig. 1. Three aircraft with initial and final viapoints assigned.

Limitations of feasible yaw rates for practical airplanes are incorporated in this model by setting

$$\omega_i \in [-\Omega_i, \Omega_i]. \tag{3}$$

Besides the initial and final state constraints, and the input constraints (3), a constraint of no–collision along solutions must be enforced. Typically, such a constraint is embodied by non intersection of safety discs of radius D_i centered at the aircraft positions, namely by $\frac{N(N-1)}{2}$ inequalities on states of type

$$V_{ij}(\xi_i, \xi_j) = (x_i - x_j)^2 + (y_i - y_j)^2 - D_i^2 \geq 0 \tag{4}$$

It should be noted that, although the Euclidean distance used in the above definition of the constraint does not appear to take into account the fact that the nonholonomic kinematics (1) induce a more complex metric on the space (compare work of Laumond and co-workers, [11], [6]), (4) is typical in current ATC systems.

In a centralized ATMS, the above optimal control problem would be solved by ground–based Air Traffic Control, using one of several available numerical techniques. Notice that, even in this simplified setup, inequality constraints on states at interior points may generate difficulties in numerical integration. In our simulations, we adopt the suboptimal strategy introduced in [2], and described below in some detail.

In decentralized ATMS schemes, each agent (aircraft) is allowed to take decisions autonomously, based on the information that is available at each time. Several models of decentralized ATC are conceivable, which may differ in the degree of cooperative/competitive behaviour of the agents, and in the information structure. For instance, a noncooperative, zero–sum game of the agents is a conservative (worst–case) approach to guarantee safety of solutions, and as such it has been studied e.g. in [13], [10], [14]. In this paper, we consider a cooperative scheme which falls within the scope of the theory of teams (cf. e.g. [9], [1]). In particular, we consider a scheme in which

- The i–th agent has information on the state and goals of all other agents which are at a distance less than an "alert" radius $R_i > D_i$;
- Each agent plans its flight according to an optimal strategy which consists in minimizing the sum of the time–to–goals of all pertinent aircraft.

Let $S_i(\tau)$ denote the set of indices of aircraft within distance R_i from the i–th aircraft at time τ, i.e. aircraft j such that

$$C_{ij}(\xi_i, \xi_j) = (x_i - x_j)^2 + (y_i - y_j)^2 - R_i^2 \leq 0.$$

The goal of the i–th agent at time τ with information S_i is therefore to minimize

$$\mathbf{J}_{i,S_i}(\tau) = \sum_{j \in S_i} \int_\tau^{T_j} dt \tag{5}$$

under the constraints

$$V_{ij}(\xi_i, \xi_j) > 0, \ \forall j \in S_i. \tag{6}$$

Obviously, when all R_i are large w.r.t. the dimension of the considered flight area, each agent solves the same problem the centralized controller would solve, and the resulting performance would be equal (albeit with N–fold computational redundancy).

When, during execution of flight maneuvers planned based on a certain information structure $I = (S_1, \ldots, S_N)$, an aircraft i with $i \notin S_j$ gets at distance R_j from aircraft j, the information structure is updated, and optimal paths are replanned according to the new cost and constraints for aircraft j. We assume that structure updates and replanning are done in real time by agents.

The system resulting from the above decentralized ATMS scheme is described by a set of continuous variables $\xi_i, \omega_i, i = 1, \ldots, N$, and a set of variables S_i that take values over discrete sets. To each different information structure I_k there corresponds a working mode for the system, i.e. dynamics (1) driven by controls $\hat{\omega}_{i,k}$ which optimize J_{i,S_i} under constraints $V_{ij} > 0, j \in S_i$. The resulting hybrid system is composed of a finite–state machine and of associated continuous–time dynamic systems, transitions among states being triggered by conditions on the continuous variables.

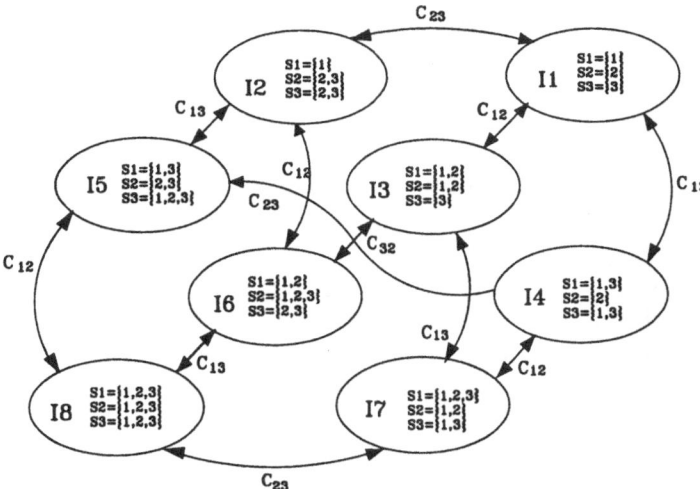

Fig. 2. A decentralized ATMS with three aircraft having equal alert radius. Each node in the graph corresponds to different costs and constraints in the agents' optimal steering problem. Optimizing controllers for such problems cause different continuous time dynamics at each node. Switching between modes is triggered when an airplane enters or exits the alert neighborhood of another ($C_{i,j}$ changes sign).

For instance, in the case with $N = 3$, $R_1 = R_2 = R_3$, there are eight possible states (modes of operation), corresponding to different information structures I_k (see Figure 2).

Remark At every state transition, each agent evaluates in real–time the optimal steering control from the current position to the goal for itself as well as for all other aircraft within its alert radius. This implies that shared information consists of present position and goal coordinates. Only the control policy evaluated by an agent for itself is then executed, as the one calculated for others may ignore part of the information available to them (as e.g. it happens in states I_5, I_6, and I_7 in Figure 2). All optimal policies coincide for large R_i's.

3   Shortest multi–agent collision–free paths

The algorithm introduced in [2], tending to optimize problem (2) with constraints (1), (3) and (4), will be considered for use by the planner, and is succinctly described below. The algorithm is first described with respect to a centralized implementation, and then adapted to the decentralized setup (5), (6).

The simplified airplane model given by (1), (3) is equivalent to what has become the well–known "Dubins' car" in robot motion planning literature, i.e. a vehicle which only goes forward and has bounded curvature. The solution of the shortest path among two via-points for such system (when a single airplane is considered) has been obtained first by Dubins ([8]), who showed that optimal paths are made of concatenations of segments, either circular with minimal radius ("C"–segment), or linear ("S"–segment), and that a shortest path can always be found among 6 candidates of type "CC-C" or type "CSC" only. Notice that computation of optimal Dubins' paths is an extremely computationally–efficient procedure. Subsequently, [12] and [5] reinterpreted this result as an application of Pontryagin's maximum principle. The latter framework is instrumental to developments presented here.

When multiple airplanes are considered, the sum of all lengths of Dubins' paths disregarding (4) is clearly a lower bound to cost (2), and one which is attained if, and only if, the unconstrained Dubins' solutions happen not to collide.

If unconstrained Dubins' solutions collide, then from the theory of optimal control with path constraints (see e.g. [7], and [4] for an application to mobile robot planning) we know that the optimal solution will be comprised of a concatenation of free and constrained arcs, i.e., arcs where $V_{ij} > 0, \forall i, j$ and arcs where $\exists (i, j) : V_{ij} = 0$, respectively. Along constrained arcs, at least two airplanes fly keeping their distance exactly equal to their safety limit.

Along free arcs, however, constraints are not active, hence their Lagrangian multipliers are all zero in the system Hamiltonian, and the problem can be reduced to N decoupled optimization problem for each airplane, whose solution is again given by paths of Dubins type.

A characterization of constrained arcs can be obtained by differentiating the constraint (4), and implies that only two types of such arcs are possible (type a and b, see Figure 3). Along arcs of type a), the velocities of the two airplanes must be parallel, and the line joining the two airplanes can only translate. Along arcs of type b), the velocities of the aircraft are symmetric with respect to the line joining the aircrafts. Two typical circular trajectories are shown (by dotted circles) in Figure 3 for type a) and b). In both cases, the constrained arc may have zero length, which we will call a_0) and b_0) such as depicted in Figure 4. A constrained arc of type a) or b) must also be a path minimizing solution for the "tandem" system comprised of the two airplanes moving in contact with either parallel or symmetric velocities. In case a), tandem arcs will be again of Dubins type.

Consider now the case of two airplanes flying in shared airspace, such that their unconstrained Dubins' paths collide (see Figure 5). A solution of this problem with a single constrained zero– length arc of either type a_0) or b_0), would be comprised of four Dubins paths $D_{ij}, i = 1, 2; j = 1, 2$, with $D_{i,1}$ joining the initial configuration of agent i with its configuration on the constrained arc, and $D_{i,2}$ joining the latter with the final configuration of the same agent. The set of all such paths for constrained arcs of

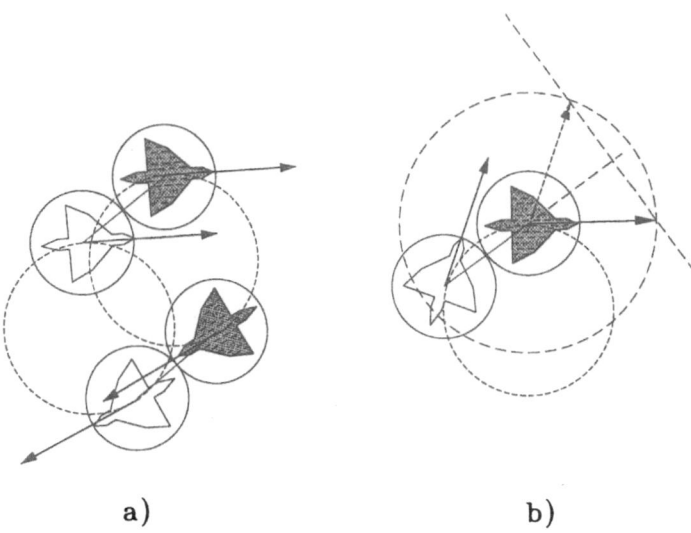

Fig. 3. Possible constrained arcs for two airplanes.

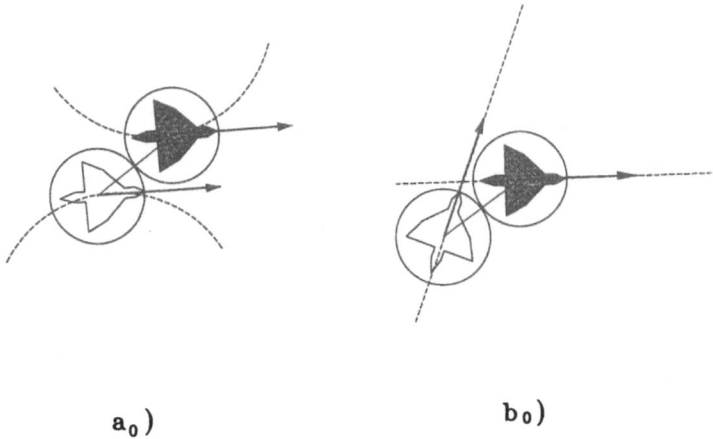

Fig. 4. Zero–length constrained arcs for two airplanes

type a) and b) can be parametrized by a quadruple in $\mathbb{R}^2 \times S^1 \times S^1$ (e.g., position and velocity direction of agent 1, and direction of the line joining the planes). Further, the constraint on the set of paths must be enforced that the constrained arc is hit simultaneously by the two agents, i.e. that $\text{length}(D_{i1}) = \text{length}(D_{j1})$.

The optimal solution within each case can be obtained by using any of several available numerical constrained optimization routines: computation is sped up considerably by using very efficient algorithms made available

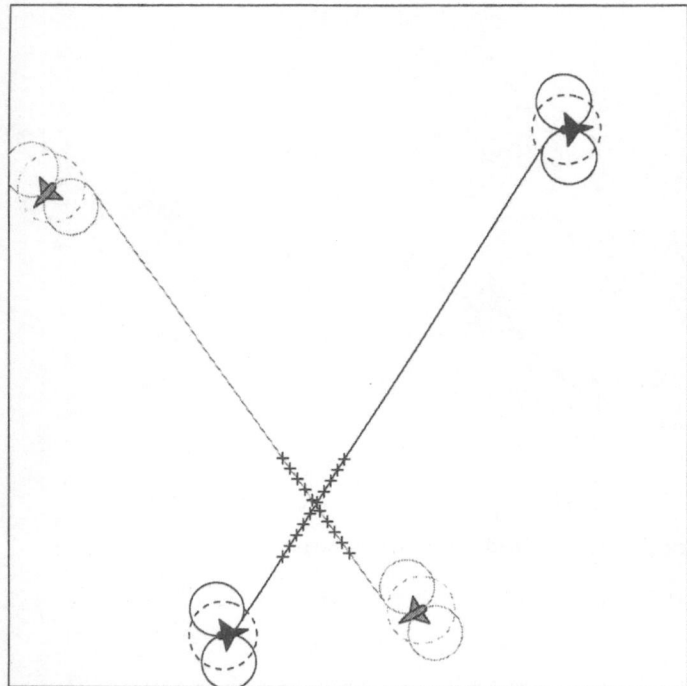

Fig. 5. Configurations of two airplanes for which unconstrained Dubins' solution collide. The dashed circle represents the safety disc; solid circles indicate maximum curvature bounds.

for evaluating Dubins' paths ([6]). A solution is guaranteed to exist for both cases. We will refer to the shortest one as to the two–agent, single–constrained zero-length arc, optimal conflict management path (OCMP21, for short). The OCMP21 solution for the example of Figure 5 is reported in Figure 6. Solutions of the two–agent problem with constrained arcs of non-zero length, or with multiple constrained arcs, are also possible in principle. An optimal solution for these cases should be searched in a larger space. However, solutions have to comply with additional requirements on lengths of intermediate free arcs, and seem to be somewhat non–generic. Further theoretical work is currently being devoted to understanding under what conditions non zero–length, and/or multiple constrained arcs may occur in an optimal multi–agent path. In principle, it may even happen that the optimal path is made by concatenating an infinite number of free and con-strained arcs (cf. the Fuller's phenomenon in optimal control).

In the present version of our planner, we do not search for solutions with multiple and/or non zero–length constrained arcs, trading optimality for

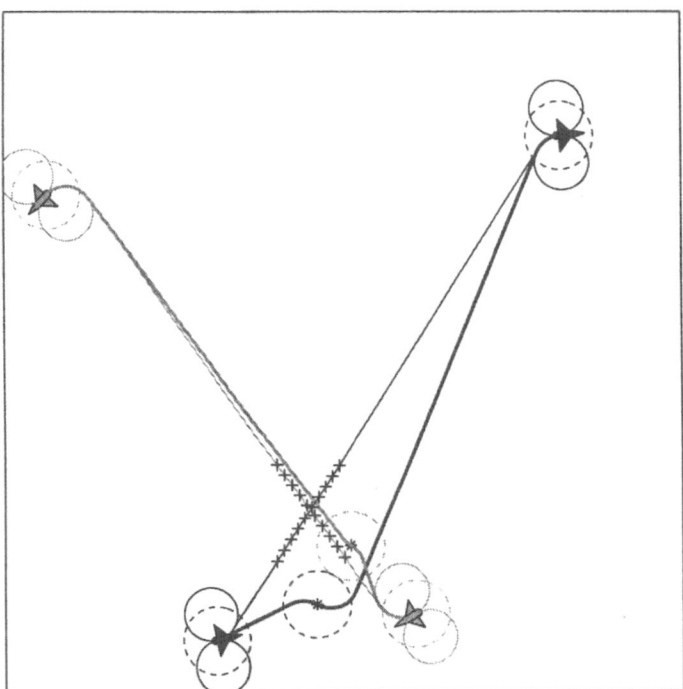

Fig. 6. The centralized OCMP21 solution of problem in 5. The total length is 92.25 (the unconstrained Dubins total length is 88.75). Safety discs contacting at the constrained arc of type b_0) are shown (dashed).

efficiency. Another motivation for such limitation is that "acrobatic" flights in tandem configuration do not sem to suit well commercial air traffic.

If three airplanes fly in a shared workspace, their possible conflicts can be managed with the following multilevel policy:

Level 0 Consider the unconstrained Dubins paths of all agents (which may be regarded as single–agent, zero–constrained arc, optimal conflict management paths, or OCMP10). If no collision occurs, the global optimum is achieved, and the algorithm stopped. Otherwise, go to next level;

Level 1 Consider the $\binom{3}{2} = 3$ OCMP21 for each pair of agents. If at least one path is collision free, choose the shortest path and stop. Otherwise, go to next level;

Level 2 Consider the three–agent, double zero–length constrained arc, optimal conflict management problem OCMP32, consisting in searching all 8–dimensional spaces of parameters identifying the first constrained arc of type ξ_1 between agent i and j, and the second constrained arc of type ξ_2 between agent k and ℓ, with ξ_1 and $\xi_2 \in \{a_0, b_0\}$; i, j, k

and $\ell \in \{1,2,3\}, i \neq j, k \neq \ell$. A solution is guaranteed to exist for all $3^2 \binom{3}{2}\left[\binom{3}{2} - 1\right] = 54$ cases; the shortest solution is OCMP32.

A three airplane conflict management solution at level two (OCMP32) is reported in Figure 7. When the number of airplanes increases, the number

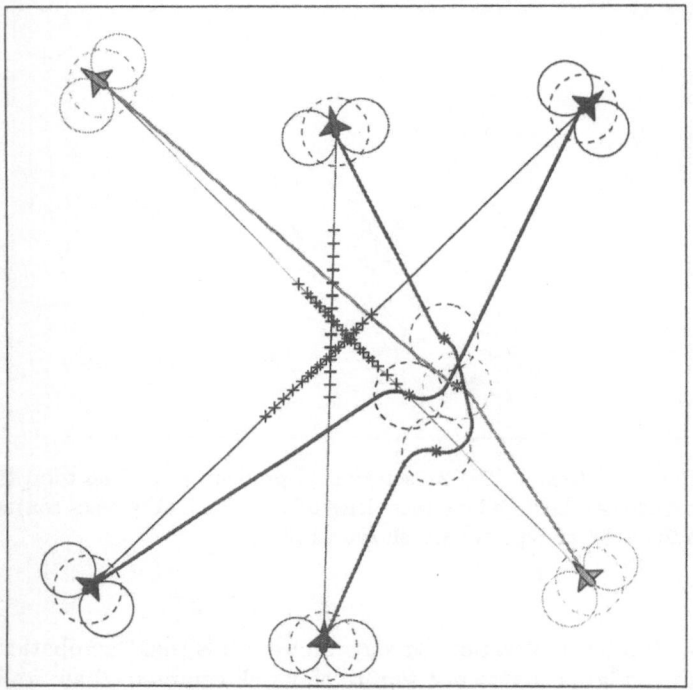

Fig. 7. An OCMP32 solution. The total length is 151.2 (the unconstrained Dubins total length is 140.0). The first constrained arc involves agents 2 and 3 (from left), while the second involves 1 and 2. Boths arcs are of type b_0).

of optimization problems to be solved grows combinatorially. However, in practice, it is hardly to be expected that conflicts between more than a few airplanes at a time have to be managed.

3.1 Decentralized implementation

The algorithm described above can be applied in a decentralized manner by simply having each agent apply its steps taking in consideration only those other agents that are within their alert disc.

The online solutions of the two–agent conflict management problem introduced in Figure 5 and Figure 6, are reported in Figure 8. It can be observed that the two aircraft initially follow their unconstrained Dubins path, until they enter each other's alert zone (this happens roughly at the third step after the start in Figure 8). At this moment, an OCMP21 is obtained by both decentralized planners. Notice that, in this two–agent problem with equal alert radius, the same problem is solved by both, although this does not hold in general. Aircrafts start following their modified paths, which differ from both the unconstarined Dubins paths and the centralized optimal paths of Figure 6. The total length of decentralized solution is 93.85.

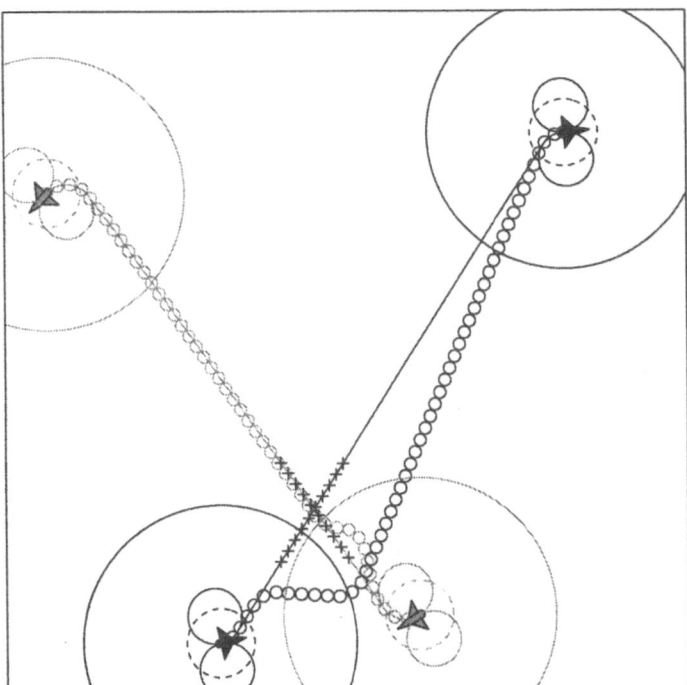

Fig. 8. Decentralized solution of the two–agent conflict management problem (trajectories traced by small circles). Alert discs are drawn in solid lines around the initial and final configurations of agents. The unconstrained Dubins' paths are superimposed for reference.

4   Decentralization: Performance and Fault Tolerance

In order to assess the effects of increasing decentralization in ATMS, we performed a number of simulations whose results are reported below.

In particular, we experimentally compared results obtained by a centralized planner with those achieved by several decentralized planners, with decreasing alert zone radius. The alert zone radius can be regarded as an inverse measure of the degree of centralization for an information structure such as that introduced in the section above.

The first set of simulations concerns performance evaluation. The performance measure, i.e. the total length cumulatively flown by all airplanes, for the problem described in Figure 9 has been calculated for three different values of the safety radius. Results of simulations are reported in Figure 10,

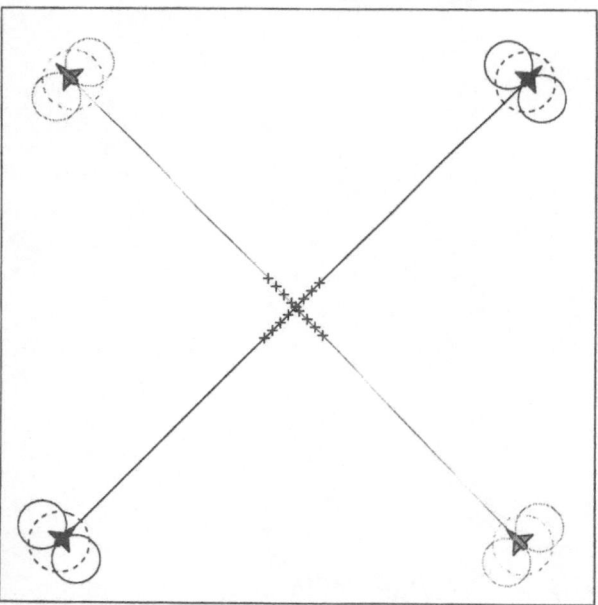

Fig. 9. Air traffic management problem simulated for performance evaluation

and show that the increase of the alert zone radius entails a rather smooth decrease of the total length flown by aircraft. As an effect of the presence of local minima in the numerical optimization process, the length does not decrease monotonically as it should be expected.

To assess fault tolerance, we simulated the same scenario and planning algorithms under degraded control conditions. In particular, some of the controllers were assumed to fail during flight. Controllers affected by failures

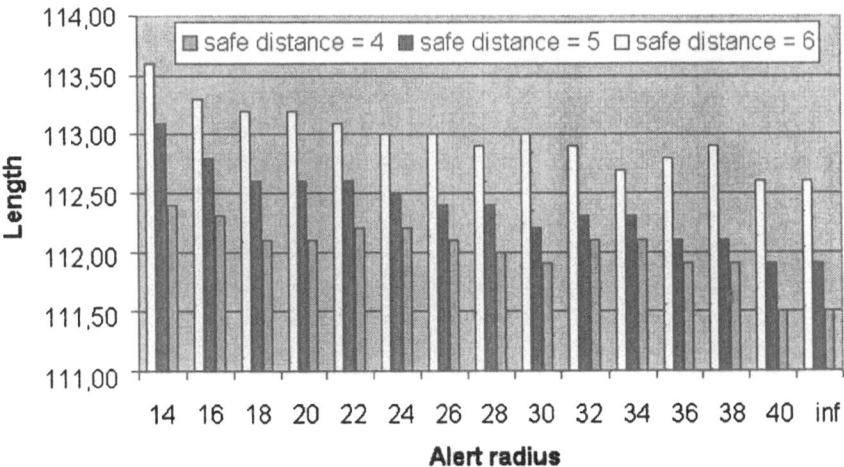

Fig. 10. Total length flown by aircraft at varying the alert zone radius.

compute their optimal plans according to the same strategy, but with via–points and information structure randomly perturbed. At the end of such crises, controllers are supposed to access correct data again, and to replan accordingly. The crisis duration is constant for all simulation runs, while they occur at random time. For centralized ATMS, all aircraft are assumed to receive random flight directions from ATC. Under decentralized ATC, other agents are able to maintain their correct operation mode, and replan in real time to try and avoid collisions. As a figure of fault tolerance of ATMS schemes, we consider the number of accidents for 100 crisis situations. Results of simulations, relative to the same initial scenario as in Figure 9, are reported in Figure 11.

Acknowledgments

This paper is largely based on material presented in [3]. The author wishes to thank Alessia Marigo, Gianfranco Parlangeli, Marco Pardini, of Univ. of Pisa, and George Pappas, Claire Tomlin, and Shankar Sastry of U.C. Berkeley for the helpful discussions and proficuous exchange of ideas. Work supported partly by grant NATO CR no. 960750 and grant ASI ARS-96-170.

References

1. M. Aicardi, F. Davoli, R. Minciardi: "Decentralized optimal control of Markov Chains with a acommon past information set", *IEEE Trans. Automat. Contr.*, vol. 32, pp. 1028–1031, 1987.

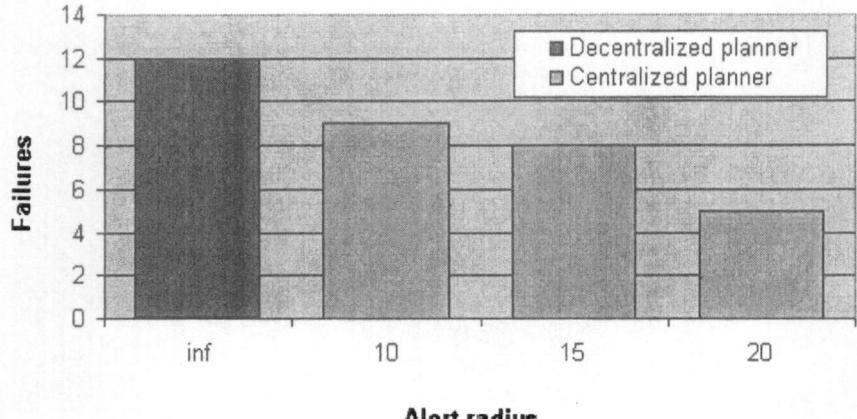

Alert radius

Fig. 11. Fault tolerance to controller failures at varying the alert zone radius. The number of accidents is relative to 100 crisis situations.

2. A. Bicchi, A. Marigo, M. Pardini, G. Parlangeli: "Shortest collision free paths of multiple Dubins' agents", Internal Report, Interdept. Research Center "E. Piaggio", 1999.
3. A. Bicchi, A. Marigo, G. Pappas, M. Pardini, G. Parlangeli, C. Tomlin, S. S. Sastry: "Decentralized Air Traffic Managemente Systems: Performance and Fault Tolerance", Proc. IFAC Int. Workshop on Motion Control, 1998.
4. A. Bicchi, G. Casalino, and C. Santilli, "Planning Shortest Bounded–Curvature Paths for a Class of Nonholonomic Vehicles among Obstacles", *Proc. IEEE Int. Conf. on Robotics and Automation*, pp. 1349–1354, 1995.
5. J.D. Boissonnat, A. Cerezo, and J. Leblond: "Shortest Paths of Bounded Curvature in the Plane", Proc. IEEE Int. Conf. on Robotics and Automation, pp.2315–2320, 1992.
6. X.N. Bui, P. Souères, J-D. Boissonnat, and J-P. Laumond: "Shortest path synthesis for Dubins nono–holonomic robots", pp. 2–7, *Proc. IEEE Int. Conf. Robotics Automat.*, 1994.
7. Chang, S.S.L.: "Optimal Control in Bounded Phase Space", Automatica, vol. I, pp.55-67, 1963.
8. L. E. Dubins: "On curves of minimal length with a constraint on average curvature and with prescribed initial and terminal positions and tangents ", American Journal of Mathematics, vol.79, pp.497–516, 1957.
9. Y.C. Ho, K.C. Chiu: " Team decision theory and information structures in optimal control problems – Part I", *IEEE Trans. Automat. Contr.*, vol. 17–1, pp. 15–21, 1972.
10. J. Košecká, C. Tomlin, G. Pappas and S. Sastry: "Generation of conflict resolution maneuvers for air traffic management", Proc. Int. Conf. Intelligent Robots and Systems, IROS'97, pp. 1598–1603, 1997.
11. J. P. Laumond: "Feasible Trajectories for Mobile Robots with Kinematic and Environment Constraints", Proc. International Conference on Intelligent Autonomous Systems, Amsterdam, 346–354, 1986.

12. H. J. Sussmann and G. Tang: "Shortest Paths for the Reeds–Shepp Car: a Worked Out Example of the Use of Geometric Techniques in Nonlinear Optimal Control", SYCON report 91-10, 1991.
13. C. Tomlin and G. Pappas and J. Kosecka and J. Lygeros and S. Sastry: "Advanced air traffic automation: a case study in distributed decentralized control", in *Control Problems in Robotics and Automation*, B. Siciliano and K. Valavanis, eds., pp. 261–295, Springer–Verlag, 1997.
14. C. Tomlin and G. J. Pappas and S. Sastry: "Conflict Resolution for Air Traffic Management: A Case Study in Multi-Agent Hybrid Systems", *IEEE Trans. Automat. Contr.*, vol. 43-4, pp. 509–521, 1998.

Index

466

Lecture Notes in Control and Information Sciences

Edited by M. Thoma

Vol. 203: Popkov, Y.S.
Macrosystems Theory and its Applications:
Equilibrium Models
344 pp. 1995 [3-540-19955-1]

Vol. 204: Takahashi, S.; Takahara, Y.
Logical Approach to Systems Theory
192 pp. 1995 [3-540-19956-X]

Vol. 205: Kotta, U.
Inversion Method in the Discrete-time
Nonlinear Control Systems Synthesis
Problems
168 pp. 1995 [3-540-19966-7]

Vol. 206: Aganovic, Z.; Gajic, Z.
Linear Optimal Control of Bilinear Systems
with Applications to Singular Perturbations
and Weak Coupling
133 pp. 1995 [3-540-19976-4]

Vol. 207: Gabasov, R.; Kirillova, F.M.;
Prischepova, S.V.
Optimal Feedback Control
224 pp. 1995 [3-540-19991-8]

Vol. 208: Khalil, H.K.; Chow, J.H.;
Ioannou, P.A. (Eds)
Proceedings of Workshop on Advances
inControl and its Applications
300 pp. 1995 [3-540-19993-4]

Vol. 209: Foias, C.; Özbay, H.;
Tannenbaum, A.
Robust Control of Infinite Dimensional
Systems: Frequency Domain Methods
230 pp. 1995 [3-540-19994-2]

Vol. 210: De Wilde, P.
Neural Network Models: An Analysis
164 pp. 1996 [3-540-19995-0]

Vol. 211: Gawronski, W.
Balanced Control of Flexible Structures
280 pp. 1996 [3-540-76017-2]

Vol. 212: Sanchez, A.
Formal Specification and Synthesis of
Procedural Controllers for Process Systems
248 pp. 1996 [3-540-76021-0]

Vol. 213: Patra, A.; Rao, G.P.
General Hybrid Orthogonal Functions and
their Applications in Systems and Control
144 pp. 1996 [3-540-76039-3]

Vol. 214: Yin, G.; Zhang, Q. (Eds)
Recent Advances in Control and Optimization
of Manufacturing Systems
240 pp. 1996 [3-540-76055-5]

Vol. 215: Bonivento, C.; Marro, G.;
Zanasi, R. (Eds)
Colloquium on Automatic Control
240 pp. 1996 [3-540-76060-1]

Vol. 216: Kulhavý, R.
Recursive Nonlinear Estimation: A Geometric
Approach
244 pp. 1996 [3-540-76063-6]

Vol. 217: Garofalo, F.; Glielmo, L. (Eds)
Robust Control via Variable Structure and
Lyapunov Techniques
336 pp. 1996 [3-540-76067-9]

Vol. 218: van der Schaft, A.
L_2 Gain and Passivity Techniques in
Nonlinear Control
176 pp. 1996 [3-540-76074-1]

Vol. 219: Berger, M.-O.; Deriche, R.;
Herlin, I.; Jaffré, J.; Morel, J.-M. (Eds)
ICAOS '96: 12th International Conference on
Analysis and Optimization of Systems -
Images, Wavelets and PDEs:
Paris, June 26-28 1996
378 pp. 1996 [3-540-76076-8]

Vol. 220: Brogliato, B.
Nonsmooth Impact Mechanics: Models,
Dynamics and Control
420 pp. 1996 [3-540-76079-2]

Vol. 221: Kelkar, A.; Joshi, S.
Control of Nonlinear Multibody Flexible Space
Structures
160 pp. 1996 [3-540-76093-8]

Vol. 222: Morse, A.S.
Control Using Logic-Based Switching
288 pp. 1997 [3-540-76097-0]